High-Performance Adaptive Control of Teleoperation Systems

Within a unified switched-control framework, this book investigates the high-performance control designs and theoretic analyses for teleoperation systems, including the joint space and task space teleoperations, the homogeneous and heterogeneous teleoperations, and the single-master single-slave and multi-master multi-slave teleoperations.

High-Performance Adaptive Control of Teleoperation Systems begins with an introduction to the concepts and challenges of networked teleoperation systems. Then, it investigates a new adaptive control framework based on auxiliary switched filters for the bilateral teleoperation systems to handle the model uncertainty and non-passive external forces. To overcome the input constraints of robotic systems, this adaptive method is also extended to the anti-windup adaptive control case. Furthermore, to apply to multi-robot remote collaboration scenarios and heterogeneous teleoperation, two tele-coordination methods and an adaptive semi-autonomous control method are respectively developed. Finally, the authors examine two finite-time control schemes and two types of improved prescribed performance controls for teleoperation systems to improve the transient-state and steady-state synchronization performances.

This title will be an essential reference for researchers and engineers interested in teleoperation, robotic systems, and nonlinear control systems. It will also prove useful to graduate students in the fields of science, engineering, and computer science.

Di-Hua Zhai is an associate professor of the School of Automation (Beijing) and the Yangtze Delta Region Academy (Jiaxing), at Beijing Institute of Technology, China. His research interests include teleoperation, intelligent robot, human-robot collaboration, switched control, optimal control, constrained control, and networked control.

Yuanqing Xia is currently a professor at Beijing Institute of Technology, China. His research interests are in the fields of cloud control systems, networked control systems, robust control and signal processing, active disturbance rejection control, unmanned system control, and flight control.

High-Performance Adaptive Control of Teleoperation Systems

Di-Hua Zhai
Yuanqing Xia

CRC Press
Taylor & Francis Group
Boca Raton London New York

CRC Press is an imprint of the
Taylor & Francis Group, an **informa** business

北京理工大学出版社
BEIJING INSTITUTE OF TECHNOLOGY PRESS

Most parts of this book are supported in part by the National Natural Science Foundation of China under Grant 62173035 and Grant 61803033, and in part by the Beijing Natural Science Foundation under Grant 4161001.

First edition published 2023
by CRC Press
6000 Broken Sound Parkway NW, Suite 300, Boca Raton, FL 33487-2742

and by CRC Press

4 Park Square, Milton Park, Abingdon, Oxon, OX14 4RN

CRC Press is an imprint of Taylor & Francis Group, LLC

Library of Congress Cataloging-in-Publication Data

Names: Zhai, Di-Hua, 1988- author. | Xia, Yuanqing, author.
Title: High-performance adaptive control of teleoperation systems / Di-Hua Zhai, Yuanqing Xia.
Description: First edition. | Boca Raton : CRC Press, [2023] | Includes bibliographical references. | Summary: "Within a unified switched-control framework, this book investigates the high-performance control designs and theoretic analyses for teleoperation systems, including the joint space and task space teleoperation, the homogeneous and heterogeneous teleoperation, and the single-master single-slave and multi-master multi-slave teleoperation. The book begins with an introduction to the concepts and challenges of networked teleoperation systems. Then, it investigates a new adaptive control framework based on auxiliary switched filters for the bilateral teleoperation systems to handle the model uncertainty and non-passive external forces. To overcome the input constraints of robotic systems, this adaptive method is also extended to the anti-windup adaptive control case. Furthermore, to apply to multi-robot remote collaboration scenarios and heterogeneous teleoperation, two tele-coordination methods and an adaptive semi-autonomous control method are respectively developed. Finally, the authors examine two finite-time control schemes and two types of improved prescribed performance control for teleoperation systems to improve the transient-state and steady-state synchronization performances. This title will be an essential reference for researchers and engineers interested in teleoperation, robotic systems, and nonlinear control systems. It would also prove useful for graduate students in the fields of science, engineering, and computer science"-- Provided by publisher.
Identifiers: LCCN 2022051005 (print) | LCCN 2022051006 (ebook) | ISBN 9781032465159 (hbk) | ISBN 9781032465173 (pbk) | ISBN 9781003382058 (ebk)
Subjects: LCSH: Adaptive control systems. | Remote control. | Robotics.
Classification: LCC TJ217 .Z49 2023 (print) | LCC TJ217 (ebook) | DDC 629.8/36--dc23/eng/20230103
LC record available at https://lccn.loc.gov/2022051005
LC ebook record available at https://lccn.loc.gov/2022051006

ISBN: 978-1-032-46515-9 (hbk)
ISBN: 978-1-032-46517-3 (pbk)
ISBN: 978-1-003-38205-8 (ebk)

DOI: 10.1201/9781003382058

Typeset in Latin Modern font
by KnowledgeWorks Global Ltd.

Dedicated to our parents.

Contents

Section II Single-Master Single-Slave Teleoperation

Section III Multi-Master Multi-Slave Teleoperation

Preface

This book addresses the adaptive control methods for teleoperation system in the presence of input and/or state and/or output constraints. The main goal of this book is to provide a unified control framework for teleoperation systems to handle the constraints in the presence of varying time delays and non-zero external forces. The specific control designs and theoretic analyses are investigated for teleoperation systems from joint space teleoperation to task space teleoperation, from homogeneous teleoperation to heterogeneous teleoperation, from single-master single-slave teleoperation to multi-master multi-slave teleoperation. Indeed, all of the targeted problems of this book are fairly new from a historical perspective and to the best of our knowledge no book or article covers such a broad range of problems within a unified control framework.

In the past half century, the study of teleoperation system has received extensive attention. Numerous control schemes have been investigated, including passivity-based control, Lyapunov-Krasovskii functional method and methods based on small-gain theorem. However, many fundamental problems are still either unexplored or less well understood. In particular, there still lacks a comprehensive framework that can systematically cope with all core issues. This motivated us to write this text.

The book is devoted to present theoretical explorations on several fundamental problems for teleoperation systems, forming the fundamental theoretical framework of teleoperations. It highlights new research on commonly encountered nonlinear teleoperation systems including adaptive control, anti-saturation control, multilateral coordinated control, adaptive semi-autonomous control, finite time control and predefined performance control. A unified switched control is developed to handle the adaptive control of teleoperation systems in the presence of input constraints, state constraints and output constraints. The presented results mark a substantial contribution to nonlinear teleoperation system theory, robotic control theory and networked control system theory.

The book has been organized into the following chapters.

In the introduction chapter, a broad overview of networked teleoperation systems is provided detailing current state-of-the-art research in this field. Chapter 2 presents basic concepts, notations and a brief background of robotics and related constructs and models that are used throughout the book. The main mathematical results, from switched control, nonlinear control and networked control, are also provided to facilitate further developments in different chapters. Chapter 3 presents the teleoperation system of a new adaptive control framework based on an auxiliary switched filter. By introducing the switched filter systems, the complete closed-loop system is modeled as a nonlinear switched delayed system, which consists of two interconnected switched and non-switched subsystems. By using the switching stability analysis in the framework of state-independent input-to-output stability, the closed-loop stability is then derived. The proposed method achieves the control of teleoperation systems in the simultaneous presence of varying time delay, model uncertainties and passive/nonpassive external forces. Chapter 4 addresses the adaptive control of the teleoperation system with actuator saturation. To unify the study of actuator saturation, passive/nonpassive external forces, asymmetric time-varying delays and unknown dynamics in the same framework, the control based on the auxiliary switched filter is developed. A specific control design is developed in the form of proportional plus damping injection plus switched filter, which simplifies the traditional one. Chapters 5 and 6 address the tele-coordinated control of multiple robots. To achieve the control objective, under the framework of auxiliary switched filter control, two tele-coordinated methods are provided, which are, respectively, the constrained coordination control and unconstrained coordination control. Chapter 7 addresses the adaptive task-space bilateral teleoperation for heterogeneous master and slave robots to guarantee stability and tracking performance, where a novel semi-autonomous teleoperation framework is developed to ensure the safety and enhance the efficiency of the slave robot. Chapter 8 addresses the adaptive finite-time control problem of nonlinear teleoperation systems in the presence of asymmetric time-varying delays. To achieve the finite-time control objective, the nonsmooth control technique is employed where the nonsmooth switched filter is investigated. On the basis of Chapter 8, Chapter 9 investigates the anti-windup finite-time control for the teleoperation system, where to study the finite-time control, actuator saturation and asymmetric time-varying delays in a unified framework, a novel non-smooth generalized switched filter is designed. Chapter 10 provides a prescribed performance control framework for teleoperation systems, which guarantees the given transient-state and steady-state synchronization performances between the master and slave robots. It overcomes the shortcomings that the predefined time control can only improve the convergence speed and do nothing to other transient state performance index like overshoot. Chapter 11 investigates a switching-based prescribed-performance-like control approach for nonlinear teleoperation systems, which balances the performance and implementability of classical predefined performance control algorithm when its hard constraint is violated and is thus applicable to a wider area.

Readers of our book are assumed to have some background in robot control and nonlinear control. We believe that this book will be of interest to university researchers, R&D engineers and graduate students in control theory and control engineering who wish to learn about the key principles, methods, algorithms and applications of networked teleoperation systems, robotic systems and nonlinear control systems. Although this book is primarily intended for students and practitioners of nonlinear teleoperation system control theory, it is also a valuable reference for those in fields such as communication engineering.

ACKNOWLEDGMENTS

The main contents of this book are the result of extensive research over the past decade. Most of the results have been reported in research papers. We would like to express our thanks to former and current students, Weizhi Lyu, Yuhan Xiong, Zigeng Yan, Sihua Zhang, Yongkang Wang and Hao Li, for their diligent and active works on the achievements of this book.

Most parts of this book are supported in part by the National Natural Science Foundation of China under Grant 62173035 and Grant 61803033, and in part by the Beijing Natural Science Foundation under Grant 4161001.

Di-Hua Zhai and **Yuanqing Xia**

Beijing Institute of Technology

I

Preliminaries

CHAPTER 1

Introduction

IN this chapter, an extensive overview of networked teleoperation systems will be carried out detailing the current state-of-art research in this field.

1.1 BACKGROUNDS

Networked teleoperation control is one of the hot research directions in the field of robot system and control since the 21st century [1–3]. Thanks to the introduction of network technology, telerobot can work in dangerous and harsh environment instead of human beings, which has been rapidly developed in various applications [4, 5]. In recent years, it can be seen in many major events, such as the accident rescue of Fukushima nuclear power plant in Japan [6], MH370 search and rescue [7], telesurgery [8], Yutu lunar rover [9].

Roughly speaking, the teleoperation system is a whole composed of operators, master robots, slave robots, working environment and communication networks [1]. Essentially, it is a typical networked nonlinear dynamic system. Therefore, it has both nonlinear and networked characteristics. Up to now, the control of teleoperation system still faces many challenges. On the one hand, the complex external environment and the strong nonlinear time-varying characteristics of the robot itself make the control design of the robot difficult. On the other hand, although the introduction of network can make teleoperation transmit data between different systems and provide convenience for operators in different places to share resources and coordinate cooperation, it also brings difficulties to teleoperation control. Teleoperation system is usually accompanied by network transmission. The instability, suddenness, vulnerability to interference and poor noise suppression ability of information transmission in the real network system, as well as the inevitable time lag, packet loss, error code and inaccuracy in the process of information transmission in the network, have exerted considerable influence on the stable control of teleoperation robot. Considering that robots are mostly nonlinear time-varying systems, the existing research results of linear networked control systems are difficult to be directly applied to the control and analysis of such systems. In fact, J. P. Richard once pointed out that the control problem of nonlinear time-delay systems is still an open problem and needs further research [10]. At present, researchers have proposed many methods for the control design and analysis of networked teleoperation system, such as methods based on

DOI: 10.1201/9781003382058-1

wave variables and scattering theory [11–13], passive methods [12,14], ISS/IOS-based (input-to-state stability/input-to-output stability) small gain methods [15–19]. However, these methods are faced with different limitations in use. The traditional wave variable method and scattering theory are only suitable for constant communication delay [2]. Passive method requires that the main input and the external input (respectively refers to the operator and the external force of the environment) meet the passive conditions, while the actual system is complex and diverse, which may not meet the passivity everywhere [19,20]. Although the existing ISS/IOS-based small gain methods well avoid the passivity requirement, the position signal of the robot also belongs to the closed-loop system state. According to the definition of ISS/IOS [21–23], when the external input of the robot is removed, the master and slave robots will gradually achieve the movement synchronization, and finally stop at the zero position. However, in the practical teleoperation system, when the external input is removed, the movement trend should be that the master and slave robots gradually achieve synchronization, and stop at this position after synchronization, instead of continuing to move synchronously to the zero position. Therefore, it is very necessary and urgent to find new effective control design and analysis methods for teleoperation system in complex network and external environment.

In this regard, this book systematically studies new methods of control design and performance analysis of networked teleoperated robots in complex external environment. Aiming at the requirements of convenience of control implementation, universality of analysis method and performance improvement, based on the idea of dynamic compensation, this book proposes an auxiliary switched filter control method for teleoperation system. For the various control tasks from joint space teleoperation to task space teleoperation, from homogeneous teleoperation to heterogeneous teleoperation, from single master and single slave teleoperation to multi-master and multi-slave teleoperation, the detailed design and performance analysis are given in the complex environment with asymmetric time-varying communication delay, passive/non-passive external input, actuator saturation, model uncertainty and external interference. Some better control performance is achieved in a wider application range. The related achievements have been used to solve the key scientific problems such as high precision control, high flexibility and autonomous obstacle avoidance performance of the networked teleoperated robots, which improves the robot's operation ability in complex and dynamic environment.

1.2 TELEOPERATION SYSTEM CONTROL

The earliest teleoperation system in modern sense can be traced back to the 1950s. At that time, to protect workers from radiation in nuclear test, Goertz and others in Argonne National Laboratory of the United States designed the first master-slave teleoperation system, in which the master and slave robots were connected by mechanical structure [4]. In 1954, they also developed a new teleoperation platform based on electric manipulator [24]. Since then, in view of the advantages of teleoperation, teleoperated robotic system has been rapidly applied to many new fields: from the submarine underwater maintenance teleoperated robot in the 1960s [25], to the teleoperated mechanical arm for high-voltage line maintenance in the 1990s [26], from Rotex space manipulator for Columbia Space Shuttle in 1993 [27] to the first teleoperated robot developed by Dr. Marescaux in 2001 [28]. In the 21st century, with the rapid development of network communication technology, remote data transmission and data exchange can be realized, which make the Internet-based teleoperated robot control system develop rapidly in various applications [25,29–31].

A typical networked teleoperation system consists of five parts, namely, human operator, master robot, slave robot, environment and communication network [1]. Teleoperation system greatly expands human's working ability in dangerous and harsh environment. The introduction of network brings convenience to teleoperation, but also brings new challenges to its control design and analysis. The following is a brief summary of related research status for the work of this book.[1]

Teleoperation control theory has experienced many important development moments since its birth in the 1950s. In the 1960s, teleoperation system began to consider the influence of communication delay [32]. At this time, supervisory control and other methods are used to solve the teleoperation control with constant communication delay [33]. Since the 1980s, more advanced control theories have been introduced and used in teleoperation system control, including Lyapunov stability method and passivity method, among which many methods are still popular in teleoperation system control [11,34]. From the end of the 1980s to the beginning of the 1990s, the virtual representation control method of teleoperation system based on two-port network [11], wave variable method [13] were developed. Teleoperation control at this stage either does not consider the influence of time delay or only considers the constant communication delay. Since the 1990s, people began to apply the Internet step by step. At this time, people began to pay attention to the teleoperation system control with time-varying communication delay [35]. The control methods used mainly include supervisory control, passivity method, sliding mode control method, control method based on Lyapunov stability theory, robust control method, predictive control method and the combination of the above control strategies.

The research on the time-delay stability of teleoperation system is almost carried out at the same time as the application research of teleoperation control. In the 1960s, Sheridan and Ferrell have built some preliminary operation experiments to verify the effect of communication delay on the control performance of teleoperation system [32,33], where the goal is to quantify the total time required to complete a prespecified task. In the experiment, the operator adopted the move-and-wait strategy to ensure that the task could be completed. The master system starts the control movement first, and then waits to observe the response of the remote slave robot. After completion, start the correction movement and wait for the slave robot to complete the response action again. Repeating the above actions until the task is completed. According to the experimental research of "move-wait" strategy, the time for teleoperation to perform specific tasks is greatly increased due to the existence

[1]It should be noted that teleoperation control has two major indicators, namely stability and transparency. Stability is the foundation of the system, and the work in this book only considers the stability of teleoperation system temporarily. Therefore, the following review of research status will only focus on stability.

of time delay, and this increasing trend is usually proportional to the time delay. To avoid this problem, Sheridan and Ferrell put forward a method of supervision and control in 1967 [33]. Since then, the time-delay problem of teleoperation system has gradually attracted people's attention.

In 1989, Anderson and Spong proposed a teleoperation control method based on scattering theory [11]. Based on the lossless transmission line theory, the master robot, slave robot and communication channel are equivalent to a two-port network, the human operator and environment are described as a port, and the force and speed of the robot are equivalent by voltage and current. Then, they also proved the asymptotic stability of teleoperation system based on scattering control method by using Lyapunov method [36]. The study found that the main reason for the instability of teleoperation system was the non-passivity of communication network. In 1991, Niemeyer and Slotine further put forward the wave variable theory [13]. The passive stability of the system is ensured by introducing wave variable controllers into the master and slave terminals, and using wave variables instead of power variables such as speed and force to transmit between the master robot and the slave robot. However, the stability criteria obtained in this period are all time-delay independent. With the development of control theory, researchers have put forward some improved passivity methods. For example, the wave variable method is suitable for the time-varying delay situation [37], integral wave variable method [38], four-channel method [39]. To solve the problem of excessive energy consumption in wave variables and scattering methods, Ryu et al. also proposed passive time-domain method [40–43]. Chopra and Spong put forward a new design method of position tracking control [12,14], which combines passive scattering theory with proportional control, and provides strict theoretical proof. At present, this method has become the theoretical analysis basis of the widely used impedance injection control method. In 2008, aiming at teleoperation system with constant communication delay, Nuno and Ortega put forward the control method of "proportional+impedance injection" for the first time [44], which has the advantage of simple control structure. In 2009, they extended this method to teleoperation system with time-varying delay [45]. At present, this method has become a very popular passive control strategy of teleoperation system [46–50]. Among them, Islam designed a "proportional+impedance injection" controller for the teleoperation system with constant value and time-varying delay in 2013 [46], and then in 2014, the model uncertainty is further considered [47]. Hashemzadeh further considered actuator saturation, and a control algorithm of "nonlinear proportional plus impedance injection" with gravity compensation is proposed for teleoperation system with constant communication delay and actuator saturation [48]. On the basis of the method proposed by Hashemzadeh, torque feedback control is further introduced to realize the hybrid position and force control [50]. At present, there are many other developments in teleoperation system control based on passivity method, see [51–57] and the references therein. It should be pointed out that Nuno et al. published a paper on Automatica in 2011, which summarized the development process of passivity methods in teleoperation system in detail [2].

Passive method requires the robot system to meet certain passive conditions, and it faces limitations in practical implementation. For example, in the popular framework of "Proportion+Impedance Injection", it is required that the external input of the robot (external force of the operator/external force of the environment) is passive relative to the speed of the robot. However, In a series of papers by Dr. Polushin [15,17–19], Polushin et al. clearly pointed out this passive assumption of external input is highly conservative, which is not valid in some cases. In order to overcome the limitation of passivity method, researchers proposed to design teleoperation system's controller in the framework of ISS/IOS stability [15,16,58]. At this time, the force applied by the operator/environment to the robot is regarded as the input of the closed-loop system, and when it is (essentially) bounded, it can be proved that the motion of the master and slave robots is bounded and synchronized. In this method, the force applied by the operator (environment) to the master (slave) robot can be non-passive, thus avoiding the conservative assumption of passive external force. Here, it should be pointed out that Polushin et al. proposed to realize the control design of teleoperation system by using ISS/IOS stability and small gain theorem method, and got a series of research results [15–19]. However, under the existing ISS/IOS small gain framework, because the complete closed-loop state contains the robot joint position information, theoretically, after the external force is removed, the master robot and the slave robot will still move to the zero position point when reaching synchronization, and finally stop at the zero position[2], which contradicts the actual teleoperation. In practice, when the external input is removed, under the action of the controller, the master and slave robots should stop moving after reaching synchronization, instead of moving to the zero position together. Therefore, the current ISS/IOS method still has limitations.

Sliding mode control technology is a very effective control method in the research of uncertain systems and nonlinear systems [59–61]. Sliding mode control can also be applied to teleoperation system with transmission delay [62,63]. In 1994, Buttolo applied sliding mode control to force feedback teleoperation system, and achieved good control performance, but it did not consider the influence of communication time delay [64]. In 1999, Park et al. proposed a new sliding mode control design scheme for teleoperation system with time-varying delay [62]. In recent years, with the maturity of terminal sliding mode control technology [65,66], researchers also apply it to teleoperation system to improve control performance. Aiming at the teleoperation system with constant time delay, the nonsingular fast terminal sliding mode control technology is adopted, and the finite time control of the teleoperation system is realized by combining with the fuzzy controller [67]. On this basis, Yang et al. further considered the transient synchronization performance of the master and slave robot systems, the barrier Lyapunov function method is introduced, and under the framework of sliding mode control, the dual control of the transient and steady synchronization performance of the teleoperation system is realized [68]. However, in practical application, the sliding mode control method will have the problem of high-frequency chattering, which is an inherent characteristic of sliding mode control and is usually hard to overcome [69].

---

[2]This phenomenon is called zero steady-state synchronization in this book.

For teleoperation system with time delay, Lyapunov-Krasovskii functional or Lyapunov-Razumikhin function can also be constructed. By using Lyapunov stability theory or Razumikhin stability theory, the stability control of the system with or without time delay can be obtained [54, 70–72]. When using this method to analyze the system, the closed-loop master-slave teleoperation system is first modeled as a dynamic system with state delay, and then the stability of the system can be proved by selecting the appropriate Lyapunov function or functional. However, when using Razumikhin's stability theorem, a delay-independent stability criterion is obtained, which is conservative to some extent [73]. Relatively speaking, Lyapunov-Krasovskii stability theorem can achieve the control performance of both stability and operational performance [54]. At present, the method based on Lyapunov-Krasovskii stability theorem has been widely used in teleoperation system [48, 74–78]. Among them, some also combine linear matrix inequalities (LMIs) technology with passivity technology [48, 75, 76, 78]. However, to obtain a less conservative stability criterion, the selection of Lyapunov-Krasovskii functional will be very complicated, usually involving double integral or even triple integral. In the process of theoretical analysis, many inequalities and the application of derivation techniques will be involved, such as Jessen inequality, Moon inequality, freedom matrix, time-delay interval decomposition, etc. The process is complicated. For different systems, the structural forms of Lyapunov-Krasovskii functional will be different, and the portability of analysis methods is not strong.

There are other control methods for teleoperation system control in complex network and external environment, such as model predictive control methods [79–84], robust control methods [85–95]. On the one hand, the model predictive control method, as an advanced control algorithm, can deal with various constraints in the control system. However, because it usually faces a huge amount of computation, it has not been widely used in teleoperation systems with high real-time requirements. At the same time, because the characteristics of "human-in-loop" in teleoperation system[3], it makes the model predictive control of teleoperation system much more complicated and difficult than the general process/motion control system. On the other hand, robust control methods (H_∞ control or μ-synthesis design) usually get conservative results for large communication delay, while for small time delay, the stability of the system may not be guaranteed. The application of robust control design sometimes needs to solve appropriate partial differential equations or inequalities, which are very difficult to solve in teleoperation system.

1.3 TELEOPERATED CONTROL WITH INPUT SATURATION

Actuator saturation widely exists in all kinds of control systems, and teleoperation systems also need to face this constraint. In order to deal with the influence of actuator saturation on control, many control methods have been proposed in the robot

[3]In the teleoperation system, the human operator of the master robot is an important component of the closed-loop system, and the external force input of the operator to the master robot has an important influence on the whole closed-loop system, so the operator factor must be considered in the control design and analysis of the teleoperation system. Teleoperation system is a typical "human-in-loop" control system.

system [96–102]. Among them, Morabito et al. studied the saturation control method of a class of general Euler-Lagrange system [103]. Loria et al. put forward a nonlinear proportional control algorithm with gravity compensation term [104], which uses saturation function to add position error term to realize nonlinear proportional input, and finally realizes bounded tracking control. Zereroglu et al. improved on this basis, and proposed an adaptive gravity compensation plus nonlinear proportional-differential control algorithm [105]. However, these control algorithms are only aimed at the saturation control of a single robot, and it is difficult to apply them to teleoperation systems with transmission delays. At present, there are some research achievements on the saturation control of teleoperation system. Lee and Ahn et al. studied the saturation control of teleoperation system with constant communication delay by assuming that the difference between any signal itself and its time-delay signal is bounded [106–108]. It should be pointed out that, in addition to the limitation of constant communication delay, the hypothesis of boundedness difference between the actual physical signal and its delayed physical signal in the existing framework [106–108] is also conservative and has no strict theoretical support. By improving the saturation function used in the algorithm, Hemzadeh et al. further extended the Loria's method [104] to teleoperation system with time-varying communication delay in 2013 [48]. However, it does not consider the uncertainty of robot model parameters, and there will be limitations in practical application. On this basis, Ganjefar et al. further considered the force feedback control, which improved the transparency of teleoperation control [50]. However, no model uncertainties and external disturbances have been considered. To deal with model uncertainty and actuator saturation at the same time, Yang et al. proposed an adaptive anti-saturation compensation controller based on neural network for teleoperation system with constant communication delay [109]. Hua et al. considered the saturation control of teleoperation system with time-varying communication delay, and designed an adaptive nonlinear proportional+nonlinear impedance injection controller based on passivity method [110]. Considering the limitations of passivity methods, it is still a theoretical difficulty to comprehensively consider the time-varying time-delay, actuator saturation, model uncertainty and external disturbances in teleoperation robot control. The related research is of great engineering significance for the practicability of teleoperation control theory.

1.4 MULTILATERAL TELE-COORDINATED CONTROL

In some complex task environments, multi-robot cooperation[4] can perform tasks that a single robot can't accomplish. For example, when controlling the removal of a heavy or fragile object from a robot, it will bring great risks if it is done independently by a single robot due to its working ability. In this case, it is a good choice to control multiple robots to accomplish this task cooperatively. In the past ten years, the coordination and cooperation control of multi-robots has received extensive attention, which can be seen from the assembly tasks of industrial automation, the material

[4]In this book, multi-robot coordination refers to the cooperation of multi-robots through the operation coordination of the robot end effector, as shown in Fig. 1.1.

handling in dangerous environment, and the maintenance tasks of spacecraft [111–118]. Many control methods are also proposed. The existing coordination and cooperation control algorithms generally include centralized hybrid position/force control [119], the Leader-Follower control [120], the master-slave control [121], the target impendence control [122] and the distributed impendence control [123].

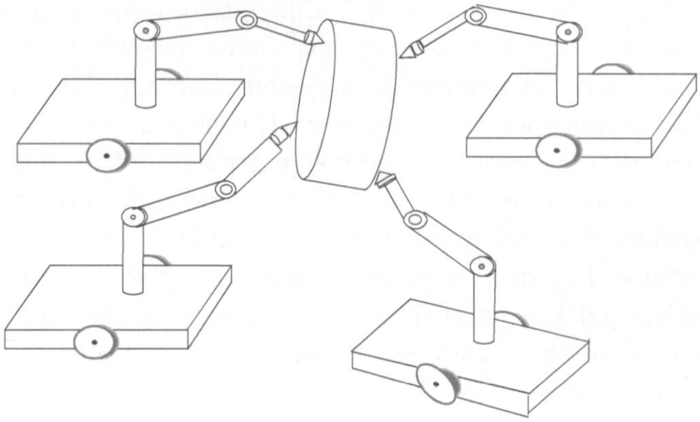

Figure 1.1 A schematic diagram of multi-robot cooperation.

Because of the unique advantages of multi-robot coordination and cooperation, researchers have also conducted related research during the development of teleoperation control theory and technology [100, 101, 124–131]. At present, the multilateral cooperative teleoperation control among multi-robots has become one of the hot spots of teleoperation system control research, including single-master multi-slave teleoperation [76, 115, 132, 133], multi-master single-slave teleoperation [3, 134]and multi-master multi-slave teleoperation [135–138].

Referring to the common multi-robot cooperation control mode, the multilateral coordination teleoperation control can be roughly divided into constrained cooperation architecture [112, 135, 136, 138] and non-constrained cooperation architecture [139]. On the one hand, in constrained cooperative control, slave robots are physically connected together to perform tasks, and a hybrid motion/force control strategy is adopted. All slave robots and their operating objects form a closed-chain mechanism, and the controller design is based on a unified dynamic model of robots and operating objects. In the control process, it is necessary to maintain a specific kinematic relationship formed by the precise internal force control applied from the robot. However, when this method realizes coordination and cooperation, it needs to measure or estimate the internal force, which will increase the economic and technical burden of application. Li and Su used this control method to realize the coordinated control of two-master and two-slave teleoperation robots [76]. Due to the introduction of communication, the multilateral cooperative teleoperation control based on constrained cooperative strategy is more complicated than the common multi-robot cooperative control. To deal with communication delay, Li et al. designed the controller by constructing Lyapunov-Krasovskii functional and applying Lyapunov stability theory [76]. The analysis process is cumbersome, and the communication delay is strictly limited. The upper bound of the change rate of the delay must be less than 1, which is highly conservative. On the other hand, under the unconstrained cooperation framework, each pair of master and slave robots will be independently controlled, with the master robot being the leader, controlled by the operator, and the slave robot being the follower, tracking the leader (master robot) through effective control strategies, and finally achieving coordination and cooperation. Because this method does not control the force on the robot, the operator needs to monitor the movements of other slave robots more carefully to avoid collision and conflicting movements. Compared with constrained collaborative control, unconstrained collaborative control method puts forward higher requirements for coordination and cooperation among operators, which will increase the workload of operators. In addition, because the coordinated teleoperation system consists of two or more pairs of master and slave robots, if the information between each pair of master and slave robots is transmitted through different communication channels, different transmission delays may occur, which will also bring difficulties to the implementation of unconstrained cooperative teleoperation control algorithm. Because of these factors, the unconstrained cooperative teleoperation control is still facing difficulties in practical implementation.

1.5 HETEROGENEOUS TELEOPERATION CONTROL

In the past few decades, teleoperation system has gradually become one of the useful tools to replace human beings in dangerous and harsh environments [1, 14, 140, 141]. A bilateral teleoperation system consists of master and slave robots, in which the master and slave robots interact with each other through communication networks. When the telerobot system is operated by the operator, the controlled coupling between the master robot and the slave robot is used by the slave robot to perform tasks. However, for most teleoperation systems, there is a long distance between the master robot and the slave robot in space, which may make it difficult for operators to obtain the (environmental) information from the slave robot in real time. This lack of information, coupled with the operator's own cognitive limitations, will lead to unpredictable (dangerous) situations, which will greatly limit the ability of teleoperation robot system [142]. In order to overcome this shortcoming, there are currently two main solutions. One is to develop intelligent robots, which can independently complete control tasks. However, limited by the technical level, it is still difficult for such robots to complete tasks independently, especially some complicated control tasks [143]. The other is the semi-autonomous mode of heterogeneous teleoperation with human intervention [82, 144–149]. Different from the general teleoperation system, in this kind of task, the master robot and the slave robot are usually heterogeneous, and the slave robot should have some redundant degrees of freedom. Using the redundant degrees of freedom of the slave robot, by designing the slave controller with semi-autonomous function, the joint position of the slave robot can be reconfigured [150], and finally the heterogeneous teleoperation control with semi-autonomous functions such as obstacle avoidance can be completed. The latter is also the current mainstream method for this type of control problem [151, 152]. Robotic systems based on the idea of semi-autonomous control have been used in rehabilitation robot [153], search and rescue robot [154], underwater robot [155], etc.

For semi-autonomous teleoperation system, the system is generally composed of non-redundant master robot and slave robot with redundant degrees of freedom [156]. Because redundant joints can be used to perform subtasks (such as obstacle avoidance, singularity avoidance, joint angular limit, etc.), it will not affect the robot task space movement [150]. Therefore, a slave robot with redundant joints can provide greater flexibility, manipulability and mobility for operators to complete some complex tasks. However, the use of kinematic redundant robots also brings difficulties to the teleoperation control design. Fortunately, many researchers have studied the teleoperation system control of slave robots with redundant degrees of freedom, see [134, 151, 152, 156, 157] and the references therein. Among them, Li et al. studied the control of asymmetric teleoperation system with two masters and one slave [134]. By applying the space transformation theory, the control task of the slave robot is decomposed into the motion space of the grasping object and the internal force space of the grasping object according to different requirements for separate control, and two master robots are used to control the motion and internal force of the slave robot respectively. However the heterogeneous framework [134] is difficult to meet the requirements of the slave robot for obstacle avoidance, singularity avoidance and joint angular limits. Nash et al. studied the control of teleoperation system in which both the master and slave robots have redundant joints [152], but its algorithm requires that the master and slave robots have the same degree of freedom, and the influence of communication delay is not considered in the control design. The control of asymmetric teleoperation system with constant communication delay is developed [151]. The robot system consists of two dual non-redundant master robots and a slave robot with redundant joints. In this system, although the slave robot has redundant joints, it is completely controlled by the operator through two master robots (two master robots cooperatively control different coordinates of the slave robot). Liu et al. studied the semi-autonomous control of single master and single slave teleoperation system, which used a redundant slave robot to enhance the efficiency of complex teleoperation tasks, and proposed a multi-class subtask controller [156, 157]. As two important research achievements of semi-autonomous teleoperation system, the Liu's works [156, 157] also face limitations in application. The control design and stability analysis proposed in 2013 [156] only considers the constant communication delays, while although the further work [157] considered the control of time-varying communication time-delay, its algorithm needs to use accurate information of time-delay change rate, and it can only realize the control of teleoperation system with nominal dynamic model in the case of non-passive operator/environmental external force.

How to develop a practical semi-autonomous control method of heterogeneous teleoperation system with time-varying communication delay is still a difficult problem to be solved urgently.

1.6 HIGH-PERFORMANCE CONTROL OF TELEOPERATION SYSTEMS

The control system performance is often judged according to the three principles of "stability, accuracy and speed". Stability is the basis of all control, and it is the same in teleoperation system control. The existing teleoperation control methods pay full attention to the closed-loop stability and give strict theoretical analyses [1,2,12,14,19,54,124,134,158–160]. After realizing the stable control of teleoperation system control, people naturally want to work hard on accuracy (steady-state error, overshoot) and rapidity (synchronization speed of master and slave robots) to pursue high-performance control with better control performance. Taking the application of robot-assisted surgery system in minimally invasive spinal surgery as an example [28,161,162]. In the process of minimally invasive spinal surgery, the accuracy of the operation position is extremely high, the operation space of the spine is relatively small, there are many unnecessary obstacles, and the operation robot moves freely, so the control accuracy and stability of the operation robot (the speed and acceleration of the robot should be smooth without sudden change) are required to be high [163]. The surgical robot, as a slave robot, will follow the movement of the master robot operated by the doctor under the action of the control law. Based on the classical control theory, the evaluation of the following performance of the slave robot mainly considers its transient tracking performance and steady tracking performance. Transient tracking performance mainly includes tracking response speed and tracking response overshoot. The steady-state tracking performance is mainly the magnitude of the steady-state tracking error. In practice, faster response speed, smaller tracking overshoot and steady-state tracking error are required. It is still very difficult to realize the high-performance control of the strong nonlinear system such as telerobot by further coupling the communication network and complex external environment. In recent years, after the successful application of non-smooth control, predefined performance control and other nonlinear control technologies, the high-performance control of teleoperation system has made some progress [164–169].

To realize the master-slave fast synchronization, researchers put forward the finite-time control strategy of teleoperation system by using the nonsingular fast terminal sliding mode technology [67,96,110]. Yang et al. propose a new fast nonsingular terminal sliding surface [67], where a finite-time fuzzy control algorithm is designed in the presence of system uncertainty and external disturbance, which realizes the finite-time control of teleoperation system with constant communication delay. However, it assumes that the master and slave robots are not affected by external forces; that is, they are in free motion, and still face strong application limitations. Yang et al. further considered the limited time control of teleoperation in the case of non-free movement [170] and, based on the sliding mode control method, realized the rapid synchronization control between the master and slave robots and the quantitative analysis of synchronization time under the constant communication delay. Since Yang et al. [67,170] only considered the constant communication delay, Hua et al. proposed a new finite-time speed observer [110]. Under the framework of passivity method, an adaptive control strategy was designed based on the observed speed signal to ensure the synchronization performance of the system with time-varying communication delay and actuator saturation. It should be pointed out that although the Hua's work [110] realized the limited time observation of unknown velocity signals, it only realized the general asymptotic control for the synchronous control of the master and slave robots that need to consider the time-varying communication delay. Recently, some methods are also developed to finite-time master-slave synchronization control of teleoperation system with time-varying communication delay, see [171–175] and the references therein. Nevertheless, the convergence time in most finite-time controllers usually depends on the initial states of the robot and is often an unbounded function of the states, which is not convenient for engineering practice. To overcome this drawback, the fixed time or predefined time control is further developed [176–180].

The finite time control algorithm can only guarantee the transient performance of the system to a certain extent, and cannot quantitatively give information such as synchronization speed and tracking overshoot. In order to make up for the shortage of limited time control strategy, in recent years, the predefined performance control technology[5] has been applied to realize higher-performance teleoperation control [68, 109, 188–192]. Under the predefined performance control framework, the transient and steady-state tracking performances are transformed into explicit constraints by the given predefined performance function, and finally used in controller design. Moreover, the minimum speed, maximum overshoot and maximum steady-state error of master-slave synchronization can be quantitatively analyzed and controlled [193]. Yang et al. designed an adaptive fuzzy controller based on PPC strategy, which realized the high-performance control of teleoperation system [188]. In 2015, they further considered the actuator input saturation constraint [109]. By comprehensively using BLF strategy and nonsingular terminal sliding mode control technology, the fast and high-performance synchronous control of teleoperation system is developed [68]. It should be noted that the predefined performance control designs [68, 109, 180, 188–192] still face many challenges in practical application. On the one hand, they only consider the constant communication delay, and can't deal with the more general time-varying communication delay. On the other hand, those algorithms [68, 109] are only applicable to the case where both the master and slave robots are in free motion, while the algorithm [188] can realize the prescribed performance control in the case of non-free motion, but it requires that the external force of the operator/environment received by the master and slave robots meet the strict impedance input form, which is highly conservative. Besides BLF and PPC methods, the funnel method is also a third means to achieve predefined performance control, which can achieve high-performance control of teleoperation systems [194].

At this time, the high-performance control of teleoperation system in complex network and external environment is still a theoretical difficulty that needs to be solved urgently.

[5]Common predefined performance control methods include prescribed performance control (PPC) [181–183] method and barrier Lyapunov function (BLF) method [184–187].

Bibliography

[1] P. F. Hokayem and M. W. Spong, "Bilateral teleoperation: an historical survey," *Automatica*, vol. 42, no. 12, pp. 2035–2057, 2006.

[2] E. Nuno, L. Basanez, and R. Ortega, "Passivity-based control for bilateral teleoperation: a tutorial," *Automatica*, vol. 47, no. 3, pp. 485–495, 2011.

[3] F. Hashemzadeh, M. Sharifi, and M. Tavakoli, "Nonlinear trilateral teleoperation stability analysis subjected to time-varying delays," *Control Engineering Practice*, vol. 56, pp. 123–135, 2016.

[4] T. B. Sheridan, "Telerobotics," *Automatica*, vol. 25, no. 4, pp. 487–507, 1989.

[5] A. F. Villaverde, A. Barreiro, and C. Raimundez, "Passive position error correction in Internet-based teleoperation," *Automatica*, vol. 46, no. 11, pp. 1884–1890, 2010.

[6] K. Nagatani, S. Kiribayashi, Y. Okada, K. Otake, K. Yoshida, S. Tadokoro, T. Nishimura, T. Yoshida, E. Koyanagi, M. Fukushima *et al.*, "Emergency response to the nuclear accident at the fukushima daiichi nuclear power plants using mobile rescue robots," *Journal of Field Robotics*, vol. 30, no. 1, pp. 44–63, 2013.

[7] R. Bogue, "Underwater robots: a review of technologies and applications," *Industrial Robot: An International Journal*, vol. 42, no. 3, pp. 182–191, 2015.

[8] A. M. Okamura, "Methods for haptic feedback in teleoperated robot-assisted surgery," *Industrial Robot: An International Journal*, vol. 31, no. 6, pp. 499–508, 2004.

[9] H. Liu, "An overview of the space robotics progress in china," *International Symposium on Artificial Intelligence, Robotics and Automation in Space*, vol. 14, pp. 1–5, 2014.

[10] J.-P. Richard, "Time-delay systems: an overview of some recent advances and open problems," *Automatica*, vol. 39, no. 10, pp. 1667–1694, 2003.

[11] R. J. Anderson and M. W. Spong, "Bilateral control of teleoperators with time delay," *IEEE Transactions on Automatic control*, vol. 34, no. 5, pp. 494–501, 1989.

[12] N. Chopra, M. W. Spong, R. Ortega, and N. E. Barabanov, "On tracking performance in bilateral teleoperation," *IEEE Transactions on Robotics*, vol. 22, no. 4, pp. 861–866, 2006.

[13] G. Niemeyer and J.-J. Slotine, "Stable adaptive teleoperation," *IEEE Journal of Oceanic Engineering*, vol. 16, no. 1, pp. 152–162, 1991.

[14] N. Chopra, M. W. Spong, and R. Lozano, "Synchronization of bilateral teleoperators with time delay," *Automatica*, vol. 44, no. 8, pp. 2142–2148, 2008.

[15] I. G. Polushin and H. J. Marquez, "Stabilization of bilaterally controlled teleoperators with communication delay: an ISS approach," *International Journal of Control*, vol. 76, no. 8, pp. 858–870, 2003.

[16] I. G. Polushin, A. Tayebi, and H. J. Marquez, "Control schemes for stable teleoperation with communication delay based on ios small gain theorem," *Automatica*, vol. 42, no. 6, pp. 905–915, 2006.

[17] I. G. Polushin, P. X. Liu, and C.-H. Lung, "A control scheme for stable force-reflecting teleoperation over ip networks," *IEEE Transactions on Systems, Man, and Cybernetics, Part B (Cybernetics)*, vol. 36, no. 4, pp. 930–939, 2006.

[18] I. G. Polushin, P. X. Liu, and C.-H. Lung, "A force-reflection algorithm for improved transparency in bilateral teleoperation with communication delay," *IEEE/ASME Transactions on Mechatronics*, vol. 12, no. 3, pp. 361–374, 2007.

[19] I. G. Polushin, S. N. Dashkovskiy, A. Takhmar, and R. V. Patel, "A small gain framework for networked cooperative force-reflecting teleoperation," *Automatica*, vol. 49, no. 2, pp. 338–348, 2013.

[20] S. Islam, P. X. Liu, and A. El Saddik, "New stability and tracking criteria for a class of bilateral teleoperation systems," *Information Sciences*, vol. 278, pp. 868–882, 2014.

[21] E. D. Sontag and Y. Wang, "Notions of input to output stability," *Systems & Control Letters*, vol. 38, no. 4-5, pp. 235–248, 1999.

[22] E. Sontag and Y. Wang, "Lyapunov characterizations of input to output stability," *SIAM Journal on Control and Optimization*, vol. 39, no. 1, pp. 226–249, 2000.

[23] E. D. Sontag, *Input to State Stability: Basic Concepts and Results.* Springer, 2008.

[24] R. C. Goertz and W. M. Thompson, "Electronically controlled manipulator," *Nucleonics*, vol. 12, no. 11, 1954.

[25] M. Ferre, R. Aracil, C. Balaguer, M. Buss, and C. Melchiorri, *Advances in Telerobotics.* Springer, 2007.

[26] M. Nakashima, K. Yano, Y. Maruyama, and H. Yakabe, "The hot line work robot system 'Phase II' and its human-robot interface 'MOS'," in *Proceedings of the IEEE/RSJ International Conference on Intelligent Robots and Systems*, vol. 2. IEEE, 1995, pp. 116–123.

[27] G. Hirzinger, B. Brunner, J. Dietrich, and J. Heindl, "Sensor-based space robotics-ROTEX and its telerobotic features," *IEEE Transactions on Robotics and Automation*, vol. 9, no. 5, pp. 649–663, 1993.

[28] S. E. Butner and M. Ghodoussi, "Transforming a surgical robot for human telesurgery," *IEEE Transactions on Robotics and Automation*, vol. 19, no. 5, pp. 818–824, 2003.

[29] G. T. Sung and I. S. Gill, "Robotic laparoscopic surgery: a comparison of the da vinci and zeus systems," *Urology*, vol. 58, no. 6, pp. 893–898, 2001.

[30] K. Qian, A. Song, J. Bao, and H. Zhang, "Small teleoperated robot for nuclear radiation and chemical leak detection," *International Journal of Advanced Robotic Systems*, vol. 9, no. 3, p. 70, 2012.

[31] L. Cragg and H. Hu, "Application of mobile agents to robust teleoperation of Internet robots in nuclear decommissioning," in *Proceedings of the IEEE International Conference on Industrial Technology*. IEEE, 2003, pp. 1214–1219.

[32] T. B. Sheridan and W. R. Ferrell, "Remote manipulative control with transmission delay," *IEEE Transactions on Human Factors in Electronics*, no. 1, pp. 25–29, 1963.

[33] W. R. Ferrell and T. B. Sheridan, "Supervisory control of remote manipulation," *IEEE Spectrum*, vol. 4, no. 10, pp. 81–88, 1967.

[34] F. Miyazaki, S. Matsubayashi, T. Yoshimi, and S. Arimoto, "A new control methodology toward advanced teleoperation of master-slave robot systems," in *Proceedings of the IEEE International Conference on Robotics and Automation*. IEEE, 1986, pp. 997–1002.

[35] Y. Yokokohji, T. Imaida, and T. Yoshikawa, "Bilateral teleoperation under time-varying communication delay," in *Proceedings of the IEEE/RSJ International Conference on Intelligent Robots and Systems*. IEEE, 1999, pp. 1854–1859.

[36] R. J. Anderson and M. W. Spong, "Asymptotic stability for force reflecting teleoperators with time delay," *The International Journal of Robotics Research*, vol. 11, no. 2, pp. 135–149, 1992.

[37] M. Alise, R. G. Roberts, D. W. Repperger, C. A. Moore, and S. Tosunoglu, "On extending the wave variable method to multiple-DOF teleoperation systems," *IEEE/ASME Transactions on Mechatronics*, vol. 14, no. 1, pp. 55–63, 2009.

[38] E. J. Rodriguez-Seda, D. Lee, and M. W. Spong, "Experimental comparison study of control architectures for bilateral teleoperators," *IEEE Transactions on Robotics*, vol. 25, no. 6, pp. 1304–1318, 2009.

[39] B. Yalcin and K. Ohnishi, "Stable and transparent time-delayed teleoperation by direct acceleration waves," *IEEE Transactions on Industrial Electronics*, vol. 57, no. 9, pp. 3228–3238, 2009.

[40] J.-H. Ryu, D.-S. Kwon, and B. Hannaford, "Control of a flexible manipulator with noncollocated feedback: time-domain passivity approach," *IEEE Transactions on Robotics*, vol. 20, no. 4, pp. 776–780, 2004.

[41] B. Hannaford and J.-H. Ryu, "Time-domain passivity control of haptic interfaces," *IEEE transactions on Robotics and Automation*, vol. 18, no. 1, pp. 1–10, 2002.

[42] J.-H. Ryu, D.-S. Kwon, and B. Hannaford, "Stable teleoperation with time-domain passivity control," *IEEE Transactions on Robotics and Automation*, vol. 20, no. 2, pp. 365–373, 2004.

[43] J.-H. Ryu, J. Artigas, and C. Preusche, "A passive bilateral control scheme for a teleoperator with time-varying communication delay," *Mechatronics*, vol. 20, no. 7, pp. 812–823, 2010.

[44] E. Nuno, R. Ortega, N. Barabanov, and L. Basanez, "A globally stable pd controller for bilateral teleoperators," *IEEE Transactions on Robotics*, vol. 24, no. 3, pp. 753–758, 2008.

[45] E. Nuno, L. Basanez, R. Ortega, and M. W. Spong, "Position tracking for non-linear teleoperators with variable time delay," *The International Journal of Robotics Research*, vol. 28, no. 7, pp. 895–910, 2009.

[46] S. Islam, P. X. Liu, and A. El Saddik, "Teleoperation systems with symmetric and unsymmetric time varying communication delay," *IEEE Transactions on Instrumentation and Measurement*, vol. 62, no. 11, pp. 2943–2953, 2013.

[47] S. Islam, P. X. Liu, and A. El Saddik, "Nonlinear control for teleoperation systems with time varying delay," *Nonlinear Dynamics*, vol. 76, no. 2, pp. 931–954, 2014.

[48] F. Hashemzadeh, I. Hassanzadeh, and M. Tavakoli, "Teleoperation in the presence of varying time delays and sandwich linearity in actuators," *Automatica*, vol. 49, no. 9, pp. 2813–2821, 2013.

[49] D.-H. Zhai and Y. Xia, "Adaptive control for teleoperation system with varying time delays and input saturation constraints," *IEEE Transactions on Industrial Electronics*, vol. 63, no. 11, pp. 6921–6929, 2016.

[50] S. Ganjefar, S. Rezaei, and F. Hashemzadeh, "Position and force tracking in nonlinear teleoperation systems with sandwich linearity in actuators and time-varying delay," *Mechanical Systems and Signal Processing*, vol. 86, pp. 308–324, 2017.

[51] E. Nuno, R. Ortega, and L. Basanez, "An adaptive controller for nonlinear teleoperators," *Automatica*, vol. 46, no. 1, pp. 155–159, 2010.

[52] E. Nuno Ortega, R. Ortega, and L. Basanez Villaluenga, "Erratum to "an adaptive controller for nonlinear teleoperators" [Automatica 46 (2010) 155-159]," *Automatica*, vol. 47, no. 5, pp. 1093–1094, 2011.

[53] E. Nuno, L. Basanez, and M. Prada, "Asymptotic stability of teleoperators with variable time-delays," in *Proceedings of the IEEE International Conference on Robotics and Automation*. IEEE, 2009, pp. 4332–4337.

[54] C.-C. Hua and P. X. Liu, "Delay-dependent stability criteria of teleoperation systems with asymmetric time-varying delays," *IEEE Transactions on Robotics*, vol. 26, no. 5, pp. 925–932, 2010.

[55] H. Li and K. Kawashima, "Achieving stable tracking in wave-variable-based teleoperation," *IEEE/ASME Transactions on Mechatronics*, vol. 19, no. 5, pp. 1574–1582, 2013.

[56] L. Bate, C. D. Cook, and Z. Li, "Reducing wave-based teleoperator reflections for unknown environments," *IEEE Transactions on Industrial Electronics*, vol. 58, no. 2, pp. 392–397, 2009.

[57] V. Chawda and M. K. O'Malley, "Position synchronization in bilateral teleoperation under time-varying communication delays," *IEEE/ASME Transactions on Mechatronics*, vol. 20, no. 1, pp. 245–253, 2014.

[58] C.-C. Hua and X. P. Liu, "A new coordinated slave torque feedback control algorithm for network-based teleoperation systems," *IEEE/ASME Transactions on Mechatronics*, vol. 18, no. 2, pp. 764–774, 2012.

[59] Y.-J. Huang, T.-C. Kuo, and S.-H. Chang, "Adaptive sliding-mode control for nonlinearsystems with uncertain parameters," *IEEE Transactions on Systems, Man, and Cybernetics, Part B (Cybernetics)*, vol. 38, no. 2, pp. 534–539, 2008.

[60] Y. Feng, F. Han, and X. Yu, "Chattering free full-order sliding-mode control," *Automatica*, vol. 50, no. 4, pp. 1310–1314, 2014.

[61] H. Khalil, *Nonlinear Systems*. New Jersey: Prentice-Hall, 2002.

[62] J. H. Park and H. C. Cho, "Sliding-mode controller for bilateral teleoperation with varying time delay," in *Proceedings of the IEEE/ASME International Conference on Advanced Intelligent Mechatronics*. IEEE, 1999, pp. 311–316.

[63] J. H. Park and H. C. Cho, "Sliding mode control of bilateral teleoperation systems with force-reflection on the Internet," in *Proceedings of the IEEE/RSJ International Conference on Intelligent Robots and Systems*. IEEE, 2000, pp. 1187–1192.

[64] P. Buttolo, P. Braathen, and B. Hannaford, "Sliding control of force reflecting teleoperation: Preliminary studies," *Presence: Teleoperators & Virtual Environments*, vol. 3, no. 2, pp. 158–172, 1994.

[65] Z. Man, A. P. Paplinski, and H. R. Wu, "A robust mimo terminal sliding mode control scheme for rigid robotic manipulators," *IEEE Transactions on Automatic Control*, vol. 39, no. 12, pp. 2464–2469, 1994.

[66] L. Wang, T. Chai, and L. Zhai, "Neural-network-based terminal sliding-mode control of robotic manipulators including actuator dynamics," *IEEE Transactions on Industrial Electronics*, vol. 56, no. 9, pp. 3296–3304, 2009.

[67] Y. Yang, C. Hua, and X. Guan, "Adaptive fuzzy finite-time coordination control for networked nonlinear bilateral teleoperation system," *IEEE Transactions on Fuzzy Systems*, vol. 22, no. 3, pp. 631–641, 2013.

[68] Y. Yang, C. Hua, and X. Guan, "Finite time control design for bilateral teleoperation system with position synchronization error constrained," *IEEE Transactions on Cybernetics*, vol. 46, no. 3, pp. 609–619, 2015.

[69] A. Levant, "Chattering analysis," *IEEE Transactions on Automatic Control*, vol. 55, no. 6, pp. 1380–1389, 2010.

[70] Y. Li, R. Johansson, and Y. Yin, "Acceleration feedback control for nonlinear teleoperation systems with time delays," *International Journal of Control*, vol. 88, no. 3, pp. 507–516, 2015.

[71] E. Nuno and L. Basanez, "Nonlinear bilateral teleoperation: stability analysis," in *Proceedings of the IEEE International Conference on Robotics and Automation.* IEEE, 2009, pp. 3718–3723.

[72] E. Nuno, L. Basanez, R. Ortega, and G. Obregon-Pulido, "Position tracking using adaptive control for bilateral teleoperators with time-delays," in *Proceedings of the IEEE International Conference on Robotics and Automation.* IEEE, 2010, pp. 5370–5375.

[73] S. M. Magdi, *Switched Time-Delay Systems: Stability and Control.* Springer, 2010.

[74] B.-Y. Kim and H.-S. Ahn, "A design of bilateral teleoperation systems using composite adaptive controller," *Control Engineering Practice*, vol. 21, no. 12, pp. 1641–1652, 2013.

[75] Z. Li, X. Cao, and N. Ding, "Adaptive fuzzy control for synchronization of nonlinear teleoperators with stochastic time-varying communication delays," *IEEE Transactions on Fuzzy Systems*, vol. 19, no. 4, pp. 745–757, 2011.

[76] Z. Li and C.-Y. Su, "Neural-adaptive control of single-master–multiple-slaves teleoperation for coordinated multiple mobile manipulators with time-varying communication delays and input uncertainties," *IEEE Transactions on Neural Networks and Learning Systems*, vol. 24, no. 9, pp. 1400–1413, 2013.

[77] X. Yang, C. Hua, J. Yan, and X. Guan, "New stability criteria for networked teleoperation system," *Information Sciences*, vol. 233, pp. 244–254, 2013.

[78] E. Nuno, L. Basanez, C. Lopez-Franco, and N. Arana-Daniel, "Stability of nonlinear teleoperators using pd controllers without velocity measurements," *Journal of the Franklin Institute*, vol. 351, no. 1, pp. 241–258, 2014.

[79] S. Sirouspour and A. Shahdi, "Model predictive control for transparent teleoperation under communication time delay," *IEEE Transactions on Robotics*, vol. 22, no. 6, pp. 1131–1145, 2006.

[80] J. Gangloff, R. Ginhoux, M. de Mathelin, L. Soler, and J. Marescaux, "Model predictive control for compensation of cyclic organ motions in teleoperated laparoscopic surgery," *IEEE Transactions on Control Systems Technology*, vol. 14, no. 2, pp. 235–246, 2006.

[81] H. Chen, P. Huang, and Z. Liu, "Mode switching-based symmetric predictive control mechanism for networked teleoperation space robot system," *IEEE/ASME Transactions on Mechatronics*, vol. 24, no. 6, pp. 2706–2717, 2019.

[82] M. Rubagotti, T. Taunyazov, B. Omarali, and A. Shintemirov, "Semi-autonomous robot teleoperation with obstacle avoidance via model predictive control," *IEEE Robotics and Automation Letters*, vol. 4, no. 3, pp. 2746–2753, 2019.

[83] N. Piccinelli and R. Muradore, "A bilateral teleoperation with interaction force constraint in unknown environment using non linear model predictive control," *European Journal of Control*, vol. 62, pp. 185–191, 2021.

[84] S. Hu, E. Babaians, M. Karimi, and E. Steinbach, "Nmpc-mp: Real-time nonlinear model predictive control for safe motion planning in manipulator teleoperation," in *Proceedings of the IEEE/RSJ International Conference on Intelligent Robots and Systems*. IEEE, 2021, pp. 8309–8316.

[85] G. Tadmor, "The standard h_∞ problem in systems with a single input delay," *IEEE Transactions on Automatic Control*, vol. 45, no. 3, pp. 382–397, 2000.

[86] J. Yan and S. E. Salcudean, "Teleoperation controller design using h_∞-optimization with application to motion-scaling," *IEEE Transactions on Control Systems Technology*, vol. 4, no. 3, pp. 244–258, 1996.

[87] G. M. Leung, B. A. Francis, and J. Apkarian, "Bilateral controller for teleoperators with time delay via μ-synthesis," *IEEE Transactions on Robotics and Automation*, vol. 11, no. 1, pp. 105–116, 1995.

[88] Z. Chen, F. Huang, W. Sun, J. Gu, and B. Yao, "RBF-neural-network-based adaptive robust control for nonlinear bilateral teleoperation manipulators with uncertainty and time delay," *IEEE/ASME Transactions on Mechatronics*, vol. 25, no. 2, pp. 906–918, 2019.

[89] Y.-C. Liu, P. N. Dao, and K. Y. Zhao, "On robust control of nonlinear teleoperators under dynamic uncertainties with variable time delays and without

relative velocity," *IEEE Transactions on Industrial Informatics*, vol. 16, no. 2, pp. 1272–1280, 2019.

[90] M. Sanchez, D. Cruz-Ortiz, M. Ballesteros, I. Salgado, and I. Chairez, "Output feedback robust control for teleoperated manipulator robots with different workspace," *Expert Systems with Applications*, vol. 206, p. 117838, 2022.

[91] P. M. Kebria, A. Khosravi, S. Nahavandi, P. Shi, and R. Alizadehsani, "Robust adaptive control scheme for teleoperation systems with delay and uncertainties," *IEEE Transactions on Cybernetics*, vol. 50, no. 7, pp. 3243–3253, 2019.

[92] H. Chen and Z. Liu, "Time-delay prediction–based smith predictive control for space teleoperation," *Journal of Guidance, Control, and Dynamics*, vol. 44, no. 4, pp. 872–879, 2021.

[93] H. Yang, L. Liu, and Y. Wang, "Observer-based sliding mode control for bilateral teleoperation with time-varying delays," *Control Engineering Practice*, vol. 91, p. 104097, 2019.

[94] S. A. M. Dehghan, H. R. Koofigar, H. Sadeghian, and M. Ekramian, "Observer-based adaptive force–position control for nonlinear bilateral teleoperation with time delay," *Control Engineering Practice*, vol. 107, p. 104679, 2021.

[95] C. Yang, G. Peng, L. Cheng, J. Na, and Z. Li, "Force sensorless admittance control for teleoperation of uncertain robot manipulator using neural networks," *IEEE Transactions on Systems, Man, and Cybernetics: Systems*, vol. 51, no. 5, pp. 3282–3292, 2019.

[96] H. Zhang, A. Song, H. Li, and S. Shen, "Novel adaptive finite-time control of teleoperation system with time-varying delays and input saturation," *IEEE Transactions on Cybernetics*, vol. 51, no. 7, pp. 3724–3737, 2019.

[97] A. Zakerimanesh, F. Hashemzadeh, and M. Tavakoli, "Task-space synchronisation of nonlinear teleoperation with time-varying delays and actuator saturation," *International Journal of Control*, vol. 93, no. 6, pp. 1328–1344, 2020.

[98] X. Yang, J. Yan, C. Hua, and X. Guan, "Effects of quantization and saturation on performance in bilateral teleoperator," *International Journal of Robust and Nonlinear Control*, vol. 30, no. 1, pp. 121–141, 2020.

[99] S. A. Deka, D. M. Stipanovic, and T. Kesavadas, "Stable bilateral teleoperation with bounded control," *IEEE Transactions on Control Systems Technology*, vol. 27, no. 6, pp. 2351–2360, 2018.

[100] A. Zakerimanesh, F. Hashemzadeh, A. Torabi, and M. Tavakoli, "A cooperative paradigm for task-space control of multilateral nonlinear teleoperation with bounded inputs and time-varying delays," *Mechatronics*, vol. 62, p. 102255, 2019.

[101] Y. Yang, J. Li, C. Hua, and X. Guan, "Adaptive synchronization control design for flexible telerobotics with actuator fault and input saturation," *International Journal of Robust and Nonlinear Control*, vol. 28, no. 3, pp. 1016–1034, 2018.

[102] L. Zhao, L. Liu, Y. Wang, and H. Yang, "Active disturbance rejection control for teleoperation systems with actuator saturation," *Asian Journal of Control*, vol. 21, no. 2, pp. 702–713, 2019.

[103] F. Morabito, A. R. Teel, and L. Zaccarian, "Nonlinear antiwindup applied to Euler-Lagrange systems," *IEEE Transactions on Robotics and Automation*, vol. 20, no. 3, pp. 526–537, 2004.

[104] A. Loria, R. Kelly, R. Ortega, and V. Santibanez, "On global output feedback regulation of Euler-Lagrange systems with bounded inputs," *IEEE Transactions on Automatic Control*, vol. 42, no. 8, pp. 1138–1143, 1997.

[105] E. Zergeroglu, W. Dixon, A. Behal, and D. Dawson, "Adaptive set-point control of robotic manipulators with amplitude-limited control inputs," *Robotica*, vol. 18, no. 2, pp. 171–181, 2000.

[106] S.-J. Lee and H.-S. Ahn, "Synchronization of bilateral teleoperation systems with input saturation," in *Proceedings of the International Conference on Control, Automation and Systems.* IEEE, 2010, pp. 1357–1361.

[107] S.-J. Lee and H.-S. Ahn, "A study on bilateral teleoperation with input saturation and systems," in *Proceedings of the 11th International Conference on Control, Automation and Systems.* IEEE, 2011, pp. 161–166.

[108] S.-J. Lee and H.-S. Ahn, "Controller designs for bilateral teleoperation with input saturation," *Control Engineering Practice*, vol. 33, pp. 35–47, 2014.

[109] Y. Yang, C. Ge, H. Wang, X. Li, and C. Hua, "Adaptive neural network based prescribed performance control for teleoperation system under input saturation," *Journal of the Franklin Institute*, vol. 352, no. 5, pp. 1850–1866, 2015.

[110] C. Hua, Y. Yang, and P. X. Liu, "Output-feedback adaptive control of networked teleoperation system with time-varying delay and bounded inputs," *IEEE/ASME Transactions on Mechatronics*, vol. 20, no. 5, pp. 2009–2020, 2014.

[111] P. F. Hokayem, D. M. Stipanović, and M. W. Spong, "Semiautonomous control of multiple networked lagrangian systems," *International Journal of Robust and Nonlinear Control*, vol. 19, no. 18, pp. 2040–2055, 2009.

[112] I. Elhajj, C. M. Kit, Y. H. Liu, N. Xi, A. Goradia, and T. Fukuda, "Telecoordinated control of multi-robot systems via the internet," in *Proceeding of IEEE International Conference on Robotics and Automation.* IEEE, 2003, pp. 1646–1652.

[113] Y. Kume, Y. Hirata, Z.-D. Wang, and K. Kosuge, "Decentralized control of multiple mobile manipulators handling a single object in coordination," in *Proceedings of the IEEE/RSJ International Conference on Intelligent Robots and Systems.* IEEE, 2002, pp. 2758–2763.

[114] N. Xi, T.-J. Tarn, and A. K. Bejczy, "Intelligent planning and control for multi-robot coordination: an event-based approach," *IEEE Transactions on Robotics and Automation*, vol. 12, no. 3, pp. 439–452, 1996.

[115] E. J. Rodríguez-Seda, J. J. Troy, C. A. Erignac, P. Murray, D. M. Stipanovic, and M. W. Spong, "Bilateral teleoperation of multiple mobile agents: coordinated motion and collision avoidance," *IEEE Transactions on Control Systems Technology*, vol. 18, no. 4, pp. 984–992, 2009.

[116] Z. Li, S. S. Ge, and Z. Wang, "Robust adaptive control of coordinated multiple mobile manipulators," *Mechatronics*, vol. 18, no. 5-6, pp. 239–250, 2008.

[117] Z. Li, J. Li, and Y. Kang, "Adaptive robust coordinated control of multiple mobile manipulators interacting with rigid environments," *Automatica*, vol. 46, no. 12, pp. 2028–2034, 2010.

[118] T. Sugar and V. Kumar, "Decentralized control of cooperating mobile manipulators," in *Proceedings of the IEEE International Conference on Robotics and Automation.* IEEE, 1998, pp. 2916–2921.

[119] V. Perdereau and M. Drouin, "Hybrid external control for two robot coordinated motion," *Robotica*, vol. 14, no. 2, pp. 141–153, 1996.

[120] C. O. Alford and S. M. Belyeu, "Coordinated control of two robot arms," in *Proceedings of the IEEE International Conference on Robotics and Automation.* IEEE, 1984, pp. 468–473.

[121] K. I. Kim and Y. F. Zheng, "Two strategies of position and force control for two industrial robots handling a single object," *Robotics and Autonomous Systems*, vol. 5, no. 4, pp. 395–403, 1989.

[122] S. Schneider and R. H. Cannon, "Object impedance control for cooperative manipulation: theory and experimental results," in *Proceedings of the International Conference on Robotics and Automation.* IEEE, 1989, pp. 1076–1083.

[123] J. Szewczyk, F. Plumet, and P. Bidaud, "Planning and controlling cooperating robots through distributed impedance," *Journal of Robotic Systems*, vol. 19, no. 6, pp. 283–297, 2002.

[124] M. Shahbazi, S. F. Atashzar, and R. V. Patel, "A systematic review of multilateral teleoperation systems," *IEEE Transactions on Haptics*, vol. 11, no. 3, pp. 338–356, 2018.

[125] D. Sun, Q. Liao, X. Gu, C. Li, and H. Ren, "Multilateral teleoperation with new cooperative structure based on reconfigurable robots and type-2 fuzzy logic," *IEEE Transactions on Cybernetics*, vol. 49, no. 8, pp. 2845–2859, 2018.

[126] U. Tumerdem and N. Yilmaz, "A unifying framework for transparency optimized controller design in multilateral teleoperation with time delays," *Control Engineering Practice*, vol. 117, p. 104931, 2021.

[127] J.-H. Ryu, Q. Ha-Van, and A. Jafari, "Multilateral teleoperation over communication time delay using the time-domain passivity approach," *IEEE Transactions on Control Systems Technology*, vol. 28, no. 6, pp. 2705–2712, 2019.

[128] M. Minelli, F. Ferraguti, N. Piccinelli, R. Muradore, and C. Secchi, "An energy-shared two-layer approach for multi-master-multi-slave bilateral teleoperation systems," in *Proceedings of the International Conference on Robotics and Automation*. IEEE, 2019, pp. 423–429.

[129] M. Farahmandrad, S. Ganjefar, H. A. Talebi, and M. Bayati, "A novel cooperative teleoperation framework for nonlinear time-delayed single-master/multi-slave system," *Robotica*, vol. 38, no. 3, pp. 475–492, 2020.

[130] U. Ahmad, Y.-J. Pan, and H. Shen, "Robust control design for teleoperation of multiple mobile manipulators under time delays," *International Journal of Robust and Nonlinear Control*, vol. 30, no. 16, pp. 6454–6472, 2020.

[131] P. M. Kebria, A. Khosravi, S. Nahavandi, F. Bello, and S. Krishnan, "Robust adaptive synchronisation of a single-master multi-slave teleoperation system over delayed communication," in *Proceeding of IEEE International Conference on Industrial Technology*. IEEE, 2019, pp. 193–198.

[132] R. Mohajerpoor, I. Sharifi, H. A. Talebi, and S. M. Rezaei, "Adaptive bilateral teleoperation of an unknown object handled by multiple robots under unknown communication delay," in *Proceeding of IEEE/ASME International Conference on Advanced Intelligent Mechatronics*. IEEE, 2013, pp. 1158–1163.

[133] D. Lee and M. W. Spong, "Bilateral teleoperation of multiple cooperative robots over delayed communication networks: theory," in *Proceedings of the IEEE International Conference on Robotics and Automation*. IEEE, 2005, pp. 360–365.

[134] Z. Li, L. Ding, H. Gao, G. Duan, and C.-Y. Su, "Trilateral teleoperation of adaptive fuzzy force/motion control for nonlinear teleoperators with communication random delays," *IEEE Transactions on Fuzzy Systems*, vol. 21, no. 4, pp. 610–624, 2012.

[135] Z. Li, Y. Xia, and F. Sun, "Adaptive fuzzy control for multilateral cooperative teleoperation of multiple robotic manipulators under random network-induced delays," *IEEE Transactions on Fuzzy Systems*, vol. 22, no. 2, pp. 437–450, 2013.

[136] D.-H. Zhai and Y. Xia, "Adaptive fuzzy control of multilateral asymmetric teleoperation for coordinated multiple mobile manipulators," *IEEE Transactions on Fuzzy Systems*, vol. 24, no. 1, pp. 57–70, 2015.

[137] B. Khademian and K. Hashtrudi-Zaad, "A framework for unconditional stability analysis of multimaster/multislave teleoperation systems," *IEEE Transactions on Robotics*, vol. 29, no. 3, pp. 684–694, 2013.

[138] S. Sirouspour, "Modeling and control of cooperative teleoperation systems," *IEEE Transactions on Robotics*, vol. 21, no. 6, pp. 1220–1225, 2005.

[139] N. Y. Chong, T. Kotoku, K. Ohba, K. Komoriya, N. Matsuhira, and K. Tanie, "Remote coordinated controls in multiple telerobot cooperation," in *Proceedings of the IEEE International Conference on Robotics and Automation*. IEEE, 2000, pp. 3138–3143.

[140] A. Franchi, C. Secchi, H. I. Son, H. H. Bulthoff, and P. R. Giordano, "Bilateral teleoperation of groups of mobile robots with time-varying topology," *IEEE Transactions on Robotics*, vol. 28, no. 5, pp. 1019–1033, 2012.

[141] O. Tokatli, P. Das, R. Nath, L. Pangione, A. Altobelli, G. Burroughes, E. T. Jonasson, M. F. Turner, and R. Skilton, "Robot-assisted glovebox teleoperation for nuclear industry," *Robotics*, vol. 10, no. 3, p. 85, 2021.

[142] B. Doroodgar, Y. Liu, and G. Nejat, "A learning-based semi-autonomous controller for robotic exploration of unknown disaster scenes while searching for victims," *IEEE Transactions on Cybernetics*, vol. 44, no. 12, pp. 2719–2732, 2014.

[143] L. Righetti, N. Sharkey, R. Arkin, D. Ansell, M. Sassoli, C. Heyns, P. Asaro, and P. Lee, "Autonomous weapon systems: technical, military, legal and humanitarian aspects," *Proceedings of the International Committee of the Red Cross, Geneva, Switzerland*, pp. 26–28, 2014.

[144] W. Li, Z. Li, Y. Liu, L. Ding, J. Wang, H. Gao, and Z. Deng, "Semi-autonomous bilateral teleoperation of six-wheeled mobile robot on soft terrains," *Mechanical Systems and Signal Processing*, vol. 133, p. 106234, 2019.

[145] G. Gonzalez, M. Agarwal, M. V. Balakuntala, M. M. Rahman, U. Kaur, R. M. Voyles, V. Aggarwal, Y. Xue, and J. Wachs, "Deserts: Delay-tolerant semi-autonomous robot teleoperation for surgery," in *Proceeding of IEEE International Conference on Robotics and Automation*. IEEE, 2021, pp. 12 693–12 700.

[146] M. M. Rahman, N. Sanchez-Tamayo, G. Gonzalez, M. Agarwal, V. Aggarwal, R. M. Voyles, Y. Xue, and J. Wachs, "Transferring dexterous surgical skill knowledge between robots for semi-autonomous teleoperation," in *Proceeding of IEEE International Conference on Robot and Human Interactive Communication*. IEEE, 2019, pp. 1–6.

[147] C. Li, Z. Jiang, Z. Li, C. Fan, and H. Liu, "A novel semi-autonomous teleoperation method for the tiangong-2 manipulator system," *IEEE Access*, vol. 7, pp. 164 453–164 467, 2019.

[148] S. Saparia, A. Schimpe, and L. Ferranti, "Active safety system for semi-autonomous teleoperated vehicles," in *Proceeding of IEEE Intelligent Vehicles Symposium Workshops*. IEEE, 2021, pp. 141–147.

[149] R. Jia, N. Koh, N. Leone, M. Singh, Z. Wu, and P. Vela, "Teleoperation of semi-autonomous robots through uncertain environments," in *Proceeding of Opportunity Research Scholars Symposium*. IEEE, 2022, pp. 31–37.

[150] B. Siciliano, "Kinematic control of redundant robot manipulators: a tutorial," *Journal of Intelligent and Robotic Systems*, vol. 3, no. 3, pp. 201–212, 1990.

[151] P. Malysz and S. Sirouspour, "A kinematic control framework for single-slave asymmetric teleoperation systems," *IEEE Transactions on Robotics*, vol. 27, no. 5, pp. 901–917, 2011.

[152] N. Nath, E. Tatlicioglu, and D. M. Dawson, "Teleoperation with kinematically redundant robot manipulators with sub-task objectives," *Robotica*, vol. 27, no. 7, pp. 1027–1038, 2009.

[153] E. Ettelt, R. Furtwängler, U. D. Hanebeck, and G. Schmidt, "Design issues of a semi-autonomous robotic assistant for the health care environment," *Journal of Intelligent and Robotic Systems*, vol. 22, no. 3, pp. 191–209, 1998.

[154] B. Doroodgar, M. Ficocelli, B. Mobedi, and G. Nejat, "The search for survivors: Cooperative human-robot interaction in search and rescue environments using semi-autonomous robots," in *Proceeding of IEEE International Conference on Robotics and Automation*. IEEE, 2010, pp. 2858–2863.

[155] J.-H. Li, B.-H. Jun, P.-M. Lee, and S.-W. Hong, "A hierarchical real-time control architecture for a semi-autonomous underwater vehicle," *Ocean Engineering*, vol. 32, no. 13, pp. 1631–1641, 2005.

[156] Y.-C. Liu and N. Chopra, "Control of semi-autonomous teleoperation system with time delays," *Automatica*, vol. 49, no. 6, pp. 1553–1565, 2013.

[157] Y.-C. Liu, "Task-space bilateral teleoperation systems for heterogeneous robots with time-varying delays," *Robotica*, vol. 33, no. 10, pp. 2065–2082, 2015.

[158] A. A. Ghavifekr, A. R. Ghiasi, and M. A. Badamchizadeh, "Discrete-time control of bilateral teleoperation systems: a review," *Robotica*, vol. 36, no. 4, pp. 552–569, 2018.

[159] P. M. Kebria, H. Abdi, M. M. Dalvand, A. Khosravi, and S. Nahavandi, "Control methods for Internet-based teleoperation systems: a review," *IEEE Transactions on Human-Machine Systems*, vol. 49, no. 1, pp. 32–46, 2018.

[160] J. S. Lee, Y. Ham, H. Park, and J. Kim, "Challenges, tasks, and opportunities in teleoperation of excavator toward human-in-the-loop construction automation," *Automation in Construction*, vol. 135, pp. 1–13(104 119), 2022.

[161] Y. Liu, J. Zhao, M. Fan, Y. Lv, W. Liu, and W. Tian, "Clinical factors affecting the accuracy of a CT-based active infrared navigation system," *International Journal of Medical Robotics and Computer Assisted Surgery*, vol. 12, pp. 568–571, 2016.

[162] H. Jin, Y. Hu, W. Tian, P. Zhang, J. Zhang, and B. Li, "Safety analysis and control of a robotic spinal surgical system," *Mechatronics*, vol. 24, no. 1, pp. 55–65, 2014.

[163] R. V. Patel, S. F. Atashzar, and M. Tavakoli, "Haptic feedback and force-based teleoperation in surgical robotics," *Proceedings of the IEEE*, vol. 110, no. 7, pp. 1012–1027, 2022.

[164] Z. Wang, Y. Sun, and B. Liang, "Synchronization control for bilateral teleoperation system with position error constraints: a fixed-time approach," *ISA transactions*, vol. 93, pp. 125–136, 2019.

[165] Z. Wang, Z. Chen, and B. Liang, "Fixed-time velocity reconstruction scheme for space teleoperation systems: Exp barrier Lyapunov function approach," *Acta Astronautica*, vol. 157, pp. 92–101, 2019.

[166] C. Hua, Y. Yang, X. Yang, and X. Guan, *Analysis and Design for Networked Teleoperation System*. Springer, 2019.

[167] D. Huang, B. Li, Y. Li, and C. Yang, "Cooperative manipulation of deformable objects by single-leader-dual-follower teleoperation," *IEEE Transactions on Industrial Electronics*, vol. 69, no. 12.

[168] Y. Yang, C. Hua, and J. Li, "Composite adaptive guaranteed performances synchronization control for bilateral teleoperation system with asymmetrical time-varying delays," *IEEE Transactions on Cybernetics*, vol. 52, no. 6, pp. 5486–5497, 2020.

[169] C. Liu, H. Wang, X. Liu, and Y. Zhou, "Adaptive finite-time fuzzy funnel control for nonaffine nonlinear systems," *IEEE Transactions on Systems, Man, and Cybernetics: Systems*, vol. 51, no. 5, pp. 2894–2903, 2019.

[170] Y. Yang, C. Hua, and X. Guan, "Finite-time synchronization control for bilateral teleoperation under communication delays," *Robotics and Computer-Integrated Manufacturing*, vol. 31, pp. 61–69, 2015.

[171] Z. Wang, Z. Chen, Y. Zhang, X. Yu, X. Wang, and B. Liang, "Adaptive finite-time control for bilateral teleoperation systems with jittering time delays," *International Journal of Robust and Nonlinear Control*, vol. 29, no. 4, pp. 1007–1030, 2019.

[172] H. Zhang, A. Song, H. Li, D. Chen, and L. Fan, "Adaptive finite-time control scheme for teleoperation with time-varying delay and uncertainties," *IEEE Transactions on Systems, Man, and Cybernetics: Systems*, vol. 52, no. 3, pp. 1552–1566, 2022.

[173] J. Bao, H. Wang, and P. X. Liu, "Finite-time synchronization control for bilateral teleoperation systems with asymmetric time-varying delay and input dead zone," *IEEE/ASME Transactions on Mechatronics*, vol. 26, no. 3, pp. 1570–1580, 2020.

[174] Y. Yang, C. Hua, and J. Li, "A novel delay-dependent finite-time control of telerobotics system with asymmetric time-varying delays," *IEEE Transactions on Control Systems Technology*, vol. 30, no. 3, pp. 985–996, 2021.

[175] H. Shen and Y.-J. Pan, "Improving tracking performance of nonlinear uncertain bilateral teleoperation systems with time-varying delays and disturbances," *IEEE/ASME Transactions on Mechatronics*, vol. 25, no. 3, pp. 1171–1181, 2019.

[176] S. Zhang, S. Yuan, X. Yu, L. Kong, Q. Li, and G. Li, "Adaptive neural network fixed-time control design for bilateral teleoperation with time delay," *IEEE Transactions on Cybernetics*, vol. 52, no. 9, pp. 9756–9769, 2022.

[177] S. Guo, Z. Liu, J. Yu, P. Huang, and Z. Ma, "Adaptive practical fixed-time synchronization control for bilateral teleoperation system with prescribed performance," *IEEE Transactions on Circuits and Systems II: Express Briefs*, vol. 69, no. 3, pp. 1243–1247, 2021.

[178] J.-Z. Xu, M.-F. Ge, T.-F. Ding, C.-D. Liang, and Z.-W. Liu, "Neuro-adaptive fixed-time trajectory tracking control for human-in-the-loop teleoperation with mixed communication delays," *IET Control Theory & Applications*, vol. 14, no. 19, pp. 3193–3203, 2020.

[179] G.-H. Xu, F. Qi, Q. Lai, and H. H.-C. Iu, "Fixed time synchronization control for bilateral teleoperation mobile manipulator with nonholonomic constraint and time delay," *IEEE Transactions on Circuits and Systems II: Express Briefs*, vol. 67, no. 12, pp. 3452–3456, 2020.

[180] J.-Z. Xu, M.-F. Ge, G. Ling, F. Liu, and J. H. Park, "Hierarchical predefined-time control of teleoperation systems with state and communication constraints," *International Journal of Robust and Nonlinear Control*, vol. 31, no. 18, pp. 9652–9675, 2021.

[181] C. P. Bechlioulis, Z. Doulgeri, and G. A. Rovithakis, "Neuro-adaptive force/position control with prescribed performance and guaranteed contact maintenance," *IEEE Transactions on Neural Networks*, vol. 21, no. 12, pp. 1857–1868, 2010.

[182] C. P. Bechlioulis and G. A. Rovithakis, "Robust partial-state feedback prescribed performance control of cascade systems with unknown nonlinearities," *IEEE Transactions on Automatic Control*, vol. 56, no. 9, pp. 2224–2230, 2011.

[183] A. K. Kostarigka and G. A. Rovithakis, "Adaptive dynamic output feedback neural network control of uncertain mimo nonlinear systems with prescribed performance," *IEEE Transactions on Neural Networks and Learning Systems*, vol. 23, no. 1, pp. 138–149, 2011.

[184] B. Ren, S. S. Ge, K. P. Tee, and T. H. Lee, "Adaptive neural control for output feedback nonlinear systems using a barrier Lyapunov function," *IEEE Transactions on Neural Networks*, vol. 21, no. 8, pp. 1339–1345, 2010.

[185] K. P. Tee, S. S. Ge, and E. H. Tay, "Barrier Lyapunov functions for the control of output-constrained nonlinear systems," *Automatica*, vol. 45, no. 4, pp. 918–927, 2009.

[186] K. P. Tee, B. Ren, and S. S. Ge, "Control of nonlinear systems with time-varying output constraints," *Automatica*, vol. 47, no. 11, pp. 2511–2516, 2011.

[187] W. He and S. S. Ge, "Vibration control of a flexible beam with output constraint," *IEEE Transactions on Industrial Electronics*, vol. 62, no. 8, pp. 5023–5030, 2015.

[188] Y. Yang, C. Hua, and X. Guan, "Synchronization control for bilateral teleoperation system with prescribed performance under asymmetric time delay," *Nonlinear Dynamics*, vol. 81, no. 1, pp. 481–493, 2015.

[189] M. S. Mahmoud and M. Maaruf, "Prescribed performance output feedback synchronisation control of bilateral teleoperation system with actuator nonlinearities," *International Journal of Systems Science*, vol. 52, no. 15, pp. 3115–3127, 2021.

[190] Z. Wang, B. Liang, Y. Sun, and T. Zhang, "Adaptive fault-tolerant prescribed-time control for teleoperation systems with position error constraints," *IEEE Transactions on Industrial Informatics*, vol. 16, no. 7, pp. 4889–4899, 2019.

[191] Z. Wang, H.-K. Lam, B. Xiao, Z. Chen, B. Liang, and T. Zhang, "Event-triggered prescribed-time fuzzy control for space teleoperation systems subject to multiple constraints and uncertainties," *IEEE Transactions on Fuzzy Systems*, vol. 29, no. 9, pp. 2785–2797, 2020.

[192] Y. Li, Z. Liu, Z. Wang, Y. Yin, and B. Zhao, "Adaptive control of teleoperation systems with prescribed tracking performance: a BLF-based approach," *International Journal of Control*, vol. 95, no. 6, pp. 1600–1610, 2022.

[193] D.-H. Zhai and Y. Xia, "A prescribed-performance-like control for improving tracking performance of networked robots," *International Journal of Robust and Nonlinear Control*, vol. 31, no. 7, pp. 2546–2571, 2021.

[194] Y. Zhao, P. X. Liu, H. Wang, and J. Bao, "Funnel-bounded synchronization control for bilateral teleoperation with asymmetric communication delays," *Nonlinear Dynamics*, vol. 107, pp. 1–14, 2022.

CHAPTER 2

Preliminaries of teleoperation control

THIS chapter presents basic concepts, notations and brief background of robotics and related constructs and models that are used throughout the book. Main mathematical results from switched control, nonlinear control and networked control are also provided to facilitate further developments in different chapters.

2.1 INTRODUCTION

In this chapter, the fundamental mathematical concepts and analysis tools in robotics and control theory are summarized, which will be used in control design and stability analysis of teleoperation system in the subsequent chapters. Much of the material is described in classical control theory textbooks and robotics books as standard form. Thus, some standard definitions, theorems, lemmas, and corollaries, which are available in references, are sometimes given without a proof.

2.2 ROBOT MODELS

In this section, the robot dynamics are given, where the modeling methods and its characteristics are introduced.

2.2.1 Lagrange's equations

In the past decades, the research on robot dynamics has been paid a lot of attention. Many modeling methods including Newton's equations of motion and Lagrange method are developed. Since these methods have been described in great detail in the books of robotics and theoretical mechanics, this book only gives a brief introduction.

Generally speaking, the modeling based on Newton's equations of motion will become complicated with the increase of robot joints. In comparison, Lagrange modeling method is simpler. The method is named after the French mathematician Joseph Louis de La Grange, and its first report can be traced back to his famous book Mécanique Analytique in 1788.[1]

[1]This method is also often called Euler-Lagrange method.

DOI: 10.1201/9781003382058-2

In the framework of Lagrange equation, the modeling of a robot can be briefly described as follows. As shown in Fig. 2.1, for a robotic manipulator with n degree of freedom (DOF), its total energy $\mathcal{E}$ can be expressed as the sum of kinetic energy $\mathcal{K}$ and potential energy $\mathcal{U}$, i.e.,

$$\mathcal{E}(\boldsymbol{q}, \dot{\boldsymbol{q}}) = \mathcal{K}(\boldsymbol{q}, \dot{\boldsymbol{q}}) + \mathcal{U}(\boldsymbol{q})$$

where $\boldsymbol{q} = [q_1, q_2, \cdots, q_n]^{\mathrm{T}}$. Assuming that the potential energy $\mathcal{U}$ is only caused by conservative force, such as gravitational potential energy and elastic potential energy stored in springs.

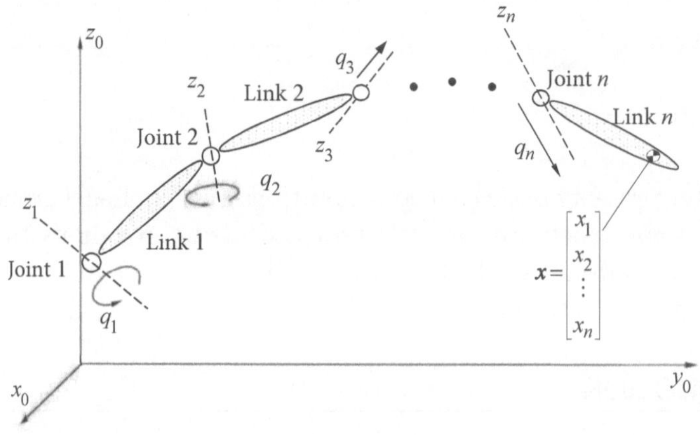

Figure 2.1 An n-DOF robotic manipulator.

For an n-DOF robotic manipulator, the Lagrange operator $\mathcal{L}(\boldsymbol{q}, \dot{\boldsymbol{q}})$ is defined as the difference between kinetic energy $\mathcal{K}$ and potential energy $\mathcal{U}$, i.e.,

$$\mathcal{L}(\boldsymbol{q}, \dot{\boldsymbol{q}}) = \mathcal{K}(\boldsymbol{q}, \dot{\boldsymbol{q}}) - \mathcal{U}(\boldsymbol{q}) \tag{2.1}$$

Then, the Lagrange equation of motion can be described as

$$\frac{\mathrm{d}}{\mathrm{d}t}\left[\frac{\partial \mathcal{L}(\boldsymbol{q}, \dot{\boldsymbol{q}})}{\partial \dot{\boldsymbol{q}}}\right] - \frac{\partial \mathcal{L}(\boldsymbol{q}, \dot{\boldsymbol{q}})}{\partial \boldsymbol{q}} = \boldsymbol{\tau}$$

or equivalently,

$$\frac{\mathrm{d}}{\mathrm{d}t}\left[\frac{\partial \mathcal{L}(\boldsymbol{q}, \dot{\boldsymbol{q}})}{\partial \dot{q}_i}\right] - \frac{\partial \mathcal{L}(\boldsymbol{q}, \dot{\boldsymbol{q}})}{\partial q_i} = \tau_i, \ i = 1, 2, \cdots, n \tag{2.2}$$

where τ_i is the external force, torque (introduced by the actuator) and other (non-conservative) forces corresponding to each joint. Here, the non-conservative force mainly refers to the friction force, the resistance of the fluid to the solid and all the forces which are only related to the time and velocity signals and have nothing to do with the position.

To establish a robot's dynamics by using Lagrange equation of motion, the following basic steps can be followed.

(1) Calculate kinetic energy $\mathcal{K}(\boldsymbol{q}, \dot{\boldsymbol{q}})$;

(2) Calculate potential energy $\mathcal{U}(\boldsymbol{q})$;

(3) From (2.1), calculate the Lagrange operator $\mathcal{L}(\boldsymbol{q}, \dot{\boldsymbol{q}})$;

(4) Get the Lagrange equation from (2.2).

In what follows, let us take a 2-DOF planar serial manipulator as an example to describe the modeling process of robot based on Lagrange equation of motion.

Example 2.1. *Assuming that the 2-DOF planar serial manipulator has the structure shown in the Fig. 2.2, where m_i and l_i are the mass and length of the link i respectively, and l_{ci} is the distance from the $i-1$th joint to the ith link centroid, I_i is the moment of inertia of the ith link.*

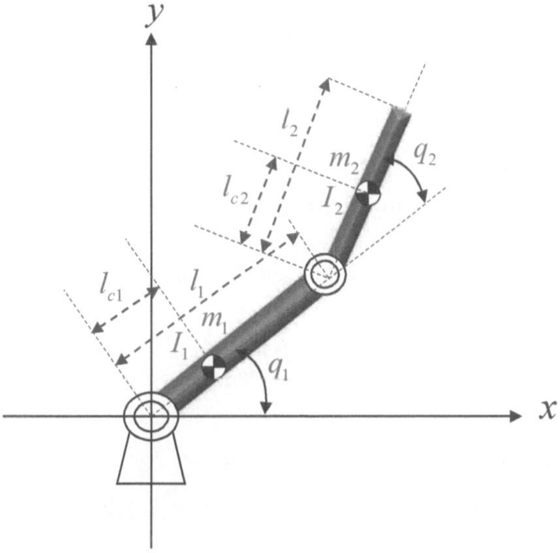

Figure 2.2 2-DOF planar serial manipulator.

From Fig. 2.2, the position of m_i $(i = 1, 2)$ is

$$\begin{bmatrix} x_{c1} \\ y_{c1} \end{bmatrix} = \begin{bmatrix} l_{c1}\cos(q_1) \\ l_{c1}\sin(q_1) \end{bmatrix}$$

$$\begin{bmatrix} x_{c2} \\ y_{c2} \end{bmatrix} = \begin{bmatrix} l_1\cos(q_1) + l_{c2}\cos(q_1+q_2) \\ l_1\sin(q_1) + l_{c2}\sin(q_1+q_2) \end{bmatrix}$$

The corresponding velocities are:

$$\boldsymbol{v}_{c1} = \begin{bmatrix} -l_{c1}\dot{q}_1\sin(q_1) \\ l_{c1}\dot{q}_1\cos(q_1) \end{bmatrix}$$

$$\boldsymbol{v}_{c2} = \begin{bmatrix} -l_1\dot{q}_1\sin(q_1) - l_{c2}(\dot{q}_1+\dot{q}_2)\sin(q_1+q_2) \\ l_1\dot{q}_1\cos(q_1) + l_{c2}(\dot{q}_1+\dot{q}_2)\cos(q_1+q_2) \end{bmatrix}$$

Then the total kinetic energy can be written as

$$\begin{aligned}
\mathcal{K}(\boldsymbol{q}, \dot{\boldsymbol{q}}) &= \frac{1}{2}m_1\boldsymbol{v}_{c1}^{\mathrm{T}}\boldsymbol{v}_{c1} + \frac{1}{2}I_1\dot{q}_1^2 + \frac{1}{2}m_2\boldsymbol{v}_{c2}^{\mathrm{T}}\boldsymbol{v}_{c2} + \frac{1}{2}I_2(\dot{q}_1+\dot{q}_2)^2 \\
&= \frac{1}{2}m_1 l_{c1}^2\dot{q}_1^2 + \frac{1}{2}I_1\dot{q}_1^2 + \frac{1}{2}m_2 l_1^2\dot{q}_1^2 + m_2 l_1 l_{c2}\dot{q}_1(\dot{q}_1+\dot{q}_2)\cos(q_2) + \\
&\quad \frac{1}{2}m_2 l_{c2}^2(\dot{q}_1+\dot{q}_2)^2 + \frac{1}{2}I_2(\dot{q}_1+\dot{q}_2)^2 \\
&= \frac{1}{2}\dot{\boldsymbol{q}}^{\mathrm{T}}\boldsymbol{M}(\boldsymbol{q})\dot{\boldsymbol{q}}
\end{aligned}$$

where $\boldsymbol{q} = [q_1, q_2]^{\mathrm{T}}$, $\boldsymbol{M}(\boldsymbol{q})$ *is the inertial matrix,*

$$\boldsymbol{M}(\boldsymbol{q}) = \begin{bmatrix} p_1 + p_2 + 2p_3\cos(q_2) & p_2 + p_3\cos(q_2) \\ p_2 + p_3\cos(q_2) & p_2 \end{bmatrix}$$

where $p_1 = m_1 l_{c1}^2 + m_2 l_1^2 + I_1$, $p_2 = m_2 l_{c2}^2 + I_2$, $p_3 = m_2 l_1 l_{c2}$.

The total potential energy can be given by

$$\begin{aligned}
\mathcal{U}(\boldsymbol{q}) &= m_1 g y_{c1} + m_2 g y_{c2} \\
&= m_1 g l_{c2}\sin(q_1) + m_2 g[l_1\sin(q_1) + l_{c2}\sin(q_1+q_2)]
\end{aligned}$$

Further, the Lagrange operator can be written as

$$\mathcal{L}(\boldsymbol{q}, \dot{\boldsymbol{q}}) = \frac{1}{2}\dot{\boldsymbol{q}}^{\mathrm{T}}\boldsymbol{M}(\boldsymbol{q})\dot{\boldsymbol{q}} - m_1 g l_{c2}\sin(q_1) - m_2 g[l_1\sin(q_1) + l_{c2}\sin(q_1+q_2)]$$

Then, it has

$$\begin{aligned}
\frac{\partial\mathcal{L}(\boldsymbol{q}, \dot{\boldsymbol{q}})}{\partial\dot{q}_1} &= m_1 l_{c1}^2\dot{q}_1 + I_1\dot{q}_1 + m_2 l_1^2\dot{q}_1 + m_2 l_1 l_{c2}(2\dot{q}_1+\dot{q}_2)\cos(q_2) + \\
&\quad m_2 l_{c2}^2(\dot{q}_1+\dot{q}_2) + I_2(\dot{q}_1+\dot{q}_2) \\
\frac{\partial\mathcal{L}(\boldsymbol{q}, \dot{\boldsymbol{q}})}{\partial\dot{q}_2} &= m_2 l_1 l_{c2}\dot{q}_1\cos(q_2) + m_2 l_{c2}^2(\dot{q}_1+\dot{q}_2) + I_2(\dot{q}_1+\dot{q}_2) \\
\frac{\partial\mathcal{L}(\boldsymbol{q}, \dot{\boldsymbol{q}})}{\partial q_1} &= 0 \\
\frac{\partial\mathcal{L}(\boldsymbol{q}, \dot{\boldsymbol{q}})}{\partial q_2} &= -m_2 l_1 l_{c2}\dot{q}_1(\dot{q}_1+\dot{q}_2)\sin(q_2) \\
\frac{\mathrm{d}}{\mathrm{dt}}\frac{\partial\mathcal{L}(\boldsymbol{q}, \dot{\boldsymbol{q}})}{\partial\dot{q}_1} &= m_1 l_{c1}^2\ddot{q}_1 + I_1\ddot{q}_1 + m_2 l_1^2\ddot{q}_1 + m_2 l_1 l_{c2}(2\ddot{q}_1+\ddot{q}_2)\cos(q_2) - \\
&\quad m_2 l_1 l_{c2}(2\dot{q}_1+\dot{q}_2)\sin(q_2)\dot{q}_2 + m_2 l_{c2}^2(\ddot{q}_1+\ddot{q}_2) + I_2(\ddot{q}_1+\ddot{q}_2) \\
\frac{\mathrm{d}}{\mathrm{dt}}\frac{\partial\mathcal{L}(\boldsymbol{q}, \dot{\boldsymbol{q}})}{\partial\dot{q}_2} &= m_2 l_1 l_{c2}\ddot{q}_1\cos(q_2) - m_2 l_1 l_{c2}\dot{q}_1\sin(q_2)\dot{q}_2 + \\
&\quad m_2 l_{c2}^2(\ddot{q}_1+\ddot{q}_2) + I_2(\ddot{q}_1+\ddot{q}_2)
\end{aligned}$$

From (2.2), it has the Lagrange equation:

$$\begin{aligned}
&\left(m_1 l_{c1}^2 + m_2 l_{c2}^2 + m_2 l_1^2 + 2 m_2 l_1 l_{c2} \cos(q_2) + I_1 + I_2\right) \ddot{q}_1 + \\
&\qquad \left(m_2 l_1 l_{c2} \cos(q_2) + m_2 l_{c2}^2 + I_2\right) \ddot{q}_2 - \\
&\qquad m_2 l_1 l_{c2} (2\dot{q}_1 + \dot{q}_2) \sin(q_2) \dot{q}_2 = \tau_1 \\
&\left(m_2 l_1 l_{c2} \cos(q_2) + m_2 l_{c2}^2 + I_2\right) \ddot{q}_1 + \left(m_2 l_{c2}^2 + I_2\right) \ddot{q}_2 - \\
&\qquad m_2 l_1 l_{c2} \dot{q}_1 \sin(q_2) \dot{q}_2 + m_2 l_1 l_{c2} \dot{q}_1 (\dot{q}_1 + \dot{q}_2) \sin(q_2) = \tau_2
\end{aligned}$$

2.2.2 Robotic dynamics

In Section 2.2.1, the modeling of robotic system based on Lagrange equation of motion method has been presented, where an example of a 2-DOF planar serial manipulator is given. In fact, this method can be used for dynamic modeling of robot with any n degree of freedom. For the convenience of reading and analysis, this section gives a compact formal model of robot dynamics based on this method.

For an n-DOF manipulator, it is assumed that its links are all rigid and the joint rotation is not affected by friction. At this time, the kinetic energy $\mathcal{K}(\boldsymbol{q}, \dot{\boldsymbol{q}})$ of the robot can be expressed as

$$\mathcal{K}(\boldsymbol{q}, \dot{\boldsymbol{q}}) = \frac{1}{2} \dot{\boldsymbol{q}}^{\mathrm{T}} \boldsymbol{M}(\boldsymbol{q}) \dot{\boldsymbol{q}} \tag{2.3}$$

where $\boldsymbol{M}(\boldsymbol{q}) \in \mathbb{R}^{n \times n}$ is the inertial matrix.

Comparatively speaking, the potential energy of robot $\mathcal{U}(\boldsymbol{q})$ is not as specific as kinetic energy, but only known as a function of joint position $\boldsymbol{q}$. Therefore, the Lagrange operator can be written as

$$\mathcal{L}(\boldsymbol{q}, \dot{\boldsymbol{q}}) = \frac{1}{2} \dot{\boldsymbol{q}}^{\mathrm{T}} \boldsymbol{M}(\boldsymbol{q}) \dot{\boldsymbol{q}} - \mathcal{U}(\boldsymbol{q}) \tag{2.4}$$

and the Lagrange motion equation

$$\frac{\mathrm{d}}{\mathrm{d}t} \left[\frac{\partial}{\partial \dot{\boldsymbol{q}}} \left[\frac{1}{2} \dot{\boldsymbol{q}}^{\mathrm{T}} \boldsymbol{M}(\boldsymbol{q}) \dot{\boldsymbol{q}} \right] \right] - \frac{\partial}{\partial \boldsymbol{q}} \left[\frac{1}{2} \dot{\boldsymbol{q}}^{\mathrm{T}} \boldsymbol{M}(\boldsymbol{q}) \dot{\boldsymbol{q}} \right] + \frac{\partial \mathcal{U}(\boldsymbol{q})}{\partial \boldsymbol{q}} = \boldsymbol{\tau}$$

According to the calculation,

$$\begin{aligned}
\frac{\partial}{\partial \dot{\boldsymbol{q}}} \left[\frac{1}{2} \dot{\boldsymbol{q}}^{\mathrm{T}} \boldsymbol{M}(\boldsymbol{q}) \dot{\boldsymbol{q}} \right] &= \boldsymbol{M}(\boldsymbol{q}) \dot{\boldsymbol{q}} \\
\frac{\mathrm{d}}{\mathrm{d}t} \left[\frac{\partial}{\partial \dot{\boldsymbol{q}}} \left(\frac{1}{2} \dot{\boldsymbol{q}}^{\mathrm{T}} \boldsymbol{M}(\boldsymbol{q}) \dot{\boldsymbol{q}} \right) \right] &= \boldsymbol{M}(\boldsymbol{q}) \ddot{\boldsymbol{q}} + \dot{\boldsymbol{M}}(\boldsymbol{q}) \dot{\boldsymbol{q}}
\end{aligned}$$

Then it holds

$$\boldsymbol{M}(\boldsymbol{q}) \ddot{\boldsymbol{q}} + \dot{\boldsymbol{M}}(\boldsymbol{q}) \dot{\boldsymbol{q}} - \frac{\partial}{\partial \boldsymbol{q}} \left[\frac{1}{2} \dot{\boldsymbol{q}}^{\mathrm{T}} \boldsymbol{M}(\boldsymbol{q}) \dot{\boldsymbol{q}} \right] + \frac{\partial \mathcal{U}(\boldsymbol{q})}{\partial \boldsymbol{q}} = \boldsymbol{\tau} \tag{2.5}$$

Equation (2.5) is usually written as

$$\boldsymbol{M}(\boldsymbol{q})\ddot{\boldsymbol{q}} + \boldsymbol{C}(\boldsymbol{q}, \dot{\boldsymbol{q}})\dot{\boldsymbol{q}} + \boldsymbol{g}(\boldsymbol{q}) = \boldsymbol{\tau} \tag{2.6}$$

where

$$\boldsymbol{C}(\boldsymbol{q}, \dot{\boldsymbol{q}})\dot{\boldsymbol{q}} = \dot{\boldsymbol{M}}(\boldsymbol{q})\dot{\boldsymbol{q}} - \frac{\partial}{\partial \boldsymbol{q}}\left[\frac{1}{2}\dot{\boldsymbol{q}}^{\mathrm{T}}\boldsymbol{M}(\boldsymbol{q})\dot{\boldsymbol{q}}\right]$$
$$\boldsymbol{g}(\boldsymbol{q}) = \frac{\partial \mathcal{U}(\boldsymbol{q})}{\partial \boldsymbol{q}}$$

The Lagrange equation (2.6) gives the general form of the dynamic equation of n-DOF robot, which is the nonlinear differential equation of state $[\boldsymbol{q}^{\mathrm{T}}, \dot{\boldsymbol{q}}^{\mathrm{T}}]^{\mathrm{T}}$. $\boldsymbol{C}(\boldsymbol{q}, \dot{\boldsymbol{q}})\dot{\boldsymbol{q}}$ is called Centrifugal and Coriolis torque vector. $\boldsymbol{g}(\boldsymbol{q})$ is gravity vector. $\boldsymbol{\tau}$ is external force vector, which generally corresponds to torque and force output by joint actuator, and is the control input in the control view.

From (2.6), the Centrifugal and Coriolis torque $\boldsymbol{C}(\boldsymbol{q}, \dot{\boldsymbol{q}})$ is not unique, but the vector $\boldsymbol{C}(\boldsymbol{q}, \dot{\boldsymbol{q}})\dot{\boldsymbol{q}}$ is unique. In theoretical analysis and engineering practice, one way to obtain $\boldsymbol{C}(\boldsymbol{q}, \dot{\boldsymbol{q}})$ is through Christoffel symbols of the first kind, that is, the coefficient $c_{ijk}(\boldsymbol{q})$:

$$c_{ijk}(\boldsymbol{q}) = \frac{1}{2}\left[\frac{\partial M_{kj}(\boldsymbol{q})}{\partial q_i} + \frac{\partial M_{ki}(\boldsymbol{q})}{\partial q_j} - \frac{\partial M_{ij}(\boldsymbol{q})}{\partial q_k}\right] \tag{2.7}$$

where $M_{ij}(\boldsymbol{q})$ is the element in the ith row and jth column of inertial matrix $\boldsymbol{M}(\boldsymbol{q})$. Then, the element $C_{kj}(\boldsymbol{q}, \dot{\boldsymbol{q}})$ in the kth row and jth column of $\boldsymbol{C}(\boldsymbol{q}, \dot{\boldsymbol{q}})$ can be written

$$C_{kj}(\boldsymbol{q}, \dot{\boldsymbol{q}}) = \sum_{i=1}^{n} c_{ijk}(\boldsymbol{q})\dot{q}_i \tag{2.8}$$

Remark 2.1. *For any fixed k, it holds $c_{ijk}(\boldsymbol{q}) = c_{jik}(\boldsymbol{q})$. This characteristic can simplify the calculation of Christopher symbols.*

Remark 2.2. *The calculation of eq.(2.8) follows:*
From the Lagrange operator (2.4), for any $k = 1, 2, \cdots, n$, it has

$$\frac{\partial \mathcal{L}(\boldsymbol{q}, \dot{\boldsymbol{q}})}{\partial \dot{q}_k} = \sum_{j=1}^{n} M_{kj}(\boldsymbol{q})\dot{q}_j$$
$$\frac{\mathrm{d}}{\mathrm{d}t}\frac{\partial \mathcal{L}(\boldsymbol{q}, \dot{\boldsymbol{q}})}{\partial \dot{q}_k} = \sum_{j=1}^{n} M_{kj}(\boldsymbol{q})\ddot{q}_j + \sum_{j=1}^{n} \frac{\mathrm{d}}{\mathrm{d}t}M_{kj}(\boldsymbol{q})\dot{q}_j$$
$$= \sum_{j=1}^{n} M_{kj}(\boldsymbol{q})\ddot{q}_j + \sum_{j=1}^{n}\sum_{i=1}^{n} \frac{\partial M_{kj}(\boldsymbol{q})}{\partial q_i}\dot{q}_i\dot{q}_j$$
$$\frac{\partial \mathcal{L}(\boldsymbol{q}, \dot{\boldsymbol{q}})}{\partial q_k} = \frac{1}{2}\sum_{i=1}^{n}\sum_{j=1}^{n} \frac{\partial M_{ij}(\boldsymbol{q})}{\partial q_k}\dot{q}_i\dot{q}_j - \frac{\partial \mathcal{U}(\boldsymbol{q})}{\partial q_k}$$

Then the Lagrange equation is

$$\sum_{j=1}^{n} M_{kj}(\boldsymbol{q})\ddot{q}_j + \sum_{i=1}^{n}\sum_{j=1}^{n}\left(\frac{\partial M_{kj}(\boldsymbol{q})}{\partial q_i} - \frac{1}{2}\frac{\partial M_{ij}(\boldsymbol{q})}{\partial q_k}\right)\dot{q}_i\dot{q}_j + \frac{\partial \mathcal{U}(\boldsymbol{q})}{\partial q_k} = \tau_k \tag{2.9}$$

By exchanging summation symbols and using the symmetry of inertia matrix, it has

$$\begin{aligned}
&\sum_{i=1}^{n}\sum_{j=1}^{n}\frac{\partial M_{kj}(\boldsymbol{q})}{\partial q_i}\dot{q}_i\dot{q}_j \\
&\quad = \frac{1}{2}\sum_{i=1}^{n}\sum_{j=1}^{n}\left(\frac{\partial M_{kj}(\boldsymbol{q})}{\partial q_i}\dot{q}_i\dot{q}_j + \frac{\partial M_{ki}(\boldsymbol{q})}{\partial q_j}\dot{q}_i\dot{q}_j\right) \\
&\quad = \frac{1}{2}\sum_{i=1}^{n}\sum_{j=1}^{n}\left(\frac{\partial M_{kj}(\boldsymbol{q})}{\partial q_i}\dot{q}_i\dot{q}_j + \frac{\partial M_{kj}(\boldsymbol{q})}{\partial q_i}\dot{q}_i\dot{q}_j\right)
\end{aligned}$$

Then the second item in eq.(2.9) can be written as

$$\begin{aligned}
&\sum_{i=1}^{n}\sum_{j=1}^{n}\left(\frac{\partial M_{kj}(\boldsymbol{q})}{\partial q_i} - \frac{1}{2}\frac{\partial M_{ij}(\boldsymbol{q})}{\partial q_k}\right)\dot{q}_i\dot{q}_j \\
&\quad = \frac{1}{2}\sum_{i=1}^{n}\sum_{j=1}^{n}\left(\frac{\partial M_{kj}(\boldsymbol{q})}{\partial q_i} + \frac{\partial M_{ki}(\boldsymbol{q})}{\partial q_j} - \frac{\partial M_{ij}(\boldsymbol{q})}{\partial q_k}\right)\dot{q}_i\dot{q}_j \\
&\quad = \sum_{i=1}^{n}\sum_{j=1}^{n} c_{ijk}(\boldsymbol{q})\dot{q}_i\dot{q}_j
\end{aligned}$$

Define $g_k(\boldsymbol{q}) = \dfrac{\partial \mathcal{U}(\boldsymbol{q})}{\partial q_k}$*, then eq.(2.9) can be simplified as*

$$\sum_{j=1}^{n} M_{kj}(\boldsymbol{q})\ddot{q}_j + \sum_{i=1}^{n}\sum_{j=1}^{n} c_{ijk}(\boldsymbol{q})\dot{q}_i\dot{q}_j + g_k(\boldsymbol{q}) = \tau_k \tag{2.10}$$

Comparing (2.10) with (2.6), it has

$$C_{kj}(\boldsymbol{q}, \dot{\boldsymbol{q}}) = \sum_{i=1}^{n} c_{ijk}(\boldsymbol{q})\dot{q}_i$$

which means that the eq.(2.8) holds.

Remark 2.3. *In eq.(2.10), the second item on the left is Centrifugal and Coriolis torque, and their distinguishing principle is: the Centrifugal torque satisfies* $\sum_{i=1}^{n} c_{iik}(\boldsymbol{q})\dot{q}_i^2$*, while the Coriolis torque satisfies* $\sum_{i=1}^{n}\sum_{j=1, j\neq i}^{n} c_{ijk}(\boldsymbol{q})\dot{q}_i\dot{q}_j$.

2.2.3 Important property

For the robotic system with the dynamics (2.6), the following classical properties are usually satisfied [1,2]:

Property 2.1. ***Inertial Matrix*** $\boldsymbol{M}(\boldsymbol{q})$*:*

(1) It is symmetrical, i.e., $\boldsymbol{M}(\boldsymbol{q}) = \boldsymbol{M}^{\mathrm{T}}(\boldsymbol{q})$.

(2) For all robots with rotating joints, it is uniformly positive definite and has upper and lower bounds. Then there exist positive constants ρ_1 *and* ρ_2 *such that*

$$\rho_1 \boldsymbol{I} \leq \boldsymbol{M}(\boldsymbol{q}) \leq \rho_2 \boldsymbol{I}$$

where $\boldsymbol{I}$ *is an identity matrix with appropriate dimension. The inverse matrix* $\boldsymbol{M}^{-1}(\boldsymbol{q})$ *exists, and is also positive definite and bounded.*

(3) For all robots with rotating joints, there exists positive constant k_{M}*, such that for any n-dimensional vector* $\boldsymbol{x}$*,* $\boldsymbol{y}$ *and* $\boldsymbol{z}$*,*

$$|\boldsymbol{M}(\boldsymbol{x})\boldsymbol{z} - \boldsymbol{M}(\boldsymbol{y})\boldsymbol{z}| \leq k_{\mathrm{M}}|\boldsymbol{x} - \boldsymbol{y}||\boldsymbol{z}|$$

where k_{M} *satisfies* $k_{\mathrm{M}} \geq n^2 \left(\max_{i,j,k,q} \left| \frac{\partial M_{ij}(\boldsymbol{q})}{\partial q_k} \right| \right)$, $M_{ij}(\boldsymbol{q})$ *is the element of* $\boldsymbol{M}(\boldsymbol{q})$ *in the ith row and jth column.*

Property 2.2. ***Centrifugal and Coriolis torque*** $\boldsymbol{C}(\boldsymbol{q}, \dot{\boldsymbol{q}})\dot{\boldsymbol{q}}$*:*

(1) It is the quadratic form of $\dot{\boldsymbol{q}}$*, and for any* $\boldsymbol{q}$*, it holds* $\boldsymbol{C}(\boldsymbol{q}, \boldsymbol{0}) = \boldsymbol{0}$.

(2) For a robot, $\boldsymbol{C}(\boldsymbol{q}, \dot{\boldsymbol{q}})$ *may not be unique, but* $\boldsymbol{C}(\boldsymbol{q}, \dot{\boldsymbol{q}})\dot{\boldsymbol{q}}$ *must be unique.*

(3) For any given vectors $\boldsymbol{x}$ *and* $\boldsymbol{y}$ *with appropriate dimensional, the matrix* $\boldsymbol{C}(\boldsymbol{q}, \dot{\boldsymbol{q}})$ *satisfies*

$$\boldsymbol{C}(\boldsymbol{q}, \boldsymbol{x})\boldsymbol{y} = \boldsymbol{C}(\boldsymbol{q}, \boldsymbol{y})\boldsymbol{x}$$

(4) For robots with all rotating joints, there exists bounded constant positive scalar k_c*, such that for any* $\boldsymbol{q}$*,* $\boldsymbol{x}$ *and* $\boldsymbol{y}$*, it has*

$$|\boldsymbol{C}(\boldsymbol{q}, \boldsymbol{x})\boldsymbol{y}| \leq k_c|\boldsymbol{x}||\boldsymbol{y}|$$

Property 2.3. *Gravity vector* $\boldsymbol{g}(\boldsymbol{q})$ *is bounded, i.e.,* $|\boldsymbol{g}(\boldsymbol{q})| \leq \alpha_g(\boldsymbol{q})$*, where* $\alpha_g(\boldsymbol{q})$ *is a scalar function. Specifically, for robots with all rotating joints,* $\alpha_g(\boldsymbol{q})$ *is a constant independent of* $\boldsymbol{q}$.

The relationship between those components is given below.

Property 2.4. *(**Skew-symmetric**) The matrix* $\dot{\boldsymbol{M}}(\boldsymbol{q}) - 2\boldsymbol{C}(\boldsymbol{q}, \dot{\boldsymbol{q}})$ *is skew symmetry, i.e., for any dimensioned* $\boldsymbol{y}$*, it has*

$$\boldsymbol{y}^{\mathrm{T}} \left(\dot{\boldsymbol{M}}(\boldsymbol{q}) - 2\boldsymbol{C}(\boldsymbol{q}, \dot{\boldsymbol{q}}) \right) \boldsymbol{y} = 0$$

or equivalently,

$$\dot{\boldsymbol{M}}(\boldsymbol{q}) = \boldsymbol{C}(\boldsymbol{q}, \dot{\boldsymbol{q}}) + \boldsymbol{C}^{\mathrm{T}}(\boldsymbol{q}, \dot{\boldsymbol{q}})$$

Property 2.5. *(Linearization in Parameters) For the Lagrange equation in (2.6), there exist $n \times l$-dimension vector $\boldsymbol{Y}(\boldsymbol{q}, \dot{\boldsymbol{q}}, \ddot{\boldsymbol{q}})$ and l-dimension vector $\boldsymbol{\Theta}$ such that*

$$\boldsymbol{M}(\boldsymbol{q})\ddot{\boldsymbol{q}} + \boldsymbol{C}(\boldsymbol{q}, \dot{\boldsymbol{q}})\dot{\boldsymbol{q}} + \boldsymbol{g}(\boldsymbol{q}) = \boldsymbol{Y}(\boldsymbol{q}, \dot{\boldsymbol{q}}, \ddot{\boldsymbol{q}})\boldsymbol{\Theta}$$

where the function $\boldsymbol{Y}(\boldsymbol{q}, \dot{\boldsymbol{q}}, \ddot{\boldsymbol{q}})$ is called regression vector, and $\boldsymbol{\Theta}$ is the parameter vector.

2.2.4 Mathematical descriptions

Considering that the teleoperation systems have different topological structures, for convenience, this section takes the single master and single slave teleoperation system as an example, and introduces its dynamics. As shown in Fig. 2.3, a typical single master and single slave teleoperation system consists of five parts: the human operator, the master robot, the communication network, the slave robot and the environment.

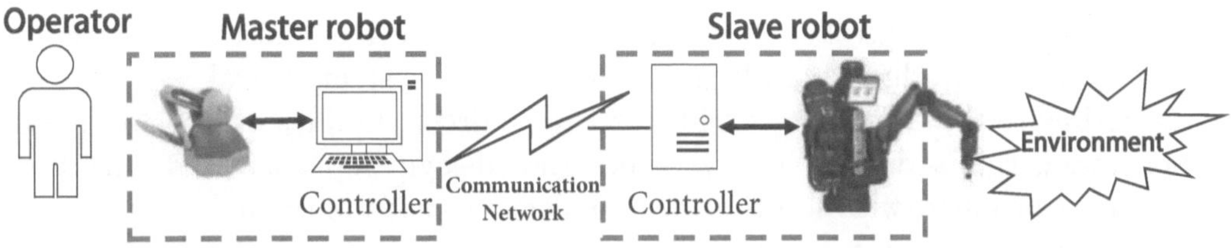

Figure 2.3 A schematic diagram of master-slave teleoperation robot system.

Without losing generality, the dynamics of the single master and single slave teleoperation system is given as follows:

$$\begin{cases} \boldsymbol{M}_m(\boldsymbol{q}_m)\ddot{\boldsymbol{q}}_m + \boldsymbol{C}_m(\boldsymbol{q}_m, \dot{\boldsymbol{q}}_m)\dot{\boldsymbol{q}}_m + \boldsymbol{g}_m(\boldsymbol{q}_m) = \boldsymbol{f}_h + \boldsymbol{\tau}_m \\ \boldsymbol{M}_s(\boldsymbol{q}_s)\ddot{\boldsymbol{q}}_s + \boldsymbol{C}_s(\boldsymbol{q}_s, \dot{\boldsymbol{q}}_s)\dot{\boldsymbol{q}}_s + \boldsymbol{g}_s(\boldsymbol{q}_s) = \boldsymbol{\tau}_s - \boldsymbol{f}_e \end{cases} \tag{2.11}$$

where the subscripts m and s represent the master robot and the slave robot respectively, the subscripts h and e represent the human operator and the environment respectively. For any $j \in \{m, s\}$, $\boldsymbol{q}_j$, $\dot{\boldsymbol{q}}_j$, $\ddot{\boldsymbol{q}}_j$ are respectively the position, velocity and acceleration of the robot joint. $\boldsymbol{M}_j(\boldsymbol{q}_j)$ is the inertia matrix of the robot, $\boldsymbol{C}_j(\boldsymbol{q}_j, \dot{\boldsymbol{q}}_j)$ is the matrix of Centrifugal/Coriolis torque, $\boldsymbol{g}_j(\boldsymbol{q}_j)$ is the gravity vector, $\boldsymbol{\tau}_j$ is the control torque, $\boldsymbol{f}_h$ and $\boldsymbol{f}_e$ are respectively the torque applied by the human operator to the master robot and the torque exerted by the environment on the slave robot.

For any $j \in \{m, s\}$, let $\boldsymbol{x}_j$ be the coordinates of the robot end effector pose in the task space, which is defined as

$$\begin{cases} \boldsymbol{x}_j = \boldsymbol{h}_j(\boldsymbol{q}_j) \\ \dot{\boldsymbol{x}}_j = \dfrac{\partial \boldsymbol{h}_j(\boldsymbol{q}_j)}{\partial \boldsymbol{q}_j}\dot{\boldsymbol{q}}_j = \boldsymbol{J}_j(\boldsymbol{q}_j)\dot{\boldsymbol{q}}_j \end{cases} \tag{2.12}$$

where $\boldsymbol{J}_j(\boldsymbol{q}_j)$ is the Jacobian matrix which reflects the transformation relationship between robot joint speed and end effector speed

In this book, it is assumed that the joints of the master and slave robots are all rotating joints, which satisfy all the classical properties in Section 2.2.3.

2.3 SYSTEM WITH TIME DELAY

2.3.1 Modeling of time delay

With the exploration of space and deep sea by human beings, the demand for remote control of robots in space or deep sea to complete special tasks is increasing. These ultra-long distance teleoperation systems usually have a time delay of 2 $\sim$ 3s, and some even reach tens of seconds or more than several minutes. Take the communication between the earth and the moon as an example, the maximum distance between the earth and the moon is 407000km, the communication delay is about 1.4 s. Considering that it also exists the delay of 1.1s in the system, the one-way loop delay between the earth and the moon is 2.5 s [3,4], which poses a serious real-time control challenge in robot teleoperation [5]. Communication delay in teleoperation system seriously threatens the control quality and even makes the system unstable. Therefore, time delay is a key technical challenge in teleoperation system control.

According to the existing control theory, the current stability analysis of teleoperation system can be divided into two types: time-delay independent and time-delay dependent. Compared with the time-delay independent stability analysis, the time-delay dependent stability analysis is generally more complex, but less conservative, so the controller designed according to this method often has better control performance. Therefore, in recent years, the delay-dependent stability analysis of teleoperation system has been widely concerned. Time-delay dependent stability analysis requires the use of time-delay information, so it is necessary to model the time-delay. The common time-delay models are summarized as follows.

(1) Constant communication delay. It is the simplest time-delay model, which assumes that all transmission times in the communication network are constant. In practice, the buffer is introduced into the control system. When the buffer length is greater than the worst-case transmission time, the communication delay can be regarded as a fixed constant. The limitation of this method is that the communication time delay adopted by the control system is often larger than the actual time delay, which will destroy the performance of the system [6].

(2) Time-varying communication delay. It is a practical time-delay model, which assumes that all transmission times in the communication network are bounded but variable, and can generally be characterized by the first-order continuously differentiable function. For example, if the time-varying communication delay is characterized as $d(t)$, there exist bounded constant positive scalars $\bar{d}$ and $\tilde{d}$ such that $0 \leq d(t) \leq \bar{d}$ and $|\dot{d}(t)| \leq \tilde{d}$. The control design will use $\bar{d}$ and $\tilde{d}$ instead of the time-delay signal itself [7,8]. Compared with constant communication delay, this model is less conservative. Teleoperation systems with high real-time requirements often use this modeling method.

(3) Random communication delay. Because of the asymmetry of communication channels and the difference of channel media, it will cause the variability and uncertainty of transmission delay, so it is also possible to use stochastic process to establish the delay model. Usually, it is assumed that the transmission delay of the communication network satisfies a certain probability distribution, such as Markov process, and the closed-loop system is designed and analyzed in the stochastic framework [9,10]. This method is more in line with the actual channel characteristics in theory. However, this method is often difficult to be used in actual control because the information of signal transmission probability distribution in actual communication network is difficult to obtain accurately.

In this book, the time-varying communication delay will be used to model the network transmission time.

2.3.2 Functional differential equation

In the practical engineering system, the mathematical model is usually described by differential equations, and the rate of change of the system state is assumed to be only related to the current system state, for example:

$$\dot{\boldsymbol{x}}(t) = \boldsymbol{f}(t, \boldsymbol{x}(t)), \ \boldsymbol{x}(t_0) = \boldsymbol{x}_0$$

That is, as long as the initial condition $\boldsymbol{x}_0$ is known, the solution of the equation can be determined by the state $\boldsymbol{x}(t)$, $T_0 \leq t < \infty$. This book pays attention to the control design and stability analysis of teleoperation system. The feedback of the state information of the master and slave robots through the network will lead to the information delay in the closed-loop teleoperation system, so that the future evolution of the state of the complete closed-loop teleoperation system is still related to the state of the system at some time in the past. So the ordinary differential equations can no longer accurately describe the model of this kind of system. For such time-delay system, the functional differential equations are often used to establish the mathematical model of the system.

Let us give a brief introduction of functional differential equations. Let the maximum upper bound of time-delay system be $\bar{d}$. For any $A > 0$ and any continuous function $\boldsymbol{\psi}(t)$, $t_0 \leq t \leq t_0 + A$, $\boldsymbol{\psi}_t \in C([t_0 - \bar{d}, t_0 + A]; \mathbb{R}^n)$ is an interval of function $\boldsymbol{\psi}(t)$. Define $\boldsymbol{\psi}_t(\theta) = \boldsymbol{\psi}(t + \theta), -\bar{d} \leq \theta \leq 0$. Then the delayed functional differential equation can be written as

$$\dot{\boldsymbol{x}}(t) = \boldsymbol{f}(t, \boldsymbol{x}_t) \tag{2.13}$$

where $\boldsymbol{x}_t(\theta) = \boldsymbol{x}(t+\theta), -\bar{d} \leq \theta \leq 0$, $\boldsymbol{x}(t) \in \mathbb{R}^n$, $\boldsymbol{f} : \mathbb{R} \times C([t_0 - \bar{d}, t_0 + A]; \mathbb{R}^n) \rightarrow \mathbb{R}^n$. The equation (2.13) shows that the differential of the state variable $\boldsymbol{x}$ at time t depends on the t and the $\boldsymbol{x}(\theta)$, $t - \bar{d} \leq \theta \leq t$. Therefore, to determine the future change law of the state, it is necessary to specify the initial value of the state variable $\boldsymbol{x}(t)$ in the time interval $[t_0 - \bar{d}, t_0]$, i.e.,

$$\boldsymbol{x}_{t_0}(\theta) = \boldsymbol{x}(t_0 + \theta) = \boldsymbol{\phi}(\theta), \ -\bar{d} \leq \theta \leq 0 \tag{2.14}$$

where $\boldsymbol{\phi}(\theta) \in C([t_0 - \bar{d}, t_0 + A]; \mathbb{R}^n)$ is given.

For the $A > 0$, if the function $\boldsymbol{x}$ is continuous in the interval $[t_0 - \bar{d}, t_0 + A)$ and satisfies the eq.(2.13), then $\boldsymbol{x}$ is one solution of eq.(2.13) in the interval $[t_0 - \bar{d}, t_0 + A)$. This solution shows that $(t, \boldsymbol{x}_t)$ is in the domain of f. If it also satisfies the initial condition equation (2.14), then $\boldsymbol{x}$ is the solution passing through (t_0, ϕ). Let us denote it as $\boldsymbol{x}(t; T_0, \boldsymbol{\phi}, \boldsymbol{f})$ or simply $\boldsymbol{x}(t; T_0, \boldsymbol{\phi})$ or $\boldsymbol{x}(t_0, \boldsymbol{\phi})$.

2.3.3 Switched systems

With the introduction of auxiliary switched filter, the closed-loop teleoperation system can be described by the switched dynamic equations, so that the closed-loop system can be designed and analyzed by using the analysis methods of switched system. The following is a brief introduction of the switched system.

The switched system consists of a group of subsystems (continuous time dynamic subsystems or discrete time dynamic subsystems) and their corresponding switching rules. This book only considers continuous time switched systems, and the typical mathematical models of such systems are as follows:

$$\dot{\boldsymbol{x}}(t) = \boldsymbol{f}(t, \boldsymbol{x}(t), r(t)) \tag{2.15}$$

where $\boldsymbol{x}(t) \in \mathbb{R}^n$ is the state, $\boldsymbol{f} : \mathbb{R}_+ \times \mathbb{R}^n \times \mathcal{S} \to \mathbb{R}^n$ is assumed to satisfy the local Lipschitz condition, and for any $t \geq t_0$ and $i \in \mathcal{S}$, it has $\boldsymbol{f}(t, 0, i) \equiv 0$.

In eq.(2.15), $r : [t_0, \infty) \to \mathcal{S}$ is the switching signal, which is a right-continuous function with piecewise constant value about t. $\mathcal{S} := \{1, 2, \cdots, N\}$ is the switching state space. For switching signal $r(t)$, let $\{t_k\}_{k \geq 0}$ be the switching time sequences, and define $r(t_k) = i_k$. Then in the interval $[t_k, t_{k+1})$, the i_k-th subsystem will be activated. In Fig. 2.4, a simple structural diagram of the switched system is given.[2]

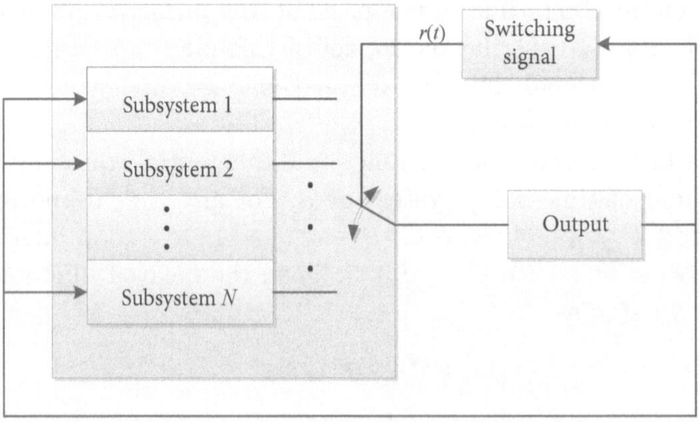

Figure 2.4 A simple diagram of switched system.

[2]When $r(t)$ is modeled as deterministic switching signal and random switching signal (such as Markov switching) respectively, the corresponding systems are called deterministic switched system and random switched system. According to the design of switched filter in the following chapters, this book only considers the deterministic switched system or switched system for short.

From Fig. 2.4, the switched system is a multi-model structure, and its dynamics are jointly determined by the dynamics of each subsystem and the switching rules. The mathematical model of each subsystem of the switched system is as follows

$$\dot{\boldsymbol{x}}(t) = \boldsymbol{f}(t, \boldsymbol{x}(t), i), \ i \in \mathcal{S}$$

which calls a mode of switched system (2.15).

Specific to the teleoperation system, when we use the auxiliary switched filter control method, the robot's dynamics can be generalized by switched nonlinear time-delay systems as follows:

$$\begin{cases} \dot{\boldsymbol{x}}(t) = \boldsymbol{f}(t, \boldsymbol{x}_t, \boldsymbol{u}(t), r(t)), \\ \boldsymbol{z}(t) = \boldsymbol{h}(\boldsymbol{x}_t, r(t)), \end{cases} \quad t \geq t_0 = 0 \tag{2.16}$$

where $\boldsymbol{z}(t) \in \mathbb{R}^w$ is the output, $\boldsymbol{z}_t = \{\boldsymbol{z}(t+\theta) : -\bar{d} \leq \theta \leq 0\} \in C([-\bar{d}, 0]; \mathbb{R}^w)$, $\boldsymbol{u}(t) \in \mathbb{R}^m$ is the input, which can be used to express the effects of operators and environment on robots in the modeling of teleoperation system in the following chapters.

2.3.4 Stability of time-delayed system

To perform the stability analysis of teleoperation system in the following chapters, this section gives some stability definitions and criteria. For the general switched nonlinear delayed systems in (2.16), let us assume that the initial states of the system is $r(t_0) = i_0 \in \mathcal{S}$, $\boldsymbol{x}_0 = \{\boldsymbol{x}(\theta) : -\bar{d} \leq \theta \leq 0\} \in C([-\bar{d}, 0]; \mathbb{R}^n)$, $\boldsymbol{z}_0 = \boldsymbol{h}(\boldsymbol{x}_0, i_0) = \{\boldsymbol{z}(\theta) : -\bar{d} \leq \theta \leq 0\} \in C([-\bar{d}, 0]; \mathbb{R}^w)$, and the input $\boldsymbol{u}$ is locally essential bounded[3].

Let us introduce the following useful definitions.

Definition 2.1. [11] *If the continuous function $\alpha : [0, a) \to [0, \infty)$ is strictly increasing, and $\alpha(0) = 0$, then α belongs to the $\mathcal{K}$-class function. If $a = \infty$, and $\alpha(r) \to \infty$ when $r \to \infty$, then α is the $\mathcal{K}_\infty$-class function. For the continuous function $\beta : [0, a) \times [0, \infty) \to [0, \infty)$, if for every fixed s, the mapping $\beta(r, s)$ is a $\mathcal{K}$-class function of r, and for every fixed r, $\beta(r, s)$ is a decreasing function of s, with $\beta(r, s) \to 0$ when $s \to \infty$, then β belongs to the $\mathcal{KL}$-class function.*

Definition 2.2. [12] *A function $\alpha : \mathbb{R}_+ \to \mathbb{R}_+$ is called generalized $\mathcal{K}$-class function ($\mathcal{GK}$), if it is continuous with $\alpha(0) = 0$, and for $r_1 > r_2$, it holds*

$$\begin{cases} \alpha(r_1) > \alpha(r_2), & \alpha(r_1) \neq 0 \\ \alpha(r_1) = \alpha(r_2), & \alpha(r_1) = 0 \end{cases} \tag{2.17}$$

Obviously, a classical $\mathcal{K}$-class function is also the $\mathcal{GK}$-class function. Further, a function $\beta : \mathbb{R}_+ \times \mathbb{R}_+ \to \mathbb{R}_+$ is generalized $\mathcal{KL}$-class function ($\mathcal{GKL}$), if for any fixed $t \geq 0$, the function $\beta(s, t)$ is $\mathcal{GK}$-class function for s, and for any fixed $s \geq 0$, it is decreasing function and equals 0 when $t = T \leq \infty$. Here, T is the settling time of the $\mathcal{GKL}$ $\beta(\cdot, \cdot)$.

[3] If the function $\boldsymbol{\nu}$ satisfies $\operatorname{ess\,sup}_{t \geq 0} |\boldsymbol{\nu}(t)| < \infty$, then it is essential bounded. For any given constants $0 \leq a < b$, $\boldsymbol{\nu}_{[a,b)} : [0, \infty) \to \mathbb{R}^m$ is defined as: $\boldsymbol{\nu}_{[a,b)}(t) = \boldsymbol{\nu}(t)$ when $t \in [a, b)$, $\boldsymbol{\nu}_{[a,b)}(t) = \boldsymbol{0}$ when $t \notin [a, b)$. A function $\boldsymbol{\nu}$ is called locally essential bounded, if for any $T > 0$, $\boldsymbol{\nu}_{[0,T)}$ is essential bounded

To overcome the shortcomings of the traditional ISS/IOS methods, all subsequent chapters of this book will design and analyze the closed-loop control in the framework of (finite time) state-independent input-output stability. The definitions of stability are stated as follows:

Definition 2.3. [13] *The system (2.16) is state-independent input-to-output practical stable (SIIOpS), if there exists $\gamma(\cdot) \in \mathcal{K}$ and $\beta(\cdot,\cdot) \in \mathcal{KL}$ such that*

$$|\boldsymbol{z}(t)| \leq \beta(\|\boldsymbol{z}_0\|, t) + \gamma(\|\boldsymbol{u}_{[0,t)}\|_\infty) + \xi \tag{2.18}$$

where ξ is a positive bounded constant. If $\xi = 0$, then the system (2.16) is state-independent input-to-output stable (SIIOS). Further, if $\mathbf{u}(t) \equiv \mathbf{0}$, the system (2.16) is state-independent uniformly output stable (SIUOS).

Remark 2.4. *The traditional ISS/IOS methods have been used in the teleoperation system* [14, 15]*. Considering that the joint position information of the robot is one of the components of the robot system state, when the traditional input-to-state stability/input-to-output stability analysis method is used, the joint position is both the state and the output of the system* [14, 15]*. At this time, according to the definition of input-to-state stability/input-to-output stability* [13]*, when the master and slave robots are in the free motion state, the master and slave robots will have zero steady-state synchronization. However, this is not what the teleoperation robot system expects. Because in teleoperation system, the joint positions of the master and slave robots should converge to each other, and they should stop moving when synchronization is reached. Therefore, the ISS/IOS method faces limitations when applied to teleoperation system. To overcome this theoretical shortcoming, this book is the first time to propose the state-independent input-output stability method for teleoperation system.*

Remark 2.5. *It should be noted that state-independent input-output stability is not a new concept. In 2000, Sontag and Wang first put forward related concepts* [13]*. Subsequently, Dashkovskiy also mentioned the definition of SIIOS in 2011* [16]*. According to the definition* [13]*, SIIOS is different from traditional IOS. An SIIOS system must be IOS, but not necessarily vice versa. Definition 2.3 extends it to the case of time-delay system.*

Definition 2.4. *The switched nonlinear delay system (2.16) is finite-time state-independent input-to-output practical stable (FTSIIOpS), if for any $\xi \geq 0$, there exist $\gamma(\cdot) \in \mathcal{K}$ and $\beta(\cdot,\cdot) \in \mathcal{GKL}$ such that*

$$|\boldsymbol{z}(t)| \leq \beta(\|\boldsymbol{z}_0\|, t) + \gamma(\|\boldsymbol{u}_{[0,t)}\|_\infty) + \xi \tag{2.19}$$

If $\xi = 0$, the system (2.16)is finite-time state-independent input-to-output stable (FTSIIOS). Further, when the input $\boldsymbol{u} \equiv \mathbf{0}$, the system (2.16) is finite-time state-independent uniformly output stable (FTSIUOS).

At present, Razumikhin method and Lyapunov-Krasovskii method are the most commonly used methods for stability analysis of time-delay systems. Razumikhin-based stability criterion is a delay-independent criterion, which can guarantee the stability of the system with any time delay, so it is a very conservative stability criterion at the expense of system operation performance. With a certain knowledge of the time delay of teleoperation system, we can choose the time delay dependent Lyapunov-Krasovskii stability theorem, which can greatly improve the operation performance on the basis of ensuring the stability of teleoperation system within a certain time delay range. Therefore, the controller design of teleoperation system based on Lyapunov-Krasovskii method is a hot topic in this field in recent years. This book will also analyze the closed-loop teleoperation system based on Lyapunov-Krasovskii stability method, and further use the multiple Lyapunov-Krasovskii methods in the switched control framework.

The SIIOS and FTSIIOS criteria of the system (2.16) are given below, before which two lemmas are introduced.

Lemma 2.1. [17] *For any (local) absolutely continuous function $y(t)$ defined at $t \geq 0$ and satisfying $y(t) \geq 0$, if there is a continuous and positive definite function $\alpha(\cdot)$, such that*

$$D^+ y(t) \leq -\alpha(y(t))$$

holds for almost all t, then it exists $\mathcal{KL}$-class function $\beta_\alpha(\cdot,\cdot)$ such that for all $t \geq 0$,

$$y(t) \leq \beta_\alpha(y_0, t)$$

where $y_0 = y(0) \geq 0$.

Lemma 2.2. [18] *For any $r \in [0, \infty)$, let $\eta(r) = \int_0^r \frac{1}{h(\theta)} \mathrm{d}\theta$, where $h \in \mathcal{K}$ and $\eta(r) < +\infty$. Define*

$$\beta(r,s) = \begin{cases} 0, & r = 0,\ s \geq 0 \\ 0, & r \neq 0,\ s \geq \eta(r) \\ \eta(r) - s, & r \neq 0,\ s < \eta(r) \end{cases} \tag{2.20}$$

then $\beta(r,s)$ belongs to $\mathcal{GKL}$-class function.

Then for system (2.16), it has

Lemma 2.3. *When the input of system (2.16) constant equals to zero, i.e., $\boldsymbol{u} \equiv \boldsymbol{0}$, let $\alpha_1, \alpha_2 \in \mathcal{K}_\infty$, $\alpha_3 \in \mathcal{K}$, then if there exists functional $V : C([-\bar{d}, 0]; \mathbb{R}^n) \times \mathcal{S} \times \mathbb{R}_+ \to \mathbb{R}_+$ such that for any $i \in \mathcal{S}$, (2.21)~(2.23) hold*

$$\alpha_1(|\boldsymbol{z}(t)|) \leq V(\boldsymbol{x}_t, i, t) \leq \alpha_2(\|\boldsymbol{z}_t\|_{M_2}) \tag{2.21}$$

$$D^+ V(\boldsymbol{x}_t, i, t) \leq -\alpha_3(\|\boldsymbol{z}_t\|_{M_2}) \tag{2.22}$$

$$V(\boldsymbol{x}_t, r(t_k), t_k) \leq V(\boldsymbol{x}_t, r(t_{k-1}), t_k), \ \forall k \geq 1 \tag{2.23}$$

where the right-hand Dini derivative [19] *is defined by*

$$D^+ V(\boldsymbol{x}_t, i, t) = \limsup_{h \to 0^+} \frac{V(\boldsymbol{x}_{t+h}, i, t+h) - V(\boldsymbol{x}_t, i, t)}{h}$$

then the system (2.16) is SIUOS.

Proof. Firstly, to simplify the expression, denote $W_{r(t)}(t) = V(\boldsymbol{x}_t, t, r(t))$ and $W_0 = V(\boldsymbol{x}_0, 0, i_0)$. From (2.21) and (2.22), for any $t \in [t_k, t_{k+1})$,$k \geq 0$ with $t_0 = 0$, assume that $r(t) = i_k$, then

$$D^+W_{i_k}(t) \leq \alpha \circ \alpha_2(W_{i_k}(t)) \tag{2.24}$$

where $\circ$ is the composite function operator. By using Lemma 2.1, there exists the $\mathcal{KL}$-class function $\beta_1(\cdot,\cdot)$ such that

$$W_{i_k}(t) \leq \beta_1(W_{i_k}(t_k), t - t_k),\ \forall t \in [t_k, t_{k+1})$$

Considering the arbitrariness of k, it's easy to verify that

$$\begin{array}{ll}
W_{i_0}(t) \leq \beta_1(W_{i_0}(t_0), t - t_0), & \forall t \in [t_0, t_1) \\
W_{i_1}(t) \leq \beta_1(W_{i_1}(t_1), t - t_1), & \forall t \in [t_1, t_2) \\
W_{i_2}(t) \leq \beta_1(W_{i_2}(t_2), t - t_2), & \forall t \in [t_2, t_3) \\
\vdots & \vdots \\
W_{i_{k-1}}(t) \leq \beta_1(W_{i_{k-1}}(t_{k-1}), t - t_{k-1}), & \forall t \in [t_{k-1}, t_k) \\
W_{i_k}(t) \leq \beta_1(W_{i_k}(t_k), t - t_k), & \forall t \in [t_k, t_{k+1}) \\
\vdots & \vdots
\end{array}$$

Moreover, from (2.23), for any $t \in [t_{k-1}, t_k)$, it has

$$W_{i_k}(t_k) \leq W_{i_{k-1}}(t_k) \leq \beta_1(W_{i_{k-1}}(t_{k-1}), t_k - t_{k-1})$$

Then there exists the $\mathcal{KL}$-class function $\bar{\beta}(\cdot,\cdot)$ such that

$$W_{r(t)}(t) \leq \bar{\beta}(W_{i_0}(t_0), t - t_0),\ \forall t \geq t_0$$

where

$$\begin{aligned}
&\bar{\beta}(W_{i_0}(t_0), t - t_0) \\
&= \begin{cases}
\beta_1(W_{i_0}(t_0), t - t_0), & \forall t \in [t_0, t_1) \\
\beta_1(\beta_1(W_{i_0}(t_0), t_1 - t_0), t - t_1), & \forall t \in [t_1, t_2) \\
\beta_1(\beta_1(\beta_1(W_{i_0}(t_0), t_1 - t_0), t_2 - t_1), t - t_2), & \forall t \in [t_2, t_3) \\
\vdots & \\
\beta_1(\beta_1(\cdots\beta_1(W_{i_0}(t_0), t_1 - t_0)\cdots, t_k - t_{k-1}), t - t_k), & \forall t \in [t_k, t_{k+1}) \\
\vdots &
\end{cases}
\end{aligned}$$

In other word, for any $t \geq t_0$, it holds

$$|z(t)| \leq \alpha_1^{-1} \circ \bar{\beta}(\alpha_2(\|z_0\|), t - t_0) := \beta(\|z_0\|, t - t_0)$$

□

Lemma 2.4. *When the system (2.16) has the external input, i.e.,* $\boldsymbol{u} \not\equiv \mathbf{0}$*, if the hypotheses of Lemma 2.3 hold, and there further exists* $\rho \in \mathcal{K}$ *such that the condition (2.22) is changed as*

$$\|z_t\|_{M_2} \geq \rho(\|\boldsymbol{u}_{[0,t)}\|_\infty) \Rightarrow$$
$$D^+V(\boldsymbol{x}_t, i, t) \leq -\alpha_3(\|z_t\|_{M_2}) \tag{2.25}$$

then the system (2.16) is SIIOS, where the gain γ *satisfies* $\gamma(s) = \alpha_1^{-1} \circ \alpha_2 \circ \rho(s)$.

Proof. Define the set $\mathfrak{B} = \{z_t \in C([-\bar{d}, 0]; \mathbb{R}^w) : \|z_t\|_{M_2} < \rho(\|u_{[0,\infty)}\|_\infty)\}$, and let $\tau \in [0, \infty)$ denote the time when the state first enters the set $\mathfrak{B}$. In the following, let us consider the following two cases: $z_0 \in \mathfrak{B}^C$ and $z_0 \in \mathfrak{B}$.

Case 1: $z_0 \in \mathfrak{B}^C$. In this case, for any $t \in [0, \tau]$, from (2.25), it has

$$D^+V(\boldsymbol{x}_t, i, t) \leq -\alpha(\|z_t\|_{M_2})$$

Then from Lemma 2.3, there exists the $\mathcal{KL}$-class function $\beta(\cdot, \cdot)$ such that

$$|z(t)| \leq \beta(\|z_0\|, t), \ \forall t \in [0, \tau] \tag{2.26}$$

For $t > \tau$, because the time derivative of $V(\boldsymbol{x}_t, i, t)$ is non-positive outside of the set $\mathfrak{B} \subset \{z_t \in C([-\bar{d}, 0]; \mathbb{R}^w) : V(\boldsymbol{x}_t, t, i) < \alpha_2 \circ \rho(\|\boldsymbol{u}_{[0,\infty)}\|_\infty)\}$, if $z_t \in \mathfrak{B}$, then

$$V(\boldsymbol{x}_t, t, i) < \alpha_2 \circ \rho(\|\boldsymbol{u}_{[0,\infty)}\|_\infty)$$

And further

$$|z(t)| < \alpha_1^{-1} \circ \alpha_2 \circ \rho(\|\boldsymbol{u}_{[0,\infty)}\|_\infty), \ \forall t > \tau \tag{2.27}$$

Then for any $t \geq 0$ and $z_0 \in \mathfrak{B}^C$, it has

$$|z(t)| \leq \beta(\|z_0\|, t) + \alpha_1^{-1} \circ \alpha_2 \circ \rho(\|\boldsymbol{u}_{[0,\infty)}\|_\infty)$$

Case 2: $z_0 \in \mathfrak{B}$. In this case, $\tau \equiv 0$, and (2.27) holds for all $t > 0$. Then it has

$$|z(t)| \leq \beta(\|z_0\|, t) + \alpha_1^{-1} \circ \alpha_2 \circ \rho(\|\boldsymbol{u}_{[0,\infty)}\|_\infty)$$

Obviously, it also holds for $t = 0$. Then, for any $t \geq 0$ and $z_0 \in \mathfrak{B}$, it has

$$|z(t)| \leq \beta(\|z_0\|, t) + \alpha_1^{-1} \circ \alpha_2 \circ \rho(\|\boldsymbol{u}_{[0,\infty)}\|_\infty)$$

Combining **Case 1** and **Case 2**, for any $t \geq 0$ and $x_0 \in C([-\bar{d}, 0]; \mathbb{R}^n)$(or equivalently $z_0 \in C([-\bar{d}, 0]; \mathbb{R}^w)$), it holds

$$|z(t)| \leq \beta(\|z_0\|, t) + \alpha_1^{-1} \circ \alpha_2 \circ \rho(\|\boldsymbol{u}_{[0,\infty)}\|_\infty)$$

By causality, it further has

$$|z(t)| \leq \beta(\|z_0\|, t) + \alpha_1^{-1} \circ \alpha_2 \circ \rho(\|\boldsymbol{u}_{[0,t)}\|_\infty)$$

□

Lemma 2.5. *Let $\alpha_1(\cdot), \alpha_2(\cdot) \in \mathcal{K}_\infty$. If there exist the functional $V : C([-\bar{d}, 0]; \mathbb{R}^n) \times \mathcal{S} \times \mathbb{R}_+ \to \mathbb{R}_+$ and the continuous differentiable function $\alpha : \mathbb{R}_+ \to \mathbb{R}_+$ such that for ant $i \in \mathcal{S}$,*

$$\alpha_1(|\boldsymbol{z}(t)|) \leq V(\boldsymbol{x}_t, i, t) \leq \alpha_2(\|\boldsymbol{z}_t\|_{M_2}) \tag{2.28}$$

$$D^+V(\boldsymbol{x}_t, i, t) \leq -\alpha(\|\boldsymbol{z}_t\|_{M_2}) \tag{2.29}$$

$$V(\boldsymbol{x}_{t_k}, r(t_k), t_k) \leq V(\boldsymbol{x}_{t_k}, r(t_{k-1}), t_k), \ \forall k \geq 2 \tag{2.30}$$

where for $r \in [0, \infty)$ and $\theta > 0$, $\int_0^r \frac{1}{\alpha(s)} \mathrm{d}s < \infty$ and $\alpha'(\theta) > 0$, then the system (2.16) without input (i.e., $\boldsymbol{u}(t) \equiv \boldsymbol{0}$) is FTSIUOS, and the settling time satisfies

$$T_0(\boldsymbol{x}_0) \leq \int_0^{\alpha_2(\|\boldsymbol{z}_0\|)} \frac{1}{\alpha(s)} \mathrm{d}s < \infty$$

Proof. For convenience, let $W_{r(t)}(t) = V(\boldsymbol{x}_t, t, r(t))$, $W_0 = V(\boldsymbol{x}_0, 0, i_0)$. From Lemma 2.5, for any $W_i(t) \in [0, \infty)$ and $i \in \mathcal{S}$, it has

$$\eta(W_i(t)) = \int_0^{W_i(t)} \frac{1}{\alpha(s)} \mathrm{d}s < \infty$$

By Newton-Leibniz formula, for any $t \in [t_k, t_{k+1})$,$k \geq 0$, let us denote $r(t) = i_k$. Then,

$$\eta(W_{i_k}(t)) = \eta(W_{i_k}(t_k)) + \int_{t_k}^{t} D^+\eta(W_{i_k}(s)) \mathrm{d}s \tag{2.31}$$

From (2.29), when $W_{i_k}(t) \neq 0$, it has

$$D^+\eta(W_{i_k}(t)) \leq -1 \tag{2.32}$$

Then from (2.31) and (2.32), it has

$$\eta(W_{i_k}(t)) \leq \eta(W_{i_k}(t_k)) - (t - t_k), \ \forall t \in [t_k, t_{k+1}) \tag{2.33}$$

From (2.30), for any $t \in [t_{k-1}, t_{k+1})$,

$$\begin{aligned}\eta(W_{i_k}(t_k)) &\leq \eta(W_{i_{k-1}}(t_k)) \\ &\leq \eta(W_{i_{k-1}}(t_{k-1})) - (t_k - t_{k-1})\end{aligned}$$

Further, it has

$$\eta(W_{i_k}(t)) \leq \eta(W_{i_{k-1}}(t_{k-1})) - (t - t_{k-1})$$

Repeat the above steps, it is easy to verify that

$$\eta(W_{r(t)}(t)) \leq \eta(W_0) - t, \ \forall t \in [0, t_{k+1})$$

Considering the arbitrariness of k, it holds

$$\eta(W_{r(t)}(t)) \leq \eta(W_0) - t, \ \forall t \geq 0$$

In fact, when $W_{r(t)}(t) = 0$, from the positive property of $W_{r(t)}(t)$, and (2.29) and (2.30), it has for all $t < T$, it holds $W_{r(t)}(T) \equiv 0$.

Define

$$\tilde{\beta}(r,s) = \begin{cases} 0, & r = 0, \ s \geq 0 \\ 0, & r \neq 0, \ s \geq \eta(r) \\ \eta(r) - s, & r \neq 0, \ s < \eta(r) \end{cases} \tag{2.34}$$

Then from Lemma 2.2, it is easy to verify $\tilde{\beta}(\cdot,\cdot) \in \mathcal{GKL}$.

Therefore,

$$\eta(W_{r(t)}(t)) \leq \tilde{\beta}(\eta(W_0), t), \ \forall t \geq 0$$

and further

$$W_{r(t)}(t) \leq \bar{\beta}(W_0, t), \ \forall t \geq 0$$

It means that

$$|\boldsymbol{z}(t)| \leq \beta(\|\boldsymbol{z}_0\|, t), \ \forall t \geq 0$$

where $\bar{\beta}(r,s) = \eta^{-1} \circ \tilde{\beta}(\eta(r), s)$, $\beta(r,s) := \alpha_1^{-1} \circ \bar{\beta}(\alpha_2(r), s)$. Obviously, $\beta(\cdot,\cdot)$ is also the $\mathcal{GKL}$-class function.

From (2.33), the settling time of the $\mathcal{GKL}$-class function $\tilde{\beta}(\cdot,\cdot)$ is given as

$$\eta(V(\boldsymbol{x}_{T_0(\boldsymbol{x}_0)}, T_0(\boldsymbol{x}_0), r(T_0(\boldsymbol{x}_0)))) - \eta(V(\boldsymbol{x}_0, 0, i_0)) \leq -T_0(\boldsymbol{x}_0)$$

i.e.,

$$T_0(\boldsymbol{x}_0) \leq \eta(V(\boldsymbol{x}_0, 0, i_0)) = \int_0^{V(\boldsymbol{x}_0,0,i_0)} \frac{1}{\alpha(s)} \mathrm{d}s \tag{2.35}$$

Based on the definitions of $\bar{\beta}(\cdot,\cdot)$ and $\beta(\cdot,\cdot)$, the settling time of $\beta(\cdot,\cdot)$ still satisfies $T_0(\boldsymbol{x}_0) \leq \int_0^{\alpha_2(\|\boldsymbol{z}_0\|)} \frac{1}{\alpha(s)} \mathrm{d}s$. □

Remark 2.6. *Specifically, in Lemma 2.5, if* $\alpha(s) = k_1 s + k_2 s^{\sigma}$, $k_1 > 0$, $k_2 > 0$, $0 < \sigma < 1$, *the settling time can be further given as* $T_0(x_0) \leq \int_0^{\alpha_2(\|\boldsymbol{z}_0\|)} \frac{1}{\alpha(s)} \mathrm{d}s = \frac{1}{k_1(1-\sigma)} \ln \frac{k_1(\alpha_2(\|\boldsymbol{z}_0\|))^{1-\sigma} + k_2}{k_2}$.

Lemma 2.6. *If there exists* $\rho(\cdot) \in \mathcal{K}_\infty$ *such that the condition (2.29) is replaced by*

$$\|\boldsymbol{z}_t\|_{M_2} \geq \rho(\|\boldsymbol{u}_{[0,\infty)}\|_\infty) \Rightarrow D^+V(\boldsymbol{x}_t, i, t) \leq -\alpha(\|\boldsymbol{z}_t\|_{M_2}) \tag{2.36}$$

Then when other conditions of Lemma 2.5 are satisfied, the system (2.16) is FTSIIOS, where $\gamma(s) = \alpha_1^{-1} \circ \alpha_2 \circ \rho(s)$.

Proof. The proof of Lemma 2.6 is similar to Lemma 2.4. □

2.4 CONCLUSION

This chapter introduces the mathematical model of robot and its classical properties, the time-delay system, the switched system, the state-independent input-output stability and its criteria, and some important definitions and conclusions, which are the model and theoretical basis of this book and will be fully used in the following chapters of this book. Because this book focuses on the research of new control technology of teleoperation system, this chapter only briefly introduces those concepts. Readers who want to study in depth can refer to the relevant literature.

Bibliography

[1] R. Kelly, V. S. Davila, and J. A. L. Perez, *Control of Robot Manipulators in Joint Space.* Springer Science & Business Media, 2005.

[2] T. H. Lee and C. J. Harris, *Adaptive Neural Network Control of Robotic Manipulators.* World Scientific, 1998.

[3] E. Krotkov, R. Simmons, F. Cozman, and S. Koenig, "Safeguarded teleoperation for lunar rovers," *SAE Transactions*, pp. 1020–1029, 1996.

[4] C. Yang, J. Song, J. Sun, Y. Miao, W. Zhang, J. She, and D. Hui, "On real-time teleoperation of lunar rover," *Science China Information Sciences*, vol. 44, no. 4, pp. 461–472, 2014.

[5] N. Chopra, M. W. Spong, S. Hirche, and M. Buss, "Bilateral teleoperation over the Internet: the time varying delay," in *Proceedings of the American Control Conference*, 2003, pp. 155–160.

[6] R. Luck and A. Ray, "An observer-based compensator for distributed delays," *Automatica*, vol. 26, no. 5, pp. 903–908, 1990.

[7] C.-C. Hua and P. X. Liu, "Delay-dependent stability criteria of teleoperation systems with asymmetric time-varying delays," *IEEE Transactions on Robotics*, vol. 26, no. 5, pp. 925–932, 2010.

[8] F. Hashemzadeh, I. Hassanzadeh, and M. Tavakoli, "Teleoperation in the presence of varying time delays and sandwich linearity in actuators," *Automatica*, vol. 49, no. 9, pp. 2813–2821, 2013.

[9] Z. Li, L. Ding, H. Gao, G. Duan, and C.-Y. Su, "Trilateral teleoperation of adaptive fuzzy force/motion control for nonlinear teleoperators with communication random delays," *IEEE Transactions on Fuzzy Systems*, vol. 21, no. 4, pp. 610–624, 2012.

[10] Y. Kang, Z. Li, X. Cao, and D. Zhai, "Robust control of motion/force for robotic manipulators with random time delays," *IEEE Transactions on Control Systems Technology*, vol. 21, no. 5, pp. 1708–1718, 2012.

[11] H. Khalil, *Nonlinear Systems.* New Jersey: Prentice-Hall, 2002.

[12] Y. Hong, Z.-P. Jiang, and G. Feng, "Finite-time input-to-state stability and applications to finite-time control design," *SIAM Journal on Control and Optimization*, vol. 48, no. 7, pp. 4395–4418, 2010.

[13] E. Sontag and Y. Wang, "Lyapunov characterizations of input to output stability," *SIAM Journal on Control and Optimization*, vol. 39, no. 1, pp. 226–249, 2000.

[14] I. G. Polushin, A. Tayebi, and H. J. Marquez, "Control schemes for stable teleoperation with communication delay based on ios small gain theorem," *Automatica*, vol. 42, no. 6, pp. 905–915, 2006.

[15] I. G. Polushin, S. N. Dashkovskiy, A. Takhmar, and R. V. Patel, "A small gain framework for networked cooperative force-reflecting teleoperation," *Automatica*, vol. 49, no. 2, pp. 338–348, 2013.

[16] S. Dashkovskiy, D. V. Efimov, and E. D. Sontag, "Input to state stability and allied system properties," *Automation and Remote Control*, vol. 72, no. 8, pp. 1579–1614, 2011.

[17] Y. Lin, E. D. Sontag, and Y. Wang, "A smooth converse Lyapunov theorem for robust stability," *SIAM Journal on Control and Optimization*, vol. 34, no. 1, pp. 124–160, 1996.

[18] Y. Kang and P. Zhao, "Finite-time stability and input-to-state stability of stochastic nonlinear systems," in *Proceedings of the 31st Chinese Control Conference.* IEEE, 2012, pp. 1529–1534.

[19] S. M. Magdi, *Switched Time-Delay Systems: Stability and Control.* Springer, 2010.

II

Single-Master Single-Slave Teleoperation

CHAPTER 3

Adaptive control of bilateral teleoperation

THIS chapter presents the teleoperation system a new adaptive control framework based on auxiliary switched filter. By introducing the switched filter systems, the complete closed-loop system is modeled as a nonlinear switched delayed system, which consists of two interconnected switched and non-switched subsystems. By using the switching stability analysis in the framework of state-independent input-to-output stability, the closed-loop stability is then derived. The proposed method achieves the control of teleoperaion system in the simultaneous presence of varying time delay, model uncertainties and passive/nonpassive external forces.

3.1 INTRODUCTION

As pointed out in Chapter 1, teleoperation system has received a lot of attention in the development process of over half a century [1–3]. Among them, how to realize the stable control of teleoperation system under the condition of non-zero communication delay and model uncertainty is a difficult problem. To solve this challenge, many advanced control methods have emerged, including passive control method [2,4–6], sliding mode control method [7,8], small gain control method [9,10] and linear matrix inequality method [11–17], etc. However, these methods still have different limitations in practical application. For example, the traditional wave variable method and scattering theory are only suitable for constant time delay. Passivity-based method requires that the master and slave inputs must meet the passive conditions. When analyzing the stability of teleoperation system with time delay, the linear inequality method needs to choose the complex Lyapunov-Krasovskii functional, and the system analysis is highly technical. In recent years, to overcome the defects of passive control methods, Polushin et al. proposed a teleoperation control method based on ISS/IOS small gain theorem [9,10,15,18]. Compared with the traditional passivity method, the small gain method does not impose passivity assumptions on operators and environmental dynamics, so it has a wider application range. It should be noted that under the definition of ISS/IOS stability framework, the joint position of the robot will also be one of the components of the complete output state of the closed-loop system. At

DOI: 10.1201/9781003382058-3

this time, when the master and slave of the teleoperation robot are in free movement, the position movement of the master and slave robots will follow the following process: approaching each other, realizing synchronization, and finally stabilizing at the zero position. However, in the actual teleoperation system, when the external force applied to the robot is removed, that is, when the master and slave robots are in free motion, the master and slave robots expect to approach each other and achieve synchronization, and stop moving after synchronization. Obviously, the zero steady-state synchronization problem of the existing ISS/IOS small gain theorem method is not the phenomenon that the actual teleoperation system wants, and needs to be improved and overcome.

This chapter presents the teleoperation system a new adaptive control framework based on auxiliary switched filter. By introducing the switched filter subsystems, the complete closed-loop system is modeled as a nonlinear switched delayed system, which consists of two interconnected switched and non-switched subsystems. By using the switching stability analysis in the framework of state-independent input-to-output stability, the closed-loop stability is then derived. The proposed method achieves the control of teleoperaion system in the simultaneous presence of varying time delay, model uncertainties and passive/nonpassive external forces.

3.2 PROBLEM FORMULATION

Let us consider a teleoperator consisting of a pair of n-DOF nonlinear robotic systems as follows:

$$\begin{cases} \boldsymbol{M}_m(\boldsymbol{q}_m)\ddot{\boldsymbol{q}}_m + \boldsymbol{C}_m(\boldsymbol{q}_m, \dot{\boldsymbol{q}}_m)\dot{\boldsymbol{q}}_m + \boldsymbol{g}_m(\boldsymbol{q}_m) = \boldsymbol{f}_h + \boldsymbol{\tau}_m \\ \boldsymbol{M}_s(\boldsymbol{q}_s)\ddot{\boldsymbol{q}}_s + \boldsymbol{C}_s(\boldsymbol{q}_s, \dot{\boldsymbol{q}}_s)\dot{\boldsymbol{q}}_s + \boldsymbol{g}_s(\boldsymbol{q}_s) = \boldsymbol{\tau}_s - \boldsymbol{f}_e \end{cases} \tag{3.1}$$

where for any $j \in \{m, s\}$, $\boldsymbol{q}_j$, $\dot{\boldsymbol{q}}_j$, $\ddot{\boldsymbol{q}}_j \in \mathbb{R}^n$. The definitions of $\boldsymbol{M}_j(\boldsymbol{q}_j) \in \mathbb{R}^{n \times n}$, $\boldsymbol{C}_j(\boldsymbol{q}_j, \dot{\boldsymbol{q}}_j) \in \mathbb{R}^{n \times n}$, $\boldsymbol{g}_j(\boldsymbol{q}_j) \in \mathbb{R}^n$, $\boldsymbol{\tau}_j \in \mathbb{R}^n$, $\boldsymbol{f}_h$ and $\boldsymbol{f}_e$ are given in (2.11).

The teleoperation system in (3.1) satisfies the Properties 2.1, 2.4 and 2.5. It is assumed that the operator torque and environmental torque meet the following conditions:

Assumption 3.1. *The human operator torque $\boldsymbol{f}_h$ and the environment torque $\boldsymbol{f}_e$ are assumed to be locally essentially bounded.*

Remark 3.1. *There are two points to explain about the Assumption 3.1. On the one hand, Assumption 3.1 is not used for the first time in this book, but it is based on the discussion in the existing literature. For example, Polushin et al. once assumed that the operator torque/environmental torque was uniformly bounded, but the upper bound was unknown [18]. In this chapter, the uniformly bounded hypothesis is extended to be locally bounded. On the other hand, unlike the traditional control method based on passivity theory, Assumtion 3.1 does not make passivity assumption for operators and environment, thus reducing the practical application limit to a certain extent [10].*

For teleoperation system (3.1), the transmission delays are $d_m(t)$ and $d_s(t)$, which respectively represent the communication delay from master robot to slave robot and from slave robot to master robot. For any $j \in \{m, s\}$, assume that

Assumption 3.2. *There exists positive constants* $\bar{d}$ *and* $\tilde{d}$ *such that* $0 \leq d_j(t) \leq \bar{d}$ *and* $0 \leq |\dot{d}_j(t)| \leq \tilde{d}$.

To describe the synchronization performance between the master and slave robots, the following position synchronization errors are introduced:

$$\begin{cases} \boldsymbol{e}_m(t) = \boldsymbol{q}_m(t) - \boldsymbol{q}_s(t - d_s(t)) \\ \boldsymbol{e}_s(t) = \boldsymbol{q}_s(t) - \boldsymbol{q}_m(t - d_m(t)) \end{cases} \tag{3.2}$$

In this chapter, the goal of teleoperation control is defined as: realizing adaptive control of teleoperation system (or realizing asymptotic bounded control of tracking error) in the presence of time-varying communication delay and unknown dynamic model.

3.3 ADAPTIVE CONTROLLER DESIGN

In this section, the subsystem decomposition incorporated with small-gain theorem is developed for single master-single-slave teleoperation. Based on the proposed auxiliary switched filters, the complete closed-loop master (or, slave) robot's dynamic is decomposed into two interconnected subsystems. By analyzing the stability of each subsystem, the SIIOS condition of the complete systems is finally derived. Note that under the discontinuous control law adopt by the proposed novel framework, the solution concepts of the closed-loop system with discontinuous right hand sides due to Filippov can be used [19,20]. Now, let us introduce the main results.

The new adaptive control method for teleoperation system (3.1) is given below.

Firstly, to simplify the expression, the following velocity synchronization error is introduced:

$$\boldsymbol{e}_{vm}(t) = \dot{\boldsymbol{q}}_m(t) - \boldsymbol{q}_{vds}(t, d_s(t))$$
$$\boldsymbol{e}_{vs}(t) = \dot{\boldsymbol{q}}_s(t) - \boldsymbol{q}_{vdm}(t, d_m(t))$$

where $\boldsymbol{q}_{vdm}(t, d_m(t)) = \dot{\boldsymbol{q}}_m(\theta)|_{\theta=t-d_m(t)}$, $\boldsymbol{q}_{vds}(t, d_s(t)) = \dot{\boldsymbol{q}}_s(\theta)|_{\theta=t-d_s(t)}$. Then, $\dot{\boldsymbol{e}}_m = \boldsymbol{e}_{vm} + \dot{d}_s(t)\boldsymbol{q}_{vds}(t, d_s(t))$, $\dot{\boldsymbol{e}}_s = \boldsymbol{e}_{vs} + \dot{d}_m(t)\boldsymbol{q}_{vdm}(t, d_m(t))$. We also define auxiliary variables:

$$\boldsymbol{\eta}_j(t) = \dot{\boldsymbol{q}}_j(t) + \lambda_j \boldsymbol{e}_j(t) \tag{3.3}$$

where $j \in \{m, s\}$, λ_j is a constant, positive scalar. From Property 2.5, we have

$$\boldsymbol{Y}_j(\boldsymbol{q}_j, \dot{\boldsymbol{q}}_j, \boldsymbol{e}_{vj}, \boldsymbol{e}_j)\boldsymbol{\Theta}_j = \boldsymbol{M}_j(\boldsymbol{q}_j)\lambda_j \boldsymbol{e}_{vj} + \boldsymbol{C}_j(\boldsymbol{q}_j, \dot{\boldsymbol{q}}_j)\lambda_j \boldsymbol{e}_j - \boldsymbol{g}_j(\boldsymbol{q}_j)$$

Substituting (3.3) into (3.1), it holds

$$\boldsymbol{M}_j(\boldsymbol{q}_j)\dot{\boldsymbol{\eta}}_j + \boldsymbol{C}_j(\boldsymbol{q}_j, \dot{\boldsymbol{q}}_j)\boldsymbol{\eta}_j = \boldsymbol{Y}_j\boldsymbol{\Theta}_j + \boldsymbol{f}_{he_j} + \boldsymbol{\tau}_j + \lambda_j \dot{d}_\iota \boldsymbol{M}_j(\boldsymbol{q}_j)\boldsymbol{q}_{vd\iota}(t, d_\iota(t)) \tag{3.4}$$

where $\boldsymbol{Y}_j = \boldsymbol{Y}_j(\boldsymbol{q}_j, \dot{\boldsymbol{q}}_j, \boldsymbol{e}_{vj}, \boldsymbol{e}_j)$, $\boldsymbol{f}_{he_m}(t) = \boldsymbol{f}_h(t)$, $\boldsymbol{f}_{he_s}(t) = -\boldsymbol{f}_e(t)$, with $\iota \in \{m, s\}$ but $\iota \neq j$.

For system (3.4), based on the design idea of dynamic compensation, the control torque of the robot is designed as follows

$$\boldsymbol{\tau}_j(t) = -K_{j1}\boldsymbol{\eta}_j(t) - \boldsymbol{Y}_j(t)\hat{\boldsymbol{\Theta}}_j(t) - K_{j2}\boldsymbol{y}_j(t) - K_{j3}\text{sgn}(\boldsymbol{\eta}_j)|\boldsymbol{q}_{vd\iota}(t, d_\iota(t))| \tag{3.5}$$

where K_{j1}, K_{j2} and K_{j3} are some positive constants.

The control torque (3.5) consists of four parts, namely, the proportional control term $-K_{j1}\boldsymbol{\eta}_j(t)$, the adaptive estimation term $-\boldsymbol{Y}_j(t)\hat{\boldsymbol{\Theta}}_j(t)$, the coordinated control torque $-K_{j3}\text{sgn}(\boldsymbol{\eta}_j)|\boldsymbol{q}_{vd\iota}(t, d_\iota(t))|$ and the dynamic compensation $-K_{j2}\boldsymbol{y}_j(t)$.

Firstly, $\hat{\boldsymbol{\Theta}}_j(t)$ is the estimation of unknown parameter $\boldsymbol{\Theta}_j$, and its dynamic update law is designed as [9]

$$\dot{\hat{\boldsymbol{\Theta}}}_j(t) = \psi_j \boldsymbol{Y}_j^{\mathrm{T}}(t)\boldsymbol{\eta}_j(t) - \vartheta_j(\hat{\boldsymbol{\Theta}}_j(t) - \boldsymbol{\Theta}_j^*) \tag{3.6}$$

where $\boldsymbol{\Theta}_j^*$ is the nominal value of $\boldsymbol{\Theta}_j$, ψ_j and ϑ_j are some positive constants.

Secondly, the coordinated torque term is mainly used to eliminate the residual term in the process of theoretical derivation. In order to simplify the expression, let $\boldsymbol{F}_j = -K_{j3}\text{sgn}(\boldsymbol{\eta}_j)|\boldsymbol{q}_{vd\iota}(t, d_\iota(t))|$.

Finally, $\boldsymbol{y}_j(t)$ in the dynamic compensation term is the state output of the dynamic compensator, and its specific design will be given through a switching filter. Let $\boldsymbol{\mathcal{K}}_j(t) = \text{diag}[\kappa_{j,1}(t), \kappa_{j,2}(t), \cdots, \kappa_{j,n}(t)]$, then $\boldsymbol{y}_j = [y_{j,1}, y_{j,2}, \cdots, y_{j,n}]^{\mathrm{T}}$ is designed as

$$\dot{y}_{j,i} = \begin{cases} -P_{j1}(y_{j,i} + e_{j,i}) - e_{vj,i} - P_{j2}\eta_{j,i} + u_{j,i}, & \kappa_{j,i}(t) = 1 \\ -P_{j1}(y_{j,i} - e_{j,i}) + e_{vj,i} - P_{j2}\eta_{j,i} + u_{j,i}, & \kappa_{j,i}(t) = -1 \end{cases} \tag{3.7}$$

where

$$u_{j,i}(t) = \begin{cases} -P_{j3}\dfrac{s_{j,i}(t)}{|s_{j,i}(t)|^2}|\boldsymbol{q}_{vd\iota}(t, d_\iota(t))|^2, & s_{j,i}(t) \neq 0 \\ 0, & \text{others} \end{cases}$$

and for any $i = 1, 2, \cdots, n$ and $j \in \{m, s\}$, $\eta_{j,i}$, $s_{j,i}$, $e_{j,i}$ and $e_{vj,i}$ are the ith component of, respectively, $\boldsymbol{\eta}_j$, $\boldsymbol{s}_j$, $\boldsymbol{e}_j$ and $\boldsymbol{e}_{vj}$. $\boldsymbol{s}_j(t) = \boldsymbol{y}_j(t) + \boldsymbol{\mathcal{K}}_j(t)\boldsymbol{e}_j(t)$. P_{j1}, P_{j2} and P_{j3} are some positive constants, $\iota \in \{m, s\}$ but $\iota \neq j$.

In (3.7), $\boldsymbol{\mathcal{K}}_j(t) = \text{diag}[\kappa_{j,1}(t), \kappa_{j,2}(t), \cdots, \kappa_{j,n}(t)]$ is the switching rule of switched filter, specifically defined as

$$\kappa_{j,i}(t) = \begin{cases} 1, & y_{j,i}(t)e_{j,i}(t) > 0 \text{ or } y_{j,i}(t)e_{j,i}(t) = 0,\ y_{j,i}(t^-)e_{j,i}(t^-) \leq 0 \\ -1, & y_{j,i}(t)e_{j,i}(t) < 0 \text{ or } y_{j,i}(t)e_{j,i}(t) = 0,\ y_{j,i}(t^-)e_{j,i}(t^-) > 0 \end{cases}$$

Remark 3.2. *Due to the introduction of $\boldsymbol{\mathcal{K}}_j(t)$, the switched filter (3.7) is a switched system. However, because $\boldsymbol{\mathcal{K}}_j(t)$ is not a standard switching signal, here we first model it as a standard switching signal. From the definition, when only $y_{j,i}(t)e_{j,i}(t) = 0$ and $y_{j,i}(t^-)e_{j,i}(t^-) > 0$, $\kappa_{j,i}(t)$ switches from 1 to -1. On the contrary, when only $y_{j,i}(t)e_{j,i}(t) = 0$ and $y_{j,i}(t^-)e_{j,i}(t^-) < 0$, $\kappa_{j,i}(t)$ switches from -1*

to 1. *Therefore, for any fixed* j *and* t, $\kappa_{j,i}(t)$ *has two values of* ± 1. *Then* $\mathcal{K}_j(t)$ *has* $2n$ *values, which are*

$$\begin{cases} \text{diag}[-1,-1,-1,\cdots,-1] \in \mathbb{R}^{n\times n} \\ \text{diag}[1,-1,-1,\cdots,-1] \in \mathbb{R}^{n\times n} \\ \text{diag}[-1,1,-1,\cdots,-1] \in \mathbb{R}^{n\times n} \\ \vdots \\ \text{diag}[-1,1,1,\cdots,1] \in \mathbb{R}^{n\times n} \\ \text{diag}[1,1,1,\cdots,1] \in \mathbb{R}^{n\times n} \end{cases} \tag{3.8}$$

Number the states in (3.8) in turn and denote them as mode 1, *mode* 2, *mode* 3, $\cdots$, *mode* 2^n. *Define discrete state set* $\mathcal{S} = \{1, 2, \cdots, 2^n\}$. *Define the function* $\bar{r}_j$: $\mathcal{K}_j(t) \to \mathcal{S}$ *as*

$$\bar{r}_j(\mathcal{K}_j(t)) = i, \ \textit{iff}\ \mathcal{K}_j(t) \textit{ is in the } i\textit{th mode}$$

where $i \in \mathcal{S}$. *For convenience, define* $r_j(t)$ *and let* $r_j(t) = \bar{r}_j(\mathcal{K}_j(t))$. *By those, one-to-one mapping between* $r_j(t)$ *and* $\mathcal{K}_j(t)$ *are obtained. Further, Let* $\{t_k^j\}_{k\geq 0}$ *be the switching times of* $r_j(t)$. *Without losing generality, assume that* $t_0^j = 0$, *then* $r_j(t)$ *can be regarded as a standard state-dependent switching signal.*

Substitute $r_j(t)$ into the system (3.7), it then has

$$\begin{aligned} \dot{\boldsymbol{y}}_j &= -P_{j1}(\boldsymbol{y}_j + \mathcal{K}_j\boldsymbol{e}_j) - \mathcal{K}_j\boldsymbol{e}_{vj} - P_{j2}\boldsymbol{\eta}_j + \boldsymbol{u}_j \\ &:= -P_{j1}\boldsymbol{y}_j - P_{j2}\boldsymbol{\eta}_j + \boldsymbol{u}_j + \boldsymbol{h}(\boldsymbol{e}_j, \boldsymbol{e}_{vj}, r_j) \end{aligned} \tag{3.9}$$

where $\boldsymbol{u}_j = [u_{j,1}, u_{j,2}, \cdots, u_{j,n}]^{\mathrm{T}}$, $\boldsymbol{h}(\boldsymbol{e}_j, \boldsymbol{e}_{vj}, r_j) = -P_{j1}\mathcal{K}_j\boldsymbol{e}_j - \mathcal{K}_j\boldsymbol{e}_{vj}$.

The system (3.9) can be regarded as a switched dynamic system with $r_j(t)$ as the switching signal.

Denoting $\tilde{\boldsymbol{\Theta}}_j(t) = \boldsymbol{\Theta}_j - \hat{\boldsymbol{\Theta}}_j(t)$ and substituting (3.5) into (3.4),

$$\begin{aligned} &\boldsymbol{M}_j(\boldsymbol{q}_j)\dot{\boldsymbol{\eta}}_j + \boldsymbol{C}_j(\boldsymbol{q}_j, \dot{\boldsymbol{q}}_j)\boldsymbol{\eta}_j \\ &\quad = \boldsymbol{Y}_j\tilde{\boldsymbol{\Theta}}_j - K_{j1}\boldsymbol{\eta}_j + \boldsymbol{f}_{he_j} - K_{j2}\boldsymbol{y}_j + \boldsymbol{F}_j + \lambda_j\dot{d}_\iota\boldsymbol{M}_j(\boldsymbol{q}_j)\boldsymbol{q}_{vd\iota}(t, d_\iota(t)) \end{aligned} \tag{3.10}$$

The whole closed-loop control system consists of subsystems (3.5), (3.6), (3.9), and (3.10). With the introduction of subsystem (3.9), the closed-loop system will also be a switched dynamic system. The adaptive control diagram of the system is shown in Fig. 3.1.

Remark 3.3. *Here, we model the closed-loop switched system. A complete closed-loop system consists of dynamic equations of the master and slave robots and corresponding controllers, so it contains two switching signals, namely* $r_m(t)$ *and* $r_s(t)$. *In order to model the complete closed-loop system into a standard switched dynamic system, referring to the existing works* [21–23], *we introduce a virtual switching signal*

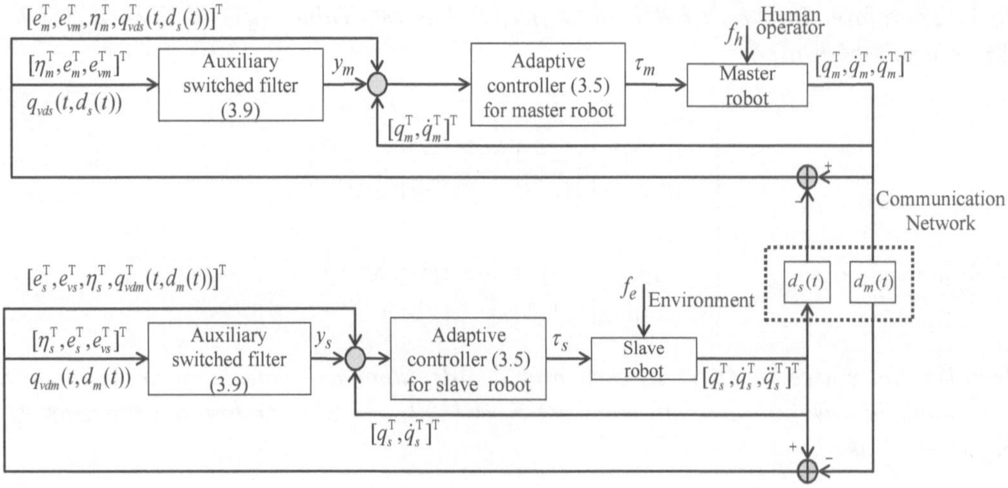

Figure 3.1 Block diagram of the proposed algorithm for single-master-single-slave teleoperation system.

$r(t)$, which is defined as

$$r(t) := (r_m(t), r_s(t)) : [0, \infty) \to \mathcal{S} \times \mathcal{S}$$

and denote the switching time sequences as $\{t_k\}_{k\geq 0} = \{t_k^m\}_{k\geq 0} \cup \{t_k^s\}_{k\geq 0}$. Then let $\chi_{m1} = \boldsymbol{q}_m$, $\chi_{m2} = \boldsymbol{e}_m$, $\chi_{m3} = \boldsymbol{\eta}_m$, $\chi_{m4} = \tilde{\boldsymbol{\Theta}}_m$, $\chi_{m5} = \boldsymbol{y}_m$, $\chi_{s1} = \boldsymbol{q}_s$, $\chi_{s2} = \boldsymbol{e}_s$, $\chi_{s3} = \boldsymbol{\eta}_s$, $\chi_{s4} = \tilde{\boldsymbol{\Theta}}_s$, $\chi_{s5} = \boldsymbol{y}_s$, and denote $\chi = [\chi_{m1}^{\mathrm{T}}, \chi_{m2}^{\mathrm{T}}, \chi_{m3}^{\mathrm{T}}, \chi_{m4}^{\mathrm{T}}, \chi_{m5}^{\mathrm{T}}, \chi_{s1}^{\mathrm{T}}, \chi_{s2}^{\mathrm{T}}, \chi_{s3}^{\mathrm{T}}, \chi_{s4}^{\mathrm{T}}, \chi_{s5}^{\mathrm{T}}]^{\mathrm{T}}$. Therefore, according to the control design, the available formula (3.11) is sorted out. This means that the complete closed-loop teleoperation system can be regarded as a switched nonlinear time-delay system with $r(t)$ as the switching signal.

It should be noted that since the closed-loop system is a discontinuous dynamic system, in the following chapters of this book, the solution of this kind of system will be considered under Filippov definition [19, 20, 24–26].

Remark 3.4. *The switched filter control method is similar to the traditional output regulation control in form, but the difference is that it introduces the switching control technology into the controller design. Compared with traditional non-switching control methods, this method has the following advantages:*

(1) it helps the controller to make full use of the status information of the system. With the introduction of $\mathcal{K}_j$, the system (3.9) can make full use of the error information.

(2) It can make the stability analysis of closed-loop system more flexible. Due to the switched design of the system (3.9), the complete closed-loop system is also a switched dynamic system. Therefore, the stability analysis can be carried out

by using the multiple Lyapunov-Krasovskii functionals method, which makes the selection of candidate Lyapunov-Krasovskii functional more flexible.

$$\dot{\chi} = \begin{bmatrix} \chi_{m3} - \lambda_m \chi_{m2} \\ \begin{pmatrix} -\lambda_m \chi_{m2} + \chi_{m3} - (1 - \dot{d}_s(t))\chi_{s3}(t - d_s(t)) + \\ (1 - \dot{d}_s(t))\lambda_s \chi_{s2}(t - d_s(t)) \end{pmatrix} \\ \begin{pmatrix} \boldsymbol{M}_m^{-1}(\chi_{m1})\Big(- \boldsymbol{C}_m(\chi_{m1}, \chi_{m3} - \lambda_m \chi_{m2})\chi_{m3} + \\ \boldsymbol{Y}_m \chi_{m4} - K_{m1}\chi_{m3} - K_{m2}\chi_{m5} + \\ \lambda_m \dot{d}_s(t)\boldsymbol{M}_m(\chi_{m1})(\chi_{s3}(t - d_s(t)) - \lambda_s \chi_{s2}(t - d_s(t))) - \\ K_{m3}\mathrm{sgn}(\chi_{m3})|\chi_{s3}(t - d_s(t)) - \lambda_s \chi_{s2}(t - d_s(t))| + \boldsymbol{f}_h\Big) \end{pmatrix} \\ -\vartheta_m \chi_{m4} - \psi_m \boldsymbol{Y}_m^{\mathrm{T}} \chi_{m3} + \vartheta_m(\boldsymbol{\Theta}_m - \boldsymbol{\Theta}_m^*) \\ \begin{pmatrix} -P_{m1}\chi_{m5} - P_{m2}\chi_{m3} + \boldsymbol{u}_m + \\ \boldsymbol{h}(\chi_{m2}, \chi_{m3} - \lambda_m \chi_{m2} - \chi_{s3}(t - d_s(t)) + \\ \lambda_s \chi_{s2}(t - d_s(t)), r_m(t)) \end{pmatrix} \\ \chi_{s3} - \lambda_m \chi_{s2} \\ \begin{pmatrix} -\lambda_s \chi_{s2} + \chi_{s3} - (1 - \dot{d}_m(t))\chi_{m3}(t - d_m(t)) + \\ (1 - \dot{d}_m(t))\lambda_m \chi_{m2}(t - d_m(t)) \end{pmatrix} \\ \begin{pmatrix} \boldsymbol{M}_s^{-1}(\chi_{s1})\Big(- \boldsymbol{C}_s(\chi_{s1}, \chi_{s3} - \lambda_s \chi_{s2})\chi_{s3} + \boldsymbol{Y}_s \chi_{s4} - \\ K_{s1}\chi_{s3} - K_{s2}\chi_{s5} + \lambda_s \dot{d}_m(t)\boldsymbol{M}_s(\chi_{s1})(\chi_{m3}(t - d_m(t)) - \\ \lambda_m \chi_{m2}(t - d_m(t))) - K_{s3}\mathrm{sgn}(\chi_{s3})|\chi_{m3}(t - d_m(t)) - \\ \lambda_m \chi_{m2}(t - d_m(t))| - \boldsymbol{f}_e\Big) \end{pmatrix} \\ -\vartheta_s \chi_{s4} - \psi_s \boldsymbol{Y}_s^{\mathrm{T}} \chi_{s3} + \vartheta_s(\boldsymbol{\Theta}_s - \boldsymbol{\Theta}_s^*) \\ \begin{pmatrix} -P_{s1}\chi_{s5} - P_{s2}\chi_{s3} + \boldsymbol{u}_s + \\ \boldsymbol{h}(\chi_{s2}, \chi_{s3} - \lambda_s \chi_{s2} - \chi_{m3}(t - d_m(t)) + \\ \lambda_m \chi_{m2}(t - d_m(t)), r_s(t)) \end{pmatrix} \end{bmatrix}$$

$$:= \boldsymbol{H}(\chi, \chi(t - d_m(t)), \chi(t - d_s(t)), \boldsymbol{\omega}, r(t)) \tag{3.11}$$

3.4 STABILITY ANALYSIS

Given that the operator/environment external force may be non-passive, in order to unify the analysis of passive and non-passive external forces, this section will discuss the stability of closed-loop system under the framework of state-independent input-output stability, and refer to Section 2.3.4 of this book for relevant definitions. In this case, let the state and output of the teleoperated robot are $\boldsymbol{x}_j = [\boldsymbol{\eta}_j^{\mathrm{T}}, \tilde{\boldsymbol{\Theta}}_j^{\mathrm{T}}, \boldsymbol{q}_j^{\mathrm{T}}, \boldsymbol{y}_j^{\mathrm{T}}, \boldsymbol{e}_j^{\mathrm{T}}]^{\mathrm{T}}$, $\boldsymbol{z}_j = [\boldsymbol{\eta}_j^{\mathrm{T}}, \tilde{\boldsymbol{\Theta}}_j^{\mathrm{T}}, \boldsymbol{y}_j^{\mathrm{T}}, \boldsymbol{e}_j^{\mathrm{T}}]^{\mathrm{T}}$. Also let $\boldsymbol{x}_{jt}(\tau) = \{\boldsymbol{x}_j(t + \tau) : \tau \in [-\bar{d}, 0]\} \in C([-\bar{d}, 0]; \mathbb{R}^{4n+l})$, $\boldsymbol{z}_{jt}(\tau) = \{\boldsymbol{z}_{1j}(t + \tau) : \tau \in [-\bar{d}, 0]\} \in C([-\bar{d}, 0]; \mathbb{R}^{3n+l})$. Then, based on the multiple Lyapunov-Krasovskii functionals method, the following conclusions are given.

Theorem 3.1. *Under the designed controller (3.5), when the control gains K_{j1}, K_{j2}, K_{j3}, P_{j1}, P_{j2}, P_{j3}, ψ_j, ϑ_j and λ_j are positive constants and satisfy*

$$\begin{cases} K_{j1} > \dfrac{K_{j2} + P_{j2} + 1}{2} + \zeta_j \\ \vartheta_j > 2\psi_j\zeta_j \\ P_{j1} > \dfrac{P_{j2} + \tilde{d} + K_{j2}}{2} + \zeta_j \\ P_{j3} \geq \dfrac{\tilde{d}}{2} \\ \zeta_j > 0 \end{cases} \tag{3.12}$$

the closed-loop teleoperation system is SIIOS. In this case, the tracking error $\mathbf{e}_j(t)$ between the master and slave robots, the joint velocity of the robot $\dot{\mathbf{q}}_j$, and the adaptive estimation error $\tilde{\mathbf{\Theta}}_j$ are all bounded.

Proof. For the closed-loop system, take Lyapunov-Krasovskii functional as

$$V_j(\boldsymbol{x}_{jt}, r_j(t)) = \sum_{i=1}^{4} V_{ji}(\boldsymbol{x}_{jt}, r_j(t))$$

where

$$V_{j1}(\boldsymbol{x}_{jt}, r_j(t)) = \frac{1}{2}\boldsymbol{\eta}_j^{\mathrm{T}}\boldsymbol{M}_j(\boldsymbol{q}_j)\boldsymbol{\eta}_j$$
$$V_{j2}(\boldsymbol{x}_{jt}, r_j(t)) = \frac{1}{2\psi_j}\tilde{\boldsymbol{\Theta}}_j^{\mathrm{T}}\tilde{\boldsymbol{\Theta}}_j$$
$$V_{j3}(\boldsymbol{x}_{jt}, r_j(t)) = \frac{1}{2}\boldsymbol{s}_j^{\mathrm{T}}(t)\boldsymbol{s}_j(t)$$
$$V_{j4}(\boldsymbol{x}_{jt}, r_j(t)) = \frac{\zeta_j}{2}\int_{-\bar{d}}^{0}\boldsymbol{z}_{jt}^{\mathrm{T}}(\tau)\left(\frac{-\tau}{\bar{d}} + \frac{2(\tau+\bar{d})}{\bar{d}}\right)\boldsymbol{z}_{jt}(\tau)\mathrm{d}\tau$$

with $\zeta_j > 0$ is a constant.

On the one hand, given that $\frac{1}{2}\int_{-\bar{d}}^{0}\boldsymbol{z}_{jt}^{\mathrm{T}}(\tau)\left(\frac{-\tau}{\bar{d}} + \frac{2(\tau+\bar{d})}{\bar{d}}\right)\boldsymbol{z}_{jt}(\tau)\mathrm{d}\tau \leq \int_{-\bar{d}}^{0}\boldsymbol{z}_{jt}^{\mathrm{T}}(\tau)\boldsymbol{z}_{jt}(\tau)\mathrm{d}\tau$, the Lyapunov-Krasovskii functional $V_j(\boldsymbol{x}_{jt}, r_j(t))$ satisfies (2.21), i.e.,

$$\alpha_1(|\boldsymbol{z}_j(t)|) \leq V_j(\boldsymbol{x}_{jt}, r_j(t)) \leq \alpha_2(\|\boldsymbol{z}_{jt}\|_{M_2}) \tag{3.13}$$

where $\alpha_1(s) = \min\left\{\rho_{j1}, \psi_j^{-1}, \frac{1}{2}\right\}s^2$, $\alpha_2(s) = \min\left\{\rho_{j2}, \psi_j^{-1}, 2\zeta_j, 1\right\}s^2$.

In addition, based on the proposed design, when only $y_{j,i}(t)\bar{e}_{j,i}(t) = 0$, $y_{j,i}(t^-)\bar{e}_{j,i}(t^-) > 0$, $\kappa_{j,i}(t)$ switches from 1 to -1. When only $y_{j,i}(t)\bar{e}_{j,i}(t) = 0$, $y_{j,i}(t^-)\bar{e}_{j,i}(t^-) < 0$, $\kappa_{j,i}(t)$ switches from -1 to 1. Then, at time $t = t_{k+1}^j$, the candidate Lyapunov-Krasovskii functional $V_j(\boldsymbol{x}_{jt}, r_j(t))$ satisfies

$$V_j(\boldsymbol{x}_{jt_{k+1}^j}, r_j(t_{k+1}^j)) \leq V_j(\boldsymbol{x}_{jt_{k+1}^j}, r_j(t_k^j)) \tag{3.14}$$

On the other hand, from system (3.10), by using Property 2.4, it holds

$$\begin{aligned} D^+V_{j1}(\boldsymbol{x}_{jt}, r_j(t)) &= \boldsymbol{\eta}_j^{\mathrm{T}}\boldsymbol{M}_j(\boldsymbol{q}_j)\dot{\boldsymbol{\eta}}_j + \frac{1}{2}\boldsymbol{\eta}_j^{\mathrm{T}}\dot{\boldsymbol{M}}_j(\boldsymbol{q}_j)\boldsymbol{\eta}_j \\ &= -K_{j1}|\boldsymbol{\eta}_j|^2 + \boldsymbol{\eta}_j^{\mathrm{T}}\boldsymbol{f}_{he_j} + \boldsymbol{\eta}_j^{\mathrm{T}}\boldsymbol{Y}_j\tilde{\boldsymbol{\Theta}}_j - K_{j2}\boldsymbol{\eta}_j^{\mathrm{T}}\boldsymbol{y}_j + \\ &\quad \boldsymbol{\eta}_j^{\mathrm{T}}\boldsymbol{F}_j + \boldsymbol{\eta}_j^{\mathrm{T}}\lambda_j\dot{d}_\iota\boldsymbol{M}_j(\boldsymbol{q}_j)\boldsymbol{q}_{vd\iota}(t, d_\iota(t)) \\ &\leq -\left(K_{j1} - \frac{K_{j2}}{2} - \frac{1}{2}\right)|\boldsymbol{\eta}_j|^2 + \frac{K_{j2}}{2}|\boldsymbol{y}_j|^2 + \\ &\quad \frac{1}{2}\|\boldsymbol{f}_{he_j[0,\infty)}\|_\infty^2 + \boldsymbol{\eta}_j^{\mathrm{T}}\boldsymbol{Y}_j\tilde{\boldsymbol{\Theta}}_j \end{aligned} \tag{3.15}$$

From (3.6), it has

$$\begin{aligned} D^+V_{j2}(\boldsymbol{x}_{jt}, r_j(t)) &= -\tilde{\boldsymbol{\Theta}}_j^{\mathrm{T}}\boldsymbol{Y}_j^{\mathrm{T}}\boldsymbol{\eta}_j - \frac{\vartheta_j}{\psi_j}\tilde{\boldsymbol{\Theta}}_j^{\mathrm{T}}\left(\boldsymbol{\Theta}_j - \hat{\boldsymbol{\Theta}}_j\right) + \frac{\vartheta_j}{\psi_j}\tilde{\boldsymbol{\Theta}}_j^{\mathrm{T}}\left(\boldsymbol{\Theta}_j - \boldsymbol{\Theta}_j^*\right) \\ &\leq -\tilde{\boldsymbol{\Theta}}_j^{\mathrm{T}}\boldsymbol{Y}_j^{\mathrm{T}}\boldsymbol{\eta}_j - \frac{\vartheta_j}{2\psi_j}|\tilde{\boldsymbol{\Theta}}_j|^2 + \frac{\vartheta_j}{2\psi_j}|\boldsymbol{\Theta}_j - \boldsymbol{\Theta}_j^*|^2 \end{aligned} \tag{3.16}$$

Based on the definition of $\boldsymbol{s}_j(t)$, from (3.7), for any $t \in [t_k^j, t_{k+1}^j)$, $k \geq 0$, it has

$$\dot{\boldsymbol{s}}_j(t) = -P_{j1}\boldsymbol{s}_j(t) - P_{j2}\boldsymbol{\eta}_j(t) + \boldsymbol{u}_j(t) + \dot{d}_\iota(t)\boldsymbol{\mathcal{K}}_j(t)\boldsymbol{q}_{vd\iota}(t, d_\iota(t))$$

where $\iota \in \{m, s\}$ but $\iota \neq j$.

Therefore for any $t \in [t_k^j, t_{k+1}^j)$,

$$\begin{aligned} D^+V_{j3}(\boldsymbol{x}_{jt}, r_j(t)) \leq& -\left(P_{j1} - \frac{P_{j2}}{2}\right)|\boldsymbol{s}_j(t)|^2 + \frac{P_{j2}}{2}|\boldsymbol{\eta}_j(t)|^2 + \boldsymbol{s}_j^{\mathrm{T}}(t)\boldsymbol{u}_j(t) + \\ & \boldsymbol{s}_j^{\mathrm{T}}(t)\boldsymbol{\mathcal{K}}_j(t)\dot{d}_\iota(t)\boldsymbol{q}_{vd\iota}(t, d_\iota(t)) \end{aligned}$$

From the definition of $\boldsymbol{u}_j(t)$, when $P_{j3} \geq \frac{\tilde{d}}{2}$, it has

$$\boldsymbol{s}_j^{\mathrm{T}}(t)\boldsymbol{u}_j(t) + \boldsymbol{s}_j^{\mathrm{T}}(t)\boldsymbol{\mathcal{K}}_j(t)\dot{d}_\iota(t)\boldsymbol{q}_{vd\iota}(t, d_\iota(t)) \leq \frac{\tilde{d}}{2}|\boldsymbol{s}_j(t)|^2$$

From the definitions of $\boldsymbol{s}_j(t)$ and $\boldsymbol{\mathcal{K}}_j(t)$, it holds $\frac{1}{2}|\boldsymbol{s}_j(t)|^2 \leq |\boldsymbol{y}_j(t)|^2 + |\boldsymbol{e}_j|^2 \leq |\boldsymbol{s}_j(t)|^2$. Further,

$$D^+V_{j3}(\boldsymbol{x}_{jt}, r_j(t)) \leq -\left(P_{j1} - \frac{P_{j2} + \tilde{d}}{2}\right)\left(|\boldsymbol{y}_j(t)|^2 + |\boldsymbol{e}_j|^2\right) + \frac{P_{j2}}{2}|\boldsymbol{\eta}_j(t)|^2 \tag{3.17}$$

Finally, similar to the previous works [27,28], it has

$$\begin{aligned} D^+V_{j4}(\boldsymbol{x}_{jt}, r_j(t)) &= \zeta_j\boldsymbol{z}_j^{\mathrm{T}}\boldsymbol{z}_j - \frac{1}{2}\zeta_j\boldsymbol{z}_{jt}^{\mathrm{T}}(-\bar{d})\boldsymbol{z}_{jt}(-\bar{d}) - \frac{\zeta_j}{2\bar{d}}\int_{-\bar{d}}^0 \boldsymbol{z}_{jt}^{\mathrm{T}}(\tau)\boldsymbol{z}_{jt}(\tau)\mathrm{d}\tau \\ &\leq \zeta_j\boldsymbol{z}_j^{\mathrm{T}}\boldsymbol{z}_j - \frac{\zeta_j}{2\bar{d}}\int_{-\bar{d}}^0 \boldsymbol{z}_{jt}^{\mathrm{T}}(\tau)\boldsymbol{z}_{jt}(\tau)\mathrm{d}\tau \end{aligned} \tag{3.18}$$

Combine (3.12), (3.15)~(3.18), for any $t \in [t_k^j, t_{k+1}^j)$, there exist positive constants μ_{j1} and μ_{j2} such that

$$\begin{aligned}
& D^+V_j(\boldsymbol{x}_{jt}, r_j(t)) \\
& \leq -\left(K_{j1} - \frac{K_{j2}+P_{j2}+1}{2} - \zeta_j\right)|\boldsymbol{\eta}_j|^2 - \left(\frac{\vartheta_j}{2\psi_j} - \zeta_j\right)|\tilde{\boldsymbol{\Theta}}_j|^2 - \\
& \left(P_{j1} - \frac{P_{j2}+\tilde{d}}{2} - \frac{K_{j2}}{2} - \zeta_j\right)|\boldsymbol{y}_j(t)|^2 - \left(P_{j1} - \frac{P_{j2}+\tilde{d}}{2} - \zeta_j\right)|\boldsymbol{e}_j|^2 - \\
& \frac{\zeta_j}{2\bar{d}}\int_{-\bar{d}}^{0} \boldsymbol{z}_{jt}^{\mathrm{T}}(\tau)\boldsymbol{z}_{jt}(\tau)\mathrm{d}\tau + \frac{\vartheta_j}{2\psi_j}|\boldsymbol{\Theta}_j - \boldsymbol{\Theta}_j^*|^2 + \frac{1}{2}\|\boldsymbol{f}_{he_j[0,\infty)}\|_\infty^2 \\
& \leq \mu_{j1}|\boldsymbol{z}_j|^2 - \frac{\zeta_j}{2\bar{d}}\int_{-\bar{d}}^{0} \boldsymbol{z}_{jt}^{\mathrm{T}}(\tau)\boldsymbol{z}_{jt}(\tau)\mathrm{d}\tau + \frac{\vartheta_j}{2\psi_j}|\boldsymbol{\Theta}_j - \boldsymbol{\Theta}_j^*|^2 + \frac{1}{2}\|\boldsymbol{f}_{he_j[0,\infty)}\|_\infty^2 \\
& \leq \mu_{j2}\|\boldsymbol{z}_{jt}\|_{M_2}^2 + \frac{\vartheta_j}{2\psi_j}|\boldsymbol{\Theta}_j - \boldsymbol{\Theta}_j^*|^2 + \frac{1}{2}\|\boldsymbol{f}_{he_j[0,\infty)}\|_\infty^2
\end{aligned} \tag{3.19}$$

From (3.13), (3.14) and (3.19), by using Lemma 2.4, it's easy to verify that the closed-loop teleoperation system is SIIOS, with the output be $\boldsymbol{z}_j(t)$. From the definition of $\boldsymbol{z}_j(t)$, when $\boldsymbol{f}_h$ and $\boldsymbol{f}_e$ satisfy Assumption 3.1, with the proposed control, the tracking error $\boldsymbol{e}_j$, the adaptive estimation error $\tilde{\boldsymbol{\Theta}}_j$ and the joint velocity $\dot{\boldsymbol{q}}_j(t)$ are all bounded. □

Remark 3.5. *Compared with the existing research results, the adaptive control based on switched filter and its analysis method proposed in this section have the following characteristics:*

(1) *Different from the traditional passive methods* [2, 4, 5, 29], *in the scheme proposed in this book, operator torque $\boldsymbol{f}_h(t)$ and environmental torque $\boldsymbol{f}_e(t)$ only need to be locally bounded. Considering the passivity assumption of operator/environment, there are many limitations in practical application* [9, 10, 18], *the control and analysis methods proposed in this section provide a new solution to overcome these limitations.*

(2) *As described in Nuno's work* [30], *if the auxiliary variable $\boldsymbol{\eta}_j$ in the definition (3.3) is $\boldsymbol{\eta}_j = \dot{\boldsymbol{q}}_j + \lambda_j \boldsymbol{q}_j$, the designed controller can only make the joint position of the master and slave robots converge to the zero position and realize synchronization at the zero position. To overcome this defect, under the constant communication time delay, Nuno and others put forward a new method* [30], *in which the position and speed errors are used to replace the position and speed signals of the robot in the control torque. In this section, a new adaptive control method based on auxiliary switching filter is proposed, which can be applied to the case of time-varying communication delay, and further broadens the applicable scope of existing theories.*

(3) *Under the framework of state-independent input-output stability, the complete output signal $\boldsymbol{z}_j$ does not contain the robot joint position information, so the problem of zero-steady-state synchronization can be avoided.*

(4) This section only considers the ordinary time-varying communication delay, and the proposed analysis method can't deal with the more realistic random delay [16, 31]*, which needs further consideration in the future work.*

3.5 EXPANSION TO SINGLE-MASTER MULTI-SLAVE TELEOPERATION

The adaptive control method of teleoperation system based on switched filter proposed in Sections 3.3 and 3.4 can be easily extended to single master multi-slave teleoperation control. Consider the following 1-master N-slave teleoperation system:

$$\begin{cases} \boldsymbol{M}_m^0(\boldsymbol{q}_m^0)\ddot{\boldsymbol{q}}_m^0 + \boldsymbol{C}_m^0(\boldsymbol{q}_m^0, \dot{\boldsymbol{q}}_m^0)\dot{\boldsymbol{q}}_m^0 + \boldsymbol{g}_m^0(\boldsymbol{q}_m^0) = \boldsymbol{\tau}_m^0 + \boldsymbol{f}_h \\ \boldsymbol{M}_s^i(\boldsymbol{q}_s^i)\ddot{\boldsymbol{q}}_s^i + \boldsymbol{C}_s^i(\boldsymbol{q}_s^i, \dot{\boldsymbol{q}}_s^i)\dot{\boldsymbol{q}}_s^i + \boldsymbol{g}_s^i(\boldsymbol{q}_s^i) = \boldsymbol{\tau}_s^i - \boldsymbol{f}_e^i \end{cases} \tag{3.20}$$

where the superscript $i = 1, 2, \cdots, N$ represents the i th slave robot, $\boldsymbol{M}_j^i(\cdot) \in \mathbb{R}^{n\times n}$, $\boldsymbol{C}_j^i(\cdot,\cdot) \in \mathbb{R}^{n\times n}$ and $\boldsymbol{g}_j^i(\cdot) \in \mathbb{R}^n$ have the same definitions as the single master and single slave case and satisfy Properties 2.1~2.5. Noting that, in Property 2.1, ρ_{j1} and ρ_{j2} should be replaced by ρ_{ji1} and ρ_{ji2}, respectively, $(j, i) \in \mathcal{S}_1 := \{(m, 0), (s, 1), \cdots, (s, N)\}$. Similarly, for convenience, let $\boldsymbol{f}_{he_m}^0 = \boldsymbol{f}_h$, $\boldsymbol{f}_{he_s}^i = -\boldsymbol{f}_e^i$.

Consider a simple topology, as shown in Fig. 3.2. For N slave robots, assume that slave robot 1 is the leader and other slave robots are followers. There are only communication links between the leader and each follower and between the master robot and the leader. Our control goal is to make the "leader" follow the movement of the main robot, and the movement of each "follower" is consistent with that of the "leader".

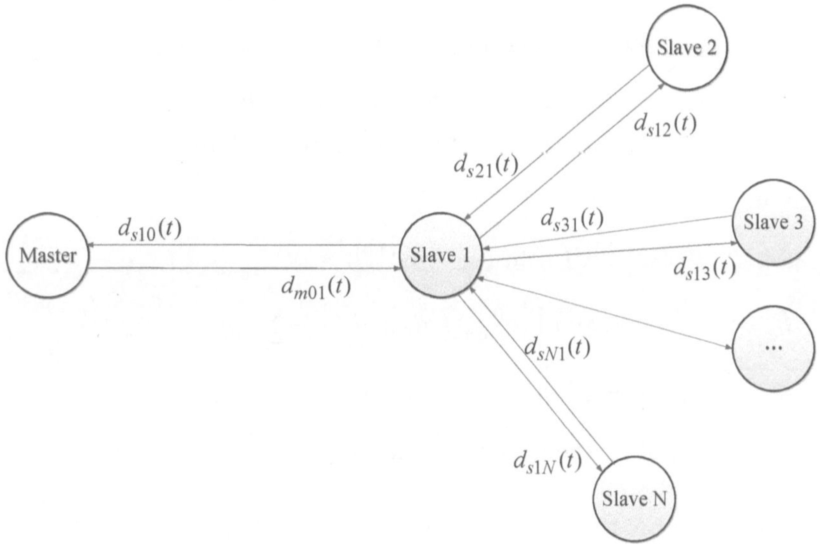

Figure 3.2 1-master N-slave teleoperation system.

To this end, define position synchronization error $\boldsymbol{e}_j^i = [e_{j,1}^i, e_{j,2}^i, \cdots, e_{j,n}^i]^{\mathrm{T}}$ as

$$
\begin{aligned}
\boldsymbol{e}_m^0(t) &= \boldsymbol{q}_m^0(t) - \boldsymbol{q}_s^1(t - d_{s10}(t)) \\
\boldsymbol{e}_s^0(t) &= \boldsymbol{q}_s^1(t) - \boldsymbol{q}_m^0(t - d_{m01}(t)) \\
\boldsymbol{e}_s^1(t) &= \boldsymbol{q}_s^1(t) - \frac{1}{N-1}\sum_{i=2}^{N} \boldsymbol{q}_s^i(t - d_{si1}(t)) \\
\boldsymbol{e}_s^k(t) &= \boldsymbol{q}_s^k(t) - \boldsymbol{q}_s^1(t - d_{s1k}(t)),\ \forall k = 2, 3, \cdots, n
\end{aligned}
$$

where all communication delays and their rate of change are assumed to be bounded; that is, Assumption 3.2 is satisfied.

Similarly, define velocity synchronization error$\boldsymbol{e}_{vj}^i = [e_{vj,1}^i, e_{vj,2}^i, \cdots, e_{vj,n}^i]^{\mathrm{T}}$:

$$
\begin{aligned}
\boldsymbol{e}_{vm}^0(t) &= \dot{\boldsymbol{q}}_m^0(t) - \boldsymbol{q}_{vds}^1(t, d_{s10}(t)) \\
\boldsymbol{e}_{vs}^0(t) &= \dot{\boldsymbol{q}}_s^1(t) - \boldsymbol{q}_{vdm}^0(t, d_{m01}(t)) \\
\boldsymbol{e}_{vs}^1(t) &= \dot{\boldsymbol{q}}_s^1(t) - \frac{1}{N-1}\sum_{i=2}^{N} \boldsymbol{q}_{vds}^i(t, d_{si1}(t)) \\
\boldsymbol{e}_{vs}^k(t) &= \dot{\boldsymbol{q}}_s^k(t) - \boldsymbol{q}_{vds}^1(t, d_{s1k}(t)),\ k = 2, 3, \cdots, n
\end{aligned}
$$

where $\boldsymbol{q}_{vdm}^0(t, d_{m01}(t)) = \dot{\boldsymbol{q}}_m^0(\theta)|_{\theta=t-d_{m01}(t)}$, $\boldsymbol{q}_{vds}^1(t, d_{s10}(t)) = \dot{\boldsymbol{q}}_s^1(\theta)|_{\theta=t-d_{s10}(t)}$, $\boldsymbol{q}_{vds}^i(t, d_{si1}(t)) = \dot{\boldsymbol{q}}_s^i(\theta)|_{\theta=t-d_{si1}(t)}$, $\boldsymbol{q}_{vds}^1(t, d_{s1k}(t)) = \dot{\boldsymbol{q}}_s^1(\theta)|_{\theta=t-d_{s1k}(t)}$.

For all $(j,k) \in \mathcal{S}_1 \setminus \{(s,1)\}$, define the auxiliary variables

$$
\begin{aligned}
\boldsymbol{\eta}_j^k(t) &= \dot{\boldsymbol{q}}_j^k(t) + \lambda_{jk}\boldsymbol{e}_j^k(t) \\
\boldsymbol{\eta}_s^1(t) &= \dot{\boldsymbol{q}}_s^1(t) + \lambda_{s10}\boldsymbol{e}_s^0(t) + \lambda_{s11}\boldsymbol{e}_s^1(t)
\end{aligned}
$$

Similar to the control design in Section 3.3, let

$$
\begin{cases}
\boldsymbol{\tau}_j^k = -K_{jk1}\boldsymbol{\eta}_j^k - \boldsymbol{Y}_j^k\hat{\boldsymbol{\Theta}}_j^k - K_{jk2}\boldsymbol{y}_j^k + \boldsymbol{F}_j^k \\
\boldsymbol{\tau}_s^1 = -K_{s11}\boldsymbol{\eta}_s^1 - \boldsymbol{Y}_s^1\hat{\boldsymbol{\Theta}}_s^1 - K_{s12}\boldsymbol{y}_s^0 - K_{s13}\boldsymbol{y}_s^1 + \boldsymbol{F}_s^1
\end{cases} \tag{3.21}
$$

where $k = 2, 3, \cdots, N$, $j \in \{m, s\}$,

$$
\begin{aligned}
\boldsymbol{F}_m^0 &= -K_{m04}\mathrm{sgn}(\boldsymbol{\eta}_m^0)|\boldsymbol{q}_{vds}^1(t, d_{s10}(t))| \\
\boldsymbol{F}_s^1 &= -K_{s14}\mathrm{sgn}(\boldsymbol{\eta}_s^1)\left(\left|\boldsymbol{q}_{vdm}^0(t, d_{m01}(t))\right| + \sum_{i=2}^{N}\left|\boldsymbol{q}_{vds}^i(t, d_{si1}(t))\right|\right) \\
\boldsymbol{F}_s^k &= -K_{sk4}\mathrm{sgn}(\boldsymbol{\eta}_s^k)|\boldsymbol{q}_{vds}^1(t, d_{s1k}(t))|
\end{aligned}
$$

$\boldsymbol{y}_j^i(t)$ is the state output of the designed auxiliary switched filter, and its design will be given later.

Then for all $(j,k) \in \mathcal{S}_1 \setminus \{(s,1)\}$, it has

$$
\begin{aligned}
&\boldsymbol{M}_j^k(\boldsymbol{q}_j^k)\dot{\boldsymbol{\eta}}_j^k + \boldsymbol{C}_j^k(\boldsymbol{q}_j^k, \dot{\boldsymbol{q}}_j^k)\boldsymbol{\eta}_j^k \\
&\qquad = \boldsymbol{Y}_j^k\tilde{\boldsymbol{\Theta}}_j^k - K_{jk1}\boldsymbol{\eta}_j^k + \boldsymbol{f}_{he_j}^k - K_{jk2}\boldsymbol{y}_j^k + \boldsymbol{F}_j^k + \boldsymbol{\Xi}_j^k
\end{aligned} \tag{3.22}
$$

$$
\begin{aligned}
&\boldsymbol{M}_s^1(\boldsymbol{q}_s^1)\dot{\boldsymbol{\eta}}_s^1 + \boldsymbol{C}_s^1(\boldsymbol{q}_s^1, \dot{\boldsymbol{q}}_s^1)\boldsymbol{\eta}_s^1 \\
&\qquad = \boldsymbol{Y}_s^1\tilde{\boldsymbol{\Theta}}_s^1 - K_{s11}\boldsymbol{\eta}_s^1 - \boldsymbol{f}_e^1 - K_{s12}\boldsymbol{y}_s^0 - K_{s13}\boldsymbol{y}_s^1 + \boldsymbol{F}_s^1 + \boldsymbol{\Xi}_s^1
\end{aligned} \tag{3.23}
$$

where

$$\begin{aligned}
\Xi_m^0 &= \lambda_{m0}\dot{d}_{s10}\boldsymbol{M}_m^0(\boldsymbol{q}_m^0)\boldsymbol{q}_{vds}^1(t, d_{s10}(t)) \\
\Xi_s^r &= \lambda_{sr}\dot{d}_{s1r}\boldsymbol{M}_s^r(\boldsymbol{q}_s^r)\boldsymbol{q}_{vds}^1(t, d_{s1r}(t)),\ r = 2, 3, \cdots, N \\
\Xi_s^1 &= \lambda_{s10}\dot{d}_{m01}\boldsymbol{M}_s^1(\boldsymbol{q}_s^1)\boldsymbol{q}_{vdm}^0(t, d_{m01}(t)) + \frac{\lambda_{s11}}{N-1}\sum_{i=2}^{N}\dot{d}_{si1}\boldsymbol{M}_s^1(\boldsymbol{q}_s^1)\boldsymbol{q}_{vds}^i(t, d_{si1}(t))
\end{aligned}$$

$$\boldsymbol{Y}_j^k\boldsymbol{\Theta}_j^k = \boldsymbol{M}_j^k(\boldsymbol{q}_j^k)\lambda_{jk}\boldsymbol{e}_{vj}^k + \boldsymbol{C}_j^k(\boldsymbol{q}_j^k, \dot{\boldsymbol{q}}_j^k)\lambda_{jk}\boldsymbol{e}_j^k - \boldsymbol{g}_j^k(\boldsymbol{q}_j^k)$$

$$\boldsymbol{Y}_s^1\boldsymbol{\Theta}_s^1 = \boldsymbol{M}_s^1(\boldsymbol{q}_s^1)\left(\lambda_{s10}\boldsymbol{e}_{vs}^0 + \lambda_{s11}\boldsymbol{e}_{vs}^1\right) + \boldsymbol{C}_s^1(\boldsymbol{q}_s^1, \dot{\boldsymbol{q}}_s^1)\left(\lambda_{s10}\boldsymbol{e}_s^0 + \lambda_{s11}\boldsymbol{e}_s^1\right) - \boldsymbol{g}_s^1(\boldsymbol{q}_s^1)$$

For this 1-master N-slave teleoperation system, the adaptive update law is designed as follows

$$\dot{\hat{\boldsymbol{\Theta}}}_j^i(t) = \psi_{ji}\boldsymbol{Y}_j^{i\mathrm{T}}(t)\boldsymbol{\eta}_j^i(t) - \vartheta_{ji}\left(\hat{\boldsymbol{\Theta}}_j^i(t) - \boldsymbol{\Theta}_j^{i*}\right)$$

Next, the design of auxiliary switched filter is given. Similar to that case of single master and single slave, for all $(j, i) \in \mathcal{S}_1 \cup \{(s, 0)\}$ and $k = 1, 2, \cdots, n$, denote the switching rule as $\boldsymbol{\mathcal{K}}_j^i(t) = \mathrm{diag}\{\kappa_{j,1}^i(t), \kappa_{j,2}^i(t), \cdots, \kappa_{j,n}^i(t)\}$, where

$$\kappa_{j,k}^i(t) = \begin{cases} 1, & y_{j,k}^i(t)e_{j,k}^i(t) > 0 \text{ or } y_{j,k}^i(t)e_{j,k}^i(t) = 0,\ y_{j,k}^i(t^-)e_{j,k}^i(t^-) \le 0 \\ -1, & y_{j,k}^i(t)e_{j,k}^i(t) < 0 \text{ or } y_{j,k}^i(t)e_{j,k}^i(t) = 0,\ y_{j,k}^i(t^-)e_{j,k}^i(t^-) > 0 \end{cases}$$

Then the auxiliary switched filter $\boldsymbol{y}_j^i = [y_{j,1}^i, y_{j,2}^i, \cdots, y_{j,n}^i]^{\mathrm{T}}$ is given as

$$\dot{y}_{j,r}^i = \begin{cases} -P_{ji1}(y_{j,r}^i + e_{j,r}^i) - e_{vj,r}^i - P_{ji2}\eta_{j,r}^i + u_{j,r}^i, & \kappa_{j,r}^i(t) = 1 \\ -P_{ji1}(y_{j,r}^i - e_{j,r}^i) + e_{vj,r}^i - P_{ji2}\eta_{j,r}^i + u_{j,r}^i, & \kappa_{j,r}^i(t) = -1 \end{cases}$$

where $\boldsymbol{\eta}_s^0(t) = \boldsymbol{\eta}_s^1(t)$, $\eta_{j,r}^i$ is the rth element of $\boldsymbol{\eta}_j^i = [\eta_{j,1}^i, \eta_{j,2}^i, \cdots, \eta_{j,n}^i]^{\mathrm{T}}$. For all $k = 2, 3, \cdots, N$, it has

$$u_{m,r}^0(t) = \begin{cases} -P_{m03}\dfrac{s_{m,r}^0(t)}{|s_{m,r}^0(t)|^2}|\boldsymbol{q}_{vds}^1(t, d_{s10}(t))|^2, & s_{m,r}^0(t) \ne 0 \\ 0, & s_{m,r}^0(t) = 0 \end{cases}$$

$$u_{s,r}^0(t) = \begin{cases} -P_{s03}\dfrac{s_{s,r}^0(t)}{|s_{s,r}^0(t)|^2}|\boldsymbol{q}_{vdm}^0(t, d_{m01}(t))|^2, & s_{s,r}^0(t) \ne 0 \\ 0, & s_{s,r}^0(t) = 0 \end{cases}$$

$$u_{s,r}^1(t) = \begin{cases} -P_{s13}\dfrac{s_{s,r}^1(t)}{|s_{s,r}^1(t)|^2}\sum_{i=2}^N|\boldsymbol{q}_{vds}^i(t, d_{si1}(t))|^2, & s_{s,r}^1(t) \ne 0 \\ 0, & s_{s,r}^1(t) = 0 \end{cases}$$

$$u_{s,r}^k(t) = \begin{cases} -P_{sk3}\dfrac{s_{s,r}^k(t)}{|s_{s,r}^k(t)|^2}|\boldsymbol{q}_{vds}^1(t, d_{s1k}(t))|^2, & s_{s,r}^k(t) \ne 0 \\ 0, & s_{s,r}^k(t) = 0 \end{cases}$$

where $\boldsymbol{s}_j^i(t) = [s_{j,1}^i(t), s_{j,2}^i(t), \cdots, s_{j,n}^i(t)]^{\mathrm{T}} = \boldsymbol{y}_j^i(t) + \boldsymbol{\mathcal{K}}_j^i(t)\boldsymbol{e}_j^i(t)$.

From Remark 3.2, we can construct a standard switching signal $r_j^i(t)$ so that it corresponds to $\boldsymbol{\mathcal{K}}_j^i(t)$ one by one. Further, we have

$$\begin{aligned}\dot{\boldsymbol{y}}_j^i &= -P_{ji1}(\boldsymbol{y}_j^i + \boldsymbol{\mathcal{K}}_j^i \boldsymbol{e}_j^i) - \boldsymbol{\mathcal{K}}_j^i \boldsymbol{e}_{vj}^i - P_{ji2}\boldsymbol{\eta}_j^i + \boldsymbol{u}_j^i \\ &:= -P_{ji1}\boldsymbol{y}_j^i - P_{ji2}\boldsymbol{\eta}_j^i + \boldsymbol{u}_j^i + \boldsymbol{g}(\boldsymbol{e}_j^i, \boldsymbol{e}_{vj}^i, r_j^i)\end{aligned} \tag{3.24}$$

where $\boldsymbol{u}_j^i = [u_{j,1}^i, u_{j,2}^i, \cdots, u_{j,n}^i]^{\mathrm{T}}$, $\boldsymbol{g}(\boldsymbol{e}_j^i, \boldsymbol{e}_{vj}^i, r_j^i) = -P_{ji1}\boldsymbol{\mathcal{K}}_j^i \boldsymbol{e}_j^i - \boldsymbol{\mathcal{K}}_j^i \boldsymbol{e}_{vj}^i$.

Remark 3.6. *Let $\{t_k^{ji}\}_{k\geq 0}$ be the switching times of $r_j^i(t)$. Then for any $t \in [t_k^{ji}, t_{k+1}^{ji})$,*

$$\dot{\boldsymbol{s}}_j^i(t) = -P_{ji1}\boldsymbol{s}_j^i(t) - P_{ji2}\boldsymbol{\eta}_j^i(t) + \boldsymbol{u}_j^i(t) + \boldsymbol{\Gamma}_{ji}(t)$$

where $(j,i) \in \mathcal{S}_1$, and

$$\begin{aligned}\boldsymbol{\Gamma}_{m0} &= \boldsymbol{\mathcal{K}}_m^0(t)\dot{d}_{s10}(t)\boldsymbol{q}_{vds}^1(t, d_{s10}(t)) \\ \boldsymbol{\Gamma}_{s0} &= \boldsymbol{\mathcal{K}}_s^0(t)\dot{d}_{m01}(t)\boldsymbol{q}_{vdm}^0(t, d_{m01}(t)) \\ \boldsymbol{\Gamma}_{s1} &= \frac{1}{N-1}\boldsymbol{\mathcal{K}}_s^1(t)\sum_{i=2}^{N}\dot{d}_{si1}(t)\boldsymbol{q}_{vds}^i(t, d_{si1}(t)) \\ \boldsymbol{\Gamma}_{sk} &= \boldsymbol{\mathcal{K}}_s^k(t)\dot{d}_{s1k}(t)\boldsymbol{q}_{vds}^1(t, d_{s1k}(t)),\ k = 2, 3, \cdots, N\end{aligned}$$

Similar to Theorem 3.1, for the master robot and all slave robots except the slave robot 1, take Lyapunov-Krasovskii functional as

$$\begin{aligned}V_{ji}(\boldsymbol{x}_{jt}^i, r_j^i(t)) = &\frac{1}{2}\boldsymbol{\eta}_j^{i\mathrm{T}}\boldsymbol{M}_j^i(\boldsymbol{q}_j^i)\boldsymbol{\eta}_j^i + \frac{1}{2}\tilde{\boldsymbol{\Theta}}_j^{i\mathrm{T}}\psi_{ji}^{-1}\tilde{\boldsymbol{\Theta}}_j^i + \frac{1}{2}\boldsymbol{s}_j^{i\mathrm{T}}(t)\boldsymbol{s}_j^i(t) + \\ &\frac{\zeta_{ji}}{2}\int_{-\bar{d}}^{0}\boldsymbol{z}_{jt}^{i\mathrm{T}}(\tau)\left(\frac{-\tau}{\bar{d}} + \frac{2(\tau+\bar{d})}{\bar{d}}\right)\boldsymbol{z}_{jt}^i(\tau)\mathrm{d}\tau\end{aligned}$$

where $(j,i) \in \mathcal{S}_1 \setminus \{(s,1)\}$, $\boldsymbol{x}_j^i = \left[\left(\boldsymbol{\eta}_j^i\right)^{\mathrm{T}}, \left(\tilde{\boldsymbol{\Theta}}_j^i\right)^{\mathrm{T}}, \left(\boldsymbol{q}_j^i\right)^{\mathrm{T}}, \left(\boldsymbol{y}_j^i\right)^{\mathrm{T}}, \left(\boldsymbol{e}_j^i\right)^{\mathrm{T}}\right]^{\mathrm{T}}$, $\boldsymbol{z}_j^i = \left[\left(\boldsymbol{\eta}_j^i\right)^{\mathrm{T}}, \left(\tilde{\boldsymbol{\Theta}}_j^i\right)^{\mathrm{T}}, \left(\boldsymbol{y}_j^i\right)^{\mathrm{T}}, \left(\boldsymbol{e}_j^i\right)^{\mathrm{T}}\right]^{\mathrm{T}}$.

For the slave robot 1, take Lyapunov-Krasovskii functional as

$$\begin{aligned}V_{s1}(\boldsymbol{x}_{st}^1, r_s^1(t)) = &\frac{1}{2}\boldsymbol{\eta}_s^{1\mathrm{T}}\boldsymbol{M}_s^1(\boldsymbol{q}_s^1)\boldsymbol{\eta}_s^1 + \frac{1}{2}\tilde{\boldsymbol{\Theta}}_s^{1\mathrm{T}}\psi_{s1}^{-1}\tilde{\boldsymbol{\Theta}}_s^1 + \frac{1}{2}\bar{\boldsymbol{s}}_s^{1\mathrm{T}}(t)\bar{\boldsymbol{s}}_s^1(t) + \\ &\frac{\zeta_{s1}}{2}\int_{-\bar{d}}^{0}\boldsymbol{z}_{st}^{1\mathrm{T}}(\tau)\left(\frac{-\tau}{\bar{d}} + \frac{2(\tau+\bar{d})}{\bar{d}}\right)\boldsymbol{z}_{st}^1(\tau)\mathrm{d}\tau\end{aligned}$$

where $\bar{\boldsymbol{s}}_s^1 = [\boldsymbol{s}_s^{0\mathrm{T}}, \boldsymbol{s}_s^{1\mathrm{T}}]^{\mathrm{T}}$, $\boldsymbol{x}_s^1 = [\boldsymbol{\eta}_s^{1\mathrm{T}}, \tilde{\boldsymbol{\Theta}}_s^{1\mathrm{T}}, \boldsymbol{q}_s^{1\mathrm{T}}, \boldsymbol{y}_s^{0\mathrm{T}}, \boldsymbol{y}_s^{1\mathrm{T}}, \boldsymbol{e}_s^{0\mathrm{T}}, \boldsymbol{e}_s^{1\mathrm{T}}]^{\mathrm{T}}$, $\boldsymbol{z}_s^1 = [\boldsymbol{\eta}_s^{1\mathrm{T}}, \tilde{\boldsymbol{\Theta}}_s^{1\mathrm{T}}, \boldsymbol{y}_s^{0\mathrm{T}}, \boldsymbol{y}_s^{1\mathrm{T}}, \boldsymbol{e}_s^{0\mathrm{T}}, \boldsymbol{e}_s^{1\mathrm{T}}]^{\mathrm{T}}$.

From Theorem 3.1, it is easy to prove that there is an appropriate control gain to make the closed-loop teleoperation system SIIOS.

3.6 SIMULATION STUDY

In this section, a simulation on single-master-two-slave framework will be presented to demonstrate the effectiveness of the obtained results. All the robots are assumed to be 2-DOF serial links, and their corresponding nonlinear dynamics are given in Nuno's work [30], where the parameters are chosen as: $m_{m01} = 3.79Kg$, $m_{m02} = 0.65Kg$, $l_{m01} = l_{m02} = 0.38m$; $m_{s11} = 3.9473Kg$, $m_{s12} = 0.6232Kg$, $m_{s21} = 3.2409Kg$, $m_{s22} = 0.3185Kg$, $l_{s11} = l_{s12} = 0.38m$ and, $l_{s21} = l_{s22} = 0.38m$. Thus, one can obtain $\rho_{m01} = 0.0533$, $\rho_{m012} = 0.9632$, $\rho_{s11} = 0.0530$, $\rho_{s12} = 0.9669$, $\rho_{s21} = 0.0323$, $\rho_{s22} = 0.6656$. In what follows, the estimated value in parametrization is set to be $\boldsymbol{\Theta}_j^i = [\alpha_{ji}, \beta_{ji}, \sigma_{ji}, l_{ji2}^{-1}\sigma_{ji}, l_{ji1}^{-1}(\alpha_{ji} - \sigma_{ji})]^T$, where $\alpha_{ji} = l_{ji2}^2 m_{ji2} + l_{ji1}^2(m_{ji1} + m_{ji2})$, $\beta_{ji} = l_{ji1}l_{ji2}m_{ji2}$ and $\sigma_{ji} = l_{ji2}^2 m_{ji2}$.

For the simulations, the human exerts a force $\boldsymbol{f}_h^0$ on the master manipulator's tip, while the two slave manipulators' interaction with the environments $\boldsymbol{f}_e^1$ and $\boldsymbol{f}_e^2$ are also done with their tips, respectively. Hence, $\boldsymbol{F}_h^0 = \boldsymbol{J}_m^{0T}(\boldsymbol{q}_m^0)\boldsymbol{f}_h^0$, $\boldsymbol{F}_e^1 = \boldsymbol{J}_s^{1T}(\boldsymbol{q}_s^1)\boldsymbol{f}_e^1$, $\boldsymbol{F}_e^2 = \boldsymbol{J}_s^{2T}(\boldsymbol{q}_s^2)\boldsymbol{f}_e^2$, where for all $(j,i) \in \mathcal{S}_1$, $\boldsymbol{J}_j^i(\boldsymbol{q}_j^i)$ are the Jacobians matrices given by:

$$\boldsymbol{J}_j^i(\boldsymbol{q}_j^i) = \begin{bmatrix} -l_{ji1}s_1 - l_{ji2}s_{12} & -l_{ji2}s_{12} \\ l_{ji1}c_1 + l_{ji2}c_{12} & l_{ji2}c_{12} \end{bmatrix}$$

where $s_1 = \sin(q_{j1}^i)$, $s_{12} = \sin(q_{j1}^i + q_{j2}^i)$, $c_1 = \cos(q_{j1}^i)$, $c_{12} = \cos(q_{j1}^i + q_{j2}^i)$.

In what follows, assume that $\boldsymbol{\Theta}_j^{i*} = 0.9\boldsymbol{\Theta}_j^i$. The time delays are assumed to be $d_{m01}(t) = 0.1 + 0.5\cos^2(t)$, $d_{s10}(t) = 0.2 + 0.6\sin^2(t)$, $d_{s21}(t) = 0.2 + 0.55\sin^2(t)$, $d_{s12}(t) = 0.1 + 0.5\sin^2(t)$. Hence, $\bar{d} = 0.8$, $\tilde{d} = 1.2$. From the theoretical analysis, one can get the following controllers' parameters: $\lambda_{m0} = 1$, $\lambda_{s10} = 5$, $\lambda_{s11} = 0.5$, $\lambda_{s2} = 2$, $K_{m01} = 5.2628$, $K_{m02} = 0.2474$, $K_{m04} = 1.3370$, $K_{s11} = 5.2723$, $K_{s12} = 0.2473$, $K_{s13} = 0.2473$, $K_{s14} = 6.3907$, $K_{s21} = 5.2628$, $K_{s22} = 0.2474$, $K_{s24} = 1.8078$, $P_{m01} = 2.6$, $P_{m02} = 4.4536 * 10^{-6}$, $P_{m03} = 23.7083$, $P_{s01} = 2.6$, $P_{s02} = 5.8897 * 10^{-5}$, $P_{s03} = 23.7083$, $P_{s11} = 2.6$, $P_{s12} = 5.8897 * 10^{-5}$, $P_{s13} = 23.7083$, $P_{s21} = 2.6$, $P_{s22} = 5.8879 * 10^{-5}$, $P_{s23} = 23.7083$, $\psi_{m0} = 0.2$, $\psi_{s1} = 1.5$, $\psi_{s2} = 1.2$, $\vartheta_{m0} = 0.6$, $\vartheta_{s1} = 4$, $\vartheta_{s2} = 3$. The initial conditions for the master/slave manipulators, EFSs as well as adaptive controllers are assumed to be: $\boldsymbol{q}_m^0(t) = [-2, 2]^T$, $\boldsymbol{q}_s^1(t) = [0, 0]^T$, $\boldsymbol{q}_s^2(t) = [1, -1]^T$, $\dot{\boldsymbol{q}}_j^i(t) = \ddot{\boldsymbol{q}}_j^i(t) = [0, 0]^T$, $\boldsymbol{y}_m^0(t) = [-5, 10]^T$, $\boldsymbol{y}_s^0(t) = [-5, -15]^T$, $\boldsymbol{y}_s^1(t) = [-10, 5]^T$, $\boldsymbol{y}_s^2(t) = [5, -5]^T$, $\dot{\boldsymbol{y}}_j^i(t) = \ddot{\boldsymbol{y}}_j^i(t) = [0, 0]^T$, $\hat{\boldsymbol{\Theta}}_m^0(t) = [2, -3, 5, 6, 3]^T$, $\hat{\boldsymbol{\Theta}}_s^1(t) = [5, 4, -3, 15, -3]^T$, $\hat{\boldsymbol{\Theta}}_s^2(t) = [-6, 5, -2, 12, 6]^T$, for all $t \in [-\tilde{d}, 0]$.

3.6.1 Stability verification

The first set of simulation is used to verify: Under the activation of bounded external forces, when we move the master robot, does the slave follow the master? In fact, when we set the human and environment forces as shown in Fig. 3.3, the simulation results are shown in Figs. 3.4–3.9. Among them, Fig. 3.4 gives the master and slave manipulators' positions in joint space. From which, the slave manipulator 1 can track the master well, while the slave 2 tracks the slave 1 well. Further, according to the theoretical analysis, the complete master (or, slave) system is SIIOS. It means that

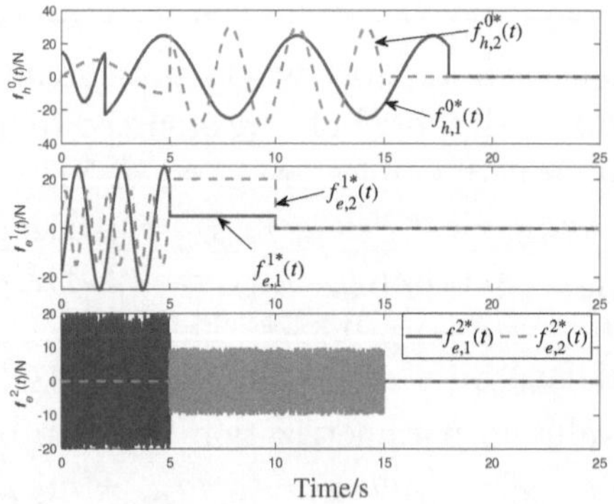

Figure 3.3 Generalized human and environment forces.

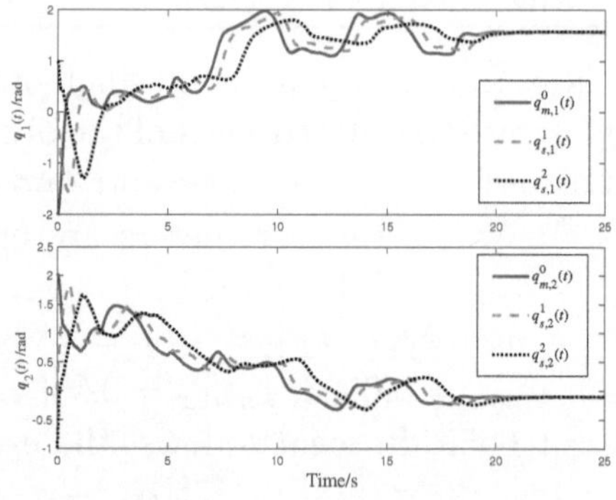

Figure 3.4 The response curves of the joint positions.

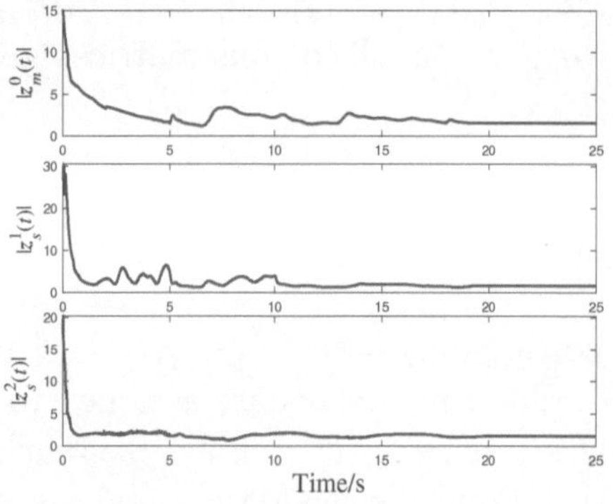

Figure 3.5 The norm of complete output $z_j^i(t)$.

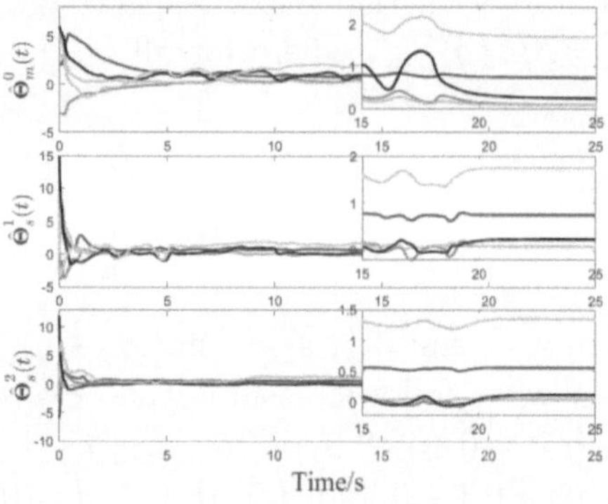

Figure 3.6 The estimated values of parameter $\Theta_j^i(t)$.

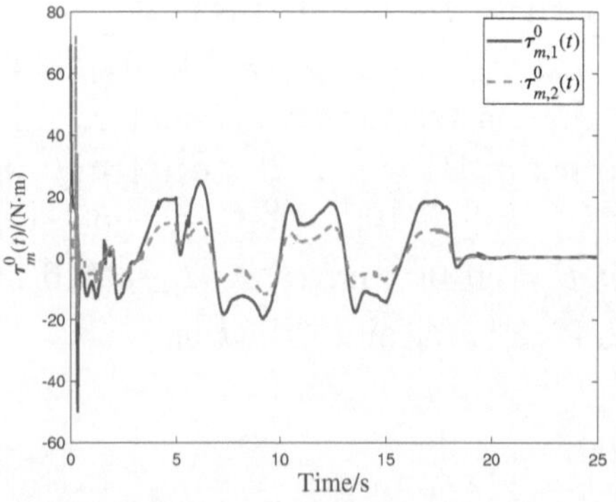

Figure 3.7 The control inputs of master robot.

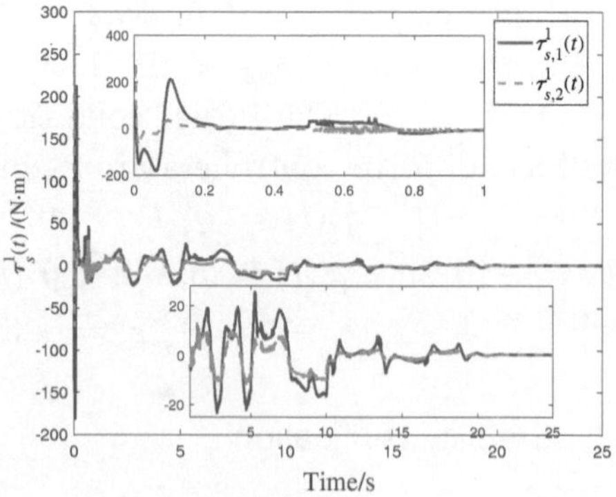

Figure 3.8 The control inputs of slave robot 1.

under the bounded inputs, the complete outputs $z_j^i(t)$ are also bounded. This fact can be observed easily from Fig. 3.5.

As a component of those complete outputs, the parameter estimation errors for each dynamics will also remain bounded, where Fig. 3.6 gives the response curves of estimated parameters (*Colors BLUE, RED, GREEN, BLACK and CYAN represent, respectively,* $\hat{\Theta}_{j,1}^i(t)$, $\hat{\Theta}_{j,2}^i(t)$, $\hat{\Theta}_{j,3}^i(t)$, $\hat{\Theta}_{j,4}^i(t)$ *and* $\hat{\Theta}_{j,5}^i(t)$).

Finally, the control inputs are given in Figs. 3.7–3.9. Note that, due to the existence of the external force $\boldsymbol{f}_e^2(t)$ in Fig. 3.3, the control inputs suffer some chattering, especially in Fig. 3.9. Additionally, since we do not consider the saturation control, the control inputs presented in Figs. 3.7–3.9 are somehow large for practical implementation. This will be a topic for future research.

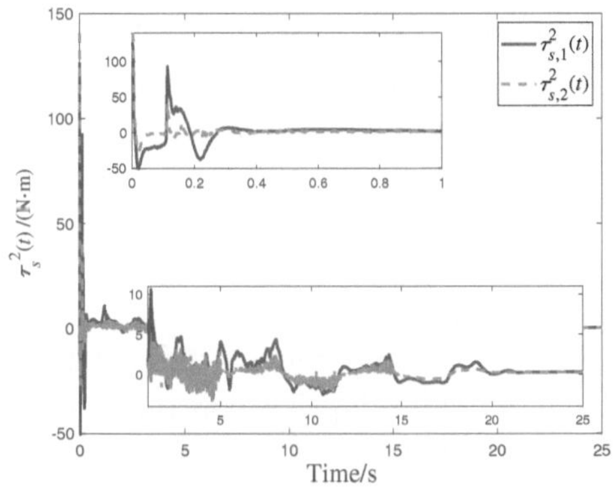

Figure 3.9 The control inputs of slave robot 2.

3.6.2 Contact stability verification

To further verify the tracking performance, another set of simulations are given. In this set of simulation, the contact stability will be evaluated. The human force is the same as the one in Fig. 3.3, while the slave manipulator 2 is assumed to run in free space, i.e., $\boldsymbol{f}_e^2 \equiv [0,0]^T$. For the slave manipulator 1, we will implement a high stiff wall in the slave environment in the xz-plant at the cartesian coordinate, $y = 0.1m$, which is modeled as a spring-damper system (with the spring and damping gains being $20000N/m$ and $20Ns/m$) reacting only along the y direction. The simulation results are shown in Figs. 3.10 and 3.11. Fig. 3.11 presents the simulations of the Cartesian positions for all, the master, the slave 1/2 manipulators, where the contact force $\boldsymbol{f}_{e2}^1(t)$ due to environment is given in Fig. 3.10. From which, the slave manipulator 1 and 2 can track the master well. Note that when the slave manipulator 1 comes in contact with the wall, stability is maintained and when the slave 1 departs from the wall, it can track the master quickly.

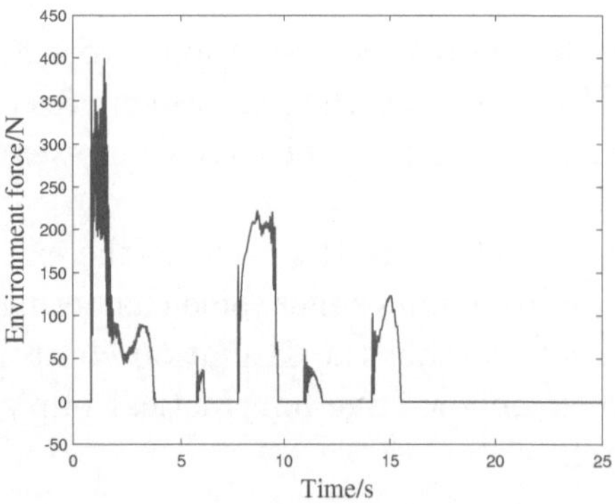

Figure 3.10 Environmental Y-force applied to the slave robot 1 when the slave robot 1 interacts with a virtual wall located in xz-plant at $y = 0.1m$.

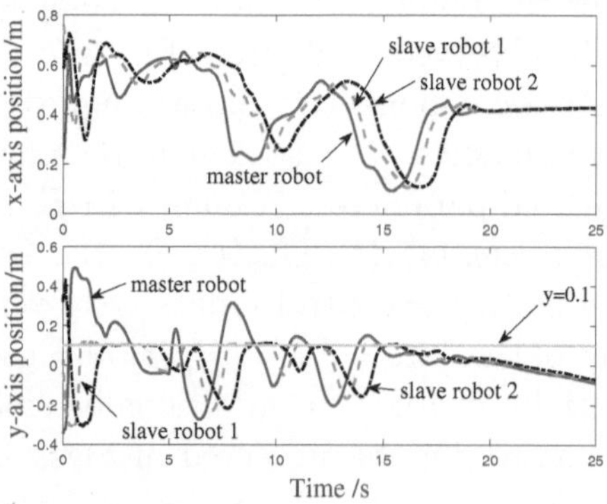

Figure 3.11 xy-position of the master, slave robot 1 and slave robot 2 at cartesian coordinate when the slave robot 1 interacts with a virtual wall located in xz-plant at $y = 0.1m$.

3.6.3 Comparison study

In the last set of simulation, a comparison study with the classical ISS/IOS approach [9] is performed, where the adaptive control for single-master-single-slave teleoperation system is investigated, and the time-free stability criteria are obtained. In the classical IOS framework [9], the state and the output of the complete closed-loop system are, respectively, $\boldsymbol{x} = [\boldsymbol{q}_m^T, \dot{\boldsymbol{q}}_m, \tilde{\boldsymbol{\Theta}}_m^T, \boldsymbol{e}_s^T, \dot{\boldsymbol{q}}_s^T, \tilde{\boldsymbol{\Theta}}_s^T]^T$ and $\boldsymbol{y} = [\boldsymbol{q}_m^T, \dot{\boldsymbol{q}}_m, \boldsymbol{e}_s^T, \dot{\boldsymbol{q}}_s^T]^T$, where the joint position $\boldsymbol{q}_m$ is one element of the state as well as the output. Roughly speaking, from the definitions of ISS (IOS), for any bounded external inputs, the state (the output) will converge to near zero [32]. Specially, in the unperturbed system, the closed-loop system with the proposed controller is asymptotically stable, thus $\boldsymbol{q}_m$ and $\boldsymbol{e}_s$ will approach the origin asymptotically, and further $\boldsymbol{q}_s$ will also approach the origin for any bounded time delays. However, in the study of teleoperation system, this consequence is not desirable because without any perturbation, the master and slave robots are expected to converge to each other, not the zero position, which has been overcome in this chapter. To verify the effectiveness, let us consider 2-DOF single-master-single-slave teleoperation system, where the master robot and the slave robot are chosen to be the same as the master and slave robot 1 in the first set of simulation. Moreover, assume that the human force $\boldsymbol{f}_h(t)$ is the same as the one in Fig. 3.3, while the slave robot is run in free motion with $\boldsymbol{f}_e(t) \equiv [0, 0]^T$. Then, the simulation results are shown in Figs. 3.12 and 3.13. On the one hand, from Fig. 3.12, one can find that the synchronization between the master and slave robots is achieved near the origin, which is in accordance with the above analysis and is an undesired coordinated behavior. On the other hand, Fig. 3.13 gives the tracking performance under the developed SIIOS framework of this article, where if human force equals zero, the master and slave robots can converge to each other, not the origin.

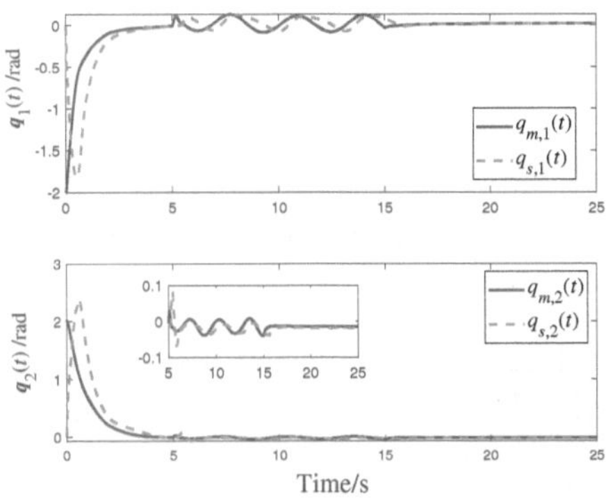

Figure 3.12 The joint positions under the classical IOS framework [9].

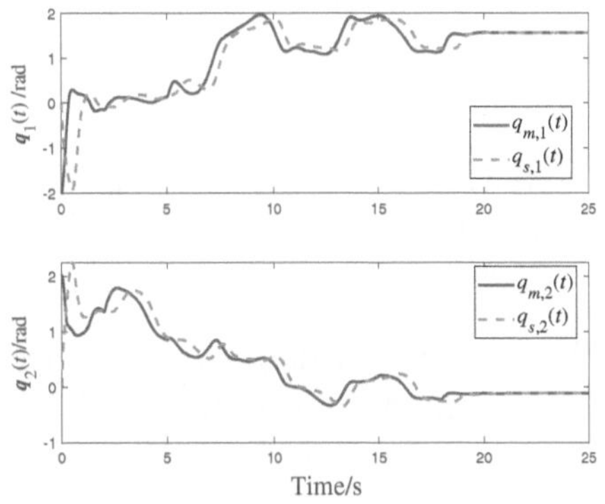

Figure 3.13 The joint positions under the new SIIOS framework.

3.7 CONCLUSION

In this chapter, we have presented a new IOS-based control framework for nonlinear teleoperation systems, in which a subsystem decomposition method based on switched filter has been developed. The designed switched filter is driven by local tracking errors, and local/remote velocities. By designing the proper adaptive nonlinear controllers with some well-defined auxiliary variables, the complete closed-loop master/slave systems are modeled by two interconnected subsystems, i.e., the auxiliary variable subsystem and the error filter subsystem. By applying the Lyapunov-Krasovskii method to the two subsystems, the state-independent input-to-output stability of each subsystem are verified. Utilizing the classical small-gain theorem, the time-dependent state-independent input-to-output stability criteria for the complete closed-loop systems are finally obtained. As an extended application, the proposed control framework is also proved to be suitable for the control of single-master-multi-slave teleoperation system, and this extension is very simple. Finally, the effectiveness of the proposed scheme is verified by the numerical example.

Bibliography

[1] P. F. Hokayem and M. W. Spong, "Bilateral teleoperation: an historical survey," *Automatica*, vol. 42, no. 12, pp. 2035–2057, 2006.

[2] E. Nuno, L. Basanez, and R. Ortega, "Passivity-based control for bilateral teleoperation: a tutorial," *Automatica*, vol. 47, no. 3, pp. 485–495, 2011.

[3] M. Ferre, R. Aracil, C. Balaguer, M. Buss, and C. Melchiorri, *Advances in Telerobotics*. Springer, 2007.

[4] E. Nuno, R. Ortega, N. Barabanov, and L. Basanez, "A globally stable pd controller for bilateral teleoperators," *IEEE Transactions on Robotics*, vol. 24, no. 3, pp. 753–758, 2008.

[5] F. Hashemzadeh, I. Hassanzadeh, and M. Tavakoli, "Teleoperation in the presence of varying time delays and sandwich linearity in actuators," *Automatica*, vol. 49, no. 9, pp. 2813–2821, 2013.

[6] E. Slawinski and V. Mut, "Pd-like controllers for delayed bilateral teleoperation of manipulators robots," *International Journal of Robust and Nonlinear Control*, vol. 25, no. 12, pp. 1801–1815, 2015.

[7] L.-G. García-Valdovinos, V. Parra-Vega, and M. A. Arteaga, "Observer-based sliding mode impedance control of bilateral teleoperation under constant unknown time delay," *Robotics and Autonomous Systems*, vol. 55, no. 8, pp. 609–617, 2007.

[8] R. Moreau, M. T. Pham, M. Tavakoli, M. Le, and T. Redarce, "Sliding-mode bilateral teleoperation control design for master–slave pneumatic servo systems," *Control Engineering Practice*, vol. 20, no. 6, pp. 584–597, 2012.

[9] I. G. Polushin, A. Tayebi, and H. J. Marquez, "Control schemes for stable teleoperation with communication delay based on ios small gain theorem," *Automatica*, vol. 42, no. 6, pp. 905–915, 2006.

[10] I. G. Polushin, S. N. Dashkovskiy, A. Takhmar, and R. V. Patel, "A small gain framework for networked cooperative force-reflecting teleoperation," *Automatica*, vol. 49, no. 2, pp. 338–348, 2013.

[11] D.-H. Zhai and Y. Xia, "Robust saturation-based control of bilateral teleoperation without velocity measurements," *International Journal of Robust and Nonlinear Control*, vol. 25, no. 15, pp. 2582–2607, 2015.

[12] C.-C. Hua and P. X. Liu, "Delay-dependent stability criteria of teleoperation systems with asymmetric time-varying delays," *IEEE Transactions on Robotics*, vol. 26, no. 5, pp. 925–932, 2010.

[13] Z. Li and C.-Y. Su, "Neural-adaptive control of single-master–multiple-slaves teleoperation for coordinated multiple mobile manipulators with time-varying communication delays and input uncertainties," *IEEE Transactions on Neural Networks and Learning Systems*, vol. 24, no. 9, pp. 1400–1413, 2013.

[14] Z. Li, Y. Xia, and X. Cao, "Adaptive control of bilateral teleoperation with unsymmetrical time-varying delays," *International Journal of Innovative Computing, Information and Control*, vol. 9, no. 2, pp. 753–767, 2013.

[15] C.-C. Hua and X. P. Liu, "A new coordinated slave torque feedback control algorithm for network-based teleoperation systems," *IEEE/ASME Transactions on Mechatronics*, vol. 18, no. 2, pp. 764–774, 2012.

[16] Z. Li, L. Ding, H. Gao, G. Duan, and C.-Y. Su, "Trilateral teleoperation of adaptive fuzzy force/motion control for nonlinear teleoperators with communication random delays," *IEEE Transactions on Fuzzy Systems*, vol. 21, no. 4, pp. 610–624, 2012.

[17] Z. Li, X. Cao, Y. Tang, R. Li, and W. Ye, "Bilateral teleoperation of holonomic constrained robotic systems with time-varying delays," *IEEE Transactions on Instrumentation and Measurement*, vol. 62, no. 4, pp. 752–765, 2013.

[18] I. G. Polushin and H. J. Marquez, "Stabilization of bilaterally controlled teleoperators with communication delay: an ISS approach," *International Journal of Control*, vol. 76, no. 8, pp. 858–870, 2003.

[19] S.-J. Kim and I.-J. Ha, "Existence of caratheodory solutions in nonlinear systems with discontinuous switching feedback controllers," *IEEE Transactions on Automatic Control*, vol. 49, no. 7, pp. 1167–1171, 2004.

[20] A. F. Filippov, *Differential Equations with Discontinuous Righthand Sides.* Springer, 1988.

[21] Y. Kang, D.-H. Zhai, G.-P. Liu, Y.-B. Zhao, and P. Zhao, "Stability analysis of a class of hybrid stochastic retarded systems under asynchronous switching," *IEEE Transactions on Automatic Control*, vol. 59, no. 6, pp. 1511–1523, 2014.

[22] Y. Kang, D.-H. Zhai, G.-P. Liu, and Y.-B. Zhao, "On input-to-state stability of switched stochastic nonlinear systems under extended asynchronous switching," *IEEE transactions on cybernetics*, vol. 46, no. 5, pp. 1092–1105, 2015.

[23] D. Zhai, Y. Kang, P. Zhao, and Y.-B. Zhao, "Stability of a class of switched stochastic nonlinear systems under asynchronous switching," *International Journal of Control, Automation and Systems*, vol. 10, no. 6, pp. 1182–1192, 2012.

[24] S. M. Magdi, *Switched Time-Delay Systems: Stability and Control.* Springer, New York, USA, 2010.

[25] Z. Sun and S. S. Ge, *Stability Theory of Switched Dynamical Systems.* Springer, 2011.

[26] D. Liberzon, *Switching in Systems and Control.* Springer, 2003.

[27] P. Pepe, I. Karafyllis, and Z.-P. Jiang, "On the liapunov–krasovskii methodology for the iss of systems described by coupled delay differential and difference equations," *Automatica*, vol. 44, no. 9, pp. 2266–2273, 2008.

[28] P. Pepe and Z.-P. Jiang, "A Lyapunov-Krasovskii methodology for ISS and iISS of time-delay systems," *Systems & Control Letters*, vol. 55, no. 12, pp. 1006–1014, 2006.

[29] N. Chopra, M. W. Spong, and R. Lozano, "Synchronization of bilateral teleoperators with time delay," *Automatica*, vol. 44, no. 8, pp. 2142–2148, 2008.

[30] E. Nuno, R. Ortega, and L. Basanez, "An adaptive controller for nonlinear teleoperators," *Automatica*, vol. 46, no. 1, pp. 155–159, 2010.

[31] Z. Li, X. Cao, and N. Ding, "Adaptive fuzzy control for synchronization of nonlinear teleoperators with stochastic time-varying communication delays," *IEEE Transactions on Fuzzy Systems*, vol. 19, no. 4, pp. 745–757, 2011.

[32] E. D. Sontag and Y. Wang, "Notions of input to output stability," *Systems & Control Letters*, vol. 38, no. 4-5, pp. 235–248, 1999.

CHAPTER 4

Anti-saturation teleoperation control

THIS chapter builds on Chapter 3 and further addresses the adaptive control of teleoperation system with input saturation constraints. To unify the study of actuator saturation, passive/nonpassive external forces, asymmetric time-varying delays, and unknown dynamics in the same framework, the adaptive anti-saturation control based on auxiliary switched filter is developed. A specific control design is developed in the form of proportional plus damping injection plus switched filter, which simplifies the traditional design while providing better saturation control performance.

4.1 INTRODUCTION

Teleoperation system can realize the long-distance interaction between operator and working robot, so it can replace human working in dangerous and harsh environment. In the past few decades, related system control issues have received extensive attention [1–10]. Among them, an important topic is how to solve the influence of time delay introduced by communication network on system performance and control. To solve this problem, many control analysis methods have been proposed so far, such as the constant time-delay cases [11–14] and the time-varying time-delay cases [15–18]. Among them, passivity method is a popular method [19]. Considering the limitation of passivity method itself, in recent years, the research on non-passive external forces has also received some attention [20–24]. However, none of the above results take into account the saturation constraint of the actual actuator.

In the actual control system, the output capacity of the actuator is always limited, which may lead to some unexpected response characteristics, thus affecting the system performance [25–27]. Therefore, the saturation constraint of the actuator must be considered in the control. At present, there are many effective control methods for robot system. Loria et al. designed a nonlinear proportional weighting force compensation controller for a single robot system, in which the position error term only acts on the controller after a saturation function link [28]. On this basis, Zergeroglu et al. proposed an improved controller with adaptive gravity compensation by using

DOI: 10.1201/9781003382058-4

adaptive control technology [29]. Because they don't consider the communication delay, they are difficult to use in teleoperation system. In recent years, researchers have also paid attention to the problem of saturation control of teleoperation system. There are related research results in both constant communication time-delay and time-varying time-delay cases [30–32]. In view of the saturation control of teleoperation system with time-varying communication delay, Hashemzadeh et al. proposed a control design scheme of nonlinear proportional plus impedance injection weighting force compensation based on passivity method, and gave the closed-loop stability analysis based on Lyapunov-Krasovskii functional method [32]. However, there are two major constraints. Firstly, the passivity method adopted has strict limitations. Secondly, it does not consider the uncertainty and disturbance of the system. Therefore, it still faces difficulties in practical use. To deal with the model uncertainty and actuator saturation at the same time, Yang et al. proposed an adaptive anti-saturation compensation controller based on neural network [33]. However, it only considers the case of constant communication delay. In view of the fact that the time delay in the actual communication channel is not fixed, Hua et al. consider the saturation adaptive control of teleoperation system with time-varying communication time delay, and design an adaptive saturation proportional plus saturation impedance injection controller [34]. However, it should be noted that the work [33] is still based on passive control. Therefore, there are still many difficulties to be solved for the saturation adaptive control of teleoperation system.

This chapter addresses the adaptive control of teleoperation system with actuator saturation. To unify the study of actuator saturation, passive/nonpassive external forces, asymmetric time-varying delays and unknown dynamics in the same framework, the control based on auxiliary switched filter is developed. A specific control design is developed in the form of proportional plus damping injection plus switched filter, which simplifies the traditional one.

4.2 PROBLEM FORMULATION

Let us consider the following teleoperation system:

$$\begin{cases} \boldsymbol{M}_m(\boldsymbol{q}_m)\ddot{\boldsymbol{q}}_m + \boldsymbol{C}_m(\boldsymbol{q}_m, \dot{\boldsymbol{q}}_m)\dot{\boldsymbol{q}}_m + \boldsymbol{g}_m(\boldsymbol{q}_m) + \boldsymbol{f}_m(\dot{\boldsymbol{q}}_m) + \boldsymbol{D}_m \\ \qquad = \boldsymbol{J}_m^{\mathrm{T}}(\boldsymbol{q}_m)\boldsymbol{f}_h + \boldsymbol{\tau}_m \\ \boldsymbol{M}_s(\boldsymbol{q}_s)\ddot{\boldsymbol{q}}_s + \boldsymbol{C}_s(\boldsymbol{q}_s, \dot{\boldsymbol{q}}_s)\dot{\boldsymbol{q}}_s + \boldsymbol{g}_s(\boldsymbol{q}_s) + \boldsymbol{f}_s(\dot{\boldsymbol{q}}_s) + \boldsymbol{D}_s = -\boldsymbol{J}_s^{\mathrm{T}}(\boldsymbol{q}_s)\boldsymbol{f}_e + \boldsymbol{\tau}_s \end{cases} \tag{4.1}$$

where for any $j \in \{m, s\}$, $\boldsymbol{q}_j \in \mathbb{R}^n$, $\boldsymbol{M}_j(\boldsymbol{q}_j) \in \mathbb{R}^{n\times n}$, $\boldsymbol{C}_j(\boldsymbol{q}_j, \dot{\boldsymbol{q}}_j) \in \mathbb{R}^{n\times n}$, $\boldsymbol{g}_j(\boldsymbol{q}_j) \in \mathbb{R}^n$ and $\boldsymbol{\tau}_j \in \mathbb{R}^n$ are defined in (2.11). $\boldsymbol{f}_j(\dot{\boldsymbol{q}}_j)$ is the friction torque, $\boldsymbol{D}_j$ is the bounded external disturbance, $\boldsymbol{J}_j(\boldsymbol{q}_j)$ is the Jacobian matrix that reflects the transformation relationship between robot joint speed and end effector motion speed, $\boldsymbol{f}_h$ and $\boldsymbol{f}_e$ are the external force of the operator and the external force of the environment, respectively.

The end-effector positions is given as

$$\begin{cases} \boldsymbol{X}_j = \boldsymbol{h}_j(\boldsymbol{q}_j) \\ \dot{\boldsymbol{X}}_j = \boldsymbol{J}_j(\boldsymbol{q}_j)\dot{\boldsymbol{q}}_j \end{cases}$$

Similar to the existing work [35], the environment and operator dynamics are assumed to be

$$\begin{cases} \boldsymbol{f}_h = \boldsymbol{f}_h^* - \boldsymbol{M}_h\ddot{\boldsymbol{X}}_m - \boldsymbol{B}_h\dot{\boldsymbol{X}}_m - \boldsymbol{K}_h\boldsymbol{X}_m \\ \boldsymbol{f}_e = \boldsymbol{f}_e^* + \boldsymbol{M}_e\ddot{\boldsymbol{X}}_s + \boldsymbol{B}_e\dot{\boldsymbol{X}}_s + \boldsymbol{K}_e\boldsymbol{X}_s \end{cases} \tag{4.2}$$

where $\boldsymbol{M}_h(\boldsymbol{M}_e)$, $\boldsymbol{B}_h(\boldsymbol{B}_e)$ and $\boldsymbol{K}_h(\boldsymbol{K}_e)$ are unknown non-negative constant diagonal matrices, corresponding to the mass coefficient, damping coefficient and stiffness coefficient in the operator (environment) dynamics model respectively. $\boldsymbol{f}_h^*$ and $\boldsymbol{f}_e^*$ represent the exogenous forces of the operator and the environment respectively, which are assumed to be locally essential bounded in this section.

Substituting (4.2) into (4.1)

$$\begin{cases} \boldsymbol{\mathcal{M}}_m(\boldsymbol{q}_m)\ddot{\boldsymbol{q}}_m + \boldsymbol{\mathcal{C}}_m(\boldsymbol{q}_m, \dot{\boldsymbol{q}}_m)\dot{\boldsymbol{q}}_m + \boldsymbol{K}_h\boldsymbol{h}_m(\boldsymbol{q}_m) + \boldsymbol{f}_m(\dot{\boldsymbol{q}}_m) + \boldsymbol{g}_m(\boldsymbol{q}_m) + \boldsymbol{D}_m \\ \qquad = \boldsymbol{J}_m^{\mathrm{T}}(\boldsymbol{q}_m)\boldsymbol{f}_h^* + \boldsymbol{\tau}_m \\ \boldsymbol{\mathcal{M}}_s(\boldsymbol{q}_s)\ddot{\boldsymbol{q}}_s + \boldsymbol{\mathcal{C}}_s(\boldsymbol{q}_s, \dot{\boldsymbol{q}}_s)\dot{\boldsymbol{q}}_s + \boldsymbol{K}_e\boldsymbol{h}_s(\boldsymbol{q}_s) + \boldsymbol{f}_s(\dot{\boldsymbol{q}}_s) + \boldsymbol{g}_s(\boldsymbol{q}_s) + \boldsymbol{D}_s \\ \qquad = -\boldsymbol{J}_s^{\mathrm{T}}(\boldsymbol{q}_s)\boldsymbol{f}_e^* + \boldsymbol{\tau}_s \end{cases} \tag{4.3}$$

where for any $j \in \{m, s\}$,

$$\begin{aligned} \boldsymbol{\mathcal{M}}_j(\boldsymbol{q}_j) &= \boldsymbol{M}_j(\boldsymbol{q}_j) + \boldsymbol{J}_j^{\mathrm{T}}(\boldsymbol{q}_j)\boldsymbol{M}_{he_j}\boldsymbol{J}_j(\boldsymbol{q}_j) \\ \boldsymbol{\mathcal{C}}_j(\boldsymbol{q}_j, \dot{\boldsymbol{q}}_j) &= \boldsymbol{C}_j(\boldsymbol{q}_j, \dot{\boldsymbol{q}}_j) + \boldsymbol{J}_j^{\mathrm{T}}(\boldsymbol{q}_j)\boldsymbol{B}_{he_j}\boldsymbol{J}_j(\boldsymbol{q}_j) + \boldsymbol{J}_j^{\mathrm{T}}(\boldsymbol{q}_j)\boldsymbol{M}_{he_j}\dot{\boldsymbol{J}}_j(\boldsymbol{q}_j) \end{aligned}$$

with $\boldsymbol{M}_{he_m} = \boldsymbol{M}_h$, $\boldsymbol{M}_{he_s} = \boldsymbol{M}_e$, $\boldsymbol{B}_{he_m} = \boldsymbol{B}_h$, $\boldsymbol{B}_{he_s} = \boldsymbol{B}_e$.

From Property 2.1 and Property 2.4, the robot system shown in (4.3) satisfies the following properties [36]:

Property 4.1. *There exist positive constants ρ_{j1} and ρ_{j2} such that $\rho_{j1}\boldsymbol{I} \leq \boldsymbol{\mathcal{M}}_j(\boldsymbol{q}_j) \leq \rho_{j2}\boldsymbol{I}$, where $\boldsymbol{I}$ is an identity matrix with appropriate dimension.*

Property 4.2. *For any $\boldsymbol{x} \in \mathbb{R}^n$, it holds*

$$\boldsymbol{x}^{\mathrm{T}}(\dot{\boldsymbol{\mathcal{M}}}_j(\boldsymbol{q}_j) - 2\boldsymbol{\mathcal{C}}_j(\boldsymbol{q}_j, \dot{\boldsymbol{q}}_j))\boldsymbol{x} = -2\boldsymbol{x}^{\mathrm{T}}\boldsymbol{J}_j^{\mathrm{T}}(\boldsymbol{q}_j)\boldsymbol{B}_{he_j}\boldsymbol{J}_j(\boldsymbol{q}_j)\boldsymbol{x}$$

For the unknown uncertainty of the system, it is assumed that:

Assumption 4.1. *For any $j \in \{m, s\}$, it has*

$$\begin{aligned} &|\boldsymbol{K}_{he_j}\boldsymbol{h}_j(\boldsymbol{q}_j)| \leq k_{j1} \\ &|\boldsymbol{f}_j(\dot{\boldsymbol{q}}_j)| \leq k_{j2} + k_{j3}|\dot{\boldsymbol{q}}_j| \\ &|\boldsymbol{g}_j(\boldsymbol{q}_j)| \leq k_{j4} \\ &|\boldsymbol{J}_j^{\mathrm{T}}(\boldsymbol{q}_j)\boldsymbol{J}_j(\boldsymbol{q}_j)\dot{\boldsymbol{q}}_j| \leq k_{j6}|\dot{\boldsymbol{q}}_j| \end{aligned}$$

where $k_{j1}, k_{j2}, \cdots, k_{j6}$ are some unknown constant scalars.

In Assumption 4.1, only k_{j3} and k_{j6} are related to $|\dot{\boldsymbol{q}}_j|$. In the follow-up study in this chapter, in order to simplify the analysis, let

$$\begin{aligned} \omega_{j1} &= k_{j1} + k_{j2} + k_{j4} + k_{j5} \\ \omega_{j2} &= k_{j3} + k_{j6} \end{aligned}$$

In the actual robot system, the actuator inevitably faces the constraint of bounded output capability, i.e., $\boldsymbol{\tau}_j = [\tau_{j,1}, \tau_{j,2}, \cdots, \tau_{j,n}]^{\mathrm{T}}$ is constrained with

$$\tau_{j,i} = \begin{cases} \tau_{j,i}, & |\tau_{j,i}| \leq \tau_M \\ \tau_M \mathrm{sgn}(\tau_{j,i}), & |\tau_{j,i}| > \tau_M \end{cases}, \quad i = 1, 2, \cdots, n$$

where τ_M is the upper bound of output torque for each joint actuator of the robot. In this chapter, to simplify the analysis, it is assumed that all joints have the same output upper bound.

For any $j \in \{m, s\}$, the tracking errors of master and slave robots are defined as

$$\begin{cases} \boldsymbol{e}_j(t) = \boldsymbol{q}_j(t) - \boldsymbol{q}_\ell(t - d_\ell(t)) \\ \boldsymbol{e}_{vj}(t) = \dot{\boldsymbol{q}}_j(t) - \boldsymbol{q}_{vd\ell}(t, d_\ell(t)) \end{cases}, \quad \ell \in \{m, s\}, \ \ell \neq j \tag{4.4}$$

where $\boldsymbol{q}_{vd\ell}(t, d_\ell(t)) = \dot{\boldsymbol{q}}_\ell(\theta)|_{\theta = t - d_\ell(t)}$, the communication delay is assumed to satisfy $0 \leq d_j(t) \leq \bar{d}$ and $0 \leq |\dot{d}_j(t)| \leq \tilde{d}$, with $\bar{d}$ and $\tilde{d}$ being some positive constants.

The control goal of teleoperation system (4.1) is to realize adaptive control in the presence of the constraint of actuator saturation.

4.3 ADAPTIVE CONTROLLER WITH INPUT CONSTRAINT

To deal with actuator saturation constraint, this section introduces nonlinear saturation function $\boldsymbol{\mathcal{P}}(\cdot)$,

$$\boldsymbol{\tau}_j = \boldsymbol{\mathcal{P}}(\bar{\boldsymbol{\tau}}_j) = [\mathcal{P}_1(\bar{\tau}_{j,1}), \mathcal{P}_2(\bar{\tau}_{j,2}), \cdots, \mathcal{P}_n(\bar{\tau}_{j,n})]^{\mathrm{T}} \in \mathbb{R}^n \tag{4.5}$$

where $\bar{\boldsymbol{\tau}}_j = [\bar{\tau}_{j,1}, \bar{\tau}_{j,2}, \cdots, \bar{\tau}_{j,n}]^{\mathrm{T}}$ is the generalized controller. Nonlinear saturation function $\mathcal{P}_i(\bar{\tau}_{j,i})$ is designed as

$$\mathcal{P}_i(\bar{\tau}_{j,i}) = \tau_M(1 - \mathrm{e}^{-|\bar{\tau}_{j,i}|})\mathrm{sgn}(\bar{\tau}_{j,i}), \ i = 1, 2, \cdots, n \tag{4.6}$$

Obviously, $0 \leq |\mathcal{P}_i(\bar{\tau}_{j,i})| \leq \tau_M$.

Due to the introduction of nonlinear saturation function $\boldsymbol{\mathcal{P}}(\cdot)$, the generalized controller needs not consider actuator saturation constraints.

Substitute $\boldsymbol{\tau}_j = \boldsymbol{\mathcal{P}}(\bar{\boldsymbol{\tau}}_j)$ into (4.3),

$$\begin{aligned} \boldsymbol{\mathcal{M}}_j(\boldsymbol{q}_j)\ddot{\boldsymbol{q}}_j + \boldsymbol{\mathcal{C}}_j(\boldsymbol{q}_j, \dot{\boldsymbol{q}}_j)\dot{\boldsymbol{q}}_j + \boldsymbol{K}_{he_j}\boldsymbol{h}_j(\boldsymbol{q}_j) + \boldsymbol{f}_j(\dot{\boldsymbol{q}}_j) + \boldsymbol{g}_j(\boldsymbol{q}_j) + \boldsymbol{D}_j \\ = \boldsymbol{J}_j^{\mathrm{T}}(\boldsymbol{q}_j)\boldsymbol{f}_{he_j} + \boldsymbol{\mathcal{P}}(\bar{\boldsymbol{\tau}}_j) \end{aligned} \tag{4.7}$$

where $\boldsymbol{f}_{he_m} = \boldsymbol{f}_h^*$, $\boldsymbol{f}_{he_s} = -\boldsymbol{f}_e^*$.

To simplify the analysis, the nonlinear saturating function $\boldsymbol{\mathcal{P}}(\bar{\boldsymbol{\tau}}_j)$ will be rewritten as an affine nonlinear function with respect to $\bar{\boldsymbol{\tau}}_j$. According to the Lagrange mean value theorem, there exists $a \in [0, \bar{\tau}_{j,i}]$ such that $\dfrac{\mathrm{e}^{-\bar{\tau}_{j,i}} - 1}{\bar{\tau}_{j,i} - 0} = -\mathrm{e}^{-a}$, i.e.,

$$1 - \mathrm{e}^{-\bar{\tau}_{j,i}} = \mathrm{e}^{-a}\bar{\tau}_{j,i}, \ a \in [0, \bar{\tau}_{j,i}], \ \bar{\tau}_{j,i} > 0 \tag{4.8}$$

Similarly,

$$1 - e^{\bar{\tau}_{j,i}} = -e^{b}\bar{\tau}_{j,i},\ b \in [\bar{\tau}_{j,i}, 0],\ \bar{\tau}_{j,i} < 0 \tag{4.9}$$

Noting that, eq.(4.8) and eq.(4.9) still hold when $\bar{\tau}_{j,i} = 0$ with $a = b = 0$. Jointing (4.8), (4.9) with (4.6), there exists $c_{ji} \in [0, |\bar{\tau}_{j,i}|]$ such that

$$\mathcal{P}_i(\bar{\tau}_{j,i}) = \tau_M e^{-c_{ji}}\bar{\tau}_{j,i} \tag{4.10}$$

$e^{-c_{ji}}$ satisfies $e^{-|\bar{\tau}_{j,i}|} \le e^{-c_{ji}} \le 1$. Finally, $\boldsymbol{\mathcal{P}}(\bar{\boldsymbol{\tau}}_j)$ has the affine system:

$$\boldsymbol{\mathcal{P}}(\bar{\boldsymbol{\tau}}_j) = \tau_M \Xi_j \bar{\boldsymbol{\tau}}_j$$

where $\Xi_j = \mathrm{diag}[e^{-c_{j1}}, e^{-c_{j2}}, \cdots, e^{-c_{jn}}] \in \mathbb{R}^{n\times n}$.

To design the adaptive controller of the teleoperation system (4.7) with saturation constraint, the design scheme of auxiliary switched filter controller is still adopted, and the adaptive closed-loop control block diagram is shown in Fig. 4.1.

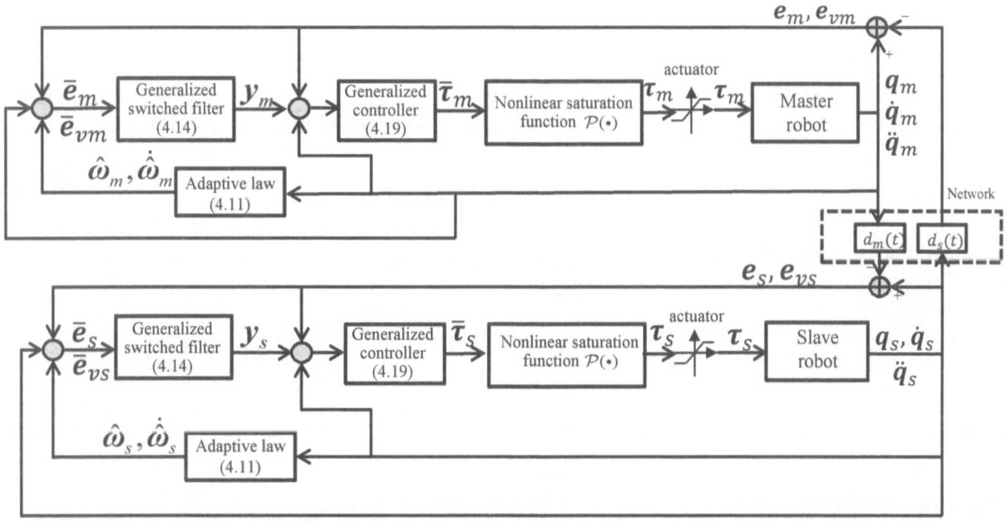

Figure 4.1 Block diagram of the proposed adaptive control algorithm for teleoperation system with input saturation.

In Fig. 4.1, $\hat{\omega}_{j1}(t)$ and $\hat{\omega}_{j2}(t)$, $j \in \{m, s\}$, are respectively the adaptive estimations of ω_{j1} and ω_{j2}, and the adaptive update laws are designed as

$$\begin{cases} \dot{\hat{\omega}}_{j1}(t) = -\psi_{j11}\hat{\omega}_{j1}(t) + \psi_{j12}P_{j1}\varepsilon_{j1}|\dot{\boldsymbol{q}}_j(t)|^2 \\ \dot{\hat{\omega}}_{j2}(t) = -\psi_{j21}\hat{\omega}_{j2}(t) + \psi_{j22}P_{j1}|\dot{\boldsymbol{q}}_j(t)|^2 \end{cases} \tag{4.11}$$

where P_{j1}, ψ_{ji1}, ψ_{ji2} $(i = 1, 2)$ and ε_{j1} are positive constants. In this section, the adaptive estimation error $\tilde{\omega}_{j1}(t) = \omega_{j1} - \hat{\omega}_{j1}(t)$ and $\tilde{\omega}_{j2}(t) = \omega_{j2} - \hat{\omega}_{j2}(t)$ are introduced.

Remark 4.1. *By solving (4.11), one can obtain*

$$\hat{\omega}_{j1}(t) = \mathrm{e}^{-\psi_{j11}t}\hat{\omega}_{j1}(0) + \int_0^t \mathrm{e}^{\psi_{j11}(w-t)}\psi_{j12}P_{j1}\varepsilon_{j1}|\dot{\boldsymbol{q}}_j(w)|^2\mathrm{d}w$$
$$\hat{\omega}_{j2}(t) = \mathrm{e}^{-\psi_{j21}t}\hat{\omega}_{j2}(0) + \int_0^t \mathrm{e}^{\psi_{j21}(w-t)}\psi_{j22}P_{j1}|\dot{\boldsymbol{q}}_j(w)|^2\mathrm{d}w$$

For any $\hat{\omega}_{j1}(0) > 0$ and $\hat{\omega}_{j2}(0) > 0$, it holds $\hat{\omega}_{j1}(t) > 0$ and $\hat{\omega}_{j2}(t) > 0$ for $0 \leq t < \infty$. In the rest of the chapter, it is assumed that $\hat{\omega}_{j1}(0) > 0$, $\hat{\omega}_{j2}(0) > 0$.

$\bar{\boldsymbol{e}}_j = [\bar{e}_{j,1}, \bar{e}_{j,2}, \cdots, \bar{e}_{j,(4n)}]^{\mathrm{T}} \in \mathbb{R}^{4n}$ is defined as

$$\bar{\boldsymbol{e}}_j = \left[\boldsymbol{e}_j^{\mathrm{T}}, \dot{\boldsymbol{q}}_j^{\mathrm{T}}, \sqrt{\varepsilon_{j1}\hat{\omega}_{j1}}\dot{\boldsymbol{q}}_j^{\mathrm{T}}, \sqrt{\hat{\omega}_{j2}}\dot{\boldsymbol{q}}_j^{\mathrm{T}}\right]^{\mathrm{T}} \tag{4.12}$$

and $\bar{\boldsymbol{e}}_{vj} = [\bar{e}_{vj,1}, \bar{e}_{vj,2}, \cdots, \bar{e}_{vj,(4n)}]^{\mathrm{T}} \in \mathbb{R}^{4n}$ is denoted by

$$\bar{\boldsymbol{e}}_{vj} = \left[\boldsymbol{e}_{vj}^{\mathrm{T}}, \ddot{\boldsymbol{q}}_j^{\mathrm{T}}, \frac{1}{2}\sqrt{\frac{\varepsilon_{j1}}{\hat{\omega}_{j1}}}\dot{\hat{\omega}}_{j1}\dot{\boldsymbol{q}}_j^{\mathrm{T}} + \sqrt{\varepsilon_{j1}\hat{\omega}_{j1}}\ddot{\boldsymbol{q}}_j^{\mathrm{T}}, \frac{1}{2}\sqrt{\frac{1}{\hat{\omega}_{j2}}}\dot{\hat{\omega}}_{j2}\dot{\boldsymbol{q}}_j^{\mathrm{T}} + \sqrt{\hat{\omega}_{j2}}\ddot{\boldsymbol{q}}_j^{\mathrm{T}}\right]^{\mathrm{T}} \tag{4.13}$$

Then in the saturation case, $\boldsymbol{y}_j = [y_{j,1}, y_{j,2}, \cdots, y_{j,4n}]^{\mathrm{T}} \in \mathbb{R}^{4n}$ the generalized switched filter. By using the concept of dynamic compensation, it is given by

$$\dot{\boldsymbol{y}}_j = -P_{j1}(\boldsymbol{y}_j + \boldsymbol{\mathcal{K}}_j\bar{\boldsymbol{e}}_j) - \boldsymbol{\mathcal{K}}_j\bar{\boldsymbol{e}}_{vj} + \boldsymbol{u}_j \tag{4.14}$$

where P_{j1} is positive constant, $\boldsymbol{u}_j = [u_{j,1}, u_{j,2}, \cdots, u_{j,4n}]^{\mathrm{T}}$ is

$$u_{j,i} = \begin{cases} -P_{j2}\mathrm{sgn}(y_{j,i} + \kappa_{j,i}\bar{e}_{j,i})|q_{vd\ell,i}(t, d_\ell(t))|, & 1 \leq i \leq n \\ 0, & i \geq n+1 \end{cases}$$

where $q_{vd\ell,i}(t, d_\ell(t))$ is the ith element of $\boldsymbol{q}_{vd\ell}(t, d_\ell(t))$. P_{j2} is a positive constant scalar, $\ell \in \{m, s\}$, $\ell \neq j$. $\boldsymbol{\mathcal{K}}_j(t) = \mathrm{diag}[\kappa_{j,1}(t), \kappa_{j,2}(t), \cdots, \kappa_{j,(4n)}(t)] \in \mathbb{R}^{4n\times 4n}$ is the switching rule, and for all $i = 1, 2, \cdots, 4n$, it is defined as

$$\kappa_{j,i}(t) = \begin{cases} 1, & y_{j,i}(t)\bar{e}_{j,i}(t) > 0 \text{ or } y_{j,i}(t)\bar{e}_{j,i}(t) = 0,\ y_{j,i}(t^-)\bar{e}_{j,i}(t^-) \leq 0 \\ -1, & y_{j,i}(t)\bar{e}_{j,i}(t) < 0 \text{ or } y_{j,i}(t)\bar{e}_{j,i}(t) = 0,\ y_{j,i}(t^-)\bar{e}_{j,i}(t^-) > 0 \end{cases} \tag{4.15}$$

Let $\mathcal{S} = \{1, 2, \cdots, 2^{4n}\}$. Similar to Remark 3.2, by introducing the map $\bar{r}_j : \boldsymbol{\mathcal{K}}_j(t) \to \mathcal{S}$, and define

$$\bar{r}_j(\boldsymbol{\mathcal{K}}_j(t)) = i, \text{ when only } \boldsymbol{\mathcal{K}}_j(t) \text{ is in the } i\text{th mode}, i \in \mathcal{S}$$

then $\boldsymbol{\mathcal{K}}_j(t)$ can be modeled as a standard switching signal, i.e., $\bar{r}_j(\boldsymbol{\mathcal{K}}_j(t))$. For convenience, let $r_j(t) = \bar{r}_j(\boldsymbol{\mathcal{K}}_j(t))$. Further, system (4.14) can be written as a standard switched system, i.e.,

$$\dot{\boldsymbol{y}}_j(t) = \boldsymbol{\mathcal{L}}(\boldsymbol{y}_j(t), \bar{\boldsymbol{e}}_j(t), \bar{\boldsymbol{e}}_{vj}(t), \boldsymbol{q}_{vd\ell}(t, d_\ell(t)), r_j(t)) \tag{4.16}$$

Remark 4.2. *In the following research in this section, the sequence* $\{t_k^j\}_{k\geq 0}$ *is used to represent the switching times of the switching signal* $r_j(t)$*, and let* $\boldsymbol{s}_j(t) = \boldsymbol{y}_j(t) + \boldsymbol{\mathcal{K}}_j(t)\bar{\boldsymbol{e}}_j(t)$*. Then,*

$$\dot{\boldsymbol{s}}_j(t) = -P_{j1}\boldsymbol{s}_j(t) + \bar{\boldsymbol{u}}_j(t), \ \forall t \in [t_k^j, t_{k+1}^j) \tag{4.17}$$

where

$$\bar{\boldsymbol{u}}_j(t) = \boldsymbol{u}_j(t) + \boldsymbol{\mathcal{K}}_j(t)\left[\dot{d}_\ell(t)\boldsymbol{q}_{vd\ell}^{\mathrm{T}}(t, d_\ell(t)), \boldsymbol{0}_{3n}^{\mathrm{T}}\right]^{\mathrm{T}}$$

$\boldsymbol{0}_{3n} \in \mathbb{R}^{3n}$ *is zero vector of* $3n$ *dimension. From the definitions of* $\boldsymbol{s}_j(t)$ *and* $\boldsymbol{\mathcal{K}}_j(t)$*, it then holds*

$$|\boldsymbol{y}_j(t)|^2 + |\bar{\boldsymbol{e}}_j(t)|^2 \leq |\boldsymbol{s}_j(t)|^2 \leq 2(|\boldsymbol{y}_j(t)|^2 + |\bar{\boldsymbol{e}}_j(t)|^2) \tag{4.18}$$

Finally, the generalized controller $\bar{\boldsymbol{\tau}}_j$ is designed as

$$\bar{\boldsymbol{\tau}}_j = -K_{j1}\dot{\boldsymbol{q}}_j - K_{j2}\boldsymbol{e}_j - \boldsymbol{K}_{j3}\boldsymbol{y}_j \tag{4.19}$$

where K_{j1} and K_{j2} are constant positive scalars, $\boldsymbol{K}_{j3} \in \mathbb{R}^{n\times 4n}$ is an appropriate non-zero matrix.

Remark 4.3. *In this section, the complete controller consists of (4.5), (4.19), (4.14) and (4.11), where a simple design form of "proportional + impedance injection + auxiliary filter compensation" is adopted. The design of auxiliary switched filter controller is based on the idea of dynamic compensation. Specifically, when the teleoperation system becomes unstable due to uncertain factors such as actuator saturation, time-varying time delay, non-zero exogenous force and so on, it will lead to changes in* $\boldsymbol{e}_j$*,* $\tilde{\omega}_{j1}$*,* $\tilde{\omega}_{j2}$ *and/or its derivative terms. At this time, based on the design of formula (4.19) and formula (4.14), these changes will first be reflected in the switched filter dynamic system (4.14), and then the control torque (4.19). In this way, the switched filter can dynamically compensate the adverse effects caused by the above uncertainty. By choosing the appropriate control gain, the stability of the closed-loop system can be ensured.*

4.4 STABILITY ANALYSIS

Considering the existence of non-zero exogenous force, the stability of the closed-loop system in this section will still be discussed under the framework of state-independent input-output stability. See Section 2.3.4 for the definition of stability.

In the framework of state-independent input-output stability, let $\boldsymbol{\chi}_j = [\boldsymbol{q}_j^{\mathrm{T}}, \tilde{\omega}_{j1}, \tilde{\omega}_{j2}, \boldsymbol{y}_j^{\mathrm{T}}, \bar{\boldsymbol{e}}_j^{\mathrm{T}}]^{\mathrm{T}} \in \mathbb{R}^{9n+2}$, $\boldsymbol{z}_j = [\tilde{\omega}_{j1}, \tilde{\omega}_{j2}, \boldsymbol{y}_j^{\mathrm{T}}, \bar{\boldsymbol{e}}_j^{\mathrm{T}}]^{\mathrm{T}} \in \mathbb{R}^{8n+2}$. And define $\boldsymbol{\chi}_{jt} = \{\boldsymbol{\chi}_j(t+\theta), \theta \in [-\bar{d}, 0]\} \in C([-\bar{d}, 0]; \mathbb{R}^{9n+2})$, $\boldsymbol{z}_{jt} = \{\boldsymbol{z}_j(t+\theta), \theta \in [-\bar{d}, 0]\} \in C([-\bar{d}, 0]; \mathbb{R}^{8n+2})$.

By using the multiple Lyapunov-Krasovskii functional method, it has

Theorem 4.1. *If the control gains K_{j1}, K_{j2}, P_{j1}, P_{j2}, ψ_{j11}, ψ_{j12}, ψ_{j21} and ψ_{j22} satisfy*

$$\begin{cases} P_{j1} > \max\left\{ \dfrac{(K_{j2} + \lambda_{\max}(\boldsymbol{K}_{j3}\boldsymbol{K}_{j3}^{\mathrm{T}}))\tau_M}{2} + \vartheta_j, \dfrac{\tau_M}{2} + \vartheta_j, 1 \right\} \\ P_{j2} \geq \tilde{d}, \ \dfrac{\psi_{j11}}{2\psi_{j12}} > \vartheta_j, \ \dfrac{\psi_{j21}}{2\psi_{j22}} > \vartheta_j \end{cases} \tag{4.20}$$

the complete closed-loop master (slave) system is SIIOS with $\boldsymbol{z}_m(t)$ ($\boldsymbol{z}_s(t)$) being the output, where $\vartheta_j > 0$ is a constant scalar, $\lambda_{\max}(\boldsymbol{K}_{j3}\boldsymbol{K}_{j3}^{\mathrm{T}})$ is the largest eigenvalue of $\boldsymbol{K}_{j3}\boldsymbol{K}_{j3}^{\mathrm{T}}$. Under the hypotheses, the position tracking error $\boldsymbol{e}_j(t)$ will maintain bounded for any bounded exogenous force $\boldsymbol{f}_{he_j}^*$.

Proof. For any $j \in \{m, s\}$, take the Lyapunov-Krasovskii functionals as

$$\begin{aligned} V_j(\chi_{jt}, r_j(t)) = &\frac{1}{2}\dot{\boldsymbol{q}}_j^{\mathrm{T}}\boldsymbol{\mathcal{M}}_j\dot{\boldsymbol{q}}_j + \frac{1}{2\psi_{j12}}\tilde{\omega}_{j1}^2 + \frac{1}{2\psi_{j22}}\tilde{\omega}_{j2}^2 + \\ &\frac{1}{2}\boldsymbol{s}_j^{\mathrm{T}}\boldsymbol{s}_j + \frac{\vartheta_j}{2}\int_{-\bar{d}}^{0} \boldsymbol{z}_{jt}^{\mathrm{T}}(\theta)\left[\frac{-\theta}{\bar{d}} + \frac{2(\theta + \bar{d})}{\bar{d}}\right]\boldsymbol{z}_{jt}(\theta)\mathrm{d}\theta \end{aligned}$$

for any fixed $r_j(t) = \mathfrak{p} \in \mathcal{S}$, from (4.18), it has

$$\begin{cases} \bar{\rho}_{j1}|\boldsymbol{z}_j|^2 \leq V_j(\chi_{jt}, \mathfrak{p}) \leq \bar{\rho}_{j2}\|\boldsymbol{z}_{jt}\|_{M_2}^2 \\ V_j(\chi_{jt_{k+1}^j}, r_j(t_{k+1}^j)) \leq V_j(\chi_{jt_{k+1}^j}, r_j(t_k^j)), \ k \geq 0 \end{cases}$$

where $\bar{\rho}_{j1} = \min\left\{\dfrac{\rho_{j1}}{2}, \dfrac{1}{2\psi_{j12}}, \dfrac{1}{2\psi_{j22}}, \dfrac{1}{2}\right\}$, $\bar{\rho}_{j2} = \max\left\{\dfrac{\rho_{j2}}{2}, \dfrac{1}{2\psi_{j12}}, \dfrac{1}{2\psi_{j22}}, 1, \vartheta_j\right\}$. In addition, it also has

$$\begin{aligned} D^+V_j(\chi_{jt}, \mathfrak{p}) = &\dot{\boldsymbol{q}}_j^{\mathrm{T}}\boldsymbol{\mathcal{M}}_j\ddot{\boldsymbol{q}}_j + \frac{1}{2}\dot{\boldsymbol{q}}_j^{\mathrm{T}}\dot{\boldsymbol{\mathcal{M}}}_j\dot{\boldsymbol{q}}_j + \frac{1}{\psi_{j12}}\tilde{\omega}_{j1}\dot{\tilde{\omega}}_{j1} + \frac{1}{\psi_{j22}}\tilde{\omega}_{j2}\dot{\tilde{\omega}}_{j2} + \\ &\boldsymbol{s}_j^{\mathrm{T}}\dot{\boldsymbol{s}}_j + \vartheta_j\boldsymbol{z}_j^{\mathrm{T}}\boldsymbol{z}_j - \vartheta_j\boldsymbol{z}_{jt}^{\mathrm{T}}(-\bar{d})\boldsymbol{z}_{jt}(-\bar{d}) - \frac{\vartheta_j}{2\bar{d}}\int_{-\bar{d}}^{0}\boldsymbol{z}_{jt}^{\mathrm{T}}(\theta)\boldsymbol{z}_{jt}(\theta)\mathrm{d}\theta \end{aligned} \tag{4.21}$$

On the one hand, from (4.7) and (4.19), by using Property 4.2 and Assumption 4.1, there exists $\varepsilon_{j1} > 0$ such that

$$\begin{aligned} &\dot{\boldsymbol{q}}_j^{\mathrm{T}}\boldsymbol{\mathcal{M}}_j\ddot{\boldsymbol{q}}_j + \frac{1}{2}\dot{\boldsymbol{q}}_j^{\mathrm{T}}\dot{\boldsymbol{\mathcal{M}}}_j\dot{\boldsymbol{q}}_j \\ &= \dot{\boldsymbol{q}}_j^{\mathrm{T}}\boldsymbol{J}_j^{\mathrm{T}}(\boldsymbol{q}_j)\boldsymbol{f}_{he_j} + \tau_M\dot{\boldsymbol{q}}_j^{\mathrm{T}}\Xi_j\bar{\boldsymbol{\tau}}_j - \boldsymbol{K}_{he_j}\dot{\boldsymbol{q}}_j^{\mathrm{T}}\boldsymbol{h}_j(\boldsymbol{q}_j) - \dot{\boldsymbol{q}}_j^{\mathrm{T}}\boldsymbol{f}_j(\dot{\boldsymbol{q}}_j) - \\ &\quad \dot{\boldsymbol{q}}_j^{\mathrm{T}}\boldsymbol{D}_j - \dot{\boldsymbol{q}}_j^{\mathrm{T}}\boldsymbol{g}_j(\boldsymbol{q}_j) - 2\dot{\boldsymbol{q}}_j^{\mathrm{T}}\boldsymbol{J}_j^{\mathrm{T}}(\boldsymbol{q}_j)B_{he_j}\boldsymbol{J}_j(\boldsymbol{q}_j)\dot{\boldsymbol{q}}_j \\ &\leq k_6|\dot{\boldsymbol{q}}_j|^2 + \frac{1}{4}|\boldsymbol{f}_{he_j}|^2 + \tau_M\dot{\boldsymbol{q}}_j^{\mathrm{T}}\Xi_j\bar{\boldsymbol{\tau}}_j + k_1|\dot{\boldsymbol{q}}_j| + k_2|\dot{\boldsymbol{q}}_j| + \\ &\quad k_3|\dot{\boldsymbol{q}}_j|^2 + k_5|\dot{\boldsymbol{q}}_j| + k_4|\dot{\boldsymbol{q}}_j| \end{aligned}$$

$$\leq \omega_{j1}|\dot{\boldsymbol{q}}_j| + \omega_{j2}|\dot{\boldsymbol{q}}_j|^2 - K_{j1}\tau_M \mathrm{e}^{-\max_{1\leq p\leq n}|\bar{\tau}_{jp}|}|\dot{\boldsymbol{q}}_j|^2 + \frac{K_{j2}\tau_M}{2}\dot{\boldsymbol{q}}_j^{\mathrm{T}}\boldsymbol{\Xi}_j\boldsymbol{\Xi}_j\dot{\boldsymbol{q}}_j +$$
$$\frac{K_{j2}\tau_M}{2}|\boldsymbol{e}_j|^2 + \frac{\tau_M}{2}|\boldsymbol{y}_j|^2 + \frac{\tau_M}{2}\dot{\boldsymbol{q}}_j^{\mathrm{T}}\boldsymbol{\Xi}_j\boldsymbol{K}_{j3}\boldsymbol{K}_{j3}^{\mathrm{T}}\boldsymbol{\Xi}_j\dot{\boldsymbol{q}}_j + \frac{1}{4}|\boldsymbol{f}_{he_j}|^2$$
$$\leq -K_{j1}\tau_M \mathrm{e}^{-\max_{1\leq p\leq n}|\bar{\tau}_{jp}|}|\dot{\boldsymbol{q}}_j|^2 + \frac{1}{4}|\boldsymbol{f}_{he_j}|^2 + \frac{\tau_M}{2}|\boldsymbol{y}_j|^2 + \frac{\omega_{j1}}{\varepsilon_{j1}} + \varepsilon_{j1}\omega_{j1}|\dot{\boldsymbol{q}}_j|^2 +$$
$$\omega_{j2}|\dot{\boldsymbol{q}}_j|^2 + \frac{K_{j2} + \lambda_{\max}(\boldsymbol{K}_{j3}\boldsymbol{K}_{j3}^{\mathrm{T}})}{2}\tau_M|\dot{\boldsymbol{q}}_j|^2 + \frac{K_{j2}\tau_M}{2}|\boldsymbol{e}_j|^2 \tag{4.22}$$

On the other hand, from (4.11), one can obtain

$$\frac{1}{\psi_{j12}}\tilde{\omega}_{j1}\dot{\tilde{\omega}}_{j1} + \frac{1}{\psi_{j22}}\tilde{\omega}_{j2}\dot{\tilde{\omega}}_{j2} = -\frac{1}{\psi_{j12}}\tilde{\omega}_{j1}\dot{\hat{\omega}}_{j1} - \frac{1}{\psi_{j22}}\tilde{\omega}_{j2}\dot{\hat{\omega}}_{j2}$$
$$\leq -\frac{\psi_{j11}}{2\psi_{j12}}|\tilde{\omega}_{j1}|^2 - \frac{\psi_{j21}}{2\psi_{j22}}|\tilde{\omega}_{j2}|^2 - P_{j1}\varepsilon_{j1}\tilde{\omega}_{j1}|\dot{\boldsymbol{q}}_j|^2 -$$
$$P_{j1}\tilde{\omega}_{j2}|\dot{\boldsymbol{q}}_j|^2 + \frac{\psi_{j11}}{2\psi_{j12}}|\omega_{j1}|^2 + \frac{\psi_{j21}}{2\psi_{j22}}|\omega_{j2}|^2 \tag{4.23}$$

For any fixed $\mathfrak{p} \in \mathcal{S}$, from (4.17), we have

$$\boldsymbol{s}_j^{\mathrm{T}}\dot{\boldsymbol{s}}_j = -P_{j1}\boldsymbol{s}_j^{\mathrm{T}}\boldsymbol{s}_j(t) + \boldsymbol{s}_j^{\mathrm{T}}\bar{\boldsymbol{u}}_j$$
$$\leq -P_{j1}|\boldsymbol{y}_j|^2 - P_{j1}|\bar{\boldsymbol{e}}_j|^2 + \boldsymbol{s}_j^{\mathrm{T}}\bar{\boldsymbol{u}}_j$$

For those $P_{j2} \geq \tilde{d}$, without loss of generality, it has

$$\boldsymbol{s}_j^{\mathrm{T}}\bar{\boldsymbol{u}}_j = \sum_{i=1}^{n}(s_{j,i}u_{j,i} + s_{j,i}\kappa_{j,i}\dot{d}_\ell q_{vd\ell,i}(t, d_\ell(t))) \leq 0$$

Thus,

$$\boldsymbol{s}_j^{\mathrm{T}}\dot{\boldsymbol{s}}_j \leq -P_{j1}|\boldsymbol{y}_j|^2 - P_{j1}|\boldsymbol{e}_j|^2 - P_{j1}|\dot{\boldsymbol{q}}_j|^2 - P_{j1}\varepsilon_{j1}\hat{\omega}_{j1}|\dot{\boldsymbol{q}}_j|^2 - P_{j1}\hat{\omega}_{j2}|\dot{\boldsymbol{q}}_j|^2 \tag{4.24}$$

Jointing (4.21)∼(4.24),

$$D^+V_j(\boldsymbol{\chi}_{jt}, \mathfrak{p})$$
$$\leq \vartheta_j\boldsymbol{z}_j^{\mathrm{T}}\boldsymbol{z}_j - \frac{\vartheta_j}{2\bar{d}}\int_{-\bar{d}}^{0}\boldsymbol{z}_{jt}^{\mathrm{T}}(\theta)\boldsymbol{z}_{jt}(\theta)\mathrm{d}\theta + \frac{1}{4}|\boldsymbol{f}_{he_j}|^2 -$$
$$K_{j1}\tau_M \mathrm{e}^{-\max_{1\leq p\leq n}|\bar{\tau}_{jp}|}|\dot{\boldsymbol{q}}_j|^2 + \frac{K_{j2}\tau_M}{2}|\boldsymbol{e}_j|^2 + \frac{\tau_M}{2}|\boldsymbol{y}_j|^2 +$$
$$\frac{K_{j2} + \lambda_{\max}(\boldsymbol{K}_{j3}\boldsymbol{K}_{j3}^{\mathrm{T}})}{2}\tau_M|\dot{\boldsymbol{q}}_j|^2 + \frac{\omega_{j1}}{\varepsilon_{j1}} + \varepsilon_{j1}\omega_{j1}|\dot{\boldsymbol{q}}_j|^2 +$$
$$\omega_{j2}|\dot{\boldsymbol{q}}_j|^2 - \frac{\psi_{j11}}{2\psi_{j12}}|\tilde{\omega}_{j1}|^2 - \frac{\psi_{j21}}{2\psi_{j22}}|\tilde{\omega}_{j2}|^2 - P_{j1}\varepsilon_{j1}\tilde{\omega}_{j1}|\dot{\boldsymbol{q}}_j|^2 -$$
$$P_{j1}\tilde{\omega}_{j2}|\dot{\boldsymbol{q}}_j|^2 + \frac{\psi_{j11}}{2\psi_{j12}}|\omega_{j1}|^2 + \frac{\psi_{j21}}{2\psi_{j22}}|\omega_{j2}|^2 - P_{j1}|\boldsymbol{y}_j|^2 -$$
$$P_{j1}|\boldsymbol{e}_j|^2 - P_{j1}|\dot{\boldsymbol{q}}_j|^2 - P_{j1}\varepsilon_{j1}\hat{\omega}_{j1}|\dot{\boldsymbol{q}}_j|^2 - P_{j1}\hat{\omega}_{j2}|\dot{\boldsymbol{q}}_j|^2$$

$$
\begin{aligned}
&\leq -\left(P_{j1} - \vartheta_j - \frac{K_{j2} + \lambda_{\max}(\boldsymbol{K}_{j3}\boldsymbol{K}_{j3}^{\mathrm{T}})}{2}\tau_M\right)|\dot{\boldsymbol{q}}_j|^2 - \\
&\left(P_{j1} - \frac{K_{j2}\tau_M}{2} - \vartheta_j\right)|\boldsymbol{e}_j|^2 - \left(P_{j1} - \frac{\tau_M}{2} - \vartheta_j\right)|\boldsymbol{y}_j|^2 - \\
&\left(\frac{\psi_{j11}}{2\psi_{j12}} - \vartheta_j\right)|\tilde{\omega}_{j1}|^2 - \left(\frac{\psi_{j21}}{2\psi_{j22}} - \vartheta_j\right)|\tilde{\omega}_{j2}|^2 + \frac{1}{4}|\boldsymbol{f}_{he_j}|^2 + \frac{\omega_{j1}}{\varepsilon_{j1}} + \\
&\frac{\psi_{j11}}{2\psi_{j12}}|\omega_{j1}|^2 + \frac{\psi_{j21}}{2\psi_{j22}}|\omega_{j2}|^2 - \frac{\vartheta_j}{2\bar{d}}\int_{-\bar{d}}^{0}\boldsymbol{z}_{jt}^{\mathrm{T}}(\theta)\boldsymbol{z}_{jt}(\theta)\mathrm{d}\theta \\
&\leq -\bar{\mu}_j|\boldsymbol{z}_j|^2 - \frac{\vartheta_j}{2\bar{d}}\int_{-\bar{d}}^{0}\boldsymbol{z}_{jt}^{\mathrm{T}}(\theta)\boldsymbol{z}_{jt}(\theta)\mathrm{d}\theta + \Delta_j \\
&\leq -\mu_j\|\boldsymbol{z}_{jt}\|_{M_2}^2 + \Delta_j
\end{aligned}
$$

where $\bar{\mu}_j = \min\left\{P_{j1} - \frac{(K_{j2} + \lambda_{\max}(\boldsymbol{K}_{j3}\boldsymbol{K}_{j3}^{\mathrm{T}}))\tau_M}{2} - \vartheta_j, P_{j1} - \frac{\tau_M}{2} - \vartheta_j, \frac{\psi_{j11}}{2\psi_{j12}} - \vartheta_j, \frac{\psi_{j21}}{2\psi_{j22}} - \vartheta_j\right\}$, $\mu_j = \min\left\{\bar{\mu}_j, \frac{\vartheta_j}{2\bar{d}}\right\}$, $\Delta_j = \frac{1}{4}|\boldsymbol{f}_{he_j}|^2 + \frac{\omega_{j1}}{\varepsilon_{j1}} + \frac{\psi_{j11}}{2\psi_{j12}}|\omega_{j1}|^2 + \frac{\psi_{j21}}{2\psi_{j22}}|\omega_{j2}|^2$.

There exists $\varepsilon_j \in (0,1)$ such that

$$
\|\boldsymbol{z}_{jt}\|_{M_2} \geq \frac{\sqrt{\Delta_j}}{\sqrt{\mu_j\varepsilon_j}} \Rightarrow D^+V_j(\chi_{jt}, \mathfrak{p}) \leq -\mu_j(1-\varepsilon_j)\|\boldsymbol{z}_{jt}\|_{M_2}^2
$$

Following Lemma 2.4, there exists $\beta_j(\cdot,\cdot) \in \mathcal{KL}$ such that

$$
|\boldsymbol{z}_j(t)| \leq \beta_j(\|\boldsymbol{z}_{j0}\|, t) + \frac{\bar{\rho}_{j2}}{\bar{\rho}_{j1}\sqrt{\mu_j\varepsilon_j}}\sqrt{\Delta_j} \tag{4.25}
$$

□

Remark 4.4. *For the control gains in controller (4.19), similar to the existing analysis* [32]*, it is possible to see a trade-off between the gains and the tracking performance. On the one hand, for fixed* K_{j2}, $\boldsymbol{K}_{j3}$ *and other parameters, if* K_{j1} *is increased, the tracking error* $\mathbf{e}_j$ *will contribute less to the control signal, which finally causes an increase in the settling time for the position tracking response; if* K_{j1} *is lowered,* P_{j1} *can only cover a smaller* ϑ_j *and further cause a smaller* μ_j*, which will result in slower convergence speed of* $\boldsymbol{z}_j$*. On the other hand, for any fixed* K_{j1}, K_{j2} *and other parameters, if* $\boldsymbol{K}_{j3}$ *is increased, the output of the switched filter will contribute more to the control output, which may cause more oscillation due to the signal switching in the switched filter. Finally, it is also not suggested to take a very large* K_{j2}*. Because a large* P_{j1} *should be taken to cover the large* K_{j2} *at this time, while the large* P_{j1} *may 'suspend'* $\bar{\mathbf{e}}_j$ *and* $\bar{\mathbf{e}}_{vj}$ *in switched filter (4.14) and then suppress the function of* $\boldsymbol{K}_{j3}\boldsymbol{y}_j$ *in controller (4.19), which may cause the undesired performance. To balance the control performance, the suggestion of the parameter selection is:* K_{j1} *and* $\boldsymbol{K}_{j3}$ *take some suitable small values,* K_{j2} *takes a suitable large value, while other control parameters can be selected according to Theorem 4.1.*

Remark 4.5. *A novel adaptive control framework based on switching technique has been developed in this chapter. Compared with the previous work* [32]*, there exist both advantages and disadvantages. The main advantages lie in: 1). The external forces applied on robots by both human operator and environment can be non-passive, while the algorithm* [32] *can only handle the passive case. However, as pointed out by Polushin et al.* [24]*, the passive assumption for those external forces has some strict shortcomings in practices. 2). The model parameter uncertainty has been considered, which makes the algorithm be more realistic. The main disadvantage is that the acceleration information is used in the proposed controller design, where both the master and slave robots have to equip with acceleration sensors.*

4.5 EXPERIMENTAL VALIDATION

This section will verify the effectiveness of the algorithm through experiments. Teleoperation system is shown in Fig. 4.2, which is composed of two PHANToM Premium 1.5A robotic manipulators with three degrees of freedom. In the experiment, in order to simulate unknown dynamic uncertainties and external disturbances, a rigid object with underactuated joints was artificially tied from the end of the manipulator. On the one hand, because the rigid object is tied to the slave robot, the dynamic model of the slave robot will change with strong uncertainty. On the other hand, the rigid body has an underactuated joint. When the slave robot moves with the master robot, the underactuated joint of the rigid body will also swing with the movement of the slave robot body, which introduces external disturbance.

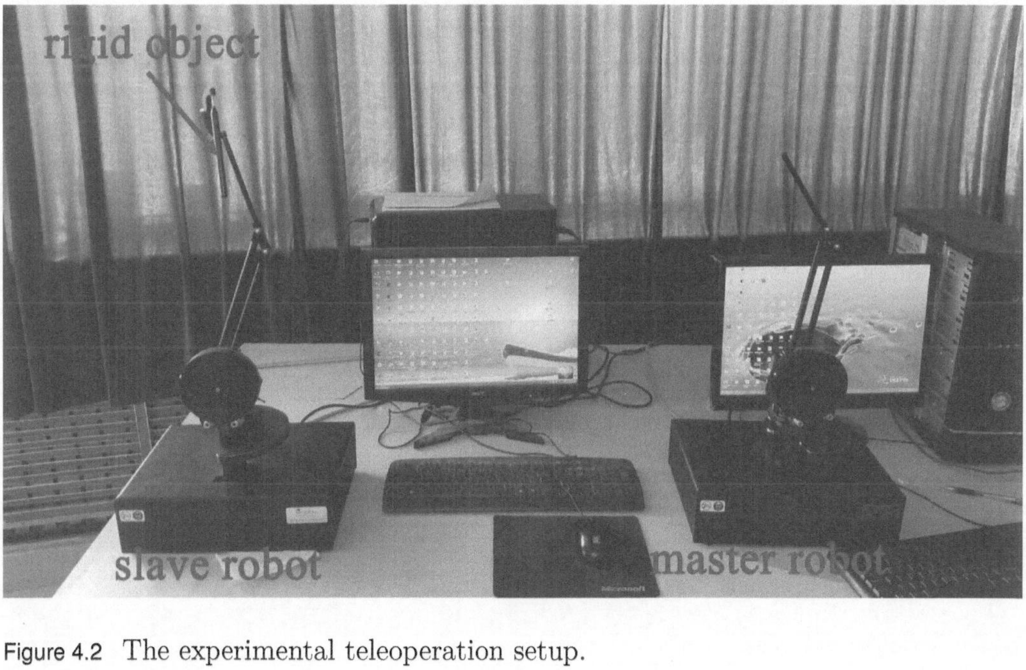

Figure 4.2 The experimental teleoperation setup.

As shown in Fig. 4.2, the sampling period of teleoperation experiment system is selected as 2 ms. The master and slave robots are connected through a LAN with a bandwidth of 100 Mb/s, and the routing delay and its variation are less than 0.1 ms. Because the routing delay is far less than the sampling period of 2 ms, in order to better verify the effectiveness of the algorithm, two additional artificial communication delays which are far greater than the actual communication delay are added to the master and slave robots through MATLAB, i.e., $d_m(t) = 0.6\sin^2(t)$ and $d_s(t) = 0.4\cos^2(2t)$. For the master and slave robots, set the upper output limit of each joint actuator as $\tau_M = 0.2$[1]. According to the Theorem 4.1, for any $j \in \{m, s\}$, the control gains are selected as $K_{j1} = 0.4$, $K_{j2} = 10$, $\boldsymbol{K}_{j3} = [\boldsymbol{I}_3, \boldsymbol{I}_3, \boldsymbol{I}_3, \boldsymbol{I}_3] \times 0.01$, $P_{m1} = 23.7$, $P_{s1} = 22.25$, $P_{j2} = 2$, $\varepsilon_{j1} = 1$, $\psi_{j12} = \psi_{j22} = 3$, $\psi_{j11} = \psi_{j21} = 33$, with $\boldsymbol{I}_3 = \text{diag}[1, 1, 1]$.

4.5.1 Stability verification

The first two sets of experiments will be used to verify the stability of the proposed algorithm.

Firstly, the slave manipulator is set to run in free motion as shown in Fig. 4.2. In this case, the experimental results are given in Fig. 4.3. Among them, Fig. 4.3(a) gives the joint position tracking between the master and slave robots, and obviously, the proposed nonlinear saturated control algorithm can guarantee the satisfactory tracking performance. Figs. 4.3(b) and 4.3(c) show the control torques of the three joints of the master and slave robots respectively.

The second set of experiments verifies the contact stability of the proposed algorithm. In the experiment, we put an obstacle in the first joint operation path of the slave manipulator. The tracking performance is then given as Fig. 4.4(a). (*Given that the master and slave robots have the same structure, only the joint positions are given here for the sake of saving space.*) From Fig. 4.4(a), when the slave manipulator comes in contact with the obstacle, stability is maintained and when the slave departs from the obstacle, it can track the master quickly. Finally, the control torques of the master and slave robots are shown in Fig. 4.4(b) and Fig. 4.4(c), respectively.

4.5.2 Comparison study

Hashemzadeh et al. investigated an effective saturation control method for teleoperation system with varying time delays [32]. This experiment verifies the adverse effect of unknown uncertainty and disturbance in robotic dynamic. The experimental setup is still given as Fig. 4.2. Because the slave robot has attached a rigid object that has an additional underactuated joint, its dynamic is of uncertainty, and it also suffers

[1] The actual output capacity of the robot actuator is greater than 0.2 N·m, and artificially set the upper output limit as $\tau_M = 0.2$ in order to simulate actuator saturation.

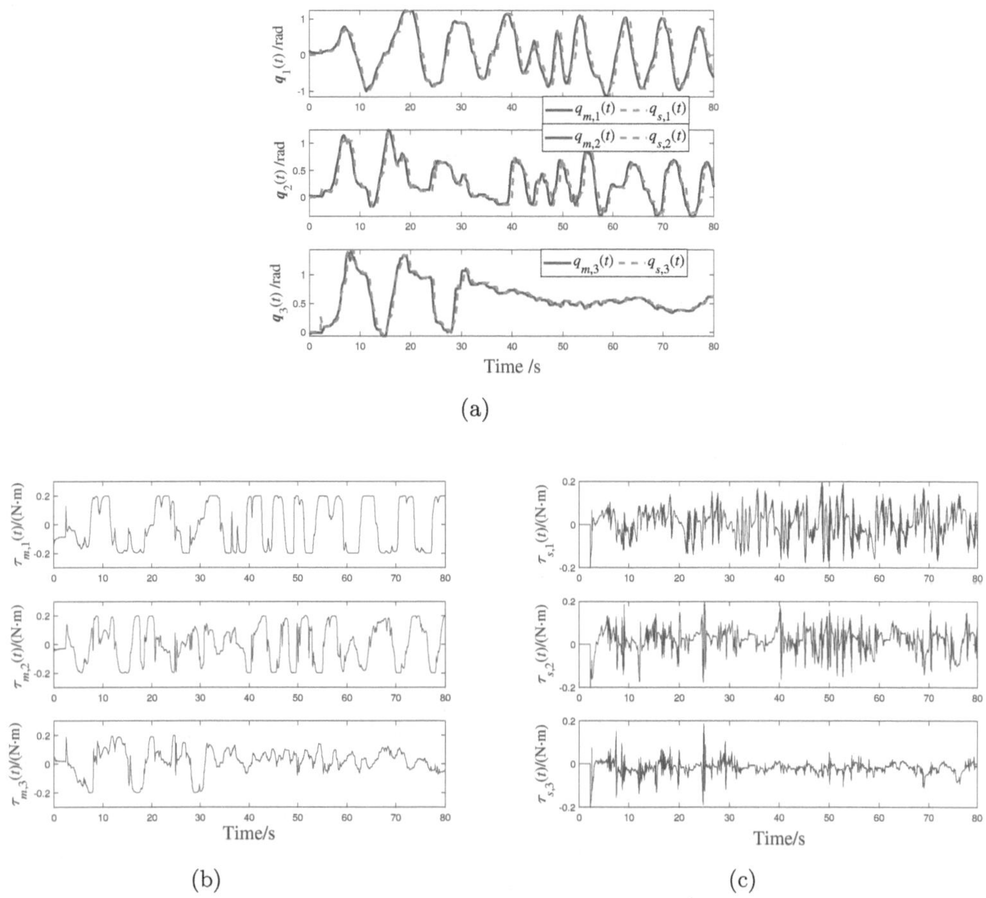

Figure 4.3 The experimental results when the slave robot runs in free motion. (a) Synchronization performance. (b) Control torque $\boldsymbol{\tau}_m$. (c) Control torque $\boldsymbol{\tau}_s$.

from unknown disturbance caused by the motion of rigid object, which have not been considered by the previous work [32].

In this experiment, to be more fair, it replaces the human operator's action in above two sets of experiments with the torque applied to master robot's motors through control. Specifically, the torque exserted on the master robot by human operator in (4.1), i.e., $\boldsymbol{J}_m^T(\boldsymbol{q}_m)\boldsymbol{f}_h$, is set to be $[0.22\sin(0.5t), 0.2\sin(0.35t), 0.6\sin(0.8t)]^T Nm$. Then under the same experimental setup (including the same communication delays and the same actuator saturation, etc.), the task-space tracking performances are shown in Fig. 4.5. (*Although the human torque has been set to be the same, there are some differences between the motion trajectories of the master robot as shown in Figs. 4.5(a and b). It is mainly because the master robot is driven simultaneously by the human torque and the control torque, while the control torques generated by the control algorithms are indeed different.*) Note that, although a poor performance is observed in Fig. 4.5(b), the selected control parameters for Hashemzadeh's method [32] in the experiment are

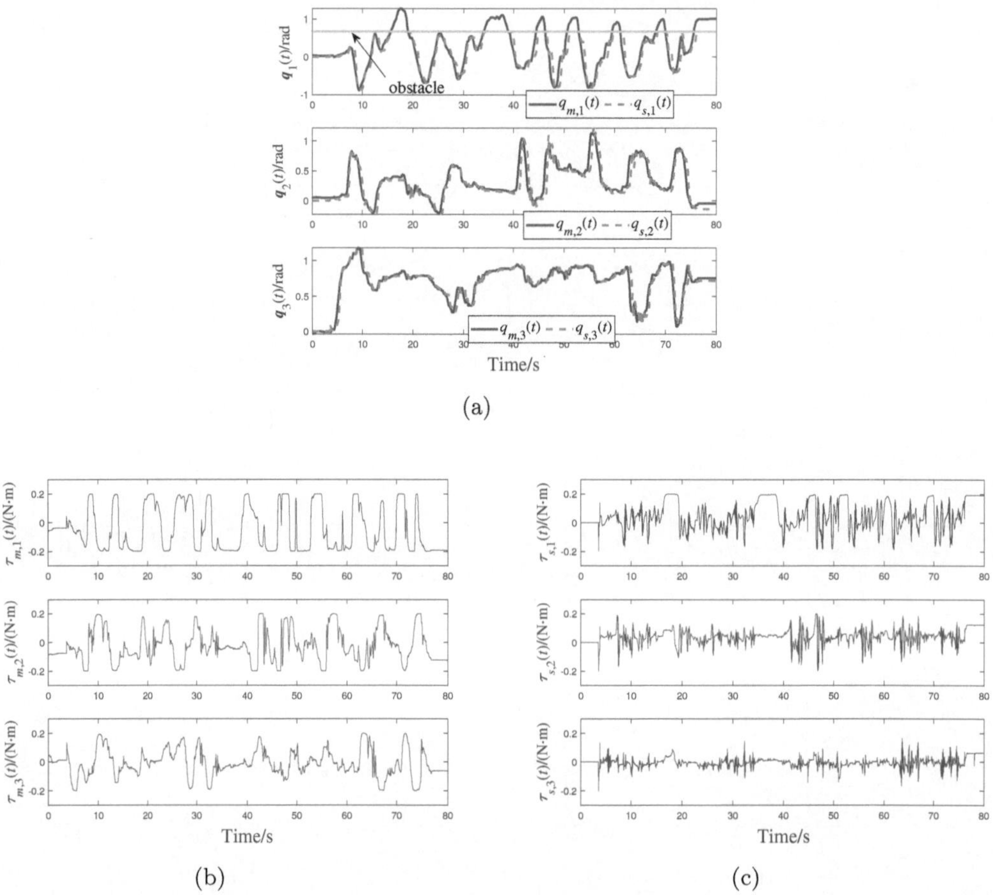

Figure 4.4 The experimental results when an obstacle is located in the first joint operation path of slave robot. (a) Synchronization performance. (b) Control torque $\boldsymbol{\tau}_m$. (c) Control torque $\boldsymbol{\tau}_s$.

capable to guarantee the satisfactory control performance when the rigid object is removed from the slave robot. Therefore from Fig. 4.5, the control performance of the Hashemzadeh's method [32] is adversely affected by the dynamic uncertainty and external disturbance, while in this scenario, a better performance is obtained by the proposed adaptive control algorithm of this chapter.

Figs. 4.6 and 4.7 show the corresponding control torques. Although the maximum allowable torque output is set to be $\tau_M = 0.2$ for both the control of this chapter and the Hashemzadeh's method [32], the slave control torque shown in Fig. 4.7 is still less than $\tau_M = 0.2$ for most of the experiment time. Given that the performance in Fig. 4.5, the Hashemzadeh's control does not make full use of output capacity of the actuators. In fact, in Hashemzadeh's method [32], it is required that $\gamma_i + N_i \leq \tau_M$, where γ_i and N_i are the upper bounds of i-th element of the gravity vector $\boldsymbol{g}_j(\boldsymbol{q}_j)$ and nonlinear proportional term, respectively. It implies that the control torque should overcome the largest gravity for all time, which thus leaves less "room" for

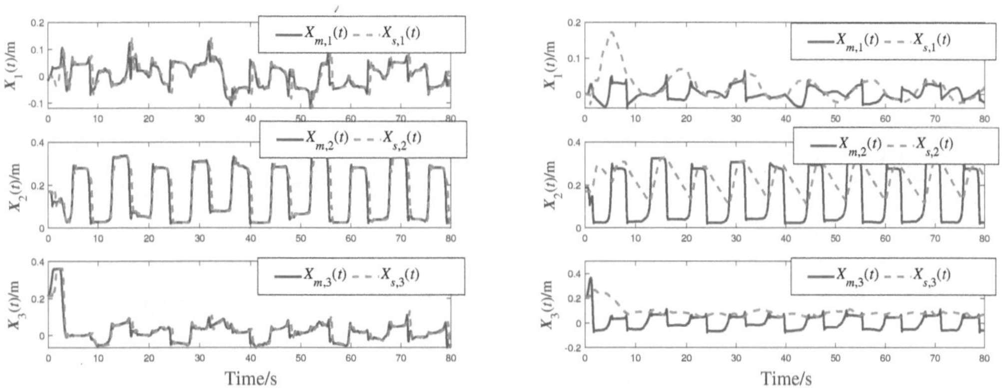

(a) (b)

Figure 4.5 Synchronization performances. (a) The method of this Chapter. (b) The Hashemzadeh's method [32].

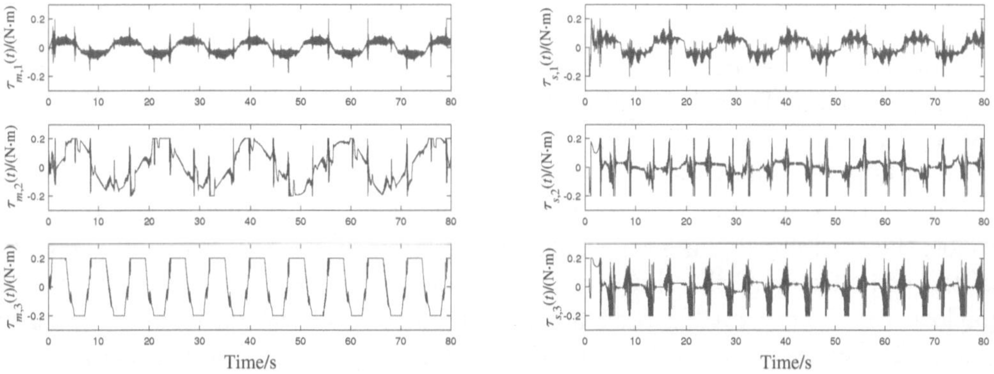

(a) (b)

Figure 4.6 Control torques under the proposed adaptive control. (a) Control torque of master robot. (b) Control torque of slave robot.

the proportional term to contribute to the control signal. Given that the gravity $\boldsymbol{g}_j(\boldsymbol{q}_j)$ is the function of $\boldsymbol{q}_j$, it does not always have the largest value. Therefore, the Hashemzadeh's control is relatively conservative. In this chapter, the nonlinear saturation function is placed on the outside of the generalized controller. This kind of design only requires that the $\gamma_i \leq \tau_M$, i.e., the actuators of each of the master and slave robots have the capacity to overcome the corresponding robot's gravity within their work spaces. Since the generalized controller does not consider any actuator saturation, the new design does not impose the constraint mentioned above and allows the controller to dynamically adjust the portion of the proportional control.

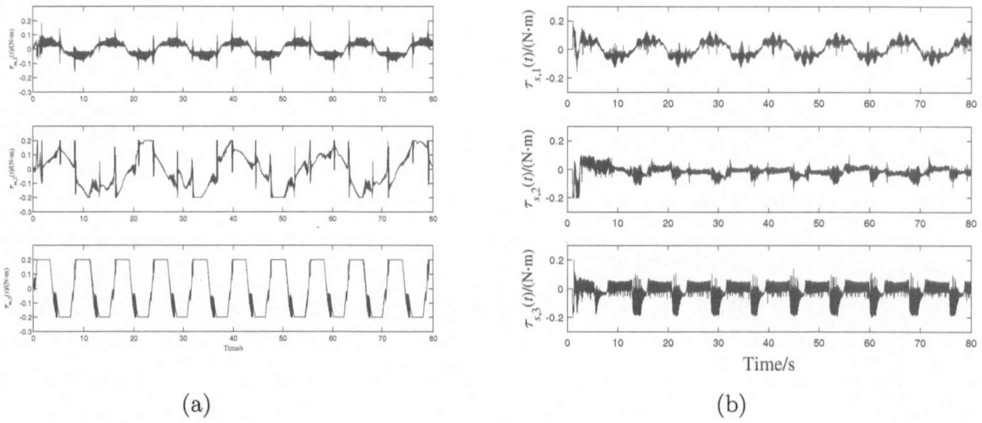

Figure 4.7 Control torques under the Hashemzadeh's control [32]. (a) Control torque of master robot. (b) Control torque of slave robot.

4.5.3 Some discussion

1. Control oscillation.

In Figs. 4.3 to 4.6, it is found that the oscillation is fewer in the control of master robot.

Firstly, it is mainly caused by the rigid object that has attached to the end-effector of the slave robot. On the one hand, when the slave robot tracks the master robot, the rigid object will swing follow the motion of the slave robot. If the motion of the end-effector of the slave robot is suddenly changed, the inertia of the rigid payload can generate a big drag force to the motion of the slave robot. To ensure the motion synchronization between the master and slave robots, the control torque of the slave robot should overcome the drag force, which raises more oscillations. On the other hand, the attached payload also introduces the model uncertainty, which has an adverse effect on the stability of the closed-loop slave system. In this scenario, given that the specific design of (4.14), the switched filter will work for the output of the control torque to ensure the stability. Since the uncertainty varies according to the motion of the slave robot, the dynamic uncertainty of slave robot is more complicated than the one in master robot. Therefore, more oscillations are found in the control torques of the slave robot.

Secondly, the joint speed should be also responsible for the more oscillations of the control torque of slave robot. In Figs. 4.3 and 4.4, the oscillation of joint speed is fewer in the master robot. *(On the one hand, since the master robot is manipulated by the human operator, while the human operator is capable of smoothing the joint speed, the oscillation of joint speed is reduced. On the other hand, the slave robot with an attached rigid payload has been set to run in free motion. The oscillation of joint speed cannot be reduced. In addition, due to the inertia, the attached rigid payload will also enlarge such oscillation. Therefore, the oscillation of joint speed is fewer in the master robot.)* Due to the designs of generalized controller (4.19) and the switched filter (4.14), the oscillation of $\dot{\boldsymbol{q}}_j$ will lead to the oscillation of control torque $\bar{\boldsymbol{\tau}}_j$. In fact, in the experiment of Subsection V-B, when we replace the human operator's action with the torque applied to master robot's motors through control, more oscillations are found in the control of the master robot.

Compared with Fig. 4.3(a) and Fig. 4.4(a), Fig. 4.5(a) has a more frequent change of the joint motion. Given that the design of the switched filter (4.14), it will cause the switching of $\boldsymbol{y}_j$. Therefore more control oscillations are found in Fig. 4.6.

In practical implementation, the following measures can be taken to reduce the oscillation. On the one hand, as discussed in Remark 4.4, one can take some small $\boldsymbol{K}_{j3}$ on the premise that the stability of the closed-loop teleoperation system is ensured. On the other hand, one can also use the low-pass filter. By letting the output of the switched filter pass through a low-pass filter before it is used for the control torque, the oscillation can be reduced. More details can be found in our future work. Due to the space limit, they are omitted here.

2. Force control.

Force tracking is another important issue for improved transparency in bilateral teleoperation, and has attracted a lot of attention [5,10,24]. However, most of them do not consider the actuator saturation. The existence of saturation will make the force control and its stability analysis more difficult. In this chapter, the main concern is to investigate a novel adaptive control framework for teleoperation system that are subjected to actuator saturation, passive/non-passive external forces, time-varying delays, and unknown dynamics. Similar to the work [32], the developed framework of this chapter does not consider the force tracking control, and therefore it cannot provide the quantitative analysis of force tracking. However, due to the master controller utilizes the position error between the master and slave robots, the developed control algorithm can help the human operator feel the environment at slave site. For example, as shown in Fig. 4.4(a) and Fig. 4.4(b), when the first joint of the slave robot comes in contact with the obstacle, the control for the first joint of the master robot is saturated ($\tau_{m1} = -\tau_M = -0.2$). At this time, the human operator feels a large drag force, which helps the human operator understand the situation of the slave robot. Of course, the force control under actuator saturation is one of our ongoing works.

4.6 CONCLUSION

In this chapter, a novel switching-based adaptive control scheme is developed to cope with actuator saturation in nonlinear teleoperation systems that are subjected to varying time delays, dynamic uncertainty and disturbance. To handle actuator saturation, a two-level control framework, i.e., the generalized controller plus a well-designed saturating function, is developed, while in implementation, the saturating function is placed on the outside of the generalized controller. Based on the concept of dynamic compensation, a special switched filter is investigated, and then the generalized controller is designed to be in the form as proportional plus damping injection plus switched filter. Thus, the closed-loop teleoperation system is modeled to be consisted of multiple subsystems including the robotic dynamic subsystem and the switched filter subsystem. Finally, to unify the study of passive and non-passive external forces, the stability of the complete system is performed in the sense of state-independent input-to-output stability, which is established by using multiple Lyapunov-Krasovskii functionals method. For any bounded exogenous forces, it is shown that the ultimate boundedness of the position tracking errors between the master and slave robots is guaranteed.

Bibliography

[1] P. F. Hokayem and M. W. Spong, "Bilateral teleoperation: an historical survey," *Automatica*, vol. 42, no. 12, pp. 2035–2057, 2006.

[2] Y. Nakajima, T. Nozaki, and K. Ohnishi, "Heartbeat synchronization with haptic feedback for telesurgical robot," *IEEE Transactions on Industrial Electronics*, vol. 61, no. 7, pp. 3753–3764, 2013.

[3] S. Livatino, L. T. De Paolis, M. D'Agostino, A. Zocco, A. Agrimi, A. De Santis, L. V. Bruno, and M. Lapresa, "Stereoscopic visualization and 3-D technologies in medical endoscopic teleoperation," *IEEE Transactions on Industrial Electronics*, vol. 62, no. 1, pp. 525–535, 2014.

[4] Y. Ye, Y.-J. Pan, and T. Hilliard, "Bilateral teleoperation with time-varying delay: A communication channel passification approach," *IEEE/ASME Transactions on Mechatronics*, vol. 18, no. 4, pp. 1431–1434, 2013.

[5] F. Hashemzadeh and M. Tavakoli, "Position and force tracking in nonlinear teleoperation systems under varying delays," *Robotica*, vol. 33, no. 4, pp. 1003–1016, 2015.

[6] B. Yalcin and K. Ohnishi, "Stable and transparent time-delayed teleoperation by direct acceleration waves," *IEEE Transactions on Industrial Electronics*, vol. 57, no. 9, pp. 3228–3238, 2009.

[7] W. R. Ferrell and T. B. Sheridan, "Supervisory control of remote manipulation," *IEEE Spectrum*, vol. 4, no. 10, pp. 81–88, 1967.

[8] T. Nozaki, T. Mizoguchi, and K. Ohnishi, "Motion expression by elemental separation of haptic information," *IEEE Transactions on Industrial Electronics*, vol. 61, no. 11, pp. 6192–6201, 2014.

[9] J. H. Cho, H. I. Son, D. G. Lee, T. Bhattacharjee, and D. Y. Lee, "Gain-scheduling control of teleoperation systems interacting with soft tissues," *IEEE Transactions on Industrial Electronics*, vol. 60, no. 3, pp. 946–957, 2012.

[10] R. J. Anderson and M. W. Spong, "Asymptotic stability for force reflecting teleoperators with time delay," *The International Journal of Robotics Research*, vol. 11, no. 2, pp. 135–149, 1992.

[11] N. Chopra, M. W. Spong, and R. Lozano, "Synchronization of bilateral teleoperators with time delay," *Automatica*, vol. 44, no. 8, pp. 2142–2148, 2008.

[12] E. Nuno, R. Ortega, and L. Basanez, "An adaptive controller for nonlinear teleoperators," *Automatica*, vol. 46, no. 1, pp. 155–159, 2010.

[13] Y. Yang, C. Hua, and X. Guan, "Adaptive fuzzy finite-time coordination control for networked nonlinear bilateral teleoperation system," *IEEE Transactions on Fuzzy Systems*, vol. 22, no. 3, pp. 631–641, 2013.

[14] Y.-C. Liu and M.-H. Khong, "Adaptive control for nonlinear teleoperators with uncertain kinematics and dynamics," *IEEE/ASME Transactions on Mechatronics*, vol. 20, no. 5, pp. 2550–2562, 2015.

[15] N. Chopra, P. Berestesky, and M. W. Spong, "Bilateral teleoperation over unreliable communication networks," *IEEE Transactions on Control Systems Technology*, vol. 16, no. 2, pp. 304–313, 2008.

[16] C.-C. Hua and P. X. Liu, "Delay-dependent stability criteria of teleoperation systems with asymmetric time-varying delays," *IEEE Transactions on Robotics*, vol. 26, no. 5, pp. 925–932, 2010.

[17] H. C. Cho, J. H. Park, K. Kim, and J.-O. Park, "Sliding-mode-based impedance controller for bilateral teleoperation under varying time-delay," in *Proceedings of the IEEE International Conference on Robotics and Automation*, vol. 1. IEEE, 2001, pp. 1025–1030.

[18] S. Islam, P. X. Liu, and A. El Saddik, "Nonlinear control for teleoperation systems with time varying delay," *Nonlinear Dynamics*, vol. 76, no. 2, pp. 931–954, 2014.

[19] E. Nuno, L. Basanez, and R. Ortega, "Passivity-based control for bilateral teleoperation: a tutorial," *Automatica*, vol. 47, no. 3, pp. 485–495, 2011.

[20] S. Islam, P. X. Liu, and A. El Saddik, "Teleoperation systems with symmetric and unsymmetric time varying communication delay," *IEEE Transactions on Instrumentation and Measurement*, vol. 62, no. 11, pp. 2943–2953, 2013.

[21] I. Sarras, E. Nuno, and L. Basanez, "An adaptive controller for nonlinear teleoperators with variable time-delays," *Journal of the Franklin Institute*, vol. 351, no. 10, pp. 4817–4837, 2014.

[22] I. G. Polushin and H. J. Marquez, "Stabilization of bilaterally controlled teleoperators with communication delay: an ISS approach," *International Journal of Control*, vol. 76, no. 8, pp. 858–870, 2003.

[23] I. G. Polushin, A. Tayebi, and H. J. Marquez, "Control schemes for stable teleoperation with communication delay based on ios small gain theorem," *Automatica*, vol. 42, no. 6, pp. 905–915, 2006.

[24] I. G. Polushin, S. N. Dashkovskiy, A. Takhmar, and R. V. Patel, "A small gain framework for networked cooperative force-reflecting teleoperation," *Automatica*, vol. 49, no. 2, pp. 338–348, 2013.

[25] M. V. Kothare, P. J. Campo, M. Morari, and C. N. Nett, "A unified framework for the study of anti-windup designs," *Automatica*, vol. 30, no. 12, pp. 1869–1883, 1994.

[26] W. Sun, Z. Zhao, and H. Gao, "Saturated adaptive robust control for active suspension systems," *IEEE Transactions on Industrial Electronics*, vol. 60, no. 9, pp. 3889–3896, 2012.

[27] H. Su, M. Z. Chen, X. Wang, and J. Lam, "Semiglobal observer-based leader-following consensus with input saturation," *IEEE Transactions on Industrial Electronics*, vol. 61, no. 6, pp. 2842–2850, 2013.

[28] A. Loria, R. Kelly, R. Ortega, and V. Santibanez, "On global output feedback regulation of Euler-Lagrange systems with bounded inputs," *IEEE Transactions on Automatic Control*, vol. 42, no. 8, pp. 1138–1143, 1997.

[29] E. Zergeroglu, W. Dixon, A. Behal, and D. Dawson, "Adaptive set-point control of robotic manipulators with amplitude-limited control inputs," *Robotica*, vol. 18, no. 2, pp. 171–181, 2000.

[30] S.-J. Lee and H.-S. Ahn, "A study on bilateral teleoperation with input saturation and systems," in *Proceedings of the 11th International Conference on Control, Automation and Systems*. IEEE, 2011, pp. 161–166.

[31] S.-J. Lee and H.-S. Ahn, "Controller designs for bilateral teleoperation with input saturation," *Control Engineering Practice*, vol. 33, pp. 35–47, 2014.

[32] F. Hashemzadeh, I. Hassanzadeh, and M. Tavakoli, "Teleoperation in the presence of varying time delays and sandwich linearity in actuators," *Automatica*, vol. 49, no. 9, pp. 2813–2821, 2013.

[33] Y. Yang, C. Ge, H. Wang, X. Li, and C. Hua, "Adaptive neural network based prescribed performance control for teleoperation system under input saturation," *Journal of the Franklin Institute*, vol. 352, no. 5, pp. 1850–1866, 2015.

[34] C. Hua, Y. Yang, and P. X. Liu, "Output-feedback adaptive control of networked teleoperation system with time-varying delay and bounded inputs," *IEEE/ASME Transactions on Mechatronics*, vol. 20, no. 5, pp. 2009–2020, 2014.

[35] P. Malysz and S. Sirouspour, "Nonlinear and filtered force/position mappings in bilateral teleoperation with application to enhanced stiffness discrimination," *IEEE Transactions on Robotics*, vol. 25, no. 5, pp. 1134–1149, 2009.

[36] T. H. Lee and C. J. Harris, *Adaptive Neural Network Control of Robotic Manipulators*. World Scientific, 1998.

III

Multi-Master Multi-Slave Teleoperation

CHAPTER 5

Adaptive tele-coordinated control of multiple mobile robots

THIS chapter addresses the tele-coordinated control problem, where an adaptive fuzzy multi-master-multi-slave teleoperation control is investigated for multiple mobile manipulators to carry a common object in a cooperative manner that subjected to asymmetric time-varying delays and model uncertainty. To achieve the control objective, under the framework of auxiliary switching filter control, the constrained tele-coordination control is developed based on the hybrid motion/force control, where all slave robots and their handled payload are considered as a closed kinematic chain. Utilizing the Lyapunov-Krasovskii method, the complete closed-loop master (slave) system is proved to be state-independent input-to-output stable.

5.1 INTRODUCTION

For some complex operation scenes, multi-robot cooperation has many advantages over single robot operation, but multi-robot cooperation also brings many difficulties to control. As discussed in Chapter 1, there are still many imperfections in the theoretical research of multilateral coordination and cooperation teleoperation control at this stage. Therefore, this chapter and Chapter 6 will give two new control schemes of multilateral teleoperation based on constrained and unconstrained coordination control methods, respectively, and using auxiliary switching filter control method.

Considering the excellent reconfigurability and adaptability of mobile manipulator, this chapter will study the coordinated teleoperation control of multi-mobile manipulator based on the hybrid motion/force control architecture. Based on the proposed auxiliary switching filter control method, an adaptive fuzzy coordination controller is designed for the multiple mobile manipulators teleoperation system with asymmetric time-varying communication delay, passive/non-passive external force, model uncertainty and external disturbance. Compared with the previous work [1], this chapter uses the state-independent input-output stability for system analysis,

DOI: 10.1201/9781003382058-5

and takes the external force of the operator as the external input, thereby reducing the number of force sensors used in the implementation of the control algorithm, and reducing the economic burden of the algorithm, and obtaining a less conservative conclusion.

5.2 PROBLEM FORMULATION

In the multilateral cooperative teleoperation system studied in this chapter, it is assumed that the system consists of N_m master robots with n_{mi} degrees of freedom $(i = 1, 2, \cdots, N_m)$ and N_s slave mobile manipulators with n_s degrees of freedom. The task of teleoperation is that the operator controls all the slave robots through the master robots to carry a rigid object cooperatively in the operation space. The control schematic diagram of the system is shown in Fig. 5.1 (for simplicity, only the architecture of 2 masters and 4 slaves is shown in the figure). The dynamics of the lth master robot and the kth slave robot are described as follows:

$$\begin{cases} \boldsymbol{M}_m^l(\boldsymbol{q}_m^l)\ddot{\boldsymbol{q}}_m^l + \boldsymbol{C}_m^l(\boldsymbol{q}_m^l, \dot{\boldsymbol{q}}_m^l)\dot{\boldsymbol{q}}_m^l + \boldsymbol{g}_m^l(\boldsymbol{q}_m^l) + \boldsymbol{D}_m^l = \boldsymbol{J}_m^{lT}(\boldsymbol{q}_m^l)\boldsymbol{f}_h^l + \boldsymbol{\tau}_m^l \\ \boldsymbol{M}_s^k(\boldsymbol{q}_s^k)\ddot{\boldsymbol{q}}_s^k + \boldsymbol{C}_s^k(\boldsymbol{q}_s^k, \dot{\boldsymbol{q}}_s^k)\dot{\boldsymbol{q}}_s^k + \boldsymbol{g}_s^k(\boldsymbol{q}_s^k) + \boldsymbol{D}_s^k = \boldsymbol{J}_s^{kT}(\boldsymbol{q}_s^k)\boldsymbol{f}_e^k + \boldsymbol{B}_s^k(\boldsymbol{q}_s^k)\boldsymbol{\tau}_s^k \end{cases} \tag{5.1}$$

where the subscripts m and s denote the master robot and the slave robot, respectively, the superscripts l and k denotes the lth and kth robot, respectively. $\boldsymbol{q}_m^l(t) \in \mathbb{R}^{n_{ml}}$ and $\boldsymbol{q}_s^k(t) \in \mathbb{R}^{n_s}$ are the robot joint vectors. For any $(j, i) \in \mathcal{S}_1 := \{(m, 1), (m, 2), \cdots, (m,\ N_m), (s, 1), (s, 2), \cdots, (s, N_s)\}$, $\boldsymbol{M}_j^i(\boldsymbol{q}_j^i) \in \mathbb{R}^{n_{ji} \times n_{ji}}$ are the positive definite inertia matrices, and $\boldsymbol{C}_j^i(\boldsymbol{q}_j^i, \dot{\boldsymbol{q}}_j^i) \in \mathbb{R}^{n_{ji} \times n_{ji}}$ are the matrices of centripetal and coriolis torque, and $\boldsymbol{g}_j^i(\boldsymbol{q}_j^i) \in \mathbb{R}^{n_{ji}}$ are the gravitational vectors, and $\boldsymbol{D}_j^i \in \mathbb{R}^{n_{ji}}$ denote the bounded unknown external disturbances, and $\boldsymbol{J}_j^i(\boldsymbol{q}_j^i) \in \mathbb{R}^{n_{ji} \times n_{ji}}$ are the Jacobian matrices, and $\boldsymbol{\tau}_j^i \in \mathbb{R}^{n_{ji}}$ are the applied torques with $n_s^k = n_s$, $\boldsymbol{B}_s^i(\boldsymbol{q}_s^i)$ is a full rank input transformation matrix for the mobile platform and the robotic manipulator and assumed to be known because it is a function of fixed geometry of the system, and $\boldsymbol{f}_h^l \in \mathbb{R}^{n_{ml}}$ and $\boldsymbol{f}_e^l \in \mathbb{R}^{n_s}$ are the forces applied by human operator

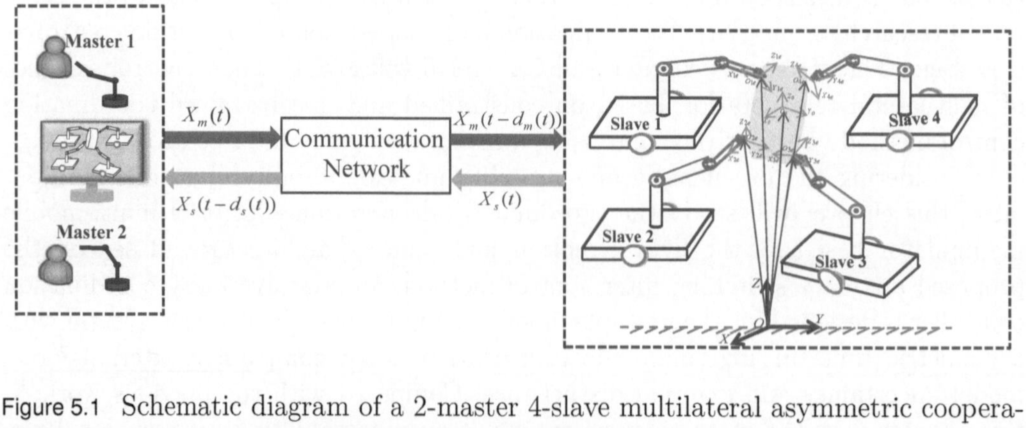

Figure 5.1 Schematic diagram of a 2-master 4-slave multilateral asymmetric cooperative teleoperation system.

and the environment on the end-effectors of the l-th master and the k-th slave robot. To facilitate the dynamic formulation, the following assumptions are made:

Assumption 5.1. [1-3] *Each slave mobile manipulator is non-redundant and operating away from singularity. The transported object is rigid. All the end-effectors of the mobile manipulators are rigidly attached to the common object so that there is no relative motion between the object and the end-effectors and the object is rigid.*

5.2.1 Dynamics of the master robots

For the ith master robot, $i = 1, 2, \cdots, N_m$, denote $\boldsymbol{X}_m^i = [\boldsymbol{X}_{1m}^{i\mathrm{T}}, \boldsymbol{X}_{2m}^{i\mathrm{T}}]^{\mathrm{T}} \in \mathbb{R}^{n_{mi}}$, where $\boldsymbol{X}_{1m}^i$ are $\boldsymbol{X}_{2m}^i$ are the position and orientation in the base frame. According to forward kinematics, they are related to the joint space vectors as $\boldsymbol{X}_m^i = \boldsymbol{h}_m^i(\boldsymbol{q}_m^i)$, $\dot{\boldsymbol{X}}_m^i = \boldsymbol{J}_m^i(\boldsymbol{q}_m^i)\dot{\boldsymbol{q}}_m^i$. For simplicity, all the master robots are assumed to be non-redundant manipulators. Without loss of generality, it has

$$\dot{\boldsymbol{q}}_m^i = \boldsymbol{J}_m^{i-1}(\boldsymbol{q}_m^i)\dot{\boldsymbol{X}}_m^i \tag{5.2}$$

Using (5.1) and (5.2), the dynamics for the ith master manipulator can be represented by

$$\boldsymbol{\mathcal{M}}_m^i(\boldsymbol{q}_m^i)\ddot{\boldsymbol{X}}_m^i + \boldsymbol{\mathcal{C}}_m^i(\boldsymbol{q}_m^i, \dot{\boldsymbol{q}}_m^i)\dot{\boldsymbol{X}}_m^i + \boldsymbol{\mathcal{G}}_m^i(\boldsymbol{q}_m^i) + \boldsymbol{\mathcal{D}}_m^i = \boldsymbol{f}_h^i + \boldsymbol{J}_m^{i-\mathrm{T}}(\boldsymbol{q}_m^i)\boldsymbol{\tau}_m^i \tag{5.3}$$

where $\boldsymbol{J}_m^{i-\mathrm{T}}(\boldsymbol{q}_m^i) = \left((\boldsymbol{J}_m^i(\boldsymbol{q}_m^i))^{-1}\right)^{\mathrm{T}}$, and

$$\begin{aligned}
\boldsymbol{\mathcal{M}}_m^i(\boldsymbol{q}_m^i) &= \boldsymbol{J}_m^{i-\mathrm{T}}(\boldsymbol{q}_m^i)\boldsymbol{M}_m^i(\boldsymbol{q}_m^i)\boldsymbol{J}_m^{i-1}(\boldsymbol{q}_m^i)\\
\boldsymbol{\mathcal{C}}_m^i(\boldsymbol{q}_m^i, \dot{\boldsymbol{q}}_m^i) &= \boldsymbol{J}_m^{i-\mathrm{T}}(\boldsymbol{q}_m^i)[\boldsymbol{C}_m^i(\boldsymbol{q}_m^i, \dot{\boldsymbol{q}}_m^i) - \boldsymbol{M}_m^i(\boldsymbol{q}_m^i)\boldsymbol{J}_m^{i-1}(\boldsymbol{q}_m^i)\dot{\boldsymbol{J}}_m^i(\boldsymbol{q}_m^i)]\boldsymbol{J}_m^{i-1}(\boldsymbol{q}_m^i)\\
\boldsymbol{\mathcal{G}}_m^i(\boldsymbol{q}_m^i) &= \boldsymbol{J}_m^{i-\mathrm{T}}(\boldsymbol{q}_m^i)\boldsymbol{g}_m^i(\boldsymbol{q}_m^i)\\
\boldsymbol{\mathcal{D}}_m^i &= \boldsymbol{J}_m^{i-\mathrm{T}}(\boldsymbol{q}_m^i)\boldsymbol{D}_m^i
\end{aligned}$$

Further, the dynamics of N_m master robots from (5.3) can then be expressed concisely as

$$\boldsymbol{\mathcal{M}}_m(\boldsymbol{q}_m)\ddot{\boldsymbol{X}}_m + \boldsymbol{\mathcal{C}}_m(\boldsymbol{q}_m, \dot{\boldsymbol{q}}_m)\dot{\boldsymbol{X}}_m + \boldsymbol{\mathcal{G}}_m(\boldsymbol{q}_m) + \boldsymbol{\mathcal{D}}_m = \boldsymbol{\mathcal{F}}_h + \boldsymbol{J}_m^{-\mathrm{T}}(\boldsymbol{q}_m)\boldsymbol{\tau}_m \tag{5.4}$$

where $\boldsymbol{q}_m = [\boldsymbol{q}_m^{1\mathrm{T}}, \boldsymbol{q}_m^{2\mathrm{T}}, \cdots, \boldsymbol{q}_m^{N_m\mathrm{T}}]^{\mathrm{T}}$, $\boldsymbol{X}_m = [\boldsymbol{X}_m^{1\mathrm{T}}, \boldsymbol{X}_m^{2\mathrm{T}}, \cdots, \boldsymbol{X}_m^{N_m\mathrm{T}}]^{\mathrm{T}}$, and

$$\begin{aligned}
\boldsymbol{\mathcal{M}}_m(\boldsymbol{q}_m) &= \text{block diag}[\boldsymbol{\mathcal{M}}_m^1(\boldsymbol{q}_m^1), \boldsymbol{\mathcal{M}}_m^2(\boldsymbol{q}_m^2), \cdots, \boldsymbol{\mathcal{M}}_m^{N_m}(\boldsymbol{q}_m^{N_m})]\\
\boldsymbol{\mathcal{C}}_m(\boldsymbol{q}_m, \dot{\boldsymbol{q}}_m) &= \text{block diag}[\boldsymbol{\mathcal{C}}_m^1(\boldsymbol{q}_m^1, \dot{\boldsymbol{q}}_m^1), \boldsymbol{\mathcal{C}}_m^2(\boldsymbol{q}_m^2, \dot{\boldsymbol{q}}_m^2), \cdots, \boldsymbol{\mathcal{C}}_m^N(\boldsymbol{q}_m^{N_m}, \dot{\boldsymbol{q}}_m^{N_m})]\\
\boldsymbol{\mathcal{G}}_m(\boldsymbol{q}_m) &= [\boldsymbol{\mathcal{G}}_m^{1\mathrm{T}}(\boldsymbol{q}_m^1), \boldsymbol{\mathcal{G}}_m^{2\mathrm{T}}(\boldsymbol{q}_m^2), \cdots, \boldsymbol{\mathcal{G}}_m^{N_m\mathrm{T}}(\boldsymbol{q}_m^{N_m})]^{\mathrm{T}}\\
\boldsymbol{\mathcal{D}}_m &= [\boldsymbol{D}_m^{1\mathrm{T}}, \boldsymbol{D}_m^{2\mathrm{T}}, \cdots, \boldsymbol{D}_m^{N_m\mathrm{T}}]^{\mathrm{T}}\\
\boldsymbol{\mathcal{F}}_h &= [\boldsymbol{F}_h^{1\mathrm{T}}, \boldsymbol{F}_h^{2\mathrm{T}}, \cdots, \boldsymbol{F}_h^{N_m\mathrm{T}}]^{\mathrm{T}}\\
\boldsymbol{J}_m^{-\mathrm{T}}(\boldsymbol{q}_m) &= \text{block diag}[\boldsymbol{J}_m^{1-\mathrm{T}}(\boldsymbol{q}_m^1), \boldsymbol{J}_m^{2-\mathrm{T}}(\boldsymbol{q}_m^2), \cdots, \boldsymbol{J}_m^{N_m-\mathrm{T}}(\boldsymbol{q}_m^{N_m})]\\
\boldsymbol{\tau}_m &= [\boldsymbol{\tau}_m^{1\mathrm{T}}, \boldsymbol{\tau}_m^{2\mathrm{T}}, \cdots, \boldsymbol{\tau}_m^{N_m\mathrm{T}}]^{\mathrm{T}}
\end{aligned}$$

5.2.2 Dynamics of the slave robots

The slave robots consist of N_s mobile manipulators with the same degree-of-freedom carrying a common object. Each slave mobile robot is assumed to be driven by two wheels. Decompose $\boldsymbol{q}_s^i$ into $\boldsymbol{q}_s^i = [\boldsymbol{q}_{sv}^{i\mathrm{T}}, \boldsymbol{q}_{sa}^{i\mathrm{T}}]^{\mathrm{T}}$, where $\boldsymbol{q}_{sv}^i \in \mathbb{R}^{n_{sv}}$ is the generalized coordinates of the mobile platform, $\boldsymbol{q}_{sa}^i \in \mathbb{R}^{n_{sa}}$ is the joint position vector of the manipulator, and have $n_s = n_{sv} + n_{sa}$. Because the wheeled mobile platform can only move in the direction perpendicular to the driving wheel, it is subject to nonholonomic constraints. The l independent nonholonomic constraints of the ith wheeled mobile manipulator can be expressed as

$$\boldsymbol{A}_s^i(\boldsymbol{q}_{sv}^i)\dot{\boldsymbol{q}}_{sv}^i = 0 \tag{5.5}$$

where $\boldsymbol{A}_s^i(\boldsymbol{q}_{sv}^i)$ is the nonholonomic constraint matrix with rank l, which is defined by $\boldsymbol{A}_s^i(\boldsymbol{q}_{sv}^i) = [\boldsymbol{A}_{s1}^{i\mathrm{T}}(\boldsymbol{q}_{sv}^i), \boldsymbol{A}_{s2}^{i\mathrm{T}}(\boldsymbol{q}_{sv}^i), \cdots, \boldsymbol{A}_{sl}^{i\mathrm{T}}(\boldsymbol{q}_{sv}^i)]^{\mathrm{T}} : \mathbb{R}^{n_{sv}} \to \mathbb{R}^{l \times n_{sv}}$. The null space of $\boldsymbol{A}_s^i(\boldsymbol{q}_{sv}^i)$ is spanned by $n_{sv} - l$ vectors, $\boldsymbol{\Delta}_s^i = \mathrm{span}\{\boldsymbol{H}_{s1}^i(\boldsymbol{q}_{sv}^i), \boldsymbol{H}_{s2}^i(\boldsymbol{q}_{sv}^i), \cdots, \boldsymbol{H}_{s(n_{sv}-l)}^i(\boldsymbol{q}_{sv}^i)\}$. Let $\boldsymbol{H}_s^i(\boldsymbol{q}_{sv}^i) = [\boldsymbol{H}_{s1}^i(\boldsymbol{q}_{sv}^i), \boldsymbol{H}_{s2}^i(\boldsymbol{q}_{sv}^i), \cdots, \boldsymbol{H}_{s(n_{sv}-l)}^i(\boldsymbol{q}_{sv}^i)] \in \mathbb{R}^{n_{sv} \times (n_{sv}-l)}$, and there exists vector $\boldsymbol{z}_{si} \in \mathbb{R}^{n_{sv}-l}$ such that

$$\dot{\boldsymbol{q}}_{sv}^i = \boldsymbol{H}_s^i(\boldsymbol{q}_{sv}^i)\dot{\boldsymbol{z}}_s^i \tag{5.6}$$

Further, $\boldsymbol{H}_s^{i\mathrm{T}}(\boldsymbol{q}_{sv}^i)\boldsymbol{A}_s^{i\mathrm{T}}(\boldsymbol{q}_{sv}^i) = 0$.

As the mobile manipulator consists of the mobile platform and the manipulator, each vector and matrix in (5.1) can be expressed as follows:

$$\boldsymbol{M}_s^i(\boldsymbol{q}_s^i) = \begin{bmatrix} \boldsymbol{M}_{sv}^i & \boldsymbol{M}_{sva}^i \\ \boldsymbol{M}_{sav}^i & \boldsymbol{M}_{sa}^i \end{bmatrix}, \ \boldsymbol{C}_s^i(\boldsymbol{q}_s^i, \dot{\boldsymbol{q}}_s^i) = \begin{bmatrix} \boldsymbol{C}_{sv}^i & \boldsymbol{C}_{sva}^i \\ \boldsymbol{C}_{sav}^i & \boldsymbol{C}_{sa}^i \end{bmatrix}$$

$$\boldsymbol{g}_s^i(\boldsymbol{q}_s^i) = \begin{bmatrix} \boldsymbol{g}_{sv}^i \\ \boldsymbol{g}_{sa}^i \end{bmatrix}, \ \boldsymbol{D}_s^i(t) = \begin{bmatrix} \boldsymbol{D}_{sv}^i \\ \boldsymbol{D}_{sa}^i \end{bmatrix}, \ \boldsymbol{\tau}_s^i(t) = \begin{bmatrix} \boldsymbol{\tau}_{sv}^i \\ \boldsymbol{\tau}_{sa}^i \end{bmatrix}$$

$$\boldsymbol{J}_s^i(\boldsymbol{q}_s^i) = \begin{bmatrix} \boldsymbol{A}_s^i & 0 \\ \boldsymbol{J}_{sv}^i & \boldsymbol{J}_{sa}^i \end{bmatrix}, \ \boldsymbol{f}_e^i = \begin{bmatrix} \boldsymbol{f}_{ev}^i \\ \boldsymbol{f}_{ee}^i \end{bmatrix}, \ \boldsymbol{B}_s^i(\boldsymbol{q}_s^i) = \begin{bmatrix} \boldsymbol{B}_{sv}^i & 0 \\ 0 & \boldsymbol{B}_{sa}^i \end{bmatrix}$$

where $\boldsymbol{f}_{ev}^i$ and $\boldsymbol{f}_{ea}^i$ are the constrained forces corresponding to nonholonomic constraints and holonomic constraints, respectively. The complete constrained force is measured by the force sensor installed on the end effector of the robotic manipulator.

From (5.6) and its derivative, the dynamics of slave robots in (5.1) can be given as

$$\mathbb{M}_s^i(\boldsymbol{\zeta}_s^i)\ddot{\boldsymbol{\zeta}}_s^i + \mathbb{C}_s^i(\boldsymbol{\zeta}_s^i, \dot{\boldsymbol{\zeta}}_s^i)\dot{\boldsymbol{\zeta}}_s^i + \mathbb{G}_s^i(\boldsymbol{\zeta}_s^i) + \mathbb{D}_s^i = \mathbb{B}_s^i\boldsymbol{\tau}_s^i + \mathbb{J}_s^{i\mathrm{T}}\boldsymbol{f}_{ee}^i$$

where $\boldsymbol{\zeta}_s^i = [\boldsymbol{z}_s^{i\mathrm{T}}, \boldsymbol{q}_{sa}^{i\mathrm{T}}]^{\mathrm{T}} \in \mathbb{R}^{n_s - l}$,

$$\mathbb{M}_s^i(\boldsymbol{\zeta}_s^i) = \begin{bmatrix} \boldsymbol{H}_s^{i\mathrm{T}}\boldsymbol{M}_{sv}^i\boldsymbol{H}_s^i & \boldsymbol{H}_s^{i\mathrm{T}}\boldsymbol{M}_{sva}^i \\ \boldsymbol{M}_{sav}^i\boldsymbol{H}_s^i & \boldsymbol{M}_{sa}^i \end{bmatrix}$$

$$\mathbb{C}_s^i(\boldsymbol{\zeta}_s^i, \dot{\boldsymbol{\zeta}}_s^i) = \begin{bmatrix} \boldsymbol{H}_s^{i\mathrm{T}}\boldsymbol{M}_{sv}^i\dot{\boldsymbol{H}}_s^i + \boldsymbol{H}_s^{i\mathrm{T}}\boldsymbol{C}_{sv}^i\boldsymbol{H}_s^i & \boldsymbol{H}_s^{i\mathrm{T}}\boldsymbol{C}_{sva}^i \\ \boldsymbol{M}_{sav}^i\dot{\boldsymbol{H}}_s^i + \boldsymbol{C}_{sav}^i\boldsymbol{H}_s^i & \boldsymbol{C}_{sa}^i \end{bmatrix}$$

$$\mathbb{G}_s^i(\boldsymbol{\zeta}_s^i) = \begin{bmatrix} \boldsymbol{H}_s^{i\mathrm{T}} \boldsymbol{G}_{sv}^i \\ \boldsymbol{G}_{sa}^i \end{bmatrix}, \ \mathbb{D}_s^i = \begin{bmatrix} \boldsymbol{H}_s^{i\mathrm{T}} \boldsymbol{D}_{sv}^i \\ \boldsymbol{D}_{sa}^i \end{bmatrix}$$

$$\mathbb{B}_s^i = \begin{bmatrix} \boldsymbol{H}_s^{i\mathrm{T}} \boldsymbol{B}_{sv}^i & 0 \\ 0 & \boldsymbol{B}_{sa}^i \end{bmatrix}, \ \mathbb{J}_s^{i\mathrm{T}} = \begin{bmatrix} \boldsymbol{H}_s^{i\mathrm{T}} \boldsymbol{J}_{sv}^{i\mathrm{T}} \\ \boldsymbol{J}_{sa}^{i\mathrm{T}} \end{bmatrix}$$

Furthermore, one can write N_s dynamics equations of slave robots into lumped form and then get

$$\mathbb{M}_s(\boldsymbol{\zeta}_s)\ddot{\boldsymbol{\zeta}}_s + \mathbb{C}_s(\boldsymbol{\zeta}_s, \dot{\boldsymbol{\zeta}}_s)\dot{\boldsymbol{\zeta}}_s + \mathbb{G}_s(\boldsymbol{\zeta}_s) + \mathbb{D}_s = \bar{\boldsymbol{\tau}}_s + \mathbb{J}_s^{\mathrm{T}} \boldsymbol{F}_e \tag{5.7}$$

where $\boldsymbol{\zeta}_s = [\boldsymbol{\zeta}_s^{1\mathrm{T}}, \boldsymbol{\zeta}_s^{2\mathrm{T}}, \cdots, \boldsymbol{\zeta}_s^{N_s\mathrm{T}}]^{\mathrm{T}}$, $\boldsymbol{F}_e = [\boldsymbol{f}_{ee}^{1\mathrm{T}}, \boldsymbol{f}_{ee}^{2\mathrm{T}}, \cdots, \boldsymbol{f}_{ee}^{N_s\mathrm{T}}]^{\mathrm{T}}$,

$$\begin{aligned}
\mathbb{M}_s(\boldsymbol{\zeta}_s) &= \text{block diag}[\mathbb{M}_s^1(\boldsymbol{\zeta}_s^1), \mathbb{M}_s^2(\boldsymbol{\zeta}_s^2), \cdots, \mathbb{M}_s^{N_s}(\boldsymbol{\zeta}_s^{N_s})] \\
\mathbb{C}_s(\boldsymbol{\zeta}_s, \dot{\boldsymbol{\zeta}}_s) &= \text{block diag}[\mathbb{C}_s^1(\boldsymbol{\zeta}_s^1, \dot{\boldsymbol{\zeta}}_s^1), \mathbb{C}_s^2(\boldsymbol{\zeta}_s^2, \dot{\boldsymbol{\zeta}}_s^2), \cdots, \mathbb{C}_s^{N_s}(\boldsymbol{\zeta}_s^{N_s}, \dot{\boldsymbol{\zeta}}_s^{N_s})] \\
\mathbb{G}_s(\boldsymbol{\zeta}_s) &= [\mathbb{G}_s^{1\mathrm{T}}(\boldsymbol{\zeta}_s^1), \mathbb{G}_s^{2\mathrm{T}}(\boldsymbol{\zeta}_s^2), \cdots, \mathbb{G}_s^{N_s\mathrm{T}}(\boldsymbol{\zeta}_s^{N_s})]^{\mathrm{T}} \\
\mathbb{D}_s &= [\mathbb{D}_s^{1\mathrm{T}}, \mathbb{D}_s^{2\mathrm{T}}, \cdots, \mathbb{D}_s^{N_s\mathrm{T}}]^{\mathrm{T}} \\
\mathbb{J}_s^{\mathrm{T}} &= \text{block diag}[\mathbb{J}_s^{1\mathrm{T}}, \mathbb{J}_s^{2\mathrm{T}}, \cdots, \mathbb{J}_s^{N_s\mathrm{T}}] \\
\bar{\boldsymbol{\tau}}_s &= [(\mathbb{B}_s^1 \boldsymbol{\tau}_s^1)^{\mathrm{T}}, (\mathbb{B}_s^2 \boldsymbol{\tau}_s^2)^{\mathrm{T}}, \cdots, (\mathbb{B}_s^{N_s} \boldsymbol{\tau}_s^{N_s})^{\mathrm{T}}]^{\mathrm{T}}
\end{aligned}$$

Denote the coordinate vector of the centroid of the rigid object grabbed by the slave robot as $\boldsymbol{X}_s \in \mathbb{R}^{n_o}$. Similar to Li's work [1], the motion of rigid body is assumed to satisfy

$$\boldsymbol{M}_o(\boldsymbol{X}_s)\ddot{\boldsymbol{X}}_s + \boldsymbol{C}_o(\boldsymbol{X}_s, \dot{\boldsymbol{X}}_s)\dot{\boldsymbol{X}}_s + \boldsymbol{g}_o(\boldsymbol{X}_s) = \boldsymbol{F}_o \tag{5.8}$$

where $\boldsymbol{M}_o(\boldsymbol{X}_s) \in \mathbb{R}^{n_o \times n_o}$, $\boldsymbol{C}_o(\boldsymbol{X}_s, \dot{\boldsymbol{X}}_s) \in \mathbb{R}^{n_o \times n_o}$ and $\boldsymbol{g}_o(\boldsymbol{X}_s) \in \mathbb{R}^{n_o}$ are the symmetric positive definite inertia matrix, centrifugal force and Coriolis force matrix and gravity vector of the object, respectively. $\boldsymbol{F}_o \in \mathbb{R}^{n_o}$ is the force on the mass center of the object. As the mobile manipulator is non-redundant, it can be known that $n_o = n_s - l$.

Define $\boldsymbol{J}_o(\boldsymbol{X}_s) \in \mathbb{R}^{N_s n_o \times n_o}$ as $\boldsymbol{J}_o(\boldsymbol{X}_s) = [\boldsymbol{J}_{o1}^{\mathrm{T}}(\boldsymbol{X}_s), \boldsymbol{J}_{o2}^{\mathrm{T}}(\boldsymbol{X}_s), \cdots, \boldsymbol{J}_{oN}^{\mathrm{T}}(\boldsymbol{X}_s)]^{\mathrm{T}}$, where $\boldsymbol{J}_{oi}(\boldsymbol{X}_s) \in \mathbb{R}^{n_o \times n_o}$ is a Jacobian matrix from the rigid object coordinate system $O_o X_o Y_o Z_o$ to the ith mobile manipulator end effector coordinate $O_{ei} X_{ei} Y_{ei} Z_{ei}$.

Then $\boldsymbol{F}_o = -\boldsymbol{J}_o^{\mathrm{T}}(\boldsymbol{X}_s)\boldsymbol{F}_e$. Further,

$$\boldsymbol{F}_e = -(\boldsymbol{J}_o^{\mathrm{T}}(\boldsymbol{X}_s))^{+}\boldsymbol{F}_o - \boldsymbol{\mathcal{F}}$$

where $\boldsymbol{\mathcal{F}}$ is the internal force in the null space of $\boldsymbol{J}_o^{\mathrm{T}}(\boldsymbol{X}_s)$, i.e., $\boldsymbol{J}_o^{\mathrm{T}}(\boldsymbol{X}_s)\boldsymbol{\mathcal{F}} = 0$.

The above formula shows that the end effector forms two orthogonal forces on the grasped object: $\boldsymbol{F}_o$ is used to push the object to move, $\boldsymbol{\mathcal{F}}$ generates internal force on the object. Similar to the previous work [4], by Lagrange multiplier $\boldsymbol{\lambda}_I$, $\boldsymbol{\mathcal{F}}$ can be parameterized to

$$\boldsymbol{\mathcal{F}} = (\boldsymbol{I} - (\boldsymbol{J}_o^{\mathrm{T}}(\boldsymbol{X}_s))^{+}\boldsymbol{J}_o^{\mathrm{T}}(\boldsymbol{X}_s))\boldsymbol{\lambda}_I$$

Let $\mathfrak{J}^{\mathrm{T}} = \boldsymbol{I} - (\boldsymbol{J}_o^{\mathrm{T}}(\boldsymbol{X}_s))^{+}\boldsymbol{J}_o^{\mathrm{T}}(\boldsymbol{X}_s)$ denote the Jacobian matrix of internal force, which satisfies $\boldsymbol{J}_o^{\mathrm{T}}(\boldsymbol{X}_s)\mathfrak{J}^{\mathrm{T}} = 0$. From (5.8), it has

$$\boldsymbol{F}_e = -(\boldsymbol{J}_o^{\mathrm{T}}(\boldsymbol{X}_s))^{+}\left(\boldsymbol{M}_o(\boldsymbol{X}_s)\ddot{\boldsymbol{X}}_s + \boldsymbol{C}_o(\boldsymbol{X}_s, \dot{\boldsymbol{X}}_s)\dot{\boldsymbol{X}}_s + \boldsymbol{g}_o(\boldsymbol{X}_s)\right) - \mathfrak{J}^{\mathrm{T}}\boldsymbol{\lambda}_I \tag{5.9}$$

Let $\boldsymbol{x}_{sie} \in \mathbb{R}^{n_s - l}$ denote the position and orientation vector of the ith end-effector. Given that $\dot{\boldsymbol{x}}_{se}^i$ is related to $\dot{\boldsymbol{\zeta}}_s^i$ by the Jacobian matrix $\mathbb{J}_s^i$,

$$\dot{\boldsymbol{x}}_{se}^i = \mathbb{J}_s^i \dot{\boldsymbol{\zeta}}_s^i$$

and the the relationship between $\dot{\boldsymbol{x}}_{se}^i$ and $\dot{\boldsymbol{X}}_s$ is given by

$$\dot{\boldsymbol{x}}_{se}^i = \boldsymbol{J}_o^i(\boldsymbol{X}_s)\dot{\boldsymbol{X}}_s$$

and the relationship between the joint velocity of the ith manipulator and the velocity of the object can be expressed as

$$\mathbb{J}_s^i \dot{\boldsymbol{\zeta}}_s^i = \boldsymbol{J}_o^i(\boldsymbol{X}_s)\dot{\boldsymbol{X}}_s$$

According to Assumption 5.1, the slave manipulators work in a nonsingular region, thus

$$\begin{cases} \dot{\boldsymbol{\zeta}}_s = \mathbb{J}_s^{+}\boldsymbol{J}_o(\boldsymbol{X}_s)\dot{\boldsymbol{X}}_s \\ \ddot{\boldsymbol{\zeta}}_s = \mathbb{J}_s^{+}\boldsymbol{J}_o(\boldsymbol{X}_s)\ddot{\boldsymbol{X}}_s + \dfrac{\mathrm{d}}{\mathrm{d}t}\left(\mathbb{J}_s^{+}\boldsymbol{J}_o(\boldsymbol{X}_s)\right)\dot{\boldsymbol{X}}_s \end{cases}$$

Let $\boldsymbol{\mathcal{L}}_s = \mathbb{J}_s^{+}\boldsymbol{J}_o(\boldsymbol{X}_s) \in \mathbb{R}^{N_s n_s \times n_o}$, combining (5.7) and (5.8), using $\boldsymbol{J}_o^{\mathrm{T}}(\boldsymbol{X}_s)\mathfrak{J}^{\mathrm{T}} = 0$, thereby having

$$\boldsymbol{\mathcal{M}}_s\ddot{\boldsymbol{X}}_s + \boldsymbol{\mathcal{C}}_s\dot{\boldsymbol{X}}_s + \boldsymbol{\mathcal{G}}_s + \boldsymbol{\mathcal{D}}_s = \boldsymbol{\mathcal{L}}_s^{\mathrm{T}}\bar{\boldsymbol{\tau}}_s \tag{5.10a}$$

$$\boldsymbol{\lambda}_I = \boldsymbol{Z}_{\lambda s}\left(\bar{\boldsymbol{\tau}}_s - \boldsymbol{C}_s^*\dot{\boldsymbol{X}}_s - \boldsymbol{G}_s^* - \mathbb{D}_s\right) \tag{5.10b}$$

where

$$\begin{aligned}
\boldsymbol{\mathcal{M}}_s &= \boldsymbol{\mathcal{L}}_s^{\mathrm{T}}\mathbb{M}_s\boldsymbol{\mathcal{L}}_s + \boldsymbol{M}_o \\
\boldsymbol{\mathcal{C}}_s &= \boldsymbol{\mathcal{L}}_s^{\mathrm{T}}\left(\mathbb{M}_s\frac{\mathrm{d}}{\mathrm{d}t}\boldsymbol{\mathcal{L}}_s + \mathbb{C}_s\boldsymbol{\mathcal{L}}_s\right) + \boldsymbol{C}_o \\
\boldsymbol{\mathcal{G}}_s &= \boldsymbol{\mathcal{L}}_s^{\mathrm{T}}\mathbb{G}_s + \boldsymbol{g}_o \\
\boldsymbol{\mathcal{D}}_s &= \boldsymbol{\mathcal{L}}_s^{\mathrm{T}}\mathbb{D}_s \\
\boldsymbol{M}_s^* &= \mathbb{M}_s + \mathbb{J}_s^{\mathrm{T}}(\boldsymbol{J}_o^{\mathrm{T}})^{+}\boldsymbol{M}_o((\boldsymbol{J}_o^{\mathrm{T}})^{+})^{\mathrm{T}}\mathbb{J}_s \\
\boldsymbol{C}_s^* &= \mathbb{M}_s\frac{\mathrm{d}}{\mathrm{d}t}\boldsymbol{\mathcal{L}}_s + \mathbb{C}_s\boldsymbol{\mathcal{L}}_s + \mathbb{J}_s^{\mathrm{T}}(\boldsymbol{J}_o^{\mathrm{T}})^{+}\boldsymbol{C}_o \\
\boldsymbol{G}_s^* &= \mathbb{G}_s + \mathbb{J}_s^{\mathrm{T}}(\boldsymbol{J}_o^{\mathrm{T}})^{+}\boldsymbol{g}_o \\
\boldsymbol{Z}_{\lambda s} &= \left(\mathfrak{J}\mathbb{J}_s(\boldsymbol{M}_s^*)^{-1}\mathbb{J}_s^{\mathrm{T}}\mathfrak{J}^{\mathrm{T}}\right)^{-1}\mathfrak{J}\mathbb{J}_s(\boldsymbol{M}_s^*)^{-1}
\end{aligned}$$

Taking $\bar{\boldsymbol{\tau}}_s = (\boldsymbol{\mathcal{L}}_s^{\mathrm{T}})^{+}\boldsymbol{\tau}_a$, then,

$$\boldsymbol{\mathcal{M}}_s\ddot{\boldsymbol{X}}_s + \boldsymbol{\mathcal{C}}_s\dot{\boldsymbol{X}}_s + \boldsymbol{\mathcal{G}}_s + \boldsymbol{\mathcal{D}}_s = \boldsymbol{\tau}_a \tag{5.11a}$$

$$\boldsymbol{\lambda}_I = \boldsymbol{Z}_{\lambda s}\left((\boldsymbol{\mathcal{L}}_s^{\mathrm{T}})^{+}\boldsymbol{\tau}_a - \boldsymbol{C}_s^{*}\dot{\boldsymbol{X}}_s - \boldsymbol{G}_s^{*} - \mathbb{D}_s\right) + \boldsymbol{\tau}_f \tag{5.11b}$$

To achieve the motion synchronization between the master and the slave robots, it is assumed that the dimension of $\boldsymbol{X}_m$ is the same as the one of $\boldsymbol{X}_s$, i.e., $\sum_{i=1}^{N_m} n_{mi} = n_o$. In general, for the typical Lagrangian dynamics shown in (5.1), they have the following famous properties [2,5]:

Property 5.1. *For all $j \in \{m, s\}$, the inertia matrix $\boldsymbol{\mathcal{M}}_j$ is symmetric, positive definite and, bounded, i.e., there exist $\rho_{j1} > 0$ and $\rho_{j2} > 0$ such that $\rho_{j1}\boldsymbol{I} \le \boldsymbol{\mathcal{M}}_j \le \rho_{j2}\boldsymbol{I}$.*

Property 5.2. *The matrix $\dot{\boldsymbol{\mathcal{M}}}_j - 2\boldsymbol{\mathcal{C}}_j$ is skew symmetric.*

Property 5.3. *All Jacobian matrices are uniformly bounded and uniformly continuous if $\boldsymbol{\zeta}_s$ and $\boldsymbol{X}_s$ arc uniformly bounded and continuous, respectively.*

5.2.3 Motion control objectives

For the multi-master multi-slave asymmetric teleoperation system (5.1), we identify the following main control objective: ① In the task space, realize the motion synchronization of the master and slave robots; ② Realize the internal force control of the slave robot.

Specifically, as shown in Fig. 5.1, there exist the asymmetric time delays $d_m(t)$ and $d_s(t)$ in the forward and backward communication channels. In this section, the delays are assumed to satisfy:

Assumption 5.2. *There exist bounded constant positive scalars $\bar{d}$ and $\tilde{d}$ such that for any $j \in \{m, s\}$, $0 \le |d_j(t)| \le \bar{d}$ and $0 \le |\dot{d}_j(t)| \le \tilde{d}$.*

Define the coordination errors between the master and slave robots as

$$\begin{cases} \boldsymbol{e}_m(t) = \boldsymbol{X}_m(t) - \boldsymbol{X}_s(t - d_s(t)) \\ \boldsymbol{e}_{vm}(t) = \dot{\boldsymbol{X}}_m(t) - \boldsymbol{X}_{vds}(t, d_s(t)) \\ \boldsymbol{e}_s(t) = \boldsymbol{X}_s(t) - \boldsymbol{X}_m(t - d_m(t)) \\ \boldsymbol{e}_{vs}(t) = \dot{\boldsymbol{X}}_s(t) - \boldsymbol{X}_{vdm}(t, d_m(t)) \end{cases} \tag{5.12}$$

The goal of motion control is to design the controller such that the position tracking errors (5.12) can converge to a certain bound. In (5.12), $\boldsymbol{X}_{vdm}(t, d_m(t)) = \dot{\boldsymbol{X}}_m(\theta)|_{\theta = t - d_m(t)}$, $\boldsymbol{X}_{vds}(t, d_s(t)) = \dot{\boldsymbol{X}}_s(\theta)|_{\theta = t - d_s(t)}$.

Denote $\boldsymbol{\lambda}_{Id}$ be the desired internal force, and define the internal force tracking error as

$$\boldsymbol{e}_\lambda = \boldsymbol{\lambda}_I - \boldsymbol{\lambda}_{Id}$$

Then the control goal of internal force is to make $\boldsymbol{e}_\lambda$ converge to any small value.

5.3 MULTILATERAL COORDINATION CONTROLLER

Firstly, for any $j \in \{m, s\}$, define the auxiliary state variables as

$$\boldsymbol{\eta}_j = \dot{\boldsymbol{X}}_j + \lambda_j \boldsymbol{e}_j \tag{5.13}$$

where λ_j is a constant positive scalar. Let $\boldsymbol{Z}_j(t) = [\boldsymbol{X}_j^{\mathrm{T}}(t), \dot{\boldsymbol{X}}_j^{\mathrm{T}}(t), \boldsymbol{X}_k^{\mathrm{T}}(t - d_k(t)), \boldsymbol{X}_{vdk}^{\mathrm{T}}(t, d_k(t))]^{\mathrm{T}}$, where $k \in \{m, s\}$ but $k \neq j$. According to the knowledge of fuzzy logic system, there are optimal approximation parameters $\boldsymbol{\Theta}_j$ such that

$$\begin{aligned} \boldsymbol{G}_j'(Z_j) &= \boldsymbol{\mathcal{M}}_j \lambda_j \boldsymbol{e}_{vj} + \boldsymbol{\mathcal{C}}_j \lambda_j \boldsymbol{e}_j - \boldsymbol{\mathcal{G}}_j - \boldsymbol{\mathcal{D}}_j \\ &:= \boldsymbol{\Theta}_j^{\mathrm{T}} \boldsymbol{\varsigma}_j(\boldsymbol{Z}_j) + \boldsymbol{\epsilon}_j(\boldsymbol{Z}_j) \end{aligned}$$

where $\boldsymbol{\epsilon}_j(\boldsymbol{Z}_j)$ is the bounded approximation error.

To simplify expression, define $\mathcal{S}_2 = \{(m, s), (s, m)\}$. Substituting (5.13) into (5.4) and (5.11a), it has

$$\begin{cases} \boldsymbol{\mathcal{M}}_m \dot{\boldsymbol{\eta}}_m + \boldsymbol{\mathcal{C}}_m \boldsymbol{\eta}_m = \boldsymbol{G}_m'(\boldsymbol{Z}_m) + \boldsymbol{\mathcal{F}}_h + \boldsymbol{J}_m^{-T} \boldsymbol{\tau}_m + \lambda_m \dot{d}_s \boldsymbol{\mathcal{M}}_m \boldsymbol{X}_{vds}(t, d_s(t)) \\ \boldsymbol{\mathcal{M}}_s \dot{\boldsymbol{\eta}}_s + \boldsymbol{\mathcal{C}}_s \boldsymbol{\eta}_s = \boldsymbol{G}_s'(\boldsymbol{Z}_s) + \boldsymbol{\tau}_a + \lambda_s \dot{d}_m \boldsymbol{\mathcal{M}}_s \boldsymbol{X}_{vdm}(t, d_m(t)) \end{cases} \tag{5.14}$$

Then, design the control inputs $\boldsymbol{\tau}_m = \boldsymbol{J}_m^{\mathrm{T}} \bar{\boldsymbol{\tau}}_m$ and $\boldsymbol{\tau}_a$ as

$$\begin{cases} \bar{\boldsymbol{\tau}}_m = -K_{m1} \boldsymbol{\eta}_m - K_{m2} \boldsymbol{y}_m - \hat{\boldsymbol{G}}_m'(\boldsymbol{Z}_m) + \boldsymbol{F}_m \\ \boldsymbol{\tau}_a = -K_{s1} \boldsymbol{\eta}_s - K_{s2} \boldsymbol{y}_s - \hat{\boldsymbol{G}}_s'(\boldsymbol{Z}_s) + \boldsymbol{F}_s \end{cases} \tag{5.15}$$

where for any $j \in \{m, s\}$, K_{ji} $(i = 1, 2)$ are some positive constant scalars, $\boldsymbol{y}_j = \boldsymbol{y}_j(t)$ is the well-designed position error filter output which will be given in the sequel, $\boldsymbol{F}_j = \boldsymbol{F}_j(t)$ are the coordinating torques and the design will also be given in the following content. $\hat{\boldsymbol{G}}_j'(\boldsymbol{Z}_j)$ is used to approximate the function $\boldsymbol{G}_j'(\boldsymbol{Z}_j)$ with $\hat{\boldsymbol{G}}_j'(\boldsymbol{Z}_j) = \hat{\boldsymbol{\Theta}}_j^{\mathrm{T}} \boldsymbol{\varsigma}_j(\boldsymbol{Z}_j)$, where $\hat{\boldsymbol{\Theta}}_j$ is the matrix of the fuzzy adaptive parameters, $\boldsymbol{\varsigma}_j(\boldsymbol{Z}_j)$ is the vector denoting the known fuzzy basis functions.

Consequently, one can get

$$\begin{cases} \boldsymbol{\mathcal{M}}_m \dot{\boldsymbol{\eta}}_m + \boldsymbol{\mathcal{C}}_m \boldsymbol{\eta}_m = -K_{m1} \boldsymbol{\eta}_m - K_{m2} \boldsymbol{y}_m + \tilde{\boldsymbol{G}}_m'(\boldsymbol{Z}_m) + \\ \qquad\qquad \boldsymbol{F}_m + \boldsymbol{\mathcal{F}}_h + \lambda_m \dot{d}_s \boldsymbol{\mathcal{M}}_m \boldsymbol{X}_{vds}(t, d_s(t)) \\ \boldsymbol{\mathcal{M}}_s \dot{\boldsymbol{\eta}}_s + \boldsymbol{\mathcal{C}}_s \boldsymbol{\eta}_s = -K_{s1} \boldsymbol{\eta}_s - K_{s2} \boldsymbol{y}_s + \tilde{\boldsymbol{G}}_s'(\boldsymbol{Z}_s) + \boldsymbol{F}_s + \\ \qquad\qquad \lambda_s \dot{d}_m \boldsymbol{\mathcal{M}}_s \boldsymbol{X}_{vdm}(t, d_m(t)) \end{cases} \tag{5.16}$$

where $\tilde{\boldsymbol{G}}_j'(\boldsymbol{Z}_j) = \boldsymbol{G}_j'(\boldsymbol{Z}_j) - \hat{\boldsymbol{G}}_j'(\boldsymbol{Z}_j) = (\boldsymbol{\Theta}_j^{\mathrm{T}} - \hat{\boldsymbol{\Theta}}_j^{\mathrm{T}}) \boldsymbol{\varsigma}_j(\boldsymbol{Z}_j) + \boldsymbol{\epsilon}_j(\boldsymbol{Z}_j) = \tilde{\boldsymbol{\Theta}}_j^{\mathrm{T}} \boldsymbol{\varsigma}_j(\boldsymbol{Z}_j) + \boldsymbol{\epsilon}_j(\boldsymbol{Z}_j)$. Assume that $|\boldsymbol{\epsilon}_j(\boldsymbol{Z}_j)| \leq \omega_j$.

In (5.16), the coordinating torque $\boldsymbol{F}_j \in \mathbb{R}^{n_o}$ is designed as

$$\begin{cases} \boldsymbol{F}_m = -|\hat{\omega}_m| \mathrm{sgn}(\boldsymbol{\eta}_m) - K_{m3} |\boldsymbol{X}_{vds}(t, d_s(t))| \mathrm{sgn}(\boldsymbol{\eta}_m) \\ \boldsymbol{F}_s = -|\hat{\omega}_s| \mathrm{sgn}(\boldsymbol{\eta}_s) - K_{s3} |\boldsymbol{X}_{vdm}(t, d_m(t))| \mathrm{sgn}(\boldsymbol{\eta}_s) \end{cases} \tag{5.17}$$

where $K_{j3} > 0$, $\hat{\omega}_j$ is the estimation of ω_j.

In this chapter, the (fuzzy logic) adaptive laws are designed as

$$\begin{cases} \dot{\hat{\boldsymbol{\Theta}}}_j = \psi_{j1}\varsigma_j(\boldsymbol{Z}_j)\boldsymbol{\eta}_j^{\mathrm{T}}(t) - \psi_{j2}\hat{\boldsymbol{\Theta}}_j(t) \\ \dot{\hat{\omega}}_j = |\boldsymbol{\eta}_j| - \psi_{j3}\hat{\omega}_j \end{cases} \tag{5.18}$$

Denote $\tilde{\omega}_j = \hat{\omega}_j - \omega_j$.

In what follows, we give the design of the position errors filter output $\boldsymbol{y}_j(t)$. Firstly, for any $j \in \{m, s\}$, define $\boldsymbol{\mathcal{K}}_j(t) = \mathrm{diag}[\kappa_{j,1}(t), \kappa_{j,2}(t), \cdots, \kappa_{j,n_o}(t)]$ with

$$\kappa_{j,i}(t) = \begin{cases} 1, & y_{j,i}(t)e_{j,i}(t) > 0 \text{ or } y_{j,i}(t)e_{j,i}(t) = 0,\ y_{j,i}(t^-)e_{j,i}(t^-) \leq 0 \\ -1, & y_{j,i}(t)e_{j,i}(t) < 0 \text{ or } y_{j,i}(t)e_{j,i}(t) = 0,\ y_{j,i}(t^-)e_{j,i}(t^-) > 0 \end{cases} \tag{5.19}$$

And let $\boldsymbol{s}_j(t) = \boldsymbol{y}_j(t) + \boldsymbol{\mathcal{K}}_j(t)\boldsymbol{e}_j(t)$. Then for all $i = 1, 2, \cdots, n_o$, the auxiliary switched filter $\boldsymbol{y}_j(t)$ is designed as

$$\dot{y}_{j,i}(t) = \\ = \begin{cases} -P_{j1}s_{j,i}(t) - e_{vj,i}(t) - P_{j2}|\boldsymbol{X}_{vdk}(t, d_k(t))|\mathrm{sgn}(s_{j,i}(t)), & \kappa_{j,i}(t) = 1 \\ -P_{j1}s_{j,i}(t) + e_{vj,i}(t) - P_{j2}|\boldsymbol{X}_{vdk}(t, d_k(t))|\mathrm{sgn}(s_{j,i}(t)), & \kappa_{j,i}(t) = -1 \end{cases} \tag{5.20}$$

where $(j, k) \in \mathcal{S}_2$, $P_{j,\ell} > 0$ $(\ell = 1, 2)$ are some scalars. $y_{j,i}(t)$ and $s_{j,i}(t)$ are the i-th elements of $\boldsymbol{y}_j(t)$ and $\boldsymbol{s}_j(t)$, respectively.

In (5.20), the filter system $\boldsymbol{y}_j(t)$ also belongs to the switched system. According to Remark 3.2, there exists a standard switching signal $r_j : \mathbb{R}_+ \to \mathcal{S}$ such that $r_j(t)$ and $\boldsymbol{\mathcal{K}}_j(t)$ satisfy a one-to-one correspondence, where $\mathcal{S} = \{1, 2, \cdots, 2^{n_o}\}$. On the basis of r_j, the standard switched dynamic system about $\boldsymbol{y}_j(t)$ can be constructed as follows

$$\begin{aligned} \dot{\boldsymbol{y}}_j(t) = & -P_{j1}\left(\boldsymbol{y}_j(t) + \boldsymbol{\mathcal{K}}_j(t)\boldsymbol{e}_j(t)\right) - \boldsymbol{\mathcal{K}}_j(t)\boldsymbol{e}_{vj}(t) - \\ & P_{j2}\mathrm{sgn}\left(\boldsymbol{y}_j(t) + \boldsymbol{\mathcal{K}}_j(t)\boldsymbol{e}_j(t)\right)|\boldsymbol{X}_{vdk}(t, d_k(t))| \\ := & \boldsymbol{h}_1(\boldsymbol{y}_j(t), \boldsymbol{e}_j(t), \boldsymbol{e}_{vj}(t), \boldsymbol{u}_j(t), r_j(t)) \end{aligned} \tag{5.21}$$

where $\boldsymbol{u}_j(t) = -P_{j2}\mathrm{sgn}(\boldsymbol{s}_j(t))|\boldsymbol{X}_{vdk}(t, d_k(t))|$.

For any fixed $\boldsymbol{\mathcal{K}}_j(t)$ from (5.21), it has

$$\dot{\boldsymbol{s}}_j(t) = -P_{j1}\boldsymbol{s}_j(t) - P_{j2}\mathrm{sgn}(\boldsymbol{s}_j(t))|\boldsymbol{X}_{vdk}(t, d_k(t))| + \boldsymbol{\mathcal{K}}_j(t)\dot{d}_k(t)\boldsymbol{X}_{vdk}(t, d_k(t)) \tag{5.22}$$

5.4 STABILITY ANALYSIS

A complete closed-loop teleoperation system consists of two subsystems (5.16) and (5.21). Define

$$\boldsymbol{x}_j = [\boldsymbol{\eta}_j^{\mathrm{T}}, \boldsymbol{X}_j^{\mathrm{T}}, \|\tilde{\boldsymbol{\Theta}}_j\|_F, \tilde{\omega}_j, \boldsymbol{y}_j^{\mathrm{T}}, \boldsymbol{e}_j^{\mathrm{T}}]^{\mathrm{T}} \in \mathbb{R}^{4n_o+2}$$
$$\boldsymbol{z}_j = [\boldsymbol{\eta}_j^{\mathrm{T}}, \boldsymbol{y}_j^{\mathrm{T}}, \boldsymbol{e}_j^{\mathrm{T}}, \|\tilde{\boldsymbol{\Theta}}_j\|_F, \tilde{\omega}_j]^{\mathrm{T}} \in \mathbb{R}^{3n_o+2}$$

and let $\boldsymbol{x}_{jt}(\tau) = \{\boldsymbol{x}_j(t+\tau)|\tau \in [-\bar{d}, 0]\} \in C([-\bar{d}, 0]; \mathbb{R}^{4n_o+2})$, $\boldsymbol{z}_{jt}(\tau) = \{\boldsymbol{z}_j(t+\tau)|\tau \in [-\bar{d}, 0]\} \in C([-\bar{d}, 0]; \mathbb{R}^{3n_o+2})$. Then, based on the multiple Lyapunov-Krasovskii functionals method, the following conclusion is obtained.

Theorem 5.1. *For any $j \in \{m, s\}$, if there exist constant positive control gains K_{j1}, K_{j2}, K_{j3}, P_{j1}, P_{j2}, ψ_{j1}, ψ_{j2} and ψ_{j3}, such that*

$$\begin{cases} K_{m1} > \dfrac{1+K_{m2}}{2} + \vartheta_m \\ K_{s1} > \dfrac{K_{s2}}{2} + \vartheta_s \\ K_{j3} \geq \lambda_j \tilde{d} \rho_{j2} \\ \psi_{j2} > 2\psi_{j1}\vartheta_j \\ \psi_{j3} > 2\vartheta_j \\ P_{j1} > \dfrac{K_{j2}}{2} + \vartheta_j \\ P_{j2} \geq \tilde{d} \end{cases} \tag{5.23}$$

with $\vartheta_j > 0$, then the complete closed-loop teleoperation system is SIIOS.

Proof. Taking the Lyapunov-Krasovskii functionals as

$$\begin{aligned} V_j(\boldsymbol{x}_{jt}, r_j(t)) = & \frac{1}{2}\boldsymbol{\eta}_j^{\mathrm{T}} \boldsymbol{\mathcal{M}}_j \boldsymbol{\eta}_j + \frac{1}{2}\mathrm{Tr}(\tilde{\boldsymbol{\Theta}}_j^{\mathrm{T}} \psi_{j1}^{-1} \tilde{\boldsymbol{\Theta}}_j) + \\ & \frac{\vartheta_j}{2}\int_{-\bar{d}}^{0} \boldsymbol{z}_{jt}^{\mathrm{T}}(\tau)\left(\frac{-\tau}{\bar{d}} + \frac{2(\tau+\bar{d})}{\bar{d}}\right)\boldsymbol{z}_{jt}(\tau)\mathrm{d}\tau + \\ & \frac{1}{2}(\hat{\omega}_j - \omega_j)^2 + \frac{1}{2}\boldsymbol{s}_j^{\mathrm{T}}(t)\boldsymbol{s}_j(t) \end{aligned}$$

where $\vartheta_j > 0$.

Considering that

$$\frac{1}{2}\int_{-\bar{d}}^{0} \boldsymbol{z}_{jt}^{\mathrm{T}}(\tau)\left(\frac{-\tau}{\bar{d}} + \frac{2(\tau+\bar{d})}{\bar{d}}\right)\boldsymbol{z}_{jt}(\tau)\mathrm{d}\tau \leq \int_{-\bar{d}}^{0} \boldsymbol{z}_{jt}^{\mathrm{T}}(\tau)\boldsymbol{z}_{jt}(\tau)\mathrm{d}\tau$$

and from the definitions of $r_j(t)$ and $\boldsymbol{\mathcal{K}}_j(t)$, for any $\ell \in \mathcal{S}$, it holds

$$\begin{cases} \bar{\rho}_{j1}|\boldsymbol{z}_j(t)|^2 \leq V_j(\boldsymbol{x}_{jt}, \ell) \leq \bar{\rho}_{j2}\|\boldsymbol{z}_{jt}\|_{M_2}^2 \\ V_j(\boldsymbol{x}_{jt_{k+1}^j}, r_j(t_{k+1}^j)) \leq V_j(\boldsymbol{x}_{jt_{k+1}^j}, r_j(t_k^j)),\ \forall k \geq 0 \end{cases}$$

where $\bar{\rho}_{j2} = \max\left\{\dfrac{\rho_{j2}}{2}, \dfrac{1}{2\psi_{j1}}, 1, \vartheta_j\right\}$, $\bar{\rho}_{j1} = \min\left\{\dfrac{\rho_{j1}}{2}, \dfrac{1}{2\psi_{j1}}, \dfrac{1}{2}\right\}$, $\{t_k^j\}_{k\geq 0}$ is the switching time sequence of switching signal $r_j(t)$

Further, it has

$$D^+V_m(\boldsymbol{x}_{mt},\ell) \tag{5.24}$$
$$\begin{aligned}
&= \boldsymbol{\eta}_m^{\mathrm{T}}\boldsymbol{\mathcal{M}}_m\dot{\boldsymbol{\eta}}_m + \frac{1}{2}\boldsymbol{\eta}_m^{\mathrm{T}}\dot{\boldsymbol{\mathcal{M}}}_m\boldsymbol{\eta}_m - \mathrm{Tr}(\tilde{\boldsymbol{\Theta}}_m^{\mathrm{T}}\psi_{m1}^{-1}\dot{\hat{\boldsymbol{\Theta}}}_m)+\\
&\quad(\hat{\omega}_m-\omega_m)\dot{\hat{\omega}}_m + \vartheta_m\boldsymbol{z}_m^{\mathrm{T}}\boldsymbol{z}_m - \vartheta_m\boldsymbol{z}_{mt}^{\mathrm{T}}(-\bar{d})\boldsymbol{z}_{mt}(-\bar{d})-\\
&\quad\frac{\vartheta_m}{2\bar{d}}\int_{-\bar{d}}^{0}\boldsymbol{z}_{mt}^{\mathrm{T}}(\tau)\boldsymbol{z}_{mt}(\tau)\mathrm{d}\tau + \boldsymbol{s}_m^{\mathrm{T}}(t)\dot{\boldsymbol{s}}_m(t)\\
&\le \boldsymbol{\eta}_m^{\mathrm{T}}\boldsymbol{\mathcal{F}}_h - K_{m1}|\boldsymbol{\eta}_m|^2 - K_{m2}\boldsymbol{\eta}_m^{\mathrm{T}}\boldsymbol{y}_m + \boldsymbol{\eta}_m^{\mathrm{T}}\tilde{\boldsymbol{G}}_m'(\boldsymbol{Z}_m)-\\
&\quad\mathrm{Tr}[\tilde{\boldsymbol{\Theta}}_m^{\mathrm{T}}\psi_{m1}^{-1}(\psi_{m1}\varsigma_m(\boldsymbol{Z}_m)\boldsymbol{\eta}_m^{\mathrm{T}} - \psi_{m2}\hat{\boldsymbol{\Theta}}_m)]+\\
&\quad(\hat{\omega}_m-\omega_m)\dot{\hat{\omega}}_m - P_{m1}|\boldsymbol{s}_m|^2 + \boldsymbol{s}_m^{\mathrm{T}}\boldsymbol{u}_m + \boldsymbol{\eta}_m^{\mathrm{T}}\boldsymbol{F}_m+\\
&\quad\boldsymbol{\eta}_m^{\mathrm{T}}\lambda_m\dot{d}_s\boldsymbol{\mathcal{M}}_m\boldsymbol{X}_{vds}(t,d_s(t)) + \boldsymbol{s}_m^{\mathrm{T}}\boldsymbol{\mathcal{K}}_m(t)\dot{d}_s\boldsymbol{X}_{vds}(t,d_s(t))+\\
&\quad\vartheta_m\boldsymbol{z}_m^{\mathrm{T}}\boldsymbol{z}_m - \frac{\vartheta_m}{2\bar{d}}\int_{-\bar{d}}^{0}\boldsymbol{z}_{mt}^{\mathrm{T}}(\tau)\boldsymbol{z}_{mt}(\tau)\mathrm{d}\tau
\end{aligned} \tag{5.25}$$

On the one hand, from Holder inequality,

$$\begin{aligned}
&\boldsymbol{\eta}_m^{\mathrm{T}}\tilde{\boldsymbol{G}}_m'(\boldsymbol{Z}_m) - \mathrm{Tr}[\tilde{\boldsymbol{\Theta}}_m^{\mathrm{T}}\psi_{m1}^{-1}(\psi_{m1}\varsigma_m(\boldsymbol{Z}_m)\boldsymbol{\eta}_m^{\mathrm{T}} - \psi_{m2}\hat{\boldsymbol{\Theta}}_m)]\\
&\quad= \boldsymbol{\eta}_m^{\mathrm{T}}\boldsymbol{\epsilon}_m(\boldsymbol{Z}_m) - \frac{\psi_{m2}}{\psi_{m1}}\mathrm{Tr}[\tilde{\boldsymbol{\Theta}}_m^{\mathrm{T}}(\boldsymbol{\Theta}_m - \hat{\boldsymbol{\Theta}}_m - \boldsymbol{\Theta}_m)]\\
&\quad\le \omega_m|\boldsymbol{\eta}_m| - \frac{\psi_{m2}}{2\psi_{m1}}\|\tilde{\boldsymbol{\Theta}}_m\|_F^2 + \frac{\psi_{m2}}{2\psi_{m1}}\|\boldsymbol{\Theta}_m\|_F^2
\end{aligned} \tag{5.26}$$

and

$$\begin{aligned}
(\hat{\omega}_m-\omega_m)\dot{\hat{\omega}}_m &= (\hat{\omega}_m-\omega_m)|\boldsymbol{\eta}_m| - \psi_{m3}(\hat{\omega}_m-\omega_m)\hat{\omega}_m\\
&\le (\hat{\omega}_m-\omega_m)|\boldsymbol{\eta}_m| - \frac{\psi_{m3}}{2}\tilde{\omega}_m^2 + \frac{\psi_{m3}}{2}\omega_m^2
\end{aligned} \tag{5.27}$$

On the other hand, when $P_{m2} \ge \tilde{d}$, it has

$$\boldsymbol{\mathcal{K}}_m(t)\dot{d}_s(t)\boldsymbol{s}_m^{\mathrm{T}}(t)\boldsymbol{X}_{vds}(t,d_s(t)) + \boldsymbol{s}_m^{\mathrm{T}}(t)\boldsymbol{u}_m(t) \le 0 \tag{5.28}$$

In addition, when $K_{m3} \ge \lambda_m\tilde{d}\rho_{m2}$,

$$\begin{aligned}
&\boldsymbol{\eta}_m^{\mathrm{T}}\boldsymbol{F}_m + \boldsymbol{\eta}_m^{\mathrm{T}}\lambda_m\dot{d}_s\boldsymbol{\mathcal{M}}_m\boldsymbol{X}_{vds}(t,d_s(t))\\
&\quad\le -|\hat{\omega}_m||\boldsymbol{\eta}_m| - K_{m3}|\boldsymbol{\eta}_m||\boldsymbol{X}_{vds}(t,d_s(t))|+\\
&\qquad\lambda_m\tilde{d}\rho_{m2}|\boldsymbol{\eta}_m||\boldsymbol{X}_{vds}(t,d_s(t))|\\
&\quad\le -|\hat{\omega}_m||\boldsymbol{\eta}_m|
\end{aligned} \tag{5.29}$$

For simplicity, let $\bar{\boldsymbol{y}}_m(t) = [\boldsymbol{y}_m^{\mathrm{T}}(t), \boldsymbol{e}_m^{\mathrm{T}}(t)]^{\mathrm{T}}$. Then, considering the definitions of $\boldsymbol{s}_m(t)$ and $\boldsymbol{\mathcal{K}}_m(t)$,

$$|\bar{\boldsymbol{y}}_m(t)|^2 \le |\boldsymbol{s}_m(t)|^2 \le 2|\bar{\boldsymbol{y}}_m(t)|^2$$

Denote $\iota_{m1} = \min\left\{P_{m1} - \frac{K_{m2}}{2} - \vartheta_m, K_{m1} - \vartheta_m - \frac{1+K_{m2}}{2}, \frac{\psi_{m2}}{2\psi_{m1}} - \vartheta_m, \frac{\psi_{m3}}{2} - \vartheta_m, \frac{\vartheta_m}{2\bar{d}}\right\} > 0$. Substituting (5.26)$\sim$(5.29) into (5.24), one thus has

$$
\begin{aligned}
&D^+V_m(\boldsymbol{x}_{mt}, \ell) \\
&\quad \leq -\left(K_{m1} - \frac{1+K_{m2}}{2}\right)|\boldsymbol{\eta}_m|^2 - \frac{\psi_{m2}}{2\psi_{m1}}\|\tilde{\boldsymbol{\Theta}}_m\|_F^2 - \frac{\psi_{m3}}{2}\tilde{\omega}_m^2 - \\
&\quad P_{m1}|\bar{\boldsymbol{y}}_m|^2 - \frac{\vartheta_m}{2\bar{d}}\int_{-\bar{d}}^{0}\boldsymbol{z}_{mt}^{\mathrm{T}}(\tau)\boldsymbol{z}_{mt}(\tau)\mathrm{d}\tau + \vartheta_m|\boldsymbol{z}_m|^2 + \frac{K_{m2}}{2}|\boldsymbol{y}_m|^2 + \\
&\quad \frac{1}{2}|\boldsymbol{\mathcal{F}}_h|^2 + \frac{\psi_{m2}}{2\psi_{m1}}\|\boldsymbol{\Theta}_m\|_F^2 + \frac{\psi_{m3}}{2}\omega_m^2 \\
&\quad \leq -\left(K_{m1} - \frac{1+K_{m2}}{2} - \vartheta_m\right)|\boldsymbol{\eta}_m|^2 - \left(\frac{\psi_{m2}}{2\psi_{m1}} - \vartheta_m\right)\|\tilde{\boldsymbol{\Theta}}_m\|_F^2 - \\
&\quad \left(\frac{\psi_{m3}}{2} - \vartheta_m\right)\tilde{\omega}_m^2 - \left(P_{m1} - \frac{K_{m2}}{2} - \vartheta_m\right)|\boldsymbol{y}_m|^2 - (P_{m1} - \vartheta_m)|\boldsymbol{e}_m|^2 - \\
&\quad \frac{\vartheta_m}{2\bar{d}}\int_{-\bar{d}}^{0}\boldsymbol{z}_{mt}^{\mathrm{T}}(\tau)\boldsymbol{z}_{mt}(\tau)\mathrm{d}\tau + \frac{1}{2}|\boldsymbol{\mathcal{F}}_h|^2 + \frac{\psi_{m2}}{2\psi_{m1}}\|\boldsymbol{\Theta}_m\|_F^2 + \frac{\psi_{m3}}{2}\omega_m^2 \\
&\quad \leq -\iota_{m1}\|\boldsymbol{z}_{mt}\|_{M_2}^2 + \frac{1}{2}|\boldsymbol{\mathcal{F}}_h|^2 + \frac{\psi_{m2}}{2\psi_{m1}}\|\boldsymbol{\Theta}_m\|_F^2 + \frac{\psi_{m3}}{2}\omega_m^2
\end{aligned}
$$

Similarly, let $\iota_{s1} = \min\left\{K_{s1} - \frac{K_{s2}}{2} - \vartheta_s, \frac{\psi_{s2}}{2\psi_{s1}} - \vartheta_s, \frac{\psi_{s3}}{2} - \vartheta_s, P_{s1} - \frac{K_{s2}}{2} - \vartheta_s, \frac{\vartheta_s}{2\bar{d}}\right\} > 0$. When $K_{s3} \geq \lambda_s\tilde{d}\rho_{s2}$ and $P_{s2} \geq \tilde{d}$, it has

$$
D^+V_s(\boldsymbol{x}_{st}, \ell) \leq -\iota_{s1}\|\boldsymbol{z}_{st}\|_{M_2}^2 + \frac{\psi_{s2}}{2\psi_{s1}}\|\boldsymbol{\Theta}_s\|_F^2 + \frac{\psi_{s3}}{2}\omega_s^2
$$

Denote $\Delta_m = \left(\frac{1}{2}\|\boldsymbol{\mathcal{F}}_{h[0,\infty)}\|_\infty^2 + \frac{\psi_{m2}}{2\psi_{m1}}\|\boldsymbol{\Theta}_m\|_F^2 + \frac{\psi_{m3}}{2}\omega_m^2\right)^{\frac{1}{2}}$, $\Delta_s = \left(\frac{\psi_{s2}}{2\psi_{s1}}\|\boldsymbol{\Theta}_s\|_F^2 + \frac{\psi_{s3}}{2}\omega_s^2\right)^{\frac{1}{2}}$, there then exists $\varrho_j \in (0,1)$ such that for any $\|\boldsymbol{z}_{jt}\|_{M_2} \geq \frac{1}{\sqrt{\iota_{j1}\varrho_j}}\Delta_j$,

$$
\begin{aligned}
D^+V_j(\boldsymbol{x}_{jt}, \ell) &\leq -\iota_{j1}(1-\varrho_j)\|\boldsymbol{z}_{jt}\|_{M_2}^2 \\
&\leq -\frac{\iota_{j1}(1-\varrho_j)}{\bar{\rho}_{j2}}V_j(\boldsymbol{x}_{jt}, \ell)
\end{aligned}
$$

where $j \in \{m, s\}$.

Finally, from Lemma 2.4, we can obtain the conclusion. □

Corollary 5.1. *Under the assumption of Theorem 5.1, when the auxiliary force control input is taken as*

$$
\boldsymbol{\tau}_f = \boldsymbol{\lambda}_{Id} - K_f\boldsymbol{e}_\lambda, \ K_f > 0 \tag{5.30}
$$

the force tracking error $\boldsymbol{e}_\lambda$ will remain bounded, and its upper bound can be adjusted by the control gain K_f.

Proof. Substituting (5.30) into (5.11b), it has

$$(1 + K_f)\boldsymbol{e}_\lambda = \boldsymbol{Z}_{\lambda s}(\boldsymbol{\mathcal{L}}_s^{\mathrm{T}})^{+}\boldsymbol{\mathcal{M}}_s\ddot{\boldsymbol{X}}_s \tag{5.31}$$

From Theorem 5.1, Assumption 5.2 and Property 5.1, it can be verified that $\dot{\boldsymbol{\eta}}_s$ is bounded. And further $\ddot{\boldsymbol{X}}_s$ is also bounded. From Property 5.3, the right terms in (5.31) will also remain bounded. There the force error is bounded. Obviously, $\boldsymbol{e}_\lambda$ can be adjusted by the control gain K_f. □

Remark 5.1. *Li et al. propose a cooperative teleoperation control method of multi-mobile robots based on neural network adaptive control* [1]*. However, it requires the human operator's force to be completely measurable, and it is regarded as a kind of "disturbance", which is compensated and cancelled by the controller. It has some application limitations. On the one hand, the robot needs to be equipped with force sensors, which increases the technical and economic burden of the equipment. On the other hand, for the teleoperation system, the external force of the human operator is generally used to drive the main robot to realize certain movements, and when it is compensated by the controller, this function may not be realized. In the switched control scheme proposed in this book, by introducing the compensation control strategy, the possible limitations in the previous work* [1] *are avoided.*

5.5 SIMULATION STUDY

Let us consider the four-master-four-slave asymmetric teleoperation system.

In the simulation, it is assumed that the master robot is four identical one-degree-of-freedom links, and its dynamics is described as $\mathcal{M}_m^i\ddot{X}_m^i + \mathcal{D}_m^i = F_h^i + \tau_m^i$, where $i = 1,2,3,4$, $\mathcal{M}_m^i$ is a positive scalar, $\mathcal{D}_m^i$ is the disturbance, F_h^i and τ_m^i are the external force and control input of the human operator, respectively.

The slave robot is also assumed to be composed of four identical mobile robotic manipulators, and the dynamics is described as $\boldsymbol{M}_s^i(\boldsymbol{q}_s^i)\ddot{\boldsymbol{q}}_s^i + \boldsymbol{C}_s^i(\boldsymbol{q}_s^i,\dot{\boldsymbol{q}}_s^i)\dot{\boldsymbol{q}}_s^i + \boldsymbol{g}_s^i(\boldsymbol{q}_s^i) + \boldsymbol{D}_s^i = \boldsymbol{B}_s^i(\boldsymbol{q}_s^i)\boldsymbol{\tau}_s^i + \boldsymbol{J}_s^{i\mathrm{T}}(\boldsymbol{q}_s^i)\boldsymbol{f}_e^i$. The ith slave robotic manipulator is shown as Fig. 5.2, and more details can be seen in the reference [4]. In this section, it is assumed that each slave robot is subject to the following constraints: $\dot{x}_s^i\sin(\theta_{si}) - \dot{y}_s^i\cos(\theta_{si}) = 0$, where $\boldsymbol{q}_{siv} = [x_s^i, y_s^i, \theta_s^i]^{\mathrm{T}}$, $\boldsymbol{q}_{sia} = [\theta_{si1}, \theta_{si2}]^{\mathrm{T}}$, $\boldsymbol{A}_s^i = [\sin(\theta_{si}), -\cos(\theta_{si}), 0]$, $\boldsymbol{\zeta}_s^i = [y_s^i, \theta_{si}, \theta_{si1}, \theta_{si2}]^{\mathrm{T}}$. The position of the ith slave robot end effector is defined as $\boldsymbol{x}_{se}^i = [x_s^{ie}, y_s^{ie}, z_s^{ie}, \beta_s^{ie}]^{\mathrm{T}}$, where β_s^{ie} is the pitch angle of the ith slave robot end effector, $x_s^{ie} = x_{sf}^i - 2l_{si2}\sin(\theta_{si2})\cos(\theta_{si}+\theta_{si1})$, $y_s^{ie} = y_{sf}^i - 2l_{si2}\sin(\theta_{si2})\sin(\theta_{si}+\theta_{si1})$, $z_s^{ie} = 2l_{si1} - 2l_{si2}\cos(\theta_{si2})$, $\beta_s^{ie} = \theta_{si} + \theta_{si1}$. The mobile platform is assumed

$$\begin{bmatrix} \dot{x}_{sf}^i \\ \dot{y}_{sf}^i \\ \dot{\theta}_{si} \end{bmatrix} = \begin{bmatrix} 1 & 0 & -d_{si}\sin(\theta_{si}) \\ 0 & 1 & d_{si}\cos(\theta_{si}) \\ 0 & 0 & 1 \end{bmatrix} \begin{bmatrix} \dot{x}_s^i \\ \dot{y}_s^i \\ \dot{\theta}_{si} \end{bmatrix}$$

$$\begin{bmatrix} \dot{x}_s^i \\ \dot{y}_s^i \\ \dot{\theta}_{si} \end{bmatrix} = \begin{bmatrix} \cos(\theta_{si}) & 0 \\ \sin(\theta_{si}) & 0 \\ 0 & 1 \end{bmatrix} \begin{bmatrix} v_{si} \\ \dot{\theta}_{si} \end{bmatrix}$$

Then the slave mobile manipulator Jacobian matrix $\mathbb{J}_s^i$ can be expressed as

$$\dot{\boldsymbol{x}}_{se}^i = \mathbb{J}_s^i[v_{si}, \dot{\theta}_{si}, \dot{\theta}_{si1}, \dot{\theta}_{si2}]^{\mathrm{T}}$$

where the elements of $\mathbb{J}_s^i$ are given by $\mathbb{J}_{s11}^i = \cos(\theta_{si})$, $\mathbb{J}_{s12}^i = -d_{si}\sin(\theta_{si}) + 2l_{si2}\sin(\theta_{si2})\sin(\theta_{si}+\theta_{si1})$, $\mathbb{J}_{s13}^i = 2l_{si2}\sin(\theta_{si2})\sin(\theta_{si}+\theta_{si1})$, $\mathbb{J}_{s14}^i = -2l_{si2}\cos(\theta_{si2})\times\cos(\theta_{si}+\theta_{si1})$, $\mathbb{J}_{s21}^i = \sin(\theta_{si})$, $\mathbb{J}_{s22}^i = d_{si}\cos(\theta_{si}) - 2l_{si2}\sin(\theta_{si2})\cos(\theta_{si}+\theta_{si1})$, $\mathbb{J}_{s23}^i = -2l_{si2}\sin(\theta_{si2})\cos(\theta_{si}+\theta_{si1})$, $\mathbb{J}_{s24}^i = -2l_{si2}\cos(\theta_{si2})\sin(\theta_{si}+\theta_{si1})$, $\mathbb{J}_{s31}^i = \mathbb{J}_{s31}^i = \mathbb{J}_{s31}^i = 0$, $\mathbb{J}_{s34}^i = 2l_{si2}\sin(\theta_{si2})$, $\mathbb{J}_{s41}^i = \mathbb{J}_{s44}^i = 0$, $\mathbb{J}_{s42}^i = \mathbb{J}_{s43}^i = 1$. Let $\boldsymbol{X}_s = [X_{s1}, X_{s2}, X_{s3}, X_{s4}]^{\mathrm{T}}$ be the positions to X-, Y-, Z-axis, and the rotation angle along Z axis, as shown in Fig. 5.1.

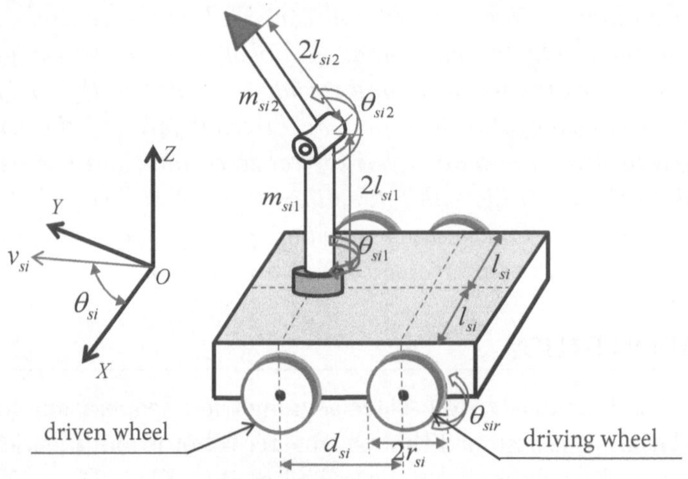

Figure 5.2 The ith slave mobile manipulator.

The dynamic equation of the object is given by

$$\boldsymbol{M}_o(\boldsymbol{X}_s)\ddot{\boldsymbol{X}}_s + \boldsymbol{g}_o(\boldsymbol{X}_s) = \sum_{i=1}^{4} \boldsymbol{J}_o^{i\mathrm{T}}(\boldsymbol{X}_s)\boldsymbol{f}_e^i$$

where $\boldsymbol{M}_o(\boldsymbol{X}_s) = \mathrm{diag}[m_o, m_o, m_o, I_o]$, $\boldsymbol{g}_o(\boldsymbol{X}_s) = [0, 0, -m_o g, 0]^{\mathrm{T}}$, $\boldsymbol{f}_e^i = [f_x^i, f_y^i, f_z^i, \tau_\beta^i]^{\mathrm{T}}$, and

$$\boldsymbol{J}_o^{1\mathrm{T}}(\boldsymbol{X}_s) = \boldsymbol{J}_o^{3\mathrm{T}}(\boldsymbol{X}_s) = \begin{bmatrix} 1 & 0 & 0 & 0 \\ 0 & 1 & 0 & 0 \\ 0 & 0 & 1 & 0 \\ l_{c1}\sin(X_{s4}) & -l_{c1}\cos(X_{s4}) & 0 & 1 \end{bmatrix}$$

$$\boldsymbol{J}_o^{2\mathrm{T}}(\boldsymbol{X}_s) = \boldsymbol{J}_o^{4\mathrm{T}}(\boldsymbol{X}_s) = \begin{bmatrix} 1 & 0 & 0 & 0 \\ 0 & 1 & 0 & 0 \\ 0 & 0 & 1 & 0 \\ -l_{c2}\sin(X_{s4}) & l_{c2}\cos(X_{s4}) & 0 & 1 \end{bmatrix}$$

In what follows, the true physical parameters of the two master manipulators are chosen as $\mathcal{M}_m^1 = 1.2$ N, $\mathcal{M}_m^2 = 0.6$ N, $\mathcal{M}_m^3 = 1.5$ N, $\mathcal{M}_m^4 = 0.8$ N, $g = 9.8$ m/s^2. In addition, similar to the existing work [4], the parameters of each slave mobile manipulators are chosen as: for any $i = 1, 2, 3, 4$, $m_{sip} = 6$ kg, $m_{si1} = m_{si2} = 1$ kg, $I_{sizp} = 19$ kg $\cdot$ m^2, $I_{siz1} = I_{siz2} = 1$ kg $\cdot$ m^2, $d_{si} = 0.6$ m, $l_{si} = r_{si} = 1$ m, $2l_{si1} = 1$ m, $2l_{si2} = 0.6$ m. The mass of the object $m_o = 1$ kg, $I_o = 1$ kg $\cdot$ m^2, $l_{c1} = l_{c2} = 0.5$ m. Given that $H_{si}^{\mathrm{T}} A_{si}^{\mathrm{T}} = 0$, let $\boldsymbol{H}_s^i = \begin{bmatrix} \cos(\theta_{si}) & 0 \\ \sin(\theta_{si}) & 0 \\ 0 & 1 \end{bmatrix}$. The disturbances for each robot are assumed to be $\mathcal{D}_m^i = \frac{2+i}{10} X_m^i \dot{X}_m^i \sin(t)$ and $\boldsymbol{D}_s^i = \begin{bmatrix} 0.2(\sin(t) + \cos(2t)) \\ 0.1\sin(2t + \frac{i}{10}\pi) \\ 0.3\cos(t) \\ 0.1(\sin(t) + \sin(2t - \frac{\pi}{3})) \\ 0.1\sin(2t) \end{bmatrix}$, where $i = 1, 2, 3, 4$.

In simulation, it is assumed that the time delays have the upper bounds $\bar{d} = 2$, $\tilde{d} = 3$. Then, by solving (5.23) in Theorem 5.1, one can get the desired control parameters: $\lambda_m = 0.5$, $\lambda_s = 1$, $\vartheta_m = 0.000\,5$, $\vartheta_s = 0.000\,1$, $K_{m1} = 13.2$, $K_{s1} = 6.6$, $K_{m2} = K_{s2} = 1$, $K_{m3} = 7.5$, $K_{s3} = 150$, $\psi_{m1} = \psi_{s1} = 10$, $\psi_{m2} = 0.011$, $\psi_{m3} = 0.002$, $\psi_{s2} = 0.003$, $\psi_{s3} = 0.002\,2$, $P_{m1} = 1.2$, $P_{s1} = 1.1$, $P_{m2} = P_{s2} = 3.5$, $\rho_{m2} = 5$, $\rho_{s2} = 50$.

The initial conditions for the complete closed-loop master/slave systems are given by $t \in [-\bar{d}, 0]$, $\boldsymbol{X}_m(t) = [0, 0, 0, 0]^{\mathrm{T}}$, $\boldsymbol{X}_s(t) = [-0.8, 0.9, 0.5, 1]^{\mathrm{T}}$, $\dot{\boldsymbol{X}}_m(t) = \dot{\boldsymbol{X}}_s(t) = \ddot{\boldsymbol{X}}_m(t) = \ddot{\boldsymbol{X}}_s(t) = [0, 0, 0, 0]^{\mathrm{T}}$, $\boldsymbol{y}_m(t) = [-20, 10, 30, -10]^{\mathrm{T}}$, $\boldsymbol{y}_s(t) = [20, 60, -40, 20]$, $\dot{\boldsymbol{y}}_m(t) = \dot{\boldsymbol{y}}_s(t) = [0, 0, 0, 0]^{\mathrm{T}}$, $\boldsymbol{\zeta}_s^1(t) = \boldsymbol{\zeta}_s^2(t) = \boldsymbol{\zeta}_s^3(t) = \boldsymbol{\zeta}_s^4(t) = [0, \frac{\pi}{2}, 0, \frac{3\pi}{2}]^{\mathrm{T}}$, $\dot{\boldsymbol{\zeta}}_s^1(t) = \dot{\boldsymbol{\zeta}}_s^2(t) = \dot{\boldsymbol{\zeta}}_s^3(t) = \dot{\boldsymbol{\zeta}}_s^4(t) = [0, 0, 0, 0]^{\mathrm{T}}$, $\hat{\boldsymbol{\Theta}}_m(t) = \begin{bmatrix} 0 & 0 & 0 & 0 \\ 0 & 0 & 0 & 0 \\ 0 & 0 & 0 & 0 \\ 0 & 0 & 0 & 0 \\ 0 & 0 & 0 & 0 \end{bmatrix}$, $\hat{\boldsymbol{\Theta}}_s(t) = \begin{bmatrix} 0 & 0 & 0 & 0 \\ 0 & 0 & 0 & 0 \\ 0 & 0 & 0 & 0 \\ 0 & 0 & 0 & 0 \\ 0 & 0 & 0 & 0 \end{bmatrix}$, $\hat{\omega}_m(t) = 3$, $\hat{\omega}_s(t) = 2$. In addition, for the fuzzy logic subsystem, $\boldsymbol{Z}_m = [\boldsymbol{X}_m^{\mathrm{T}}(t), \dot{\boldsymbol{X}}_m^{\mathrm{T}}(t), \boldsymbol{X}_{vds}^{\mathrm{T}}(t, d_s(t))]^{\mathrm{T}}$, $\boldsymbol{Z}_s = [\boldsymbol{X}_s^{\mathrm{T}}(t), \dot{\boldsymbol{X}}_s^{\mathrm{T}}(t), \boldsymbol{X}_m^{\mathrm{T}}(t - d_m(t)), \boldsymbol{X}_{vdm}^{\mathrm{T}}(t, d_m(t))]^{\mathrm{T}}$, and the membership function is designed as $\mu_{A_{i_k}^j}(Z_{ki_k}) = \mathrm{e}^{-\frac{(Z_{ki_k} - c_{i_k j})^2}{\sigma_{i_k j}^2}}$, with $\sigma_{i_k j} = 2$ and the centers $c_{i_k j}$ spaced evenly distributed to span the input space $[-2, 2]$, where $j = 1, 2, 3, 4, 5$, $k \in \{m, s\}$, $i_m = 1, 2, \cdots, 12$, $i_s = 1, 2, \cdots, 16$.

To verify the teleoperated performance under the proposed control algorithm, it is assumed that the human torques shown in Fig. 5.3 are applied respectively to the four master manipulators.

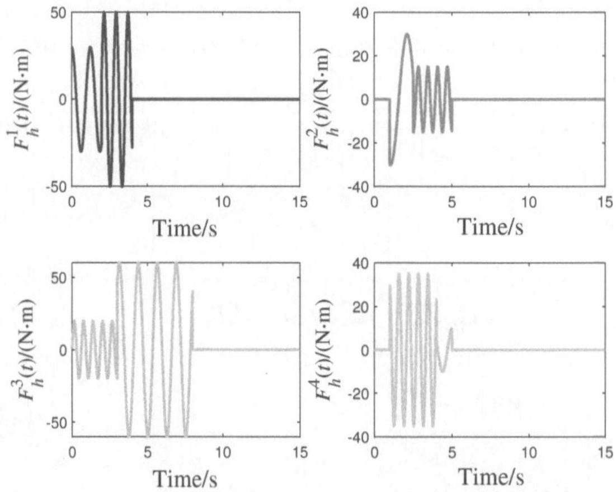

Figure 5.3 Generalized human torques F_h^i.

In the first set of simulation, the communication delays are assumed to be $d_m(t) = 0.1 + 0.5\cos^2(t)$ and $d_s(t) = 0.1 + 0.6\sin^2(t)$. According to the above simulation settings, the simulation results are shown as Figs. 5.4–5.11. Among them, Fig. 5.4 presents the position trajectories synchronization between the master and slave robots, where the corresponding tracking errors are given in Fig. 5.5. And as shown, the control algorithm can present a good position tracking performance. Figs. 5.6 and 5.7 present the estimated values of $\boldsymbol{\Theta}_j$ and ω_j. Obviously, both $\hat{\boldsymbol{\Theta}}_j(t)$ and $\hat{\omega}_j(t)$ are bounded. Considering that the boundedness of $\boldsymbol{\Theta}_j$ and ω_j, the parameter estimation errors $\tilde{\boldsymbol{\Theta}}_j(t)$ and $\tilde{\omega}_j(t)$ are also bounded. Fig. 5.8 shows the 2-norm of the output

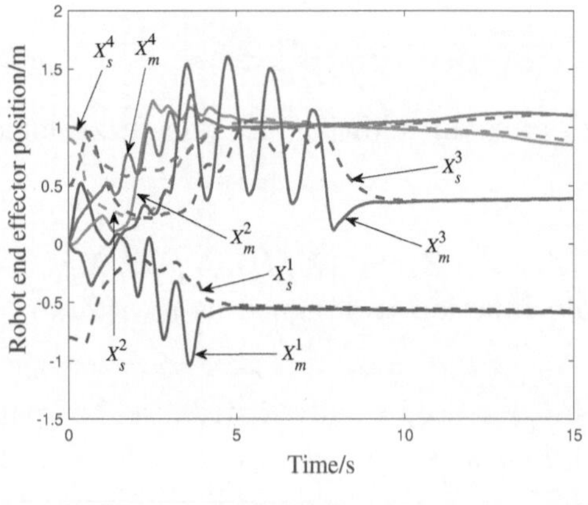

Figure 5.4 The response curves of the master and slave robots' positions.

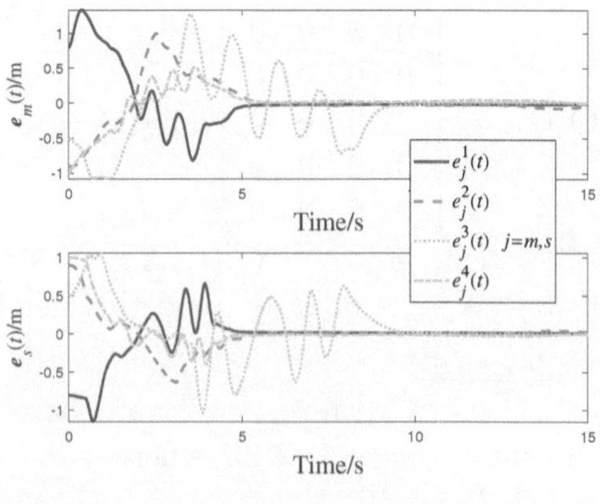

Figure 5.5 The position tracking errors between the master and slave robots.

$\bar{z}_j = [\eta_j^{\mathrm{T}}, e_j^{\mathrm{T}}, y_j^{\mathrm{T}}]$. From Figs. 5.6 to 5.8, it can be inferred that the norm of the complete output $z_j(t)$ is bounded, which is in accordance with the input- to-output stability analysis in Theorem 5.1. In the simulation, taking the desired internal force as $\boldsymbol{\lambda}_{Id} = [1, 0, 0]^{\mathrm{T}}$. Using the force controller (5.30) and taking a large value for K_f (e.g., $K_f = 10\ 000$), the actual internal force is shown as Fig. 5.9, which is in accordance with Corollary 5.1. Figs. 5.10 and 5.11 present the control inputs for the master and slave robots, respectively.

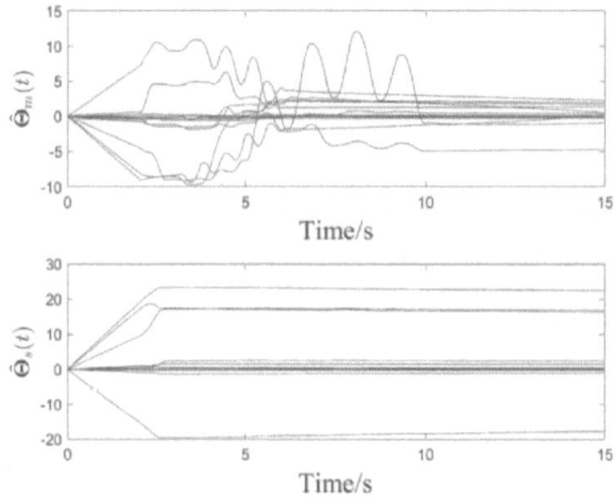

Figure 5.6 The fuzzy estimations $\hat{\Theta}_j(t)$.

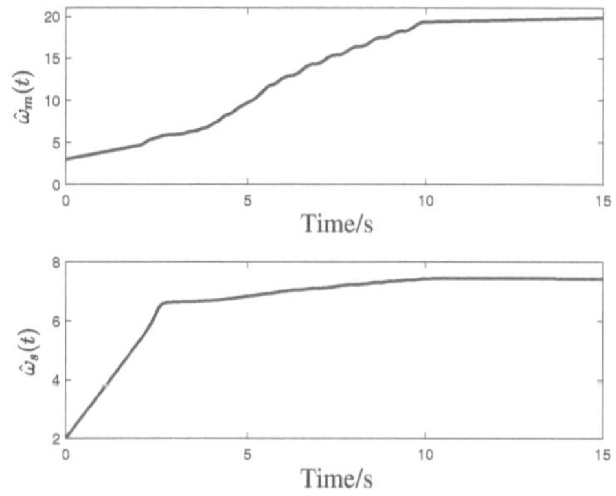

Figure 5.7 The estimation $\hat{\omega}_j(t)$.

For the comparison, to verify the tracking performance under the larger communication delays, the time delays are also set to be $d_m(t) = 0.1 + 1.8\cos^2(t)$ and $d_s(t) = 0.1 + 0.8\sin^2(t)$. In this case, the tracking performance is shown in Figs. 5.12 and 5.13. The slave robots can still track the master robots well.

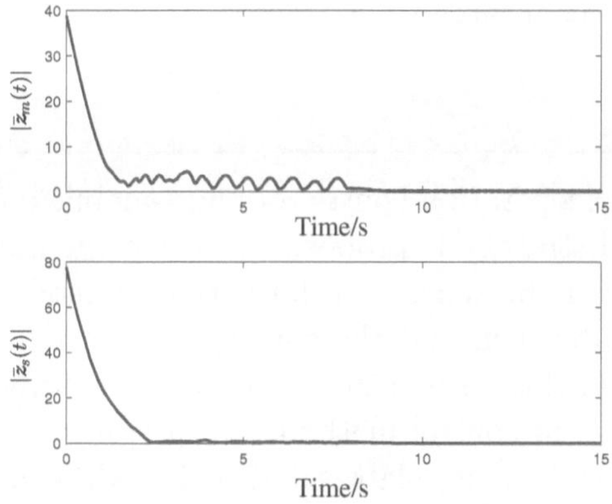

Figure 5.8 The response curves of $|\bar{z}_j(t)|$.

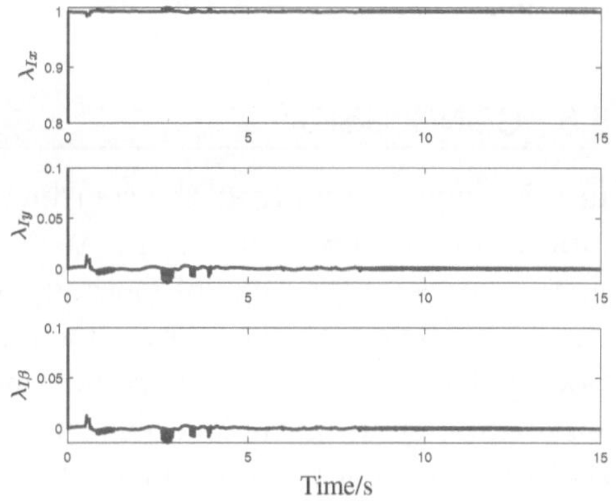

Figure 5.9 The internal force $\boldsymbol{\lambda}_I(t)$.

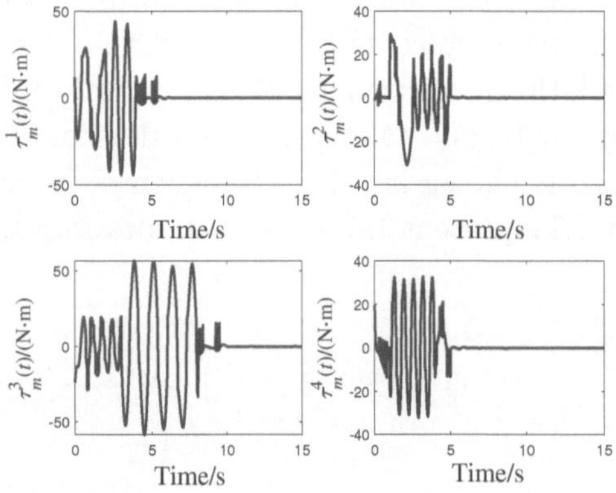

Figure 5.10 The control inputs for master robots.

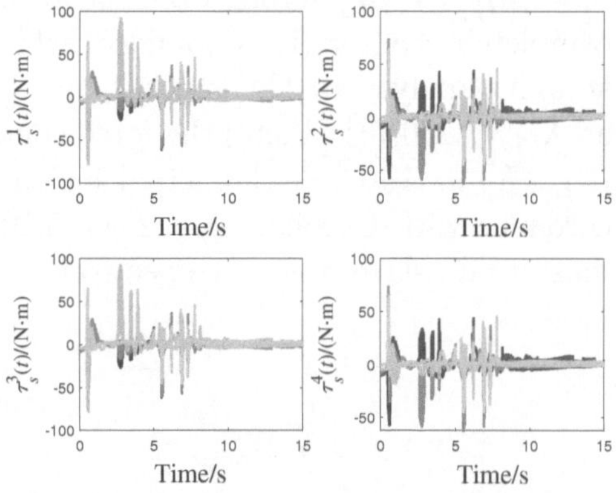

Figure 5.11 The control inputs for slave robots.

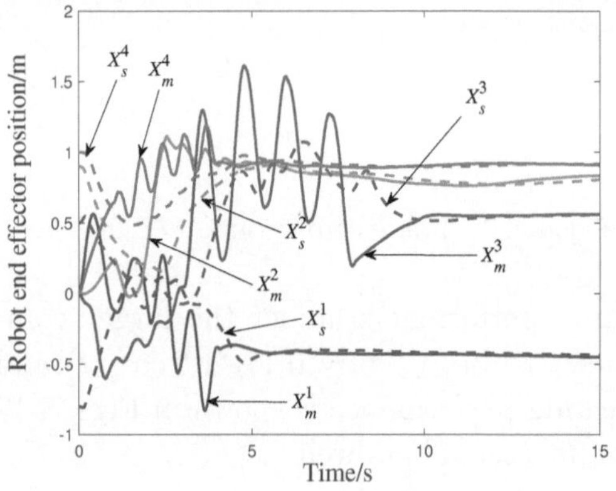

Figure 5.12 The master and slave robots' positions under large time delays.

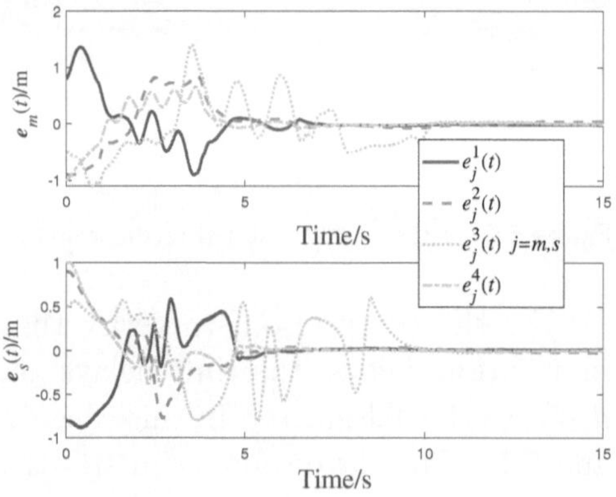

Figure 5.13 The position tracking errors under large time delays.

5.6 CONCLUSION

In this chapter, aiming at the coordinated teleoperation control of multi-mobile manipulators, an improved coordinated control strategy is proposed by using the auxiliary switched filter control method. By using the space transformation method [1], the decoupling space of motion/force is established, and the concise and clear motion model and force model are obtained. On this basis, a fuzzy adaptive controller is designed by using the auxiliary switched filter control method, which realizes the coordinated teleoperation control of multi-mobile manipulators and achieves less conservative results. All conclusions are verified by simulation.

Bibliography

[1] Z. Li and C.-Y. Su, "Neural-adaptive control of single-master–multiple-slaves teleoperation for coordinated multiple mobile manipulators with time-varying communication delays and input uncertainties," *IEEE Transactions on Neural Networks and Learning Systems*, vol. 24, no. 9, pp. 1400–1413, 2013.

[2] Z. Li, J. Li, and Y. Kang, "Adaptive robust coordinated control of multiple mobile manipulators interacting with rigid environments," *Automatica*, vol. 46, no. 12, pp. 2028–2034, 2010.

[3] J.-H. Jean and L.-C. Fu, "An adaptive control scheme for coordinated multimanipulator systems," *IEEE Transactions on Robotics and Automation*, vol. 9, no. 2, pp. 226–231, 1993.

[4] Z. Li, S. S. Ge, and Z. Wang, "Robust adaptive control of coordinated multiple mobile manipulators," *Mechatronics*, vol. 18, no. 5-6, pp. 239–250, 2008.

[5] R. Kelly, V. S. Davila, and J. A. L. Perez, *Control of Robot Manipulators in Joint Space*. Springer Science & Business Media, 2005.

CHAPTER 6

Multilateral tele-coordinated control

MOTIVATED by the significant demand for easy and efficient implementation, as well as a wider range of applications, a novel unconstrained tele-coordinated control approach is investigated for multilateral teleoperation system based on prescribed performance control and switching control technique, where the basic idea is to employ the concept of motion synchronization in each pair of master-slave robots and among all slave robots. By using the multiple Lyapunov-Krasovskii functionals method, the state-independent input-to-output practical stability of the closed-loop system is established. Compared with the existing approaches, the new design is straightforward and easier to implement and is applicable to a wider area.

6.1 INTRODUCTION

Chapter 5 presents the constrained multilateral coordinated teleoperation control method based on the auxiliary switched filter control framework. This chapter further studies the unconstrained coordinated teleoperation control design problem. Aiming at the unfavorable factors such as heavy workload of the operator and difficult manipulation in the traditional unconstrained coordinated teleoperation control, this chapter proposes a multilateral coordinated teleoperation control based on the high-performance motion synchronization of the slave robot. Inspired by the previous work [1], this chapter introduces the synchronous error control between slave robots in the traditional unconstrained coordinated control framework to ensure that the difference in operator manipulation and external uncertainties (different communication delays, model uncertainty) from robot coordination performance. Different from Sun's work [1], this chapter realizes high-performance synchronous control between slave robots by the prescribed performance control strategy. While improving the control performance, it reduces the operator's workload and control difficulty.

DOI: 10.1201/9781003382058-6

6.2 PROBLEM FORMULATION

The dynamics of a multilateral teleoperation system consisting of N pairs of n-DOF master and slave robots can be described as

$$\begin{cases} \boldsymbol{M}_m^i(\boldsymbol{q}_m^i)\ddot{\boldsymbol{q}}_m^i + \boldsymbol{C}_m^i(\boldsymbol{q}_m^i, \dot{\boldsymbol{q}}_m^i)\dot{\boldsymbol{q}}_m^i + \boldsymbol{g}_m^i(\boldsymbol{q}_m^i) \\ \qquad\qquad + \boldsymbol{F}_m^i(\boldsymbol{q}_m^i, \dot{\boldsymbol{q}}_m^i) = \boldsymbol{J}_m^{i\mathrm{T}}(\boldsymbol{q}_m^i)\boldsymbol{f}_h^i + \boldsymbol{\tau}_m^i \\ \boldsymbol{M}_s^i(\boldsymbol{q}_s^i)\ddot{\boldsymbol{q}}_s^i + \boldsymbol{C}_s^i(\boldsymbol{q}_s^i, \dot{\boldsymbol{q}}_s^i)\dot{\boldsymbol{q}}_s^i + \boldsymbol{g}_s^i(\boldsymbol{q}_s^i) \\ \qquad\qquad + \boldsymbol{F}_s^i(\boldsymbol{q}_s^i, \dot{\boldsymbol{q}}_s^i) = -\boldsymbol{J}_s^{i\mathrm{T}}(\boldsymbol{q}_s^i)\boldsymbol{f}_e^i + \boldsymbol{\tau}_s^i \end{cases} \tag{6.1}$$

where $i = 1, 2, \cdots, N$, the superscript i represents the ith robot, $\boldsymbol{F}_m^i(\boldsymbol{q}_m^i, \dot{\boldsymbol{q}}_m^i)$ and $\boldsymbol{F}_s^i(\boldsymbol{q}_s^i, \dot{\boldsymbol{q}}_s^i)$ are the total disturbance of the master and slave robots, respectively. The definitions of other components are consistent with those in (2.11). In this chapter, $\boldsymbol{f}_h^i$ and $\boldsymbol{f}_e^i$ are assumed to be locally bounded.

The generalized end-effector position $\boldsymbol{X}_j^i$ of the robot in the task space can be given by:

$$\boldsymbol{X}_j^i = \boldsymbol{h}_j^i(\boldsymbol{q}_j^i), \quad \dot{\boldsymbol{X}}_j^i = \boldsymbol{J}_j^i(\boldsymbol{q}_j^i)\dot{\boldsymbol{q}}_j^i$$

In this chapter, for the sake of simplifying the design, it is assumed that $\boldsymbol{X}_j^i \in \mathbb{R}^n$.

In practices, due to the existence of the possible Coulomb friction, the total disturbance $\boldsymbol{F}_j^i(\boldsymbol{q}_j, \dot{\boldsymbol{q}}_j)$ can be divided into the continuous component $\boldsymbol{F}_{cj}^i(\boldsymbol{q}_j, \dot{\boldsymbol{q}}_j)$ and the piecewise or non-continuous component $\boldsymbol{F}_{pj}^i(\boldsymbol{q}_j, \dot{\boldsymbol{q}}_j)$, i.e.,

$$\boldsymbol{F}_j^i(\boldsymbol{q}_j, \dot{\boldsymbol{q}}_j) = \boldsymbol{F}_{cj}^i(\boldsymbol{q}_j, \dot{\boldsymbol{q}}_j) + \boldsymbol{F}_{pj}^i(\boldsymbol{q}_j, \dot{\boldsymbol{q}}_j)$$

In what follows, it is assumed that there exists the unknown constant $D_{ji} > 0$ such that $|\boldsymbol{F}_{pj}^i(\boldsymbol{q}_j, \dot{\boldsymbol{q}}_j)| \leq D_{ji}$.

The dynamic model in (6.1) satisfies the Property 2.1 and Property 2.4, and according to the Ge's work [2], it also holds that

Property 6.1. *For any differentiable vector $\boldsymbol{\xi} \in \mathbb{R}^n$, it has*

$$\boldsymbol{J}_j^i(\boldsymbol{q}_j^i)\boldsymbol{\xi} = \boldsymbol{Y}_j^i(\boldsymbol{q}_j^i, \boldsymbol{\xi})\boldsymbol{\theta}_j^i$$

where $\boldsymbol{\theta}_j^i \in \mathbb{R}^{w_{ji}}$ is a unknown constant vector of kinematics parameters, $\boldsymbol{Y}_j^i(\boldsymbol{q}_j^i, \boldsymbol{\xi}) \in \mathbb{R}^{n \times w_{ji}}$ is called the known kinematics regressor matrix..

For the above multilateral teleoperation system, the control task is defined as follows: ensuring the position tracking between each pair of master and slave robots, and simultaneously coordinating the movements of all slave robots.

On the one hand, the positions of the master and slave robots are expected to converge with each other. In order to measure the motion tracking between each pair of master and slave robots, the following tracking errors are introduced:

$$\begin{cases} \boldsymbol{e}_m^i(t) = \boldsymbol{X}_m^i(t) - \boldsymbol{X}_s^i(t - d_s^i(t)) \\ \boldsymbol{e}_s^i(t) = \boldsymbol{X}_s^i(t) - \boldsymbol{X}_m^i(t - d_m^i(t)) \end{cases}$$

where $i = 1, 2, \cdots, N$, $d_m^i(t)$ and $d_s^i(t)$ is the transmission delay between the ith pair of master and slave robots.

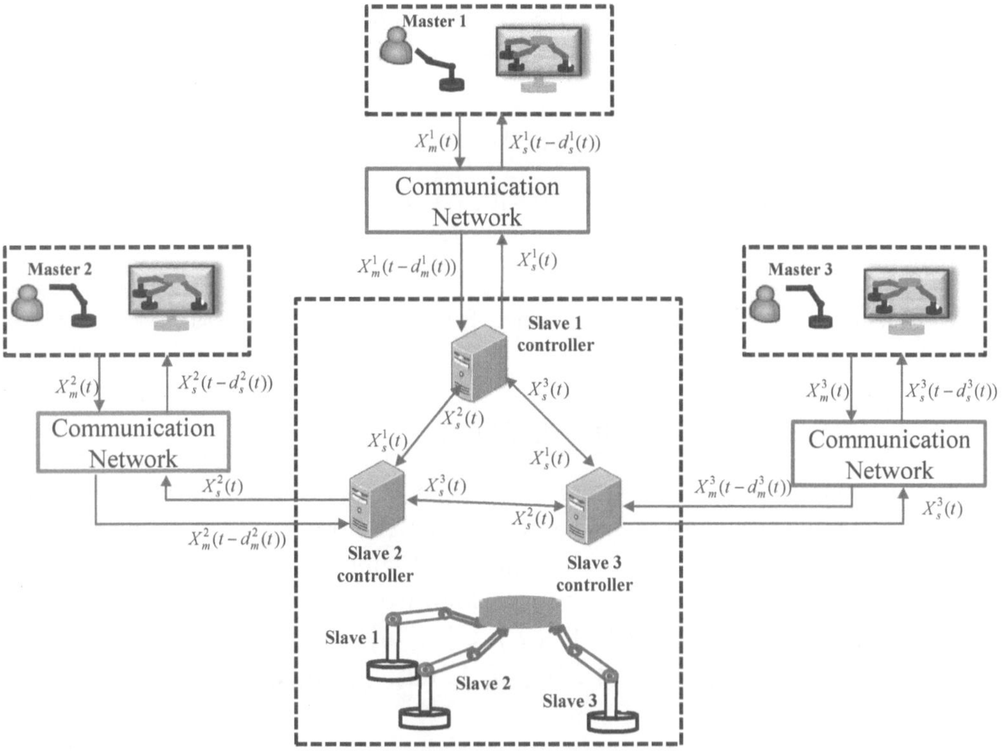

Figure 6.1 Multilateral tele-coordination system.

In this chapter, all communication delays are assumed to satisfy

Assumption 6.1. *There exist positive constant* $\bar{d} < \infty$ *and* $\tilde{d} < \infty$ *such that* $0 \leq d_j^i(t) \leq \bar{d}$, $|\dot{d}_j^i(t)| \leq \tilde{d}$, *where* $j \in \{m, s\}$, $i = 1, 2, \cdots, N$.

On the other hand, the problem of coordinating multiple slave robots in multilateral teleoperation control can be regarded as maintaining a certain kinematic relationship among all slave robots. For example, three slave robots jointly grasp a rigid object, as shown in Fig. 6.1. In order not to damage the load or the slave robot, the difference between the position/orientation of the slave robot end effector should remain constant. The synchronization error is defined as follows:

$$\begin{cases} \boldsymbol{\epsilon}_s^1(t) = \boldsymbol{X}_s^1(t) - \boldsymbol{X}_s^2(t) + \boldsymbol{a}_1 \\ \boldsymbol{\epsilon}_s^2(t) = \boldsymbol{X}_s^2(t) - \boldsymbol{X}_s^3(t) + \boldsymbol{a}_2 \\ \boldsymbol{\epsilon}_s^3(t) = \boldsymbol{X}_s^3(t) - \boldsymbol{X}_s^1(t) + \boldsymbol{a}_3 \end{cases} \tag{6.2}$$

where $\boldsymbol{a}_1$, $\boldsymbol{a}_2$ and $\boldsymbol{a}_3$ are some constant non-negative vectors. Then the tele-coordinated control should ensure that the synchronization errors $\boldsymbol{\epsilon}_s^i(t)$, $i = 1, 2, 3$, converge to zero.

Similar to the previous work [1], this chapter considers some more general co-constrained cases, introducing the following synchronization errors:

$$\begin{cases} \boldsymbol{\epsilon}_s^1(t) = c_1(t)\boldsymbol{X}_s^1(t) + c_2(t)\boldsymbol{X}_s^2(t) + \boldsymbol{a}_1 \\ \boldsymbol{\epsilon}_s^2(t) = c_2(t)\boldsymbol{X}_s^2(t) + c_3(t)\boldsymbol{X}_s^3(t) + \boldsymbol{a}_2 \\ \boldsymbol{\epsilon}_s^k(t) = c_k(t)\boldsymbol{X}_s^k(t) + c_{k+1}(t)\boldsymbol{X}_s^{k+1}(t) + \boldsymbol{a}_k \\ \boldsymbol{\epsilon}_s^N(t) = c_N(t)\boldsymbol{X}_s^N(t) + c_1(t)\boldsymbol{X}_s^1(t) + \boldsymbol{a}_N \end{cases} \tag{6.3}$$

where $k = 3, 4, \cdots, N-1$, $c_i(t)$ are bounded functions, $\boldsymbol{a}_i$ are constant nonnegative vectors.

6.3 MULTILATERAL COORDINATED CONTROLLER

For the above coordinated control tasks, this section gives the specific control scheme.

First of all, the actual robot dynamics model is not known precisely. In this section, the neural network is used to estimate and approximate the unknown robot dynamics model. Specifically, according to the knowledge of the neural network [2], the elements of $\bar{\boldsymbol{J}}_j^i(\boldsymbol{q}_j^i) = \boldsymbol{J}_j^{i\mathrm{T}}(\boldsymbol{q}_j^i)\boldsymbol{J}_j^i(\boldsymbol{q}_j^i)$, $\boldsymbol{M}_j^i(\boldsymbol{q}_j^i)$, $\boldsymbol{C}_j^i(\boldsymbol{q}_j^i, \dot{\boldsymbol{q}}_j^i)$, $\boldsymbol{G}_j^i(\boldsymbol{q}_j^i)$ and $\boldsymbol{F}_{cj}^i(\boldsymbol{q}_j^i, \dot{\boldsymbol{q}}_j^i)$, i.e., $\bar{J}_{j,k_1k_2}^i(\boldsymbol{q}_j^i)$, $M_{j,k_1k_2}^i(\boldsymbol{q}_j^i)$, $C_{j,k_1k_2}^i(\boldsymbol{q}_j^i, \dot{\boldsymbol{q}}_j^i)$, $G_{j,k_1}^i(\boldsymbol{q}_j^i)$ and $F_{cj,k_1}^i(\boldsymbol{q}_j^i, \dot{\boldsymbol{q}}_j^i)$, can be modeled as

$$\begin{aligned} \bar{J}_{j,k_1k_2}^i(\boldsymbol{q}_j^i) &= \boldsymbol{\Theta}_{Jj,k_1k_2}^{i\mathrm{T}}\boldsymbol{\varphi}_{Jj,k_1k_2}^i(\boldsymbol{Z}_{j1}^i) + o_{Jj,k_1k_2}^i(\boldsymbol{Z}_{j1}^i) \\ M_{j,k_1k_2}^i(\boldsymbol{q}_j^i) &= \boldsymbol{\Theta}_{Mj,k_1k_2}^{i\mathrm{T}}\boldsymbol{\varphi}_{Mj,k_1k_2}^i(\boldsymbol{Z}_{j1}^i) + o_{Mj,k_1k_2}^i(\boldsymbol{Z}_{j1}^i) \\ C_{j,k_1k_2}^i(\boldsymbol{q}_j^i, \dot{\boldsymbol{q}}_j^i) &= \boldsymbol{\Theta}_{Cj,k_1k_2}^{i\mathrm{T}}\boldsymbol{\varphi}_{Cj,k_1k_2}^i(\boldsymbol{Z}_{j2}^i) + o_{Cj,k_1k_2}^i(\boldsymbol{Z}_{j2}^i) \\ G_{j,k_1}^i(\boldsymbol{q}_j^i) &= \boldsymbol{\Theta}_{Gj,k_1}^{i\mathrm{T}}\boldsymbol{\varphi}_{Gj,k_1}^i(\boldsymbol{Z}_{j1}^i) + o_{Gj,k_1}^i(\boldsymbol{Z}_{j1}^i) \\ F_{cj,k_1}^i(\boldsymbol{q}_j, \dot{\boldsymbol{q}}_j) &= \boldsymbol{\Theta}_{Fj,k_1}^{i\mathrm{T}}\boldsymbol{\varphi}_{Fj,k_1}^i(\boldsymbol{Z}_{j2}^i) + o_{Fj,k_1}^i(\boldsymbol{Z}_{j2}^i) \end{aligned}$$

where $\boldsymbol{Z}_{j1}^i = \boldsymbol{q}_j^i$, $\boldsymbol{Z}_{j2}^i = [\boldsymbol{q}_j^{i\mathrm{T}}, \dot{\boldsymbol{q}}_j^{i\mathrm{T}}]^{\mathrm{T}}$, $1 \le k_1, k_2 \le n$.

In what follows, $[\cdot]$ denotes an ordinary matrix, and $\{\cdot\}$ denotes a Ge-Lee (GL) matrix explicitly, and $\bullet$ denotes a GL operator, where the properties of the GL matrix and its operator can be seen in the Ge's work [2]. Denote

$$\boldsymbol{o}_{Jj}^i(\boldsymbol{Z}_{j1}^i) = \begin{bmatrix} o_{Jj,11}^i(\boldsymbol{Z}_{j1}^i) & o_{Jj,12}^i(\boldsymbol{Z}_{j1}^i) & \cdots & o_{Jj,1n}^i(\boldsymbol{Z}_{j1}^i) \\ o_{Jj,21}^i(\boldsymbol{Z}_{j1}^i) & o_{Jj,22}^i(\boldsymbol{Z}_{j1}^i) & \cdots & o_{Jj,2n}^i(\boldsymbol{Z}_{j1}^i) \\ \vdots & \vdots & \ddots & \vdots \\ o_{Jj,n1}^i(\boldsymbol{Z}_{j1}^i) & o_{Jj,n2}^i(\boldsymbol{Z}_{j1}^i) & \cdots & o_{Jj,nn}^i(\boldsymbol{Z}_{j1}^i) \end{bmatrix}$$

$$\boldsymbol{o}_{Mj}^i(\boldsymbol{Z}_{j1}^i) = \begin{bmatrix} o_{Mj,11}^i(\boldsymbol{Z}_{j1}^i) & o_{Mj,12}^i(\boldsymbol{Z}_{j1}^i) & \cdots & o_{Mj,1n}^i(\boldsymbol{Z}_{j1}^i) \\ o_{Mj,21}^i(\boldsymbol{Z}_{j1}^i) & o_{Mj,22}^i(\boldsymbol{Z}_{j1}^i) & \cdots & o_{Mj,2n}^i(\boldsymbol{Z}_{j1}^i) \\ \vdots & \vdots & \ddots & \vdots \\ o_{Mj,n1}^i(\boldsymbol{Z}_{j1}^i) & o_{Mj,n2}^i(\boldsymbol{Z}_{j1}^i) & \cdots & o_{Mj,nn}^i(\boldsymbol{Z}_{j1}^i) \end{bmatrix}$$

$$\boldsymbol{o}_{Cj}^{i}(\boldsymbol{Z}_{j2}^{i}) = \begin{bmatrix} o_{Cj,11}^{i}(\boldsymbol{Z}_{j2}^{i}) & o_{Cj,12}^{i}(\boldsymbol{Z}_{j2}^{i}) & \cdots & o_{Cj,1n}^{i}(\boldsymbol{Z}_{j2}^{i}) \\ o_{Cj,21}^{i}(\boldsymbol{Z}_{j2}^{i}) & o_{Cj,22}^{i}(\boldsymbol{Z}_{j2}^{i}) & \cdots & o_{Cj,2n}^{i}(\boldsymbol{Z}_{j2}^{i}) \\ \vdots & \vdots & \ddots & \vdots \\ o_{Cj,n1}^{i}(\boldsymbol{Z}_{j2}^{i}) & o_{Cj,n2}^{i}(\boldsymbol{Z}_{j2}^{i}) & \cdots & o_{Cj,nn}^{i}(\boldsymbol{Z}_{j2}^{i}) \end{bmatrix}$$

$$\begin{aligned} \{\boldsymbol{\Theta}_{Jj,k_1}^{i}\} &= [\boldsymbol{\Theta}_{Jj,k_1 1}^{i}, \boldsymbol{\Theta}_{Jj,k_1 2}^{i}, \cdots, \boldsymbol{\Theta}_{Jj,k_1 n}^{i}] \\ \{\boldsymbol{\varphi}_{Jj,k_1}^{i}(\boldsymbol{Z}_{j1}^{i})\} &= [\boldsymbol{\varphi}_{Jj,k_1 1}^{i}(\boldsymbol{Z}_{j1}^{i}), \boldsymbol{\varphi}_{Jj,k_1 2}^{i}(\boldsymbol{Z}_{j1}^{i}), \cdots, \boldsymbol{\varphi}_{Jj,k_1 n}^{i}(\boldsymbol{Z}_{j1}^{i})] \end{aligned}$$

$$\begin{aligned} \{\boldsymbol{\Theta}_{Mj,k_1}^{i}\} &= [\boldsymbol{\Theta}_{Mj,k_1 1}^{i}, \boldsymbol{\Theta}_{Mj,k_1 2}^{i}, \cdots, \boldsymbol{\Theta}_{Mj,k_1 n}^{i}] \\ \{\boldsymbol{\varphi}_{Mj,k_1}^{i}(\boldsymbol{Z}_{j1}^{i})\} &= [\boldsymbol{\varphi}_{Mj,k_1 1}^{i}(\boldsymbol{Z}_{j1}^{i}), \boldsymbol{\varphi}_{Mj,k_1 2}^{i}(\boldsymbol{Z}_{j1}^{i}), \cdots, \boldsymbol{\varphi}_{Mj,k_1 n}^{i}(\boldsymbol{Z}_{j1}^{i})] \\ \{\boldsymbol{\Theta}_{Cj,k_1}^{i}\} &= [\boldsymbol{\Theta}_{Cj,k_1 1}^{i}, \boldsymbol{\Theta}_{Cj,k_1 2}^{i}, \cdots, \boldsymbol{\Theta}_{Cj,k_1 n}^{i}] \\ \{\boldsymbol{\varphi}_{C,k_1}^{i}(\boldsymbol{Z}_{j2}^{i})\} &= [\boldsymbol{\varphi}_{Cj,k_1 1}^{i}(\boldsymbol{Z}_{j2}^{i}), \boldsymbol{\varphi}_{Cj,k_1 2}^{i}(\boldsymbol{Z}_{j2}^{i}), \cdots, \boldsymbol{\varphi}_{Cj,k_1 n}^{i}(\boldsymbol{Z}_{j2}^{i})] \\ \boldsymbol{o}_{Gj}^{i}(\boldsymbol{Z}_{j1}^{i}) &= \begin{bmatrix} o_{Gj,1}^{i}(\boldsymbol{Z}_{j1}^{i}) \\ o_{Gj,2}^{i}(\boldsymbol{Z}_{j1}^{i}) \\ \vdots \\ o_{Gj,n}^{i}(\boldsymbol{Z}_{j1}^{i}) \end{bmatrix} \\ \boldsymbol{o}_{Fj}^{i}(\boldsymbol{Z}_{j2}^{i}) &= \begin{bmatrix} o_{Fj,1}^{i}(\boldsymbol{Z}_{j2}^{i}) \\ o_{Fj,2}^{i}(\boldsymbol{Z}_{j2}^{i}) \\ \vdots \\ o_{Fj,n}^{i}(\boldsymbol{Z}_{j2}^{i}) \end{bmatrix} \\ \boldsymbol{\Theta}_{Gj,k_1}^{i} &= [\boldsymbol{\Theta}_{Gj,k_1 1}^{i\mathrm{T}}, \boldsymbol{\Theta}_{Gj,k_1 2}^{i\mathrm{T}}, \cdots, \boldsymbol{\Theta}_{Gj,k_1 n}^{i\mathrm{T}}]^{\mathrm{T}} \\ \boldsymbol{\varphi}_{Gj,k_1}^{i}(\boldsymbol{Z}_{j1}^{i}) &= [\boldsymbol{\varphi}_{Gj,k_1 1}^{i\mathrm{T}}(\boldsymbol{Z}_{j1}^{i}), \boldsymbol{\varphi}_{Gj,k_1 2}^{i\mathrm{T}}(\boldsymbol{Z}_{j1}^{i}), \cdots, \boldsymbol{\varphi}_{Gj,k_1 n}^{i\mathrm{T}}(\boldsymbol{Z}_{j1}^{i})]^{\mathrm{T}} \\ \boldsymbol{\Theta}_{Fj,k_1}^{i} &= [\boldsymbol{\Theta}_{Fj,k_1 1}^{i\mathrm{T}}, \boldsymbol{\Theta}_{Fj,k_1 2}^{i\mathrm{T}}, \cdots, \boldsymbol{\Theta}_{Fj,k_1 n}^{i\mathrm{T}}]^{\mathrm{T}} \\ \boldsymbol{\varphi}_{Fj,k_1}^{i}(\boldsymbol{Z}_{j2}^{i}) &= [\boldsymbol{\varphi}_{Fj,k_1 1}^{i\mathrm{T}}(\boldsymbol{Z}_{j2}^{i}), \boldsymbol{\varphi}_{Fj,k_1 2}^{i\mathrm{T}}(\boldsymbol{Z}_{j2}^{i}), \cdots, \boldsymbol{\varphi}_{Fj,k_1 n}^{i\mathrm{T}}(\boldsymbol{Z}_{j2}^{i})]^{\mathrm{T}} \end{aligned}$$

Then from the Ge's work [2], the neural network implementation of robotic dynamics can be conveniently expressed as

$$\begin{aligned} \bar{\boldsymbol{J}}_{j}^{i}(\boldsymbol{q}_{j}^{i}) &= [\{\boldsymbol{\Theta}_{Jj}^{i}\}^{\mathrm{T}} \bullet \{\boldsymbol{\varphi}_{Jj}^{i}(\boldsymbol{Z}_{j1}^{i})\}] + \boldsymbol{o}_{Jj}^{i}(\boldsymbol{Z}_{j1}^{i}) \\ &= \begin{bmatrix} \{\boldsymbol{\Theta}_{Jj,1}^{i}\}^{\mathrm{T}} \bullet \{\boldsymbol{\varphi}_{Jj,1}^{i}(\boldsymbol{Z}_{j1}^{i})\} \\ \{\boldsymbol{\Theta}_{Jj,2}^{i}\}^{\mathrm{T}} \bullet \{\boldsymbol{\varphi}_{Jj,2}^{i}(\boldsymbol{Z}_{j1}^{i})\} \\ \{\boldsymbol{\Theta}_{Jj,3}^{i}\}^{\mathrm{T}} \bullet \{\boldsymbol{\varphi}_{Jj,3}^{i}(\boldsymbol{Z}_{j1}^{i})\} \\ \vdots \\ \{\boldsymbol{\Theta}_{Jj,n}^{i}\}^{\mathrm{T}} \bullet \{\boldsymbol{\varphi}_{Jj,n}^{i}(\boldsymbol{Z}_{j1}^{i})\} \end{bmatrix} + \boldsymbol{o}_{Jj}^{i}(\boldsymbol{Z}_{j1}^{i}) \end{aligned}$$

$$\begin{aligned}
\boldsymbol{M}_j^i(\boldsymbol{q}_j^i) &= [\{\boldsymbol{\Theta}_{Mj}^i\}^{\mathrm{T}} \bullet \{\boldsymbol{\varphi}_{Mj}^i(\boldsymbol{Z}_{j1}^i)\}] + \boldsymbol{o}_{Mj}^i(\boldsymbol{Z}_{j1}^i) \\
&= \begin{bmatrix} \{\boldsymbol{\Theta}_{Mj,1}^i\}^{\mathrm{T}} \bullet \{\boldsymbol{\varphi}_{Mj,1}^i(\boldsymbol{Z}_{j1}^i)\} \\ \{\boldsymbol{\Theta}_{Mj,2}^i\}^{\mathrm{T}} \bullet \{\boldsymbol{\varphi}_{Mj,2}^i(\boldsymbol{Z}_{j1}^i)\} \\ \{\boldsymbol{\Theta}_{Mj,3}^i\}^{\mathrm{T}} \bullet \{\boldsymbol{\varphi}_{Mj,3}^i(\boldsymbol{Z}_{j1}^i)\} \\ \vdots \\ \{\boldsymbol{\Theta}_{Mj,n}^i\}^{\mathrm{T}} \bullet \{\boldsymbol{\varphi}_{Mj,n}^i(\boldsymbol{Z}_{j1}^i)\} \end{bmatrix} + \boldsymbol{o}_{Mj}^i(\boldsymbol{Z}_{j1}^i) \\
\boldsymbol{C}_j^i(\boldsymbol{q}_j^i, \dot{\boldsymbol{q}}_j^i) &= [\{\boldsymbol{\Theta}_{Cj}^i\}^{\mathrm{T}} \bullet \{\boldsymbol{\varphi}_{Cj}^i(\boldsymbol{Z}_{j2}^i)\}] + \boldsymbol{o}_{Cj}^i(\boldsymbol{Z}_{j2}^i) \\
&= \begin{bmatrix} \{\boldsymbol{\Theta}_{Cj,1}^i\}^{\mathrm{T}} \bullet \{\boldsymbol{\varphi}_{Cj,1}^i(\boldsymbol{Z}_{j2}^i)\} \\ \{\boldsymbol{\Theta}_{Cj,2}^i\}^{\mathrm{T}} \bullet \{\boldsymbol{\varphi}_{Cj,2}^i(\boldsymbol{Z}_{j2}^i)\} \\ \{\boldsymbol{\Theta}_{Cj,3}^i\}^{\mathrm{T}} \bullet \{\boldsymbol{\varphi}_{Cj,3}^i(\boldsymbol{Z}_{j2}^i)\} \\ \vdots \\ \{\boldsymbol{\Theta}_{Cj,n}^i\}^{\mathrm{T}} \bullet \{\boldsymbol{\varphi}_{Cj,n}^i(\boldsymbol{Z}_{j2}^i)\} \end{bmatrix} + \boldsymbol{o}_{Cj}^i(\boldsymbol{Z}_{j2}^i)
\end{aligned}$$

$$\begin{aligned}
\boldsymbol{G}_j^i(\boldsymbol{q}_j^i) &= [\{\boldsymbol{\Theta}_{Gj}^i\}^{\mathrm{T}} \bullet \{\boldsymbol{\varphi}_{Gj}^i(\boldsymbol{Z}_{j1}^i)\}] + \boldsymbol{o}_{Gj}^i(\boldsymbol{Z}_{j1}^i) \\
&= \begin{bmatrix} \boldsymbol{\Theta}_{Gj,1}^{i\mathrm{T}} \boldsymbol{\varphi}_{Gj,1}^i(\boldsymbol{Z}_{j1}^i) \\ \boldsymbol{\Theta}_{Gj,2}^{i\mathrm{T}} \boldsymbol{\varphi}_{Gj,2}^i(\boldsymbol{Z}_{j1}^i) \\ \boldsymbol{\Theta}_{Gj,3}^{i\mathrm{T}} \boldsymbol{\varphi}_{Gj,3}^i(\boldsymbol{Z}_{j1}^i) \\ \vdots \\ \boldsymbol{\Theta}_{Gj,n}^{i\mathrm{T}} \boldsymbol{\varphi}_{Gj,n}^i(\boldsymbol{Z}_{j1}^i) \end{bmatrix} + \boldsymbol{o}_{Gj}^i(\boldsymbol{Z}_{j1}^i)
\end{aligned}$$

and

$$\begin{aligned}
\boldsymbol{F}_{cj}^i(\boldsymbol{q}_j^i, \dot{\boldsymbol{q}}_j^i) &= [\{\boldsymbol{\Theta}_{Fj}^i\}^{\mathrm{T}} \bullet \{\boldsymbol{\varphi}_{Fj}^i(\boldsymbol{Z}_{j2}^i)\}] + \boldsymbol{o}_{Fj}^i(\boldsymbol{Z}_{j2}^i) \\
&= \begin{bmatrix} \boldsymbol{\Theta}_{Fj,1}^{i\mathrm{T}} \boldsymbol{\varphi}_{Fj,1}^i(\boldsymbol{Z}_{j2}^i) \\ \boldsymbol{\Theta}_{Fj,2}^{i\mathrm{T}} \boldsymbol{\varphi}_{Fj,2}^i(\boldsymbol{Z}_{j2}^i) \\ \boldsymbol{\Theta}_{Fj,3}^{i\mathrm{T}} \boldsymbol{\varphi}_{Fj,3}^i(\boldsymbol{Z}_{j2}^i) \\ \vdots \\ \boldsymbol{\Theta}_{Fj,n}^{i\mathrm{T}} \boldsymbol{\varphi}_{Fj,n}^i(\boldsymbol{Z}_{j2}^i) \end{bmatrix} + \boldsymbol{o}_{Fj}^i(\boldsymbol{Z}_{j2}^i)
\end{aligned}$$

From the Ge's work [2], it is assumed that:

Assumption 6.2. *There exist unknown constant positive scalars* o_{ji1}, o_{ji2}, o_{ji3} *and* o_{ji4} *such that* $\lambda_{\max}\left(\boldsymbol{o}_{Jj}^i(\boldsymbol{Z}_{j1}^i)\right) \le o_{ji1}$, $\lambda_{\max}\left(\boldsymbol{o}_{Mj}^i(\boldsymbol{Z}_{j1}^i)\right) \le o_{ji2}$, $\lambda_{\max}\left(\boldsymbol{o}_{Cj}^i(\boldsymbol{Z}_{j2}^i)\right) \le o_{ji3}$, $|\boldsymbol{o}_{Gj}^i(\boldsymbol{Z}_{j1}^i) + \boldsymbol{o}_{Fj}^i(\boldsymbol{Z}_{j2}^i)| \le o_{ji4}$.

As shown in Fig. 6.1, each slave robot is controlled by an operator through a master robot. Due to the difference of the operator's individual control actions, the time delay of the communication channel and the uncertainty of the robot itself, in practical implementation, it is almost impossible to make the synchronization error $\boldsymbol{\epsilon}_s^i(t)$ converge to zero immediately. Before $\boldsymbol{\epsilon}_s^i(t)$ converges to a satisfactory steady-state value, they are likely to have a large value that could compromise the safety of the payload and/or the slave robot itself. Therefore, a more realistic control strategy is to introduce some kind of constraint to $\boldsymbol{\epsilon}_s^i(t)$, so that it has satisfactory transient and steady-state performance. In order to achieve this goal, this chapter adopts a description performance control strategy [3,4]. According to the references [3,4], the synchronization error $\boldsymbol{\epsilon}_s^i(t) = [\epsilon_{s,1}^i(t), \epsilon_{s,2}^i(t), \cdots, \epsilon_{s,n}^i(t)]^{\mathrm{T}}$ can be described as

$$\begin{cases} -H_{s,k}^i \varrho_{s,k}^i(t) < \epsilon_{s,k}^i(t) < \varrho_{s,k}^i(t), & \epsilon_{s,k}^i(0) \geq 0 \\ -\varrho_{s,k}^i(t) < \epsilon_{s,k}^i(t) < H_{s,k}^i \varrho_{s,k}^i(t), & \epsilon_{s,k}^i(0) \leq 0 \end{cases} \tag{6.4}$$

where $k = 1, 2, \cdots, n$, $H_{s,k}^i \in (0, 1]$, $\varrho_{s,k}^i(t)$ is the performance function, which is a bounded smooth strictly positive decreasing function and satisfies $\varrho_{s,k}^i(t) = (\varrho_{s,k0}^i - \varrho_{s,k\infty}^i)\mathrm{e}^{-l_{s,k}^i t} + \varrho_{s,k\infty}^i$, $\varrho_{s,k\infty}^i = \lim_{t\to\infty} \varrho_{s,k}^i(t) > 0$, where $l_{s,k}^i > 0$, $\varrho_{s,k0}^i = \varrho_{s,k}^i(0)$ is selected such that (6.4) is satisfied at $t = 0$, i.e., when $\epsilon_{s,k}^i(0) \geq 0$, $\varrho_{s,k0}^i > \epsilon_{s,k}^i(0)$, when $\epsilon_{s,k}^i(0) \leq 0$, $\varrho_{s,k0}^i > -\epsilon_{s,k}^i(0)$.

According to the PPC paradigm [3], let

$$\bar{\epsilon}_{s,k}^i(t) = T_k^i\left(\frac{\epsilon_{s,k}^i(t)}{\varrho_{s,k}^i(t)}\right) = \begin{cases} \ln\left(\dfrac{H_{s,k}^i + \dfrac{\epsilon_{s,k}^i(t)}{\varrho_{s,k}^i(t)}}{1 - \dfrac{\epsilon_{s,k}^i(t)}{\varrho_{s,k}^i(t)}}\right), & \epsilon_{s,k}^i(0) \geq 0 \\ \ln\left(\dfrac{1 + \dfrac{\epsilon_{s,k}^i(t)}{\varrho_{s,k}^i(t)}}{H_{s,k}^i - \dfrac{\epsilon_{s,k}^i(t)}{\varrho_{s,k}^i(t)}}\right), & \epsilon_{s,k}^i(0) < 0 \end{cases}$$

and

$$\dot{\bar{\epsilon}}_{s,k}^i(t) = \frac{\partial T_k^i\left(\dfrac{\epsilon_{s,k}^i(t)}{\varrho_{s,k}^i(t)}\right)}{\partial \dfrac{\epsilon_{s,k}^i(t)}{\varrho_{s,k}^i(t)}} \frac{1}{\varrho_{s,k}^i(t)}\left(\dot{\epsilon}_{s,k}^i(t) - \frac{\dot{\varrho}_{o,k}^i(t)}{\varrho_{s,k}^i(t)}\epsilon_{s,k}^i(t)\right)$$

then the prescribed performance in (6.4) can be guaranteed by simply preserving the boundedness of $\bar{\boldsymbol{\epsilon}}_s^i(t) = [\bar{\epsilon}_{s,1}^i(t), \bar{\epsilon}_{s,2}^i(t), \cdots, \bar{\epsilon}_{s,n}^i(t)]^{\mathrm{T}}$, where $k = 1, 2, \cdots, n$, $i = 1, 2, \cdots, N$. A graphical illustration of $T_k^i\left(\dfrac{\epsilon_{s,k}^i(t)}{\varrho_{s,k}^i(t)}\right)$ for both cases is shown in Fig. 6.2.

6.3.1 Neuroadaptive controller

Next, the specific controller design is given according to the above description.

Due to the kinematic uncertainty caused by measurement errors and possible robot flexibility, it is difficult to obtain the exact value of the Jacobian matrix $\boldsymbol{J}_j^i(\boldsymbol{q}_j^i)$ in practical control. For this purpose, this chapter uses the adaptive Jacobian control technique. Let $\hat{\boldsymbol{J}}_j^i(\boldsymbol{q}_j^i)$ denote the estimation of $\boldsymbol{J}_j^i(\boldsymbol{q}_j^i)$, and $\hat{\boldsymbol{\theta}}_j^i$ represents the estimated

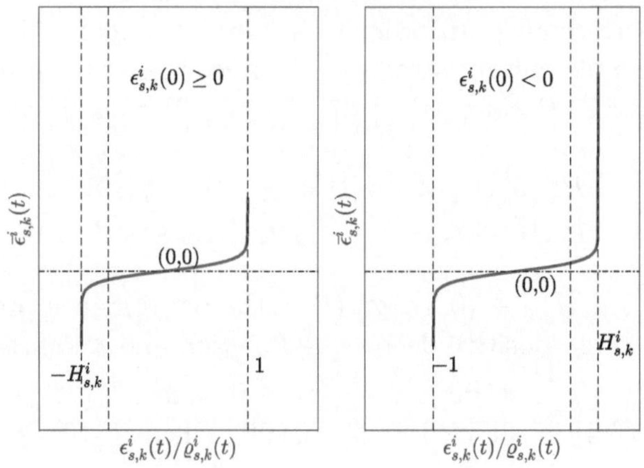

Figure 6.2 Transformation function $T_k^i\left(\dfrac{\epsilon^i_{s,k}(t)}{\varrho^i_{s,k}(t)}\right)$.

value of $\boldsymbol{\theta}_j^i$. According to the Property 6.1, the estimated task space velocity can be expressed as

$$\dot{\hat{\boldsymbol{X}}}_j^i = \hat{\boldsymbol{J}}_j^i(\boldsymbol{q}_j^i)\dot{\boldsymbol{q}}_j^i = \boldsymbol{Y}_j^i(\boldsymbol{q}_j^i, \dot{\boldsymbol{q}}_j^i)\hat{\boldsymbol{\theta}}_j^i \tag{6.5}$$

To give the controller design, define the following auxiliary variables:

$$\begin{aligned}
\boldsymbol{\sigma}_m^i &= \hat{\boldsymbol{J}}_m^{i\mathrm{T}}(\boldsymbol{q}_m^i)K_{mi1}\boldsymbol{e}_m^i \\
\boldsymbol{\sigma}_s^i &= \hat{\boldsymbol{J}}_s^{i\mathrm{T}}(\boldsymbol{q}_s^i)\left(K_{si1}\boldsymbol{e}_s^i + K_{si2}\bar{\boldsymbol{\epsilon}}_s^i\right) \\
\bar{\boldsymbol{\sigma}}_m^i &= \hat{\boldsymbol{J}}_m^{i\mathrm{T}}(\boldsymbol{q}_m^i)K_{mi1}\boldsymbol{e}_{vm}^i + \frac{\mathrm{d}}{\mathrm{d}t}\hat{\boldsymbol{J}}_m^{i\mathrm{T}}(\boldsymbol{q}_m^i)K_{mi1}\boldsymbol{e}_m^i \\
\bar{\boldsymbol{\sigma}}_s^i &= \hat{\boldsymbol{J}}_s^{i\mathrm{T}}(\boldsymbol{q}_s^i)\left(K_{si1}\boldsymbol{e}_{vs}^i + K_{si2}\dot{\bar{\boldsymbol{\epsilon}}}_s^i\right) + \frac{\mathrm{d}}{\mathrm{d}t}\hat{\boldsymbol{J}}_s^{i\mathrm{T}}(\boldsymbol{q}_s^i)\left(K_{si1}\boldsymbol{e}_s^i + K_{si2}\bar{\boldsymbol{\epsilon}}_s^i\right)
\end{aligned}$$

and

$$\begin{cases}
\boldsymbol{\eta}_m^i = \dot{\boldsymbol{q}}_m^i + \boldsymbol{\sigma}_m^i \\
\boldsymbol{\eta}_s^i = \dot{\boldsymbol{q}}_s^i + \boldsymbol{\sigma}_s^i
\end{cases} \tag{6.6}$$

where K_{mi1}, K_{si1}, K_{si2} are some positive constants.

Substituting (6.6) into (6.1), it has

$$\begin{cases}
\boldsymbol{M}_m^i(\boldsymbol{q}_m^i)\dot{\boldsymbol{\eta}}_m^i + \boldsymbol{C}_m^i(\boldsymbol{q}_m^i, \dot{\boldsymbol{q}}_m^i)\boldsymbol{\eta}_m^i \\
\qquad = \boldsymbol{M}_m^i(\boldsymbol{q}_m^i)\bar{\boldsymbol{\sigma}}_m^i + \boldsymbol{C}_m^i(\boldsymbol{q}_m^i, \dot{\boldsymbol{q}}_m^i)\boldsymbol{\sigma}_m^i - \boldsymbol{g}_m^i(\boldsymbol{q}_m^i) + \\
\qquad\quad \boldsymbol{M}_m^i(\boldsymbol{q}_m^i)\hat{\boldsymbol{J}}_m^{i\mathrm{T}}(\boldsymbol{q}_m^i)K_{mi1}\dot{d}_s^i\boldsymbol{X}_{vds}^i(t, d_s^i(t)) - \\
\qquad\quad \boldsymbol{F}_m^i(\boldsymbol{q}_m, \dot{\boldsymbol{q}}_m) + \boldsymbol{J}_m^{i\mathrm{T}}(\boldsymbol{q}_m^i)\boldsymbol{f}_h^i + \boldsymbol{\tau}_m^i \\
\boldsymbol{M}_s^i(\boldsymbol{q}_s^i)\dot{\boldsymbol{\eta}}_s^i + \boldsymbol{C}_s^i(\boldsymbol{q}_s^i, \dot{\boldsymbol{q}}_s^i)\boldsymbol{\eta}_s^i \\
\qquad = \boldsymbol{M}_s^i(\boldsymbol{q}_s^i)\bar{\boldsymbol{\sigma}}_s^i + \boldsymbol{C}_s^i(\boldsymbol{q}_s^i, \dot{\boldsymbol{q}}_s^i)\boldsymbol{\sigma}_s^i - \boldsymbol{g}_s^i(\boldsymbol{q}_s^i) + \\
\qquad\quad \boldsymbol{M}_s^i(\boldsymbol{q}_s^i)\hat{\boldsymbol{J}}_s^{i\mathrm{T}}(\boldsymbol{q}_s^i)K_{si1}\dot{d}_m^i\boldsymbol{X}_{vdm}^i(t, d_m^i(t)) - \\
\qquad\quad \boldsymbol{F}_s^i(\boldsymbol{q}_s, \dot{\boldsymbol{q}}_s) - \boldsymbol{J}_s^{i\mathrm{T}}(\boldsymbol{q}_s^i)\boldsymbol{f}_e^i + \boldsymbol{\tau}_s^i
\end{cases} \tag{6.7}$$

where $\boldsymbol{X}_{vdm}^i(t, d_m^i(t)) = \dot{\boldsymbol{X}}_m^i(\theta)|_{t=t-d_m^i(t)}$, $\boldsymbol{X}_{vds}^i(t, d_s^i(t)) = \dot{\boldsymbol{X}}_s^i(\theta)|_{t=t-d_s^i(t)}$.

Then, for the ith pair of master-slave robots ($i = 1, 2, \cdots, N$), the control torque is designed as

$$\begin{cases} \boldsymbol{\tau}_m^i = \hat{\boldsymbol{g}}_m^i(\boldsymbol{q}_m^i) + \hat{\boldsymbol{F}}_m^i(\boldsymbol{q}_m^i, \dot{\boldsymbol{q}}_m^i) - \hat{\boldsymbol{M}}_m^i(\boldsymbol{q}_m^i)\bar{\boldsymbol{\sigma}}_m^i - \hat{\boldsymbol{C}}_m^i(\boldsymbol{q}_m^i, \dot{\boldsymbol{q}}_m^i)\boldsymbol{\sigma}_m^i - \\ \quad \hat{\hat{\boldsymbol{J}}}_m^i(\boldsymbol{q}_m^i)\boldsymbol{\eta}_m^i - \hat{\omega}_{m1}^i\boldsymbol{\eta}_m^i - \hat{\omega}_{m2}^i|\hat{\boldsymbol{J}}_m^{i\mathrm{T}}(\boldsymbol{q}_m^i)\boldsymbol{X}_{vds}^i(t, d_s^i(t))|^2\boldsymbol{\eta}_m^i - \\ \quad \hat{\omega}_{m3}^i|\bar{\boldsymbol{\sigma}}_m^i|^2\boldsymbol{\eta}_m^i - \hat{\omega}_{m4}^i|\boldsymbol{\sigma}_m^i|^2\boldsymbol{\eta}_m^i - \hat{\boldsymbol{J}}_m^{i\mathrm{T}}(\boldsymbol{q}_m^i)K_{mi3}\boldsymbol{\varpi}_m^i - \\ \quad K_{mi4}\boldsymbol{\eta}_m^i - K_{mi5}\hat{\boldsymbol{J}}_m^{i\mathrm{T}}(\boldsymbol{q}_m^i)\hat{\boldsymbol{J}}_m^i(\boldsymbol{q}_m^i)\boldsymbol{\eta}_m^i \\ \boldsymbol{\tau}_s^i = \hat{\boldsymbol{g}}_s^i(\boldsymbol{q}_s^i) + \hat{\boldsymbol{F}}_s^i(\boldsymbol{q}_s^i, \dot{\boldsymbol{q}}_s^i) - \hat{\boldsymbol{M}}_s^i(\boldsymbol{q}_s^i)\bar{\boldsymbol{\sigma}}_s^i - \hat{\boldsymbol{C}}_s^i(\boldsymbol{q}_s^i, \dot{\boldsymbol{q}}_s^i)\boldsymbol{\sigma}_s^i - \hat{\hat{\boldsymbol{J}}}_s^i(\boldsymbol{q}_s^i)\boldsymbol{\eta}_s^i - \\ \quad \hat{\omega}_{s1}^i\boldsymbol{\eta}_s^i - \hat{\omega}_{s2}^i|\hat{\boldsymbol{J}}_s^{i\mathrm{T}}(\boldsymbol{q}_s^i)\boldsymbol{X}_{vdm}^i(t, d_m^i(t))|^2\boldsymbol{\eta}_s^i - \hat{\omega}_{s3}^i|\bar{\boldsymbol{\sigma}}_s^i|^2\boldsymbol{\eta}_s^i - \\ \quad \hat{\omega}_{s4}^i|\boldsymbol{\sigma}_s^i|^2\boldsymbol{\eta}_s^i - \hat{\boldsymbol{J}}_s^{i\mathrm{T}}(\boldsymbol{q}_s^i)K_{si3}\boldsymbol{\varpi}_s^i - K_{si4}\boldsymbol{\eta}_s^i - K_{si5}\hat{\boldsymbol{J}}_s^{i\mathrm{T}}(\boldsymbol{q}_s^i)\hat{\boldsymbol{J}}_s^i(\boldsymbol{q}_s^i)\boldsymbol{\eta}_s^i \end{cases} \tag{6.8}$$

where for all $j \in \{m, s\}$, $i = 1, 2, \cdots, N$, it has

$$\begin{aligned} \hat{\hat{\boldsymbol{J}}}_j^i(\boldsymbol{q}_j^i)\boldsymbol{\eta}_j^i &= [\{\hat{\boldsymbol{\Theta}}_{Jj}^i\}^{\mathrm{T}} \bullet \{\boldsymbol{\varphi}_{Jj}^i(\boldsymbol{Z}_{j1}^i)\}]\boldsymbol{\eta}_j^i \\ \hat{\boldsymbol{M}}_j^i(\boldsymbol{q}_j^i)\bar{\boldsymbol{\sigma}}_j^i &= [\{\hat{\boldsymbol{\Theta}}_{Mj}^i\}^{\mathrm{T}} \bullet \{\boldsymbol{\varphi}_{Mj}^i(\boldsymbol{Z}_{j1}^i)\}]\bar{\boldsymbol{\sigma}}_j^i \\ \hat{\boldsymbol{C}}_j^i(\boldsymbol{q}_j^i, \dot{\boldsymbol{q}}_j^i)\boldsymbol{\sigma}_j^i &= [\{\hat{\boldsymbol{\Theta}}_{Cj}^i\}^{\mathrm{T}} \bullet \{\boldsymbol{\varphi}_{Cj}^i(\boldsymbol{Z}_{j2}^i)\}]\boldsymbol{\sigma}_j^i \\ \hat{\boldsymbol{g}}_j^i(\boldsymbol{q}_j^i) &= [\{\hat{\boldsymbol{\Theta}}_{Gj}^i\}^{\mathrm{T}} \bullet \{\boldsymbol{\varphi}_{Gj}^i(\boldsymbol{Z}_{j1}^i)\}] \\ \hat{\boldsymbol{F}}_{cj}^i(\boldsymbol{q}_j^i, \dot{\boldsymbol{q}}_j^i) &= [\{\hat{\boldsymbol{\Theta}}_{Fj}^i\}^{\mathrm{T}} \bullet \{\boldsymbol{\varphi}_{Fj}^i(\boldsymbol{Z}_{j2}^i)\}] \end{aligned}$$

K_{ji3}, K_{ji4} and K_{ji5} are constant positive scalars, $\hat{\omega}_{j1}^i = \dfrac{1}{2\varepsilon_{ji1}}\hat{\upsilon}_{j1}^i$, $\hat{\omega}_{j2}^i = \dfrac{K_{ji1}\tilde{d}}{2\varepsilon_{ji1}}\hat{\upsilon}_{j2}^i$, $\hat{\omega}_{j3}^i = \dfrac{1}{2\varepsilon_{ji1}}\hat{\upsilon}_{j3}^i$, $\hat{\omega}_{j4}^i = \dfrac{1}{2\varepsilon_{ji1}}\hat{\upsilon}_{j4}^i$, $\varepsilon_{ji1} > 0$, $\upsilon_{j1}^i = o_{ji4}^2 + D_{ji}^2 + 2\varepsilon_{ji1}o_{ji1}$, $\upsilon_{j2}^i = \rho_{ji2}^2$, $\upsilon_{j3}^i = o_{ji2}^2$, $\upsilon_{j4}^i = o_{ji3}^2$. $\{\hat{\boldsymbol{\Theta}}_{Jj}^i\}$, $\{\hat{\boldsymbol{\Theta}}_{Mj}^i\}$, $\{\hat{\boldsymbol{\Theta}}_{Cj}^i\}$, $\{\hat{\boldsymbol{\Theta}}_{Gj}^i\}$, $\{\hat{\boldsymbol{\Theta}}_{Fj}^i\}$, $\hat{\upsilon}_{j1}^i$, $\hat{\upsilon}_{j2}^i$, $\hat{\upsilon}_{j3}^i$, $\hat{\upsilon}_{j4}^i$ and $\hat{\boldsymbol{\theta}}_j^i$ are respectively the estimations of $\{\boldsymbol{\Theta}_{Jj}^i\}$, $\{\boldsymbol{\Theta}_{Mj}^i\}$, $\{\boldsymbol{\Theta}_{Cj}^i\}$, $\{\boldsymbol{\Theta}_{Gj}^i\}$, $\{\boldsymbol{\Theta}_{Fj}^i\}$, υ_{j1}^i, υ_{j2}^i, υ_{j3}^i, υ_{j4}^i and $\boldsymbol{\theta}_j^i$, where the adaptive update laws are given by

$$\begin{cases} \dot{\hat{\boldsymbol{\Theta}}}_{Jj,k}^i = \boldsymbol{\Lambda}_{jki1} \bullet \{\boldsymbol{\varphi}_{Jj,k}^i(\boldsymbol{Z}_{j1}^i)\}\boldsymbol{\eta}_j^i\eta_{j,k}^i \\ \dot{\hat{\boldsymbol{\Theta}}}_{Mj,k}^i = \boldsymbol{\Lambda}_{jki2} \bullet \{\boldsymbol{\varphi}_{Mj,k}^i(\boldsymbol{Z}_{j1}^i)\}\bar{\boldsymbol{\sigma}}_j^i\eta_{j,k}^i \\ \dot{\hat{\boldsymbol{\Theta}}}_{Cj,k}^i = \boldsymbol{\Lambda}_{jki3} \bullet \{\boldsymbol{\varphi}_{Cj,k}^i(\boldsymbol{Z}_{j2}^i)\}\boldsymbol{\sigma}_j^i\eta_{j,k}^i \\ \dot{\hat{\boldsymbol{\Theta}}}_{Gj,k}^i = -\boldsymbol{\Lambda}_{jki4}\boldsymbol{\varphi}_{Gj,k}^i(\boldsymbol{Z}_{j1}^i)\eta_{j,k}^i \\ \dot{\hat{\boldsymbol{\Theta}}}_{Fj,k}^i = -\boldsymbol{\Lambda}_{jki5}\boldsymbol{\varphi}_{Fj,k}^i(\boldsymbol{Z}_{j2}^i)\eta_{j,k}^i \\ \dot{\hat{\omega}}_{j1}^i = \Lambda_{ji6}|\boldsymbol{\eta}_j^i|^2 \\ \dot{\hat{\omega}}_{j2}^i = \Lambda_{ji7}|\hat{\boldsymbol{J}}_j^{i\mathrm{T}}(\boldsymbol{q}_j^i)\boldsymbol{X}_{vd\iota}^i(t, d_\iota^i(t))|^2|\boldsymbol{\eta}_j^i|^2 \\ \dot{\hat{\omega}}_{j3}^i = \Lambda_{ji8}|\bar{\boldsymbol{\sigma}}_j^i|^2|\boldsymbol{\eta}_j^i|^2 \\ \dot{\hat{\omega}}_{j4}^i = \Lambda_{ji9}|\boldsymbol{\sigma}_j^i|^2|\boldsymbol{\eta}_j^i|^2 \\ \dot{\hat{\boldsymbol{\theta}}}_j^i = \Lambda_{ji10}\boldsymbol{Y}_j^{i\mathrm{T}}(\boldsymbol{q}_j^i, \dot{\boldsymbol{q}}_j^i)\tilde{\boldsymbol{X}}_j^i \end{cases} \tag{6.9}$$

where $k = 1, 2, \cdots, n$, $\hat{\boldsymbol{\Theta}}_{Jj,k}^i$, $\hat{\boldsymbol{\Theta}}_{Mj,k}^i$, $\hat{\boldsymbol{\Theta}}_{Cj,k}^i$, $\hat{\boldsymbol{\Theta}}_{Gj,k}^i$ and $\hat{\boldsymbol{\Theta}}_{Fj,k}^i$ are the elements of $\{\hat{\boldsymbol{\Theta}}_{Jj}^i\}$, $\{\hat{\boldsymbol{\Theta}}_{Mj}^i\}$, $\{\hat{\boldsymbol{\Theta}}_{Cj}^i\}$, $\{\hat{\boldsymbol{\Theta}}_{Gj}^i\}$ and $\{\hat{\boldsymbol{\Theta}}_{Fj}^i\}$, respectively, $\boldsymbol{\Lambda}_{jki1}$, $\boldsymbol{\Lambda}_{jki2}$, $\boldsymbol{\Lambda}_{jki3}$, $\boldsymbol{\Lambda}_{jki4}$ and $\boldsymbol{\Lambda}_{jki5}$ are symmetric positive-definite constant matrices, Λ_{ji6}, Λ_{ji7}, Λ_{ji8}, Λ_{ji9} and

Λ_{ji10} are positive constant scalars. For simplifying expression, let

$$\hat{\boldsymbol{\Theta}}^i_{j,k} = [\hat{\boldsymbol{\Theta}}^{i\mathrm{T}}_{Jj,k}, \hat{\boldsymbol{\Theta}}^{i\mathrm{T}}_{Mj,k}, \hat{\boldsymbol{\Theta}}^{i\mathrm{T}}_{Cj,k}, \hat{\boldsymbol{\Theta}}^{i\mathrm{T}}_{Gj,k}, \hat{\boldsymbol{\Theta}}^{i\mathrm{T}}_{Fj,k}]^{\mathrm{T}}$$
$$\hat{\boldsymbol{\Theta}}^i_j = [\hat{\boldsymbol{\Theta}}^{i\mathrm{T}}_{j,1}, \hat{\boldsymbol{\Theta}}^{i\mathrm{T}}_{j,2}, \cdots, \hat{\boldsymbol{\Theta}}^{i\mathrm{T}}_{j,n}]^{\mathrm{T}} \in \mathbb{R}^{n_{ji1}}$$
$$\hat{\boldsymbol{\omega}}^i_j = [\hat{\omega}^i_{j1}, \hat{\omega}^i_{j2}, \hat{\omega}^i_{j3}, \hat{\omega}^i_{j4}]^{\mathrm{T}} \in \mathbb{R}^4$$

6.3.2 Designs of filter subsystems

In (6.8), $\boldsymbol{\varpi}^i_m$ and $\boldsymbol{\varpi}^i_s$ are the low-pass filters, which are given as

$$\dot{\boldsymbol{\varpi}}^i_j = -K_{ji6}\boldsymbol{\varpi}^i_j + K_{ji7}|\boldsymbol{y}^i_j|\boldsymbol{\mathcal{R}}(\boldsymbol{\eta}^i_j) \tag{6.10}$$

where $j \in \{m, s\}$, K_{ji6} and K_{ji7} are some constant positive scalars, $\boldsymbol{\mathcal{R}}(\boldsymbol{\eta}^i_j) = [\mathcal{R}(\eta^i_{j,1}), \mathcal{R}(\eta^i_{j,2}), \cdots, \mathcal{R}(\eta^i_{j,n})]^{\mathrm{T}}$, $\mathcal{R}(\eta^i_{j,k})$ $(k = 1, 2, \cdots, n)$ is required to be strictly increasing, bounded, continuous, passing through the origin with continuous first derivative around the origin, such that $|\mathcal{R}(\eta^i_{j,k})| \leq 1$, $\mathcal{R}(-\eta^i_{j,k}) = -\mathcal{R}(\eta^i_{j,k})$.

To give the design of $\boldsymbol{y}^i_j$, denote $\tilde{\boldsymbol{X}}^i_j = \boldsymbol{X}^i_j - \hat{\boldsymbol{X}}^i_j$, and

$$\bar{\boldsymbol{e}}^i_m = [\boldsymbol{e}^{i\mathrm{T}}_m, \tilde{\boldsymbol{X}}^{i\mathrm{T}}_m, \hat{\boldsymbol{\Theta}}^{i\mathrm{T}}_m, \hat{\boldsymbol{\omega}}^{i\mathrm{T}}_m, \hat{\boldsymbol{\theta}}^{i\mathrm{T}}_m]^{\mathrm{T}} \in \mathbb{R}^{n_{mi}}$$
$$\bar{\boldsymbol{e}}^i_{vm} = [\boldsymbol{e}^{i\mathrm{T}}_{vm}, \dot{\tilde{\boldsymbol{X}}}^{i\mathrm{T}}_m, \dot{\hat{\boldsymbol{\Theta}}}^{i\mathrm{T}}_m, \dot{\hat{\boldsymbol{\omega}}}^{i\mathrm{T}}_m, \dot{\hat{\boldsymbol{\theta}}}^{i\mathrm{T}}_m]^{\mathrm{T}} \in \mathbb{R}^{n_{mi}}$$
$$\bar{\boldsymbol{e}}^i_s = [\boldsymbol{e}^{i\mathrm{T}}_s, \tilde{\boldsymbol{X}}^{i\mathrm{T}}_s, \bar{\boldsymbol{\epsilon}}^{i\mathrm{T}}_s, \hat{\boldsymbol{\Theta}}^{i\mathrm{T}}_s, \hat{\boldsymbol{\omega}}^{i\mathrm{T}}_s, \hat{\boldsymbol{\theta}}^{i\mathrm{T}}_s]^{\mathrm{T}} \in \mathbb{R}^{n_{si}}$$
$$\bar{\boldsymbol{e}}^i_{vs} = [\boldsymbol{e}^{i\mathrm{T}}_{vs}, \dot{\tilde{\boldsymbol{X}}}^{i\mathrm{T}}_s, \dot{\bar{\boldsymbol{\epsilon}}}^{i\mathrm{T}}_s, \dot{\hat{\boldsymbol{\Theta}}}^{i\mathrm{T}}_s, \dot{\hat{\boldsymbol{\omega}}}^{i\mathrm{T}}_s, \dot{\hat{\boldsymbol{\theta}}}^{i\mathrm{T}}_s]^{\mathrm{T}} \in \mathbb{R}^{n_{si}}$$

Similar to our previous work [5], for all $j \in \{m, s\}$, $\boldsymbol{y}^i_j = [y^i_{j,1}, y^i_{j,2}, \cdots, y^i_{j,n_{ji}}]^{\mathrm{T}}$ is designed as

$$\dot{y}^i_{j,k} = \begin{cases} -P_{ji1}(y^i_{j,k} + \bar{e}^i_{j,k}) - \bar{e}^i_{vj,k} + u^i_{j,k}, & \kappa^i_{j,k}(t) = 1 \\ -P_{ji1}(y^i_{j,k} - \bar{e}^i_{j,k}) + \bar{e}^i_{vj,k} + u^i_{j,k}, & \kappa^i_{j,k}(t) = -1 \end{cases} \tag{6.11}$$

where $k = 1, 2, \cdots, n_{ji}$, $P_{ji1} > 0$ is a constant scalar, $\bar{e}^i_{j,k}$ and $\bar{e}^i_{vj,k}$ are the kth element of $\bar{\boldsymbol{e}}^i_j = [\bar{e}^i_{j,1}, \bar{e}^i_{j,2}, \cdots, \bar{e}^i_{j,n_{ji}}]^{\mathrm{T}}$ and $\bar{\boldsymbol{e}}^i_{vj} = [\bar{e}^i_{vj,1}, \bar{e}^i_{vj,2}, \cdots, \bar{e}^i_{vj,n_{ji}}]^{\mathrm{T}}$, respectively. $u^i_{j,k}$ is designed as

$$u^i_{j,k} = \begin{cases} -P_{ji2}\mathrm{sgn}(y^i_{j,k} + \kappa^i_{j,k}\bar{e}^i_{j,k})|\boldsymbol{X}^i_{v d\iota}(t, d^i_\iota(t))|^2, & k = 1, 2, \cdots, n \\ 0, & k = n+1, n+2, \cdots, n_{ji} \end{cases}$$

where $\iota \in \{m, s\}$, $\iota \neq j$, $P_{ji2} > 0$ is a constant scalar. Similar to Chapter 4, $\kappa^i_{j,k}(t)$ is given by

$$\kappa^i_{j,k}(t) = \begin{cases} 1, & y^i_{j,k}(t)\bar{e}^i_{j,k}(t) > 0 \text{ or } y^i_{j,k}(t)\bar{e}^i_{j,k}(t) = 0,\ y^i_{j,k}(t^-)\bar{e}^i_{j,k}(t^-) \leq 0 \\ -1, & y^i_{j,k}(t)\bar{e}^i_{j,k}(t) < 0 \text{ or } y^i_{j,k}(t)\bar{e}^i_{j,k}(t) = 0,\ y^i_{j,k}(t^-)\bar{e}^i_{j,k}(t^-) > 0 \end{cases} \tag{6.12}$$

Rewrite (6.11) into the vector form as

$$\dot{\boldsymbol{y}}^i_j = -P_{ji1}(\boldsymbol{y}^i_j + \boldsymbol{\mathcal{K}}^i_j\bar{\boldsymbol{e}}^i_j) - \boldsymbol{\mathcal{K}}^i_j\bar{\boldsymbol{e}}^i_{vj} + \boldsymbol{u}^i_j \tag{6.13}$$

where $\boldsymbol{\mathcal{K}}^i_j = \mathrm{diag}[\kappa^i_{j,1}, \kappa^i_{j,2}, \cdots, \kappa^i_{j,n_{ji}}]$, $\boldsymbol{u}^i_j = [u^i_{j,1}, u^i_{j,2}, \cdots, u^i_{j,n_{ji}}]^{\mathrm{T}}$.

From (6.12), $\mathcal{K}_j^i(t)$ has $2^{n_{ji}}$ modes. Repeating the modeling of Remark 3.2, define $\mathcal{S}_j^i = \{1, 2, \cdots, 2^{n_{ji}}\}$, then there exists a map $\bar{r}_j^i : \mathcal{K}_j^i(t) \to \mathcal{S}_j^i$ such that

$$\bar{r}_j^i(\mathcal{K}_j^i(t)) = \ell, \text{ iff } \mathcal{K}_j^i(t) \text{ is in the } \ell\text{th mode}$$

where $\ell \in \mathcal{S}_j^i$.

For simplicity, let $r_j^i(t) = \bar{r}_j^i(\mathcal{K}_j^i(t))$. Then, (6.13) can be seen as a standard switched system with $r_j(t)$ being the switching signal. Let time sequence $\{t_{jk}^i\}_{k \geq 1}$ denote the switching time of $r_j^i(t)$, with $t_{j0}^i = t_0 = 0$.

Then by denoting $\boldsymbol{s}_j^i = \boldsymbol{y}_j^i + \mathcal{K}_j^i \bar{\boldsymbol{e}}_j^i$, when $t \in [t_{j(k-1)}^i, t_{jk}^i)$, it holds

$$\dot{\boldsymbol{s}}_j^i(t) = -P_{ji1}\boldsymbol{s}_j^i(t) + \bar{\boldsymbol{u}}_j^i(t) \tag{6.14}$$

6.3.3 Complete closed-loop teleoperation system

The complete closed-loop teleoperatiion system consists of the robot dynamics equation (6.1), the controller equation (6.8), the adaptive update law (6.9), low-pass filter (6.10) and auxiliary switching filter (6.13). Due to the switching properties of the subsystem (6.13), the complete closed-loop teleoperating system will be a switched dynamic system.

Remark 6.1. *Unlike the previously designed switched filter controller, this chapter introduces an additional low-pass filter (6.10). By using (6.10), the filtered* $\boldsymbol{y}_j^i$ *(i.e.,* $\boldsymbol{\varpi}_j^i$*) is applied to the control torque (6.8), not* $\boldsymbol{y}_j^i$ *itself. In fact, Considering the switching nature of (6.13),* $\boldsymbol{y}_j^i$ *will have some non-smooth control chattering characteristics, which is not expected from the control. The introduction of low-pass filter (6.10) can reduce this unwanted chattering.*

In the rest of the chapter, define $\tilde{\boldsymbol{\Theta}}_j^i = \boldsymbol{\Theta}_j^i - \hat{\boldsymbol{\Theta}}_j^i$, $\tilde{\boldsymbol{\omega}}_j^i = \boldsymbol{\omega}_j^i - \hat{\boldsymbol{\omega}}_j^i$, and denote

$$\begin{aligned}
\boldsymbol{x}_j^i &= [\boldsymbol{q}_j^{i\mathrm{T}}, \boldsymbol{\eta}_j^{i\mathrm{T}}, \tilde{\boldsymbol{\Theta}}_j^{i\mathrm{T}}, \tilde{\boldsymbol{\omega}}_j^{i\mathrm{T}}, \tilde{\boldsymbol{\theta}}_j^{i\mathrm{T}}, \boldsymbol{y}_j^{i\mathrm{T}}, \boldsymbol{\varpi}_j^{i\mathrm{T}}, \bar{\boldsymbol{e}}_j^{i\mathrm{T}}]^\mathrm{T} \\
\boldsymbol{z}_m^i &= [\boldsymbol{\eta}_m^{i\mathrm{T}}, \tilde{\boldsymbol{\Theta}}_m^{i\mathrm{T}}, \tilde{\boldsymbol{\omega}}_m^{i\mathrm{T}}, \tilde{\boldsymbol{\theta}}_m^{i\mathrm{T}}, \boldsymbol{y}_m^{i\mathrm{T}}, \boldsymbol{\varpi}_m^{i\mathrm{T}}, \boldsymbol{e}_m^{i\mathrm{T}}, \tilde{\boldsymbol{X}}_m^{i\mathrm{T}}]^\mathrm{T} \\
\boldsymbol{z}_s^i &= [\boldsymbol{\eta}_s^{i\mathrm{T}}, \tilde{\boldsymbol{\Theta}}_s^{i\mathrm{T}}, \tilde{\boldsymbol{\omega}}_s^{i\mathrm{T}}, \tilde{\boldsymbol{\theta}}_s^{i\mathrm{T}}, \boldsymbol{y}_s^{i\mathrm{T}}, \boldsymbol{\varpi}_s^{i\mathrm{T}}, \boldsymbol{e}_s^{i\mathrm{T}}, \tilde{\boldsymbol{X}}_s^{i\mathrm{T}}, \bar{\boldsymbol{\epsilon}}_s^{i\mathrm{T}}]^\mathrm{T}
\end{aligned}$$

To be precise, for the time delayed complete closed-loop system, denote the new states and outputs of the complete systems as

$$\begin{aligned}
\boldsymbol{x}_{jt}^i &= \{\boldsymbol{x}_j^i(t+\phi) | \phi \in [-\bar{d}, 0]\} \in C([-\bar{d}, 0]; \mathbb{R}^{2n_{ji}+3n+4+w_{ji}+n_{ji1}}) \\
\boldsymbol{z}_{mt}^i &= \{\boldsymbol{z}_m^i(t+\phi) | \phi \in [-\bar{d}, 0]\} \in C([-\bar{d}, 0]; \mathbb{R}^{n_{mi}+4n+4+w_{mi}+n_{mi1}}) \\
\boldsymbol{z}_{st}^i &= \{\boldsymbol{z}_s^i(t+\phi) | \phi \in [-\bar{d}, 0]\} \in C([-\bar{d}, 0]; \mathbb{R}^{n_{si}+5n+4+w_{si}+n_{si1}})
\end{aligned}$$

6.4 STABILITY ANALYSIS

By using the multiple Lyapunov-Krasovskii functionals method, it holds:

Theorem 6.1. *Consider the nonlinear multilateral teleoperation system (6.1) controlled by (6.8)∼(6.10) and (6.13), under the Assumption 6.1 and the initial condition* $\varrho^i_{s,k0} > |\epsilon^i_{s,k}(0)|$, *if the constant positive scalar* ε_{ji1}, ϑ_{ji1}, ϑ_{ji2}, K_{ji1}, K_{si2}, K_{ji3}, K_{ji4}, K_{ji5}, K_{ji6}, K_{ji7}, P_{ji1}, P_{ji2}, Λ_{ji6}, Λ_{ji7}, Λ_{ji8}, Λ_{ji9}, Λ_{ji10} *and the symmetric positive-definite constant matrices* $\mathbf{\Lambda}_{jki1}$, $\mathbf{\Lambda}_{jki2}$, $\mathbf{\Lambda}_{jki3}$, $\mathbf{\Lambda}_{jki4}$, $\mathbf{\Lambda}_{jki5}$ *satisfy:*

$$\min\left\{\begin{array}{c} K_{ji4}, P_{ji1} - \dfrac{nK_{ji7}}{4}, \dfrac{P_{ji1}}{2}, \\ K_{ji6} - \dfrac{K^2_{ji3}}{4K_{ji5}} - K_{ji7} \end{array}\right\} > \vartheta_{ji2} \tag{6.15}$$

and

$$P_{ji2} \geq \tilde{d} \tag{6.16}$$

then for any bounded external forces $\boldsymbol{f}^i_h$ *and* $\boldsymbol{f}^i_e$, *the complete closed-loop master/slave systems are SIIOpS. Under the hypotheses, the tracking error* $\mathbf{e}^i_j(t)$ *maintains bounded, while the synchronization error* $\boldsymbol{\epsilon}^i_s(t)$ *evolves strictly within the predefined region shown in (6.4).*

Proof. Take the Lyapunov functional as $V_{ji}(\boldsymbol{x}^i_{jt}, r^i_j(t)) = \sum_{k=1}^{6} V_{jik}$, where

$$\begin{aligned}
V_{ji1} &= \frac{1}{2}\boldsymbol{\eta}^{iT}_j \boldsymbol{M}^i_j(\boldsymbol{q}^i_j)\boldsymbol{\eta}^i_j \\
V_{ji2} &= \frac{1}{2}\sum_{k=1}^{n} \tilde{\mathbf{\Theta}}^{iT}_{Jj,k}\mathbf{\Lambda}^{-1}_{jki1}\tilde{\mathbf{\Theta}}^i_{Jj,k} + \frac{1}{2}\sum_{k=1}^{n} \tilde{\mathbf{\Theta}}^{iT}_{Mj,k}\mathbf{\Lambda}^{-1}_{jki2}\tilde{\mathbf{\Theta}}^i_{Mj,k} + \\
&\quad \frac{1}{2}\sum_{k=1}^{n} \tilde{\mathbf{\Theta}}^{iT}_{Cj,k}\mathbf{\Lambda}^{-1}_{jki3}\tilde{\mathbf{\Theta}}^i_{Cj,k} + \frac{1}{2}\sum_{k=1}^{n} \tilde{\mathbf{\Theta}}^{iT}_{Gj,k}\mathbf{\Lambda}^{-1}_{jki4}\tilde{\mathbf{\Theta}}^i_{Gj,k} + \\
&\quad \frac{1}{2}\sum_{k=1}^{n} \tilde{\mathbf{\Theta}}^{iT}_{Fj,k}\mathbf{\Lambda}^{-1}_{jki5}\tilde{\mathbf{\Theta}}^i_{Fj,k} \\
V_{ji3} &= \frac{1}{2}\tilde{\omega}^i_{j1}\Lambda^{-1}_{ji6}\tilde{\omega}^i_{j1} + \frac{1}{2}\tilde{\omega}^i_{j2}\Lambda^{-1}_{ji7}\tilde{\omega}^i_{j2} + \frac{1}{2}\tilde{\omega}^i_{j3}\Lambda^{-1}_{ji8}\tilde{\omega}^i_{j3} + \\
&\quad \frac{1}{2}\tilde{\omega}^i_{j4}\Lambda^{-1}_{ji9}\tilde{\omega}^i_{j4} + \frac{\vartheta_{ji1}}{2}\tilde{\boldsymbol{X}}^{iT}_j\tilde{\boldsymbol{X}}^i_j + \frac{\vartheta_{ji1}}{2}\tilde{\boldsymbol{\theta}}^{iT}_j\Lambda^{-1}_{ji10}\tilde{\boldsymbol{\theta}}^i_j \\
V_{ji4} &= \frac{1}{2}\boldsymbol{s}^{iT}_j\boldsymbol{s}^i_j \\
V_{ji5} &= \frac{1}{2}\boldsymbol{\varpi}^{iT}_j\boldsymbol{\varpi}^i_j \\
V_{ji6} &= \frac{\vartheta_{ji2}}{2}\int_{-\bar{d}}^{0} \boldsymbol{z}^{iT}_{jt}(\tau)\left(\frac{-\tau}{\bar{d}} + \frac{2(\tau+\bar{d})}{\bar{d}}\right)\boldsymbol{z}^i_{jt}(\tau)\mathrm{d}\tau
\end{aligned}$$

where ϑ_{ji1} and ϑ_{ji2} are some constant positive scalars.

For convenience, denote $\boldsymbol{f}^i_{he_m} = \boldsymbol{f}^i_m$, $\boldsymbol{f}^i_{he_s} = -\boldsymbol{f}^i_e$, then from Property 2.4 and (6.7), (6.8), for all $\iota \in \{m, s\}$ and $\iota \neq j$,

$$\begin{aligned}D^+V_{ji1} = &\boldsymbol{\eta}^{i\mathrm{T}}_j \boldsymbol{J}^{i\mathrm{T}}_j(\boldsymbol{q}^i_j)\boldsymbol{f}^i_{he_j} - \boldsymbol{\eta}^{i\mathrm{T}}_j \boldsymbol{F}^i_{pj}(\boldsymbol{q}^i_j, \dot{\boldsymbol{q}}^i_j) + \boldsymbol{\eta}^{i\mathrm{T}}_j[\{\tilde{\boldsymbol{\Theta}}^i_{Mj}\}^{\mathrm{T}} \bullet \{\boldsymbol{\varphi}^i_{Mj}(\boldsymbol{Z}^i_{j1})\}]\bar{\boldsymbol{\sigma}}^i_j + \\ &\boldsymbol{\eta}^{i\mathrm{T}}_j[\{\tilde{\boldsymbol{\Theta}}^i_{Cj}\}^{\mathrm{T}} \bullet \{\boldsymbol{\varphi}^i_{Cj}(\boldsymbol{Z}^i_{j2})\}]\boldsymbol{\sigma}^i_j - \boldsymbol{\eta}^{i\mathrm{T}}_j[\{\tilde{\boldsymbol{\Theta}}^i_{Gj}\}^{\mathrm{T}} \bullet \{\boldsymbol{\varphi}^i_{Gj}(\boldsymbol{Z}^i_{j1})\}] - \\ &\boldsymbol{\eta}^{i\mathrm{T}}_j[\{\tilde{\boldsymbol{\Theta}}^i_{Fj}\}^{\mathrm{T}} \bullet \{\boldsymbol{\varphi}^i_{Fj}(\boldsymbol{Z}^i_{j2})\}] + \boldsymbol{\eta}^{i\mathrm{T}}_j \boldsymbol{M}^i_j(\boldsymbol{q}^i_j)\hat{\boldsymbol{J}}^{i\mathrm{T}}_j(\boldsymbol{q}^i_j)K_{ji1}\dot{d}^i_\iota \times \\ &\boldsymbol{X}^i_{vd\iota}(t, d^i_\iota(t)) + \boldsymbol{\eta}^{i\mathrm{T}}_j \boldsymbol{o}^i_{Mj}(\boldsymbol{Z}^i_{j1})\bar{\boldsymbol{\sigma}}^i_j + \boldsymbol{\eta}^{i\mathrm{T}}_j \boldsymbol{o}^i_{Cj}(\boldsymbol{Z}^i_{j2})\boldsymbol{\sigma}^i_j + \\ &\boldsymbol{\eta}^{i\mathrm{T}}_j(\boldsymbol{o}^i_{Gj}(\boldsymbol{Z}^i_{j1}) + \boldsymbol{o}^i_{Fj}(\boldsymbol{Z}^i_{j2})) - \boldsymbol{\eta}^{i\mathrm{T}}_j \hat{\boldsymbol{J}}^i_j(\boldsymbol{q}^i_j)\boldsymbol{\eta}^i_j - \hat{\omega}^i_{j1}|\boldsymbol{\eta}^i_j|^2 - \\ &\hat{\omega}^i_{j2}|\hat{\boldsymbol{J}}^{i\mathrm{T}}_j(\boldsymbol{q}^i_j)\boldsymbol{X}^i_{vd\iota}(t, d^i_\iota(t))|^2|\boldsymbol{\eta}^i_j|^2 - \hat{\omega}^i_{j3}|\bar{\boldsymbol{\sigma}}^i_j|^2|\boldsymbol{\eta}^i_j|^2 - \hat{\omega}^i_{j4}|\boldsymbol{\sigma}^i_j|^2|\boldsymbol{\eta}^i_j|^2 - \\ &\boldsymbol{\eta}^{i\mathrm{T}}_j \hat{\boldsymbol{J}}^{i\mathrm{T}}_j(\boldsymbol{q}^i_j)K_{ji3}\boldsymbol{\varpi}^i_j - K_{ji4}|\boldsymbol{\eta}^i_j|^2 - K_{ji5}\boldsymbol{\eta}^{i\mathrm{T}}_j \hat{\boldsymbol{J}}^{i\mathrm{T}}_j(\boldsymbol{q}^i_j)\hat{\boldsymbol{J}}^i_j(\boldsymbol{q}^i_j)\boldsymbol{\eta}^i_j\end{aligned}$$

Since $\boldsymbol{\eta}^{i\mathrm{T}}_j \boldsymbol{J}^{i\mathrm{T}}_j(\boldsymbol{q}^i_j)\boldsymbol{f}^i_{he_j} \leq \boldsymbol{\eta}^{i\mathrm{T}}_j \boldsymbol{J}^{i\mathrm{T}}_j(\boldsymbol{q}^i_j)\boldsymbol{J}^i_j(\boldsymbol{q}^i_j)\boldsymbol{\eta}^i_j + \frac{1}{4}|\boldsymbol{f}^i_{he_j}|^2$,

$$\begin{aligned}&\boldsymbol{\eta}^{i\mathrm{T}}_j \boldsymbol{J}^{i\mathrm{T}}_j(\boldsymbol{q}^i_j)\boldsymbol{f}^i_{he_j} - \boldsymbol{\eta}^{i\mathrm{T}}_j \hat{\boldsymbol{J}}^i_j(\boldsymbol{q}^i_j)\boldsymbol{\eta}^i_j \\ &\quad \leq \frac{1}{4}|\boldsymbol{f}^i_{he_j}|^2 + \boldsymbol{\eta}^{i\mathrm{T}}_j[\{\tilde{\boldsymbol{\Theta}}^i_{Jj}\}^{\mathrm{T}} \bullet \{\boldsymbol{\varphi}^i_{Jj}(\boldsymbol{Z}^i_{j1})\}]\boldsymbol{\eta}^i_j + \boldsymbol{\eta}^{i\mathrm{T}}_j \boldsymbol{o}^i_{Jj}(\boldsymbol{Z}^i_{j1})\boldsymbol{\eta}^i_j\end{aligned}$$

According to Assumption 6.2,

$$\begin{aligned}&\boldsymbol{\eta}^{i\mathrm{T}}_j \boldsymbol{o}^i_{Jj}(\boldsymbol{Z}^i_{j1})\boldsymbol{\eta}^i_j + \boldsymbol{\eta}^{i\mathrm{T}}_j \boldsymbol{o}^i_{Mj}(\boldsymbol{Z}^i_{j1})\bar{\boldsymbol{\sigma}}^i_j + \boldsymbol{\eta}^{i\mathrm{T}}_j \boldsymbol{o}^i_{Cj}(\boldsymbol{Z}^i_{j2})\boldsymbol{\sigma}^i_j + \\ &\quad \boldsymbol{\eta}^{i\mathrm{T}}_j(\boldsymbol{o}^i_{Gj}(\boldsymbol{Z}^i_{j1}) + \boldsymbol{o}^i_{Fj}(\boldsymbol{Z}^i_{j2})) \\ &\quad \leq o_{ji1}|\boldsymbol{\eta}^i_j|^2 + o_{ji2}|\boldsymbol{\eta}^i_j||\bar{\boldsymbol{\sigma}}^i_j| + o_{ji3}|\boldsymbol{\eta}^i_j||\boldsymbol{\sigma}^i_j| + o_{ji4}|\boldsymbol{\eta}^i_j|\end{aligned}$$

Using the fact that $\boldsymbol{x}^{\mathrm{T}}\boldsymbol{y} \leq \frac{\varepsilon}{2}|\boldsymbol{x}|^2 + \frac{1}{2\varepsilon}|\boldsymbol{y}|^2$, $\varepsilon > 0$, then

$$-\boldsymbol{\eta}^{i\mathrm{T}}_j \hat{\boldsymbol{J}}^{i\mathrm{T}}_j(\boldsymbol{q}^i_j)K_{ji3}\boldsymbol{\varpi}^i_j - K_{ji5}\boldsymbol{\eta}^{i\mathrm{T}}_j \hat{\boldsymbol{J}}^{i\mathrm{T}}_j(\boldsymbol{q}^i_j)\hat{\boldsymbol{J}}^i_j(\boldsymbol{q}^i_j)\boldsymbol{\eta}^i_j \leq \frac{K^2_{ji3}}{4K_{ji5}}|\boldsymbol{\varpi}^i_j|^2$$

and there exists $\varepsilon_{ji1} > 0$ such that

$$\begin{aligned}&o_{ji2}|\boldsymbol{\eta}^i_j||\bar{\boldsymbol{\sigma}}^i_j| + o_{ji3}|\boldsymbol{\eta}^i_j||\boldsymbol{\sigma}^i_j| + o_{ji4}|\boldsymbol{\eta}^i_j| - \boldsymbol{\eta}^{i\mathrm{T}}_j \boldsymbol{F}^i_{pj}(\boldsymbol{q}^i_j, \dot{\boldsymbol{q}}^i_j) \\ &\quad \leq \frac{1}{2\varepsilon_{ji1}}o^2_{ji2}|\boldsymbol{\eta}^i_j|^2|\bar{\boldsymbol{\sigma}}^i_j|^2 + \frac{1}{2\varepsilon_{ji1}}o^2_{ji3}|\boldsymbol{\eta}^i_j|^2|\boldsymbol{\sigma}^i_j|^2 + \\ &\quad \frac{1}{2\varepsilon_{ji1}}o^2_{ji4}|\boldsymbol{\eta}^i_j|^2 + \frac{1}{2\varepsilon_{ji1}}D^2_{ji}|\boldsymbol{\eta}^i_j|^2 + 2\varepsilon_{ji1}\end{aligned}$$

Further, from Property 2.1,

$$\begin{aligned}&\boldsymbol{\eta}^{i\mathrm{T}}_j \boldsymbol{M}^i_j(\boldsymbol{q}^i_j)\hat{\boldsymbol{J}}^{i\mathrm{T}}_j(\boldsymbol{q}^i_j)K_{ji1}\dot{d}^i_\iota \boldsymbol{X}^i_{vd\iota}(t, d^i_\iota(t)) \\ &\quad \leq K_{ji1}\tilde{d}\rho_{ji2}|\hat{\boldsymbol{J}}^{i\mathrm{T}}_j(\boldsymbol{q}^i_j)\boldsymbol{X}^i_{vd\iota}(t, d^i_\iota(t))||\boldsymbol{\eta}^i_j| \\ &\quad \leq \frac{K_{ji1}\tilde{d}}{2\varepsilon_{ji1}}\rho^2_{ji2}|\hat{\boldsymbol{J}}^{i\mathrm{T}}_j(\boldsymbol{q}^i_j)\boldsymbol{X}^i_{vd\iota}(t, d^i_\iota(t))|^2|\boldsymbol{\eta}^i_j|^2 + \frac{K_{ji1}\tilde{d}\varepsilon_{ji1}}{2}\end{aligned}$$

Since $\omega_{j1}^{i} = \dfrac{1}{2\varepsilon_{ji1}}\left(o_{ji4}^{2} + D_{ji}^{2} + 2\varepsilon_{ji1}o_{ji1}\right)$, $\omega_{j2}^{i} = \dfrac{K_{ji1}\tilde{d}}{2\varepsilon_{ji1}}\rho_{ji2}^{2}$, $\omega_{j3}^{i} = \dfrac{1}{2\varepsilon_{ji1}}o_{ji2}^{2}$, $\omega_{j4}^{i} = \dfrac{1}{2\varepsilon_{ji1}}o_{ji3}^{2}$, then

$$
\begin{aligned}
D^{+}V_{ji1} \leq\; & \boldsymbol{\eta}_{j}^{i\mathrm{T}}[\{\tilde{\boldsymbol{\Theta}}_{Jj}^{i}\}^{\mathrm{T}} \bullet \{\boldsymbol{\varphi}_{Jj}^{i}(\boldsymbol{Z}_{j1}^{i})\}]\boldsymbol{\eta}_{j}^{i} + \boldsymbol{\eta}_{j}^{i\mathrm{T}}[\{\tilde{\boldsymbol{\Theta}}_{Mj}^{i}\}^{\mathrm{T}} \bullet \{\boldsymbol{\varphi}_{Mj}^{i}(\boldsymbol{Z}_{j1}^{i})\}]\bar{\boldsymbol{\sigma}}_{j}^{i} + \\
& \boldsymbol{\eta}_{j}^{i\mathrm{T}}[\{\tilde{\boldsymbol{\Theta}}_{Cj}^{i}\}^{\mathrm{T}} \bullet \{\boldsymbol{\varphi}_{Cj}^{i}(\boldsymbol{Z}_{j2}^{i})\}]\boldsymbol{\sigma}_{j}^{i} - \boldsymbol{\eta}_{j}^{i\mathrm{T}}[\{\tilde{\boldsymbol{\Theta}}_{Gj}^{i}\}^{\mathrm{T}} \bullet \{\boldsymbol{\varphi}_{Gj}^{i}(\boldsymbol{Z}_{j1}^{i})\}] - \\
& \boldsymbol{\eta}_{j}^{i\mathrm{T}}[\{\tilde{\boldsymbol{\Theta}}_{Fj}^{i}\}^{\mathrm{T}} \bullet \{\boldsymbol{\varphi}_{Fj}^{i}(\boldsymbol{Z}_{j2}^{i})\}] - K_{ji4}|\boldsymbol{\eta}_{j}^{i}|^{2} + \frac{K_{ji3}^{2}}{4K_{ji5}}|\boldsymbol{\varpi}_{j}^{i}|^{2} + \\
& \tilde{\omega}_{j1}^{i}|\boldsymbol{\eta}_{j}^{i}|^{2} + \tilde{\omega}_{j2}^{i}|\hat{\boldsymbol{J}}_{j}^{i\mathrm{T}}(\boldsymbol{q}_{j}^{i})\boldsymbol{X}_{vd\iota}^{i}(t, d_{\iota}^{i}(t))|^{2}|\boldsymbol{\eta}_{j}^{i}|^{2} + \tilde{\omega}_{j3}^{i}|\bar{\boldsymbol{\sigma}}_{j}^{i}|^{2}|\boldsymbol{\eta}_{j}^{i}|^{2} + \\
& \tilde{\omega}_{j4}^{i}|\boldsymbol{\sigma}_{j}^{i}|^{2}|\boldsymbol{\eta}_{j}^{i}|^{2} + \frac{K_{ji1}\tilde{d}\varepsilon_{ji1}}{2} + 2\varepsilon_{ji1} + \frac{1}{4}|\boldsymbol{f}_{he_{j}}^{i}|^{2}
\end{aligned}
$$

Further, because

$$
\begin{aligned}
\boldsymbol{\eta}_{j}^{i\mathrm{T}}[\{\tilde{\boldsymbol{\Theta}}_{Jj}^{i}\}^{\mathrm{T}} \bullet \{\boldsymbol{\varphi}_{Jj}^{i}(\boldsymbol{Z}_{j1}^{i})\}]\boldsymbol{\eta}_{j}^{i} &= \boldsymbol{\eta}_{j}^{i\mathrm{T}}\begin{bmatrix} \{\tilde{\boldsymbol{\Theta}}_{Jj,1}^{i}\}^{\mathrm{T}} \bullet \{\boldsymbol{\varphi}_{Jj,1}^{i}(\boldsymbol{Z}_{j1}^{i})\}\boldsymbol{\eta}_{j}^{i} \\ \{\tilde{\boldsymbol{\Theta}}_{Jj,2}^{i}\}^{\mathrm{T}} \bullet \{\boldsymbol{\varphi}_{Jj,2}^{i}(\boldsymbol{Z}_{j1}^{i})\}\boldsymbol{\eta}_{j}^{i} \\ \{\tilde{\boldsymbol{\Theta}}_{Jj,3}^{i}\}^{\mathrm{T}} \bullet \{\boldsymbol{\varphi}_{Jj,3}^{i}(\boldsymbol{Z}_{j1}^{i})\}\boldsymbol{\eta}_{j}^{i} \\ \vdots \\ \{\tilde{\boldsymbol{\Theta}}_{Jj,n}^{i}\}^{\mathrm{T}} \bullet \{\boldsymbol{\varphi}_{Jj,n}^{i}(\boldsymbol{Z}_{j1}^{i})\}\boldsymbol{\eta}_{j}^{i} \end{bmatrix} \\
&= \sum_{k=1}^{n}\{\tilde{\boldsymbol{\Theta}}_{Jj,k}^{i}\}^{\mathrm{T}} \bullet \{\boldsymbol{\varphi}_{Jj,k}^{i}(\boldsymbol{Z}_{j1}^{i})\}\boldsymbol{\eta}_{j}^{i}\eta_{j,k}^{i} \\
\boldsymbol{\eta}_{j}^{i\mathrm{T}}[\{\tilde{\boldsymbol{\Theta}}_{Mj}^{i}\}^{\mathrm{T}} \bullet \{\boldsymbol{\varphi}_{Mj}^{i}(\boldsymbol{Z}_{j1}^{i})\}]\bar{\boldsymbol{\sigma}}_{j}^{i} &= \boldsymbol{\eta}_{j}^{i\mathrm{T}}\begin{bmatrix} \{\tilde{\boldsymbol{\Theta}}_{Mj,1}^{i}\}^{\mathrm{T}} \bullet \{\boldsymbol{\varphi}_{Mj,1}^{i}(\boldsymbol{Z}_{j1}^{i})\}\bar{\boldsymbol{\sigma}}_{j}^{i} \\ \{\tilde{\boldsymbol{\Theta}}_{Mj,2}^{i}\}^{\mathrm{T}} \bullet \{\boldsymbol{\varphi}_{Mj,2}^{i}(\boldsymbol{Z}_{j1}^{i})\}\bar{\boldsymbol{\sigma}}_{j}^{i} \\ \vdots \\ \{\tilde{\boldsymbol{\Theta}}_{Mj,n}^{i}\}^{\mathrm{T}} \bullet \{\boldsymbol{\varphi}_{Mj,n}^{i}(\boldsymbol{Z}_{j1}^{i})\}\bar{\boldsymbol{\sigma}}_{j}^{i} \end{bmatrix} \\
&= \sum_{k=1}^{n}\{\tilde{\boldsymbol{\Theta}}_{Mj,k}^{i}\}^{\mathrm{T}} \bullet \{\boldsymbol{\varphi}_{Mj,k}^{i}(\boldsymbol{Z}_{j1}^{i})\}\bar{\boldsymbol{\sigma}}_{j}^{i}\eta_{j,k}^{i} \\
\boldsymbol{\eta}_{j}^{i\mathrm{T}}[\{\tilde{\boldsymbol{\Theta}}_{Cj}^{i}\}^{\mathrm{T}} \bullet \{\boldsymbol{\varphi}_{Cj}^{i}(\boldsymbol{Z}_{j2}^{i})\}]\boldsymbol{\sigma}_{j}^{i} &= \boldsymbol{\eta}_{j}^{i\mathrm{T}}\begin{bmatrix} \{\tilde{\boldsymbol{\Theta}}_{Cj,1}^{i}\}^{\mathrm{T}} \bullet \{\boldsymbol{\varphi}_{Cj,1}^{i}(\boldsymbol{Z}_{j2}^{i})\}\boldsymbol{\sigma}_{j}^{i} \\ \{\tilde{\boldsymbol{\Theta}}_{Cj,2}^{i}\}^{\mathrm{T}} \bullet \{\boldsymbol{\varphi}_{Cj,2}^{i}(\boldsymbol{Z}_{j2}^{i})\}\boldsymbol{\sigma}_{j}^{i} \\ \vdots \\ \{\tilde{\boldsymbol{\Theta}}_{Cj,n}^{i}\}^{\mathrm{T}} \bullet \{\boldsymbol{\varphi}_{Cj,n}^{i}(\boldsymbol{Z}_{j2}^{i})\}\boldsymbol{\sigma}_{j}^{i} \end{bmatrix} \\
&= \sum_{k=1}^{n}\{\tilde{\boldsymbol{\Theta}}_{Cj,k}^{i}\}^{\mathrm{T}} \bullet \{\boldsymbol{\varphi}_{Cj,k}^{i}(\boldsymbol{Z}_{j2}^{i})\}\boldsymbol{\sigma}_{j}^{i}\eta_{j,k}^{i} \\
\boldsymbol{\eta}_{j}^{i\mathrm{T}}[\{\tilde{\boldsymbol{\Theta}}_{Gj}^{i}\}^{\mathrm{T}} \bullet \{\boldsymbol{\varphi}_{Gj}^{i}(\boldsymbol{Z}_{j1}^{i})\}] &= \sum_{k=1}^{n}\tilde{\boldsymbol{\Theta}}_{Gj,k}^{i\mathrm{T}}\boldsymbol{\varphi}_{Gj,k}^{i}(\boldsymbol{Z}_{j1}^{i})\eta_{j,k}^{i} \\
\boldsymbol{\eta}_{j}^{i\mathrm{T}}[\{\tilde{\boldsymbol{\Theta}}_{Fj}^{i}\}^{\mathrm{T}} \bullet \{\boldsymbol{\varphi}_{Fj}^{i}(\boldsymbol{Z}_{j2}^{i})\}] &= \sum_{k=1}^{n}\tilde{\boldsymbol{\Theta}}_{Fj,k}^{i\mathrm{T}}\boldsymbol{\varphi}_{Fj,k}^{i}(\boldsymbol{Z}_{j2}^{i})\eta_{j,k}^{i}
\end{aligned}
$$

it has

$$
\begin{aligned}
D^+V_{ji1} \leq & -K_{ji4}|\boldsymbol{\eta}_j^i|^2 \sum_{k=1}^{n} \{\tilde{\boldsymbol{\Theta}}_{Jj,k}^i\}^{\mathrm{T}} \bullet \{\boldsymbol{\varphi}_{Jj,k}^i(\boldsymbol{Z}_{j1}^i)\}\boldsymbol{\eta}_j^i \eta_{j,k}^i + \\
& \frac{K_{ji3}^2}{4K_{ji5}}|\boldsymbol{\varpi}_j^i|^2 + \sum_{k=1}^{n} \{\tilde{\boldsymbol{\Theta}}_{Mj,k}^i\}^{\mathrm{T}} \bullet \{\boldsymbol{\varphi}_{Mj,k}^i(\boldsymbol{Z}_{j1}^i)\}\bar{\boldsymbol{\sigma}}_j^i \eta_{j,k}^i + \\
& \tilde{\omega}_{j1}^i|\boldsymbol{\eta}_j^i|^2 + \sum_{k=1}^{n} \{\tilde{\boldsymbol{\Theta}}_{Cj,k}^i\}^{\mathrm{T}} \bullet \{\boldsymbol{\varphi}_{Cj,k}^i(\boldsymbol{Z}_{j2}^i)\}\boldsymbol{\sigma}_j^i \eta_{j,k}^i - \\
& \sum_{k=1}^{n} \tilde{\boldsymbol{\Theta}}_{Gj,k}^{i\mathrm{T}} \boldsymbol{\varphi}_{Gj,k}^i(\boldsymbol{Z}_{j1}^i)\eta_{j,k}^i - \sum_{k=1}^{n} \tilde{\boldsymbol{\Theta}}_{Fj,k}^{i\mathrm{T}} \boldsymbol{\varphi}_{Fj,k}^i(\boldsymbol{Z}_{j2}^i)\eta_{j,k}^i + \\
& \tilde{\omega}_{j2}^i|\hat{J}_j^{i\mathrm{T}}(\boldsymbol{q}_j^i)\boldsymbol{X}_{vd\iota}^i(t, d_\iota^i(t))|^2|\boldsymbol{\eta}_j^i|^2 + \tilde{\omega}_{j3}^i|\bar{\boldsymbol{\sigma}}_j^i|^2|\boldsymbol{\eta}_j^i|^2 + \\
& \tilde{\omega}_{j4}^i|\boldsymbol{\sigma}_j^i|^2|\boldsymbol{\eta}_j^i|^2 + \frac{K_{ji1}\tilde{d}\varepsilon_{ji1}}{2} + 2\varepsilon_{ji1} + \frac{1}{4}|\boldsymbol{f}_{he_j}^i|^2
\end{aligned}
$$

For the functional V_{ji2}, from (6.9), it has

$$
\begin{aligned}
D^+V_{ji2} = & -\sum_{k=1}^{n} \tilde{\boldsymbol{\Theta}}_{Jj,k}^{i\mathrm{T}} \boldsymbol{\Lambda}_{jki1}^{-1} \dot{\hat{\boldsymbol{\Theta}}}_{Jj,k}^i - \sum_{k=1}^{n} \tilde{\boldsymbol{\Theta}}_{Mj,k}^{i\mathrm{T}} \boldsymbol{\Lambda}_{jki2}^{-1} \dot{\hat{\boldsymbol{\Theta}}}_{Mj,k}^i - \\
& \sum_{k=1}^{n} \tilde{\boldsymbol{\Theta}}_{Cj,k}^{i\mathrm{T}} \boldsymbol{\Lambda}_{jki3}^{-1} \dot{\hat{\boldsymbol{\Theta}}}_{Cj,k}^i - \sum_{k=1}^{n} \tilde{\boldsymbol{\Theta}}_{Gj,k}^{i\mathrm{T}} \boldsymbol{\Lambda}_{jki4}^{-1} \dot{\hat{\boldsymbol{\Theta}}}_{Gj,k}^i - \sum_{k=1}^{n} \tilde{\boldsymbol{\Theta}}_{Fj,k}^{i\mathrm{T}} \boldsymbol{\Lambda}_{jki5}^{-1} \dot{\hat{\boldsymbol{\Theta}}}_{Fj,k}^i \\
= & -\sum_{k=1}^{n} \{\tilde{\boldsymbol{\Theta}}_{Jj,k}^i\}^{\mathrm{T}} \bullet \boldsymbol{\varphi}_{Jj,k}^i(\boldsymbol{Z}_{j1}^i)\boldsymbol{\eta}_j^i \eta_{j,k}^i - \sum_{k=1}^{n} \{\tilde{\boldsymbol{\Theta}}_{Mj,k}^i\}^{\mathrm{T}} \bullet \{\boldsymbol{\varphi}_{Mj,k}^i(\boldsymbol{Z}_{j1}^i)\} \times \\
& \bar{\boldsymbol{\sigma}}_j^i \eta_{j,k}^i - \sum_{k=1}^{n} \{\tilde{\boldsymbol{\Theta}}_{Cj,k}^i\}^{\mathrm{T}} \bullet \{\boldsymbol{\varphi}_{Cj,k}^i(\boldsymbol{Z}_{j2}^i)\}\boldsymbol{\sigma}_j^i \eta_{j,k}^i + \\
& \sum_{k=1}^{n} \tilde{\boldsymbol{\Theta}}_{Gj,k}^{i\mathrm{T}} \boldsymbol{\varphi}_{Gj,k}^i(\boldsymbol{Z}_{j1}^i)\eta_{j,k}^i + \sum_{k=1}^{n} \tilde{\boldsymbol{\Theta}}_{Fj,k}^{i\mathrm{T}} \boldsymbol{\varphi}_{Fj,k}^i(\boldsymbol{Z}_{j2}^i)\eta_{j,k}^i
\end{aligned}
$$

while for V_{ji3}, from (6.5),

$$
\begin{aligned}
D^+V_{ji3} = & \vartheta_{ji1}\tilde{\boldsymbol{X}}_j^{i\mathrm{T}}(\dot{\boldsymbol{X}}_j^i - \dot{\hat{\boldsymbol{X}}}_j^i) - \vartheta_{ji1}\tilde{\boldsymbol{\theta}}_j^{i\mathrm{T}}\Lambda_{ji10}^{-1}\dot{\hat{\boldsymbol{\theta}}}_j^i - \tilde{\omega}_{j1}^i\Lambda_{ji6}^{-1}\dot{\hat{\omega}}_{j1}^i - \\
& \tilde{\omega}_{j2}^i\Lambda_{ji7}^{-1}\dot{\hat{\omega}}_{j2}^i - \tilde{\omega}_{j3}^i\Lambda_{ji8}^{-1}\dot{\hat{\omega}}_{j3}^i - \tilde{\omega}_{j4}^i\Lambda_{ji9}^{-1}\dot{\hat{\omega}}_{j4}^i \\
= & \vartheta_{ji1}\tilde{\boldsymbol{X}}_j^{i\mathrm{T}}\boldsymbol{Y}_j^i(\boldsymbol{q}_j^i, \dot{\boldsymbol{q}}_j^i)\tilde{\boldsymbol{\theta}}_j^i - \vartheta_{ji1}\tilde{\boldsymbol{\theta}}_j^{i\mathrm{T}}\boldsymbol{Y}_j^{i\mathrm{T}}(\boldsymbol{q}_j^i, \dot{\boldsymbol{q}}_j^i)\tilde{\boldsymbol{X}}_j^i - \tilde{\omega}_{j1}^i|\boldsymbol{\eta}_j^i|^2 - \\
& \tilde{\omega}_{j2}^i|\hat{\boldsymbol{J}}_j^{i\mathrm{T}}(\boldsymbol{q}_j^i)\boldsymbol{X}_{vd\iota}^i(t, d_\iota^i(t))|^2|\boldsymbol{\eta}_j^i|^2 - \tilde{\omega}_{j3}^i|\bar{\boldsymbol{\sigma}}_j^i|^2|\boldsymbol{\eta}_j^i|^2 - \tilde{\omega}_{j4}^i|\boldsymbol{\sigma}_j^i|^2|\boldsymbol{\eta}_j^i|^2 \\
= & -\tilde{\omega}_{j1}^i|\boldsymbol{\eta}_j^i|^2 - \tilde{\omega}_{j2}^i|\hat{\boldsymbol{J}}_j^{i\mathrm{T}}(\boldsymbol{q}_j^i)\boldsymbol{X}_{vd\iota}^i(t, d_\iota^i(t))|^2|\boldsymbol{\eta}_j^i|^2 - \\
& \tilde{\omega}_{j3}^i|\bar{\boldsymbol{\sigma}}_j^i|^2|\boldsymbol{\eta}_j^i|^2 - \tilde{\omega}_{j4}^i|\boldsymbol{\sigma}_j^i|^2|\boldsymbol{\eta}_j^i|^2
\end{aligned}
$$

For any $t \in [t^i_{jk}, t^i_{j(k+1)})$, when $P_{ji2} \geq \tilde{d}$, by considering the fact that $\boldsymbol{s}^i_j = \boldsymbol{y}^i_j + \boldsymbol{\mathcal{K}}^i_j \bar{\boldsymbol{e}}^i_j$, it also obtains

$$\begin{aligned} D^+V_{ji4} &= -P_{ji1}|\boldsymbol{s}^i_j|^2 + \sum_{k=1}^{n} \left(s^i_{j,k} u^i_{j,k} + s^i_{j,k} \kappa^i_{j,k} \dot{d}^i_\iota X^i_{vd\iota,k}(t, d^i_\iota(t)) \right) \\ &\leq -P_{ji1} \left(|\boldsymbol{y}^i_j|^2 + |\bar{\boldsymbol{e}}^i_j|^2 \right) \end{aligned}$$

For the sake of simplifying expression, denote $\breve{\boldsymbol{e}}^i_m = [\boldsymbol{e}^{i\mathrm{T}}_m, \tilde{\boldsymbol{X}}^{i\mathrm{T}}_m]^\mathrm{T}$, $\breve{\boldsymbol{e}}^i_s = [\boldsymbol{e}^{i\mathrm{T}}_s, \tilde{\boldsymbol{X}}^{i\mathrm{T}}_s, \bar{\boldsymbol{\epsilon}}^{i\mathrm{T}}_s]^\mathrm{T}$. Then for all $j \in \{m, s\}$, $\bar{\boldsymbol{e}}^i_j = [\breve{\boldsymbol{e}}^{i\mathrm{T}}_j, \hat{\boldsymbol{\Theta}}^{i\mathrm{T}}_j, \hat{\boldsymbol{\omega}}^{i\mathrm{T}}_j, \hat{\boldsymbol{\theta}}^{i\mathrm{T}}_j]^\mathrm{T}$. Thus

$$\begin{aligned} D^+V_{ji4} &\leq -P_{ji1} \left(|\boldsymbol{y}^i_j|^2 + |\bar{\boldsymbol{e}}^i_j|^2 \right) \\ &\leq -P_{ji1} \left(|\boldsymbol{y}^i_j|^2 + |\breve{\boldsymbol{e}}^i_j|^2 + |\hat{\boldsymbol{\Theta}}^i_j|^2 + |\hat{\boldsymbol{\omega}}^i_j|^2 + |\hat{\boldsymbol{\theta}}^i_j|^2 \right) \\ &\leq -P_{ji1}|\boldsymbol{y}^i_j|^2 - P_{ji1}|\breve{\boldsymbol{e}}^i_j|^2 - \frac{P_{ji1}}{2}|\tilde{\boldsymbol{\Theta}}^i_j|^2 + P_{ji1}|\boldsymbol{\Theta}^i_j|^2 - \\ &\quad \frac{P_{ji1}}{2}|\tilde{\boldsymbol{\omega}}^i_j|^2 + P_{ji1}|\boldsymbol{\omega}^i_j|^2 - \frac{P_{ji1}}{2}|\tilde{\boldsymbol{\theta}}^i_j|^2 + P_{ji1}|\boldsymbol{\theta}^i_j|^2 \end{aligned}$$

In addition, from (6.10),

$$\begin{aligned} D^+V_{ji5} &= -K_{ji6}|\boldsymbol{\varpi}^i_j|^2 + K_{ji7}|\boldsymbol{y}^i_j|\boldsymbol{\varpi}^{i\mathrm{T}}_j \mathrm{sgn}(\boldsymbol{\eta}^i_j) \\ &\leq -(K_{ji6} - K_{ji7})|\boldsymbol{\varpi}^i_j|^2 + \frac{nK_{ji7}}{4}|\boldsymbol{y}^i_j|^2 \end{aligned}$$

Finally, for V_{ji6}, it has

$$\begin{aligned} D^+V_{ji6} &= \vartheta_{ji2}\boldsymbol{z}^{i\mathrm{T}}_j \boldsymbol{z}^i_j - \frac{\vartheta_{ji2}}{2}\boldsymbol{z}^{i\mathrm{T}}_{jt}(-\bar{d})\boldsymbol{z}^i_{jt}(-\bar{d}) - \frac{\vartheta_{ji2}}{2\bar{d}}\int_{-\bar{d}}^{0} \boldsymbol{z}^{i\mathrm{T}}_{jt}(\tau)\boldsymbol{z}^i_{jt}(\tau)\mathrm{d}\tau \\ &\leq \vartheta_{ji2}\boldsymbol{z}^{i\mathrm{T}}_j \boldsymbol{z}^i_j - \frac{\vartheta_{ji2}}{2\bar{d}}\int_{-\bar{d}}^{0} \boldsymbol{z}^{i\mathrm{T}}_{jt}(\tau)\boldsymbol{z}^i_{jt}(\tau)\mathrm{d}\tau \end{aligned}$$

Consequently, for any $t \in [t^i_{jk}, t^i_{j(k+1)})$, when $P_{ji2} \geq \tilde{d}$,

$$\begin{aligned} D^+V_{ji}(\boldsymbol{x}^i_{jt}, r^i_j(t)) \leq\ & -K_{ji4}|\boldsymbol{\eta}^i_j|^2 - \left(P_{ji1} - \frac{nK_{ji7}}{4} \right)|\boldsymbol{y}^i_j|^2 - \\ & \left(K_{ji6} - \frac{K^2_{ji3}}{4K_{ji5}} - K_{ji7} \right)|\boldsymbol{\varpi}^i_j|^2 - P_{ji1}|\tilde{\boldsymbol{e}}^i_j|^2 - \\ & \frac{P_{ji1}}{2}|\tilde{\boldsymbol{\Theta}}^i_j|^2 - \frac{P_{ji1}}{2}|\tilde{\boldsymbol{\omega}}^i_j|^2 - \frac{P_{ji1}}{2}|\tilde{\boldsymbol{\theta}}^i_j|^2 + \vartheta_{ji2}\boldsymbol{z}^{i\mathrm{T}}_j\boldsymbol{z}^i_j - \\ & \frac{\vartheta_{ji2}}{2\bar{d}}\int_{-\bar{d}}^{0} \boldsymbol{z}^{i\mathrm{T}}_{jt}(\tau)\boldsymbol{z}^i_{jt}(\tau)\mathrm{d}\tau + \frac{K_{ji1}\tilde{d}\varepsilon_{ji1}}{2} + 2\varepsilon_{ji1} + \\ & \frac{1}{4}|\boldsymbol{f}^i_{he_j}|^2 + P_{ji1}|\boldsymbol{\Theta}^i_j|^2 + P_{ji1}|\boldsymbol{\omega}^i_j|^2 + P_{ji1}|\boldsymbol{\theta}^i_j|^2 \end{aligned}$$

From the definition of $\boldsymbol{z}_j^i$, it further has

$$
\begin{aligned}
D^{+}V_{ji}(\boldsymbol{x}_{jt}^i, r_j^i(t)) \leq & -\Psi_{ji1}|\boldsymbol{z}_j^i|^2 - \frac{\vartheta_{ji2}}{2\bar{d}}\int_{-\bar{d}}^{0} \boldsymbol{z}_{jt}^{i\mathrm{T}}(\tau)\boldsymbol{z}_{jt}^i(\tau)\mathrm{d}\tau + \frac{K_{ji1}\tilde{d}\varepsilon_{ji1}}{2} + \\
& 2\varepsilon_{ji1} + \frac{1}{4}|\boldsymbol{f}_{he_j}^i|^2 + P_{ji1}|\boldsymbol{\Theta}_j^i|^2 + P_{ji1}|\boldsymbol{\omega}_j^i|^2 + P_{ji1}|\boldsymbol{\theta}_j^i|^2 \\
\leq & -\Psi_{ji2}\|\boldsymbol{z}_{jt}^i\|_{M_2}^2 + \frac{K_{ji1}\tilde{d}\varepsilon_{ji1}}{2} + 2\varepsilon_{ji1} + \frac{1}{4}|\boldsymbol{f}_{he_j}^i|^2 + \\
& P_{ji1}|\boldsymbol{\Theta}_j^i|^2 + P_{ji1}|\boldsymbol{\omega}_j^i|^2 + P_{ji1}|\boldsymbol{\theta}_j^i|^2
\end{aligned} \tag{6.17}
$$

where $\Psi_{ji1} = \min\left\{K_{ji4}, P_{ji1} - \frac{nK_{ji7}}{4}, \frac{P_{ji1}}{2}, K_{ji6} - \frac{K_{ji3}^2}{4K_{ji5}} - K_{ji7}\right\} - \vartheta_{ji2} > 0$, $\Psi_{ji2} = \min\{\Psi_{ji1}, \frac{\vartheta_{ji2}}{2\bar{d}}\}$.

From Property 2.1 and the definitions of $\boldsymbol{s}_j^i$ and $\boldsymbol{\mathcal{K}}_j^i$, by considering the fact that

$$
\frac{1}{2}\int_{-\bar{d}}^{0} \boldsymbol{z}_{jt}^{i\mathrm{T}}(\tau)\left(\frac{-\tau}{\bar{d}} + \frac{2(\tau+\bar{d})}{\bar{d}}\right)\boldsymbol{z}_{jt}^i(\tau)\mathrm{d}\tau \leq \int_{-\bar{d}}^{0} \boldsymbol{z}_{jt}^{i\mathrm{T}}(\tau)\boldsymbol{z}_{jt}^i(\tau)\mathrm{d}\tau
$$

the Lyapunov-Krasovskii function satisfies

$$
\alpha_{ji1}|\boldsymbol{z}_j^i|^2 \leq V_{ji}(\boldsymbol{x}_{jt}^i, r_j^i(t)) \leq \alpha_{ji2}\|\boldsymbol{z}_{jt}^i\|_{M_2}^2 \tag{6.18}
$$

where

$$
\alpha_{ji1} = \frac{1}{2}\min\left\{\begin{array}{c}
\rho_{ji1}, \vartheta_{ji1}, \vartheta_{ji1}\Lambda_{ji10}^{-1}, 1, \Lambda_{ji6}^{-1}, \Lambda_{ji7}^{-1}, \Lambda_{ji8}^{-1}, \Lambda_{ji9}^{-1}, \\
\lambda_{\min}(\boldsymbol{\Lambda}_{j11}^{-1}), \cdots, \lambda_{\min}(\boldsymbol{\Lambda}_{jn1}^{-1}), \lambda_{\min}(\boldsymbol{\Lambda}_{j12}^{-1}), \cdots, \lambda_{\min}(\boldsymbol{\Lambda}_{jn2}^{-1}), \\
\lambda_{\min}(\boldsymbol{\Lambda}_{j13}^{-1}), \cdots, \lambda_{\min}(\boldsymbol{\Lambda}_{jn3}^{-1}), \lambda_{\min}(\boldsymbol{\Lambda}_{j14}^{-1}), \cdots, \lambda_{\min}(\boldsymbol{\Lambda}_{jn4}^{-1}), \\
\lambda_{\min}(\boldsymbol{\Lambda}_{j15}^{-1}), \cdots, \lambda_{\min}(\boldsymbol{\Lambda}_{jn5}^{-1})
\end{array}\right\}
$$

$$
\alpha_{ji2} = \frac{1}{2}\max\left\{\begin{array}{c}
\rho_{ji2}, \vartheta_{ji1}, \vartheta_{ji1}\Lambda_{ji10}^{-1}, 1, 2\vartheta_{ji2}, \Lambda_{ji6}^{-1}, \Lambda_{ji7}^{-1}, \Lambda_{ji8}^{-1}, \Lambda_{ji9}^{-1}, \\
\lambda_{\max}(\boldsymbol{\Lambda}_{j11}^{-1}), \cdots, \lambda_{\max}(\boldsymbol{\Lambda}_{jn1}^{-1}), \lambda_{\max}(\boldsymbol{\Lambda}_{j12}^{-1}), \cdots, \lambda_{\max}(\boldsymbol{\Lambda}_{jn2}^{-1}), \\
\lambda_{\max}(\boldsymbol{\Lambda}_{j13}^{-1}), \cdots, \lambda_{\max}(\boldsymbol{\Lambda}_{jn3}^{-1}), \lambda_{\max}(\boldsymbol{\Lambda}_{j14}^{-1}), \cdots, \lambda_{\max}(\boldsymbol{\Lambda}_{jn4}^{-1}), \\
\lambda_{\max}(\boldsymbol{\Lambda}_{j15}^{-1}), \cdots, \lambda_{\max}(\boldsymbol{\Lambda}_{jn5}^{-1})
\end{array}\right\}
$$

Since for all $k_1 = 1, 2, \cdots, n$, when $y_{j,k_1}^i(t)\bar{e}_{j,k_1}^i(t) = 0$ and $y_{j,k_1}^i(t^-)\bar{e}_{j,k_1}^i(t^-) > 0$, κ_{j,k_1}^i switches from 1 to -1, and when $y_{j,k_1}^i(t)\bar{e}_{j,k_1}^i(t) = 0$ and $y_{j,k_1}^i(t^-)\bar{e}_{j,k_1}^i(t^-) < 0$, κ_{j,k_1}^i switches from -1 to 1. From (6.12), at time $t = t_{j(k+1)}^i$, $k \geq 0$, it holds

$$
V_{ji}(\boldsymbol{x}_{jt_{j(k+1)}^i}^i, r_j^i(t_{j(k+1)}^i)) \leq V_{ji}(\boldsymbol{x}_{jt_{j(k+1)}^i}^i, r_j^i(t_{jk}^i)) \tag{6.19}
$$

From (6.17)~(6.19), by using Lemma 2.4, the complete closed-loop master/slave systems are SIIOpS. Under the assertion, from the definition of $\boldsymbol{z}_j^i(t)$, for any bounded $\boldsymbol{f}_h^i$ and $\boldsymbol{f}_e^i$, the signals $\boldsymbol{e}_j^i(t)$, $\tilde{\boldsymbol{\Theta}}_j^i$, $\tilde{\boldsymbol{\theta}}_j^i$, $\tilde{\boldsymbol{\omega}}_j^i$ and $\boldsymbol{\epsilon}_s^i(t)$ remain bounded. □

Remark 6.2. *For teleoperation systems with time-varying communication delays, there have been many research results [6–10]. However, only very few studies* *consider kinematics uncertainties and time-varying communication delay* [11–13]. *In this chapter, an adaptive neural network control scheme based on switched filter control is proposed for the multilateral coordinated teleoperation system, which realizes the simultaneous processing of asymmetric time-varying delay, passive/non-passive external forces, and kinematics/dynamic uncertainties.*

6.5 SIMULATION STUDY

To verify the effectiveness of the proposed algorithm, let us consider a 3-master-3-slave multilateral tele-coordination system as shown in Fig. 6.1. All the robots are assumed to be 3-DOF planar robots, whose nominal dynamics and parameters are given in the reference [14]. The tele-coordinated control task is that: the three slave robots carry a common payload, where each slave robot is controlled by a human operator via a master robot; in order not to damage the payload, the three human operators manipulate the master robots in a cooperative manner such that the end-effectors of the three slave robots maintain a relatively fixed position. Given that the existences of model uncertainties and external disturbances, it is very difficult to achieve this tele-coordinated control objective. Moreover, the differences of the human operators' actions and the communication delays in different channels will also worsen the control performance. In what follows, the simulations will be conducted under the different human operators' actions, the different communication delays, the different model uncertainties and external disturbances.

For a simulation purpose, as shown in Fig. 6.1, the communication delays are set to be $d_m^1(t) = 0.1 + 0.4\sin^2(4t)$, $d_s^1(t) = 0.3\sin^2(8t)$, $d_m^2(t) = 0.1 + 0.3\sin^2(7t)$, $d_s^2(t) = 0.2\cos^2(6t)$, $d_m^3(t) = 0.2 + 0.15\sin^2(8t)$, $d_s^3(t) = 0.3\sin^2(10t)$, which are different from each other. The external disturbances are assumed to be

$$\begin{aligned}
\boldsymbol{F}_m^1(\boldsymbol{q}_m^1, \dot{\boldsymbol{q}}_m^1) &= 0.2\dot{\boldsymbol{q}}_m^1 + 0.3\mathrm{sgn}(\dot{\boldsymbol{q}}_m^1) + \\
&\quad 0.4[\sin(q_{m,1}^1\dot{q}_{m,1}^1), \sin(q_{m,2}^1\dot{q}_{m,2}^1), \sin(q_{m,3}^1\dot{q}_{m,3}^1)]^{\mathrm{T}} \\
\boldsymbol{F}_m^2(\boldsymbol{q}_m^2, \dot{\boldsymbol{q}}_m^2) &= 0.3\dot{\boldsymbol{q}}_m^2 + 0.2\mathrm{sgn}(\dot{\boldsymbol{q}}_m^2) + \\
&\quad 0.6[\cos(q_{m,1}^2\dot{q}_{m,1}^2), \cos(q_{m,2}^2\dot{q}_{m,2}^2), \cos(q_{m,3}^2\dot{q}_{m,3}^2)]^{\mathrm{T}} \\
\boldsymbol{F}_m^3(\boldsymbol{q}_m^3, \dot{\boldsymbol{q}}_m^3) &= 0.2\dot{\boldsymbol{q}}_m^3 + 0.2\mathrm{sgn}(\dot{\boldsymbol{q}}_m^3) + \\
&\quad 0.2\cos(t)[\sin(q_{m,1}^3\dot{q}_{m,1}^3), \sin(q_{m,2}^3\dot{q}_{m,2}^3), \sin(q_{m,3}^3\dot{q}_{m,3}^3)]^{\mathrm{T}} \\
\boldsymbol{F}_s^1(\boldsymbol{q}_s^1, \dot{\boldsymbol{q}}_s^1) &= 0.5\dot{\boldsymbol{q}}_s^1 + 0.3\mathrm{sgn}(\dot{\boldsymbol{q}}_s^1) + \\
&\quad 0.6[\sin(q_{s,1}^1\dot{q}_{s,1}^1), \sin(q_{s,2}^1\dot{q}_{s,2}^1), \sin(q_{s,3}^1\dot{q}_{s,3}^1)]^{\mathrm{T}} \\
\boldsymbol{F}_s^2(\boldsymbol{q}_s^2, \dot{\boldsymbol{q}}_s^2) &= 0.4\dot{\boldsymbol{q}}_s^2 + 0.2\mathrm{sgn}(\dot{\boldsymbol{q}}_s^2) + \\
&\quad 0.3[\cos(q_{s,1}^2\dot{q}_{s,1}^2), \cos(q_{s,2}^2\dot{q}_{s,2}^2), \cos(q_{s,3}^2\dot{q}_{s,3}^2)]^{\mathrm{T}} \\
\boldsymbol{F}_s^3(\boldsymbol{q}_s^3, \dot{\boldsymbol{q}}_s^3) &= 0.3\dot{\boldsymbol{q}}_s^3 + 0.1\mathrm{sgn}(\dot{\boldsymbol{q}}_s^3) + \\
&\quad 0.5[\sin(q_{s,1}^3\dot{q}_{s,1}^3), \sin(q_{s,2}^3\dot{q}_{s,2}^3), \sin(q_{s,3}^3\dot{q}_{s,3}^3)]^{\mathrm{T}}
\end{aligned}$$

To simulate the different human operators' actions in practices, $\boldsymbol{f}_h^i = [f_{h,1}^i, f_{h,2}^i, f_{h,3}^i]^{\mathrm{T}}$ $(i = 1, 2, 3)$ are set to be as shown in Fig. 6.3(a). In addition, given that the work environment of the different robots may be different, the environment forces $\boldsymbol{f}_e^i = [f_{e,1}^i, f_{e,2}^i, f_{e,3}^i]^{\mathrm{T}}$ are also set to be different, as shown in Fig. 6.3(b).

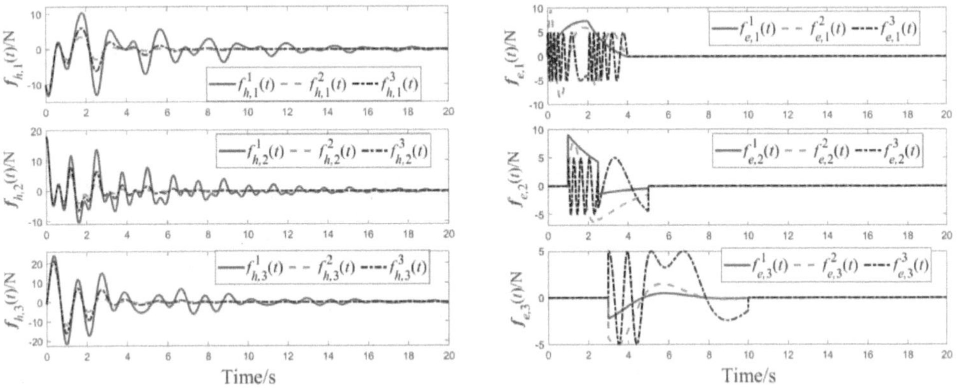

(a) (b)

Figure 6.3 The external forces $\boldsymbol{f}_h^i$ and $\boldsymbol{f}_e^i$. Forces exerted on the end-effectors of (a) the master robots by human operators and (b) the slave robots by environments.

Three sets of simulations are conducted to verify the effectiveness of the proposed algorithm.

6.5.1 Stability verification

The first simulation is used to verify the stability of the complete closed-loop tele-operation systems under the proposed algorithm. For the given control task, the synchronization errors are defined in (6.2). For simplicity, let $\boldsymbol{a}_1 = \boldsymbol{a}_2 = \boldsymbol{a}_3 \equiv \mathbf{0}$. The parameters of the predefined regions of the synchronization errors shown in (6.2) are given as: for all $i = 1, 2, 3$ and $k = 1, 2, 3$, $H_{s,1}^i = H_{s,2}^i = H_{s,3}^i = 1$, $l_{s,1}^1 = l_{s,1}^2 = l_{s,1}^3 = 0.6$, $l_{s,2}^1 = l_{s,2}^2 = l_{s,2}^3 = 0.72$, $l_{s,3}^1 = l_{s,3}^2 = l_{s,3}^3 = 0.6$, $\boldsymbol{\varrho}_{s\infty}^1 = \boldsymbol{\varrho}_{s\infty}^2 = \boldsymbol{\varrho}_{s\infty}^3 = [0.06, 0.06, 0.08]^{\mathrm{T}}$. Under the above simulation setup, from Theorem 6.1, the simulation results are given in Figs. 6.4 and 6.5.

Fig. 6.4 gives the tracking performance in each pair of master-slave robots, while Fig. 6.5 presents the trajectories of master robots and slave robots. One can find that, a satisfactory position tracking is obtained in each pair of master-slave robots by the proposed adaptive neural network controller. In Fig. 6.5, the blue solid lines are the trajectories of the synchronization errors $\boldsymbol{\epsilon}_s^i = [\epsilon_{s,1}^i, \epsilon_{s,2}^i, \epsilon_{s,3}^i]^{\mathrm{T}}$, and the dotted line is the response trajectory of the performance function (the convergence rate and steady-state error of the performance function are respectively determined by $\boldsymbol{l}_s^i = [l_{s,1}^i, l_{s,2}^i, l_{s,3}^i]^{\mathrm{T}}$ and $\boldsymbol{\varrho}_{s,\infty}^i$). It is demonstrated that for any $k = 1, 2, 3$ and $i = 1, 2, 3$, $\epsilon_{s,k}^i$ satisfy the constraint (6.4), which means that both the transient and steady-state coordination performances among three slave robots are guaranteed. In other words, the proposed tele-coordination algorithm can get the satisfactory task performance in simultaneous presence of asymmetric time-varying delays, non-passive external forces, and uncertain kinematics/dynamics.

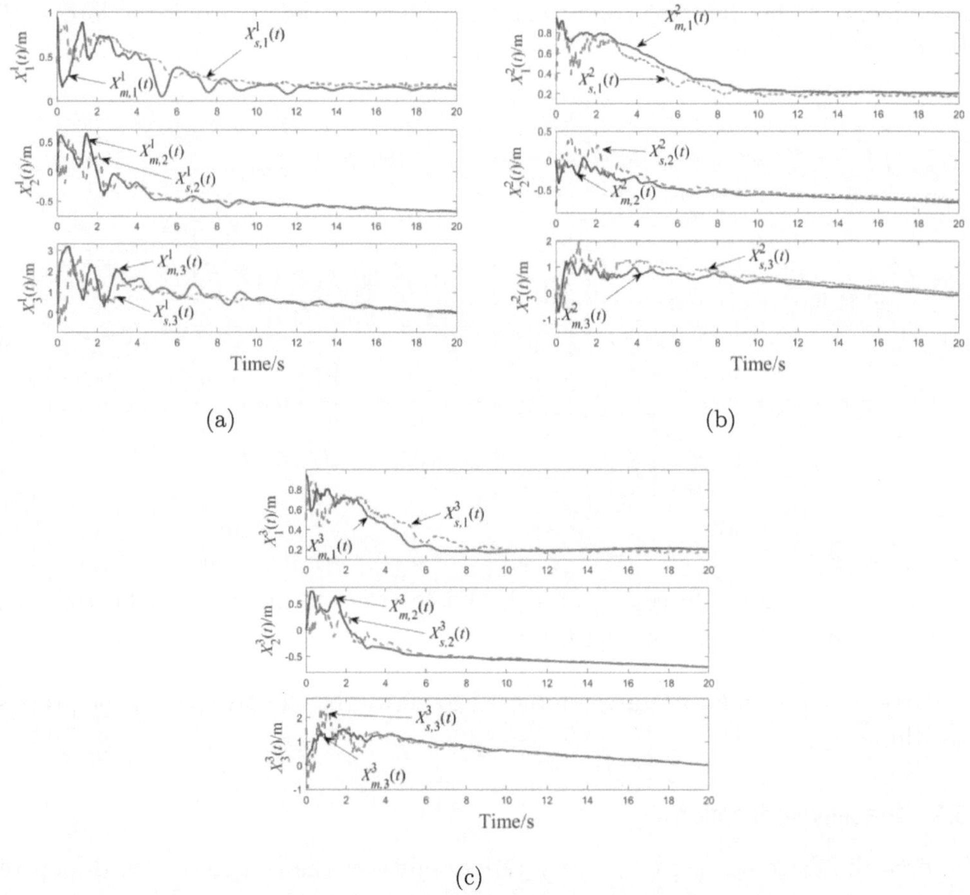

Figure 6.4 The position tracking between the master and slave robots. Position tracking between (a) the first pair of master-slave robots, (b) the second pair of master-slave robots and (c) the third pair of master-slave robots.

6.5.2 Comparison studies

Comparison study with the Li's method [15].

This set of simulations is used to illustrate that the new coordinated teleoperation algorithm proposed in this chapter has better control performance than the constrained coordinated control method. Here we take the algorithm [15] as an example, where an effective adaptive fuzzy control framework has been developed for multilateral cooperative teleoperation of multiple robotic manipulators with random communication delays. It adopts a hybrid motion/force architecture, while the algorithm proposed in this book is based on the concept of motion synchronization. Compared with the two methods, the algorithm proposed in this book is simpler in implementation. And because of the introduction of prescribed performance strategy, the new coordination algorithm has a wider range of applications. Specific to the coordination task shown in Fig. 6.1, in order not to damage the payload and/or the slave robots, the end-effector positions of the three slave robots have a high requirement,

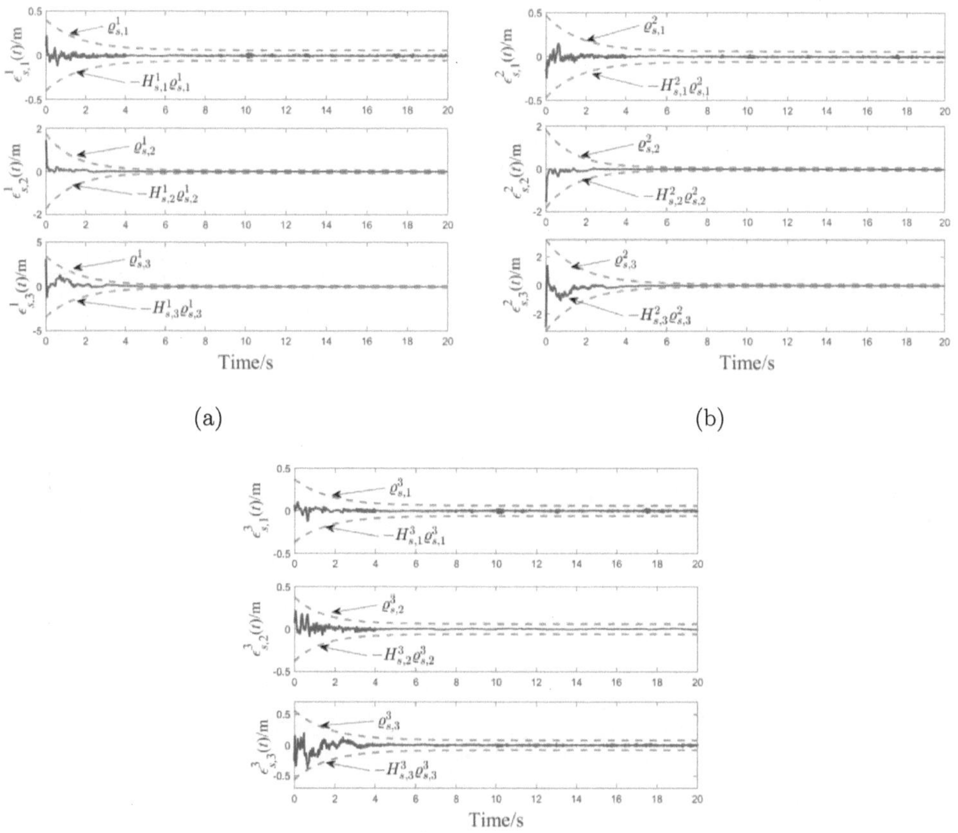

(c)

Figure 6.5 The synchronization errors $\boldsymbol{\epsilon}_s^i = [\epsilon_{s,1}^i, \epsilon_{s,2}^i, \epsilon_{s,3}^i]^{\mathrm{T}}$. Synchronization error (a) $\boldsymbol{\epsilon}_s^1$, (b)$\boldsymbol{\epsilon}_s^2$, (c)$\boldsymbol{\epsilon}_s^3$.

where the synchronization errors among the slave robots are expected to have small overshoot, fast convergence rate and small steady-state error. However, the algorithm based on the hybrid motion/force architecture [15] cannot add the above constraints in the controller, so it cannot satisfy the above control tasks. In fact, when the Li's algorithm [15] is applied, under the same simulation settings[1], the simulation results are given in Figs. 6.6 and 6.7, which show respectively the position tracking in each pair of master-slave robots and the coordination performance among slave robots. From 6.6, the Li's control method has a satisfactory tracking performance in each pair of master-slave robots. However, as shown in Fig. 6.7, it cannot guarantee the motion synchronization among the three slave robots. When applied to the above-mentioned tele-coordination tasks, it may damage the handled objects and/or slave robots, so it is not suitable for a given collaborative task and faces limitations.

[1]Since Li et al. [15] do not consider the kinematic uncertainty, to be fair, in this simulation implementation, its controller can use accurate kinematic information.

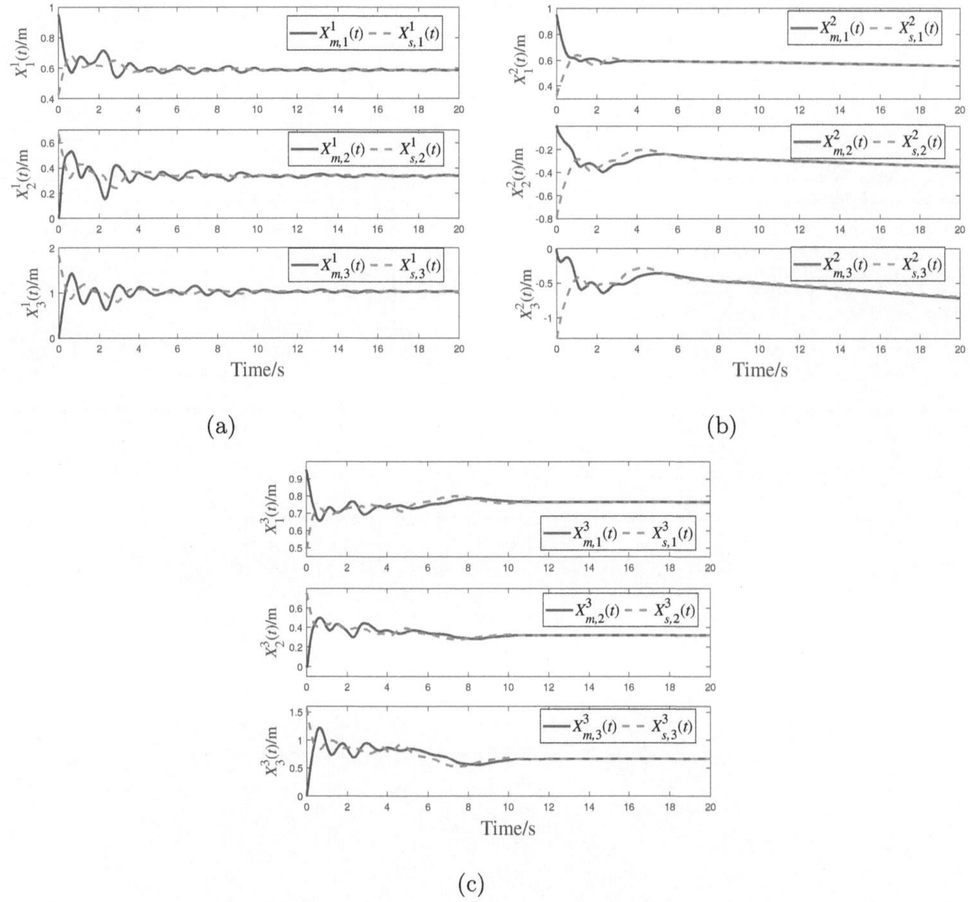

Figure 6.6 The position tracking between the master and slave robots under the Li's control method [15]. Position tracking between (a) the first pair of master-slave robots, (b) the second pair of master-slave robots and (c) the third pair of master-slave robots.

Comparison study with the Liu's method [13] This comparison study is performed to demonstrate that the proposed coordinated control has a better coordination performance than the non-constrained coordination control algorithm. Since each pair of master-slave robots is controlled independently, the coordination performance based on the non-constrained control approach can be adversely affected by the differences of the human operators' actions, the different delays in communication channels, as well as the unknown dynamics and external disturbances. To show this, this comparison study will be performed between this chapter and the Liu's method [13]. In this set of simulations, based on the non-constrained architecture, the Liu's algorithm [13] is applied to the coordinated teleoperation control of multiple robots.

Remark 6.3. *On the one hand, to our best knowledge, there have been few results addressing the multilateral tele-coordinated control that is subjected to time-delays, unknown kinematics/dynamics and external disturbances. On the other hand,*

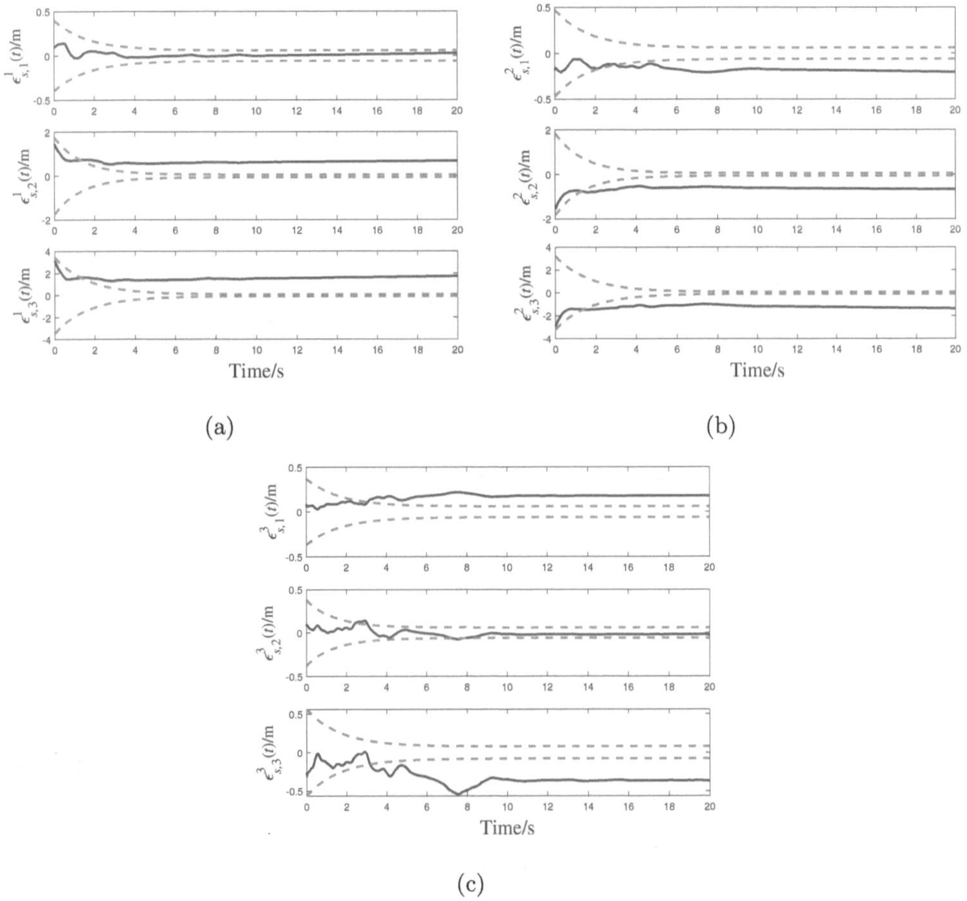

Figure 6.7 The synchronization errors $\epsilon_s^i = [\epsilon_{s,1}^i, \epsilon_{s,2}^i, \epsilon_{s,3}^i]^{\mathrm{T}}$ under the Li's control method [15]. Synchronization error (a) ϵ_s^1, (b)ϵ_s^2 and (c) ϵ_s^3.

under the architecture of non-constrained coordination control, each pair of master-slave robots is controlled independently. Therefore, the existing control approaches for single-master-single-slave teleoperation system are applicable to the multilateral tele-coordinated control based on non-constrained architecture. In Liu's framework [13], an adaptive controller is developed for bilateral teleoperation system in the presence of time delays and kinematics and dynamics uncertainties. In this comparison study, based on the non-constrained coordination architecture, the Liu's method [13] is extended to tele-coordinated control of multiple robots.

Under the same simulation setup as Subsection 6.5.1, the simulation results of Liu's method [13] are given in Figs. 6.8 and 6.9, which present respectively the position tracking errors and the synchronization errors.

Although the satisfactory position tracking is obtained as shown in Figs. 6.8, from 6.9, it has a poor performance in stabilizing the synchronization errors among the slave robots. By observing Figs. 6.5 and 6.9, a better coordination performance is obtained by this chapter.

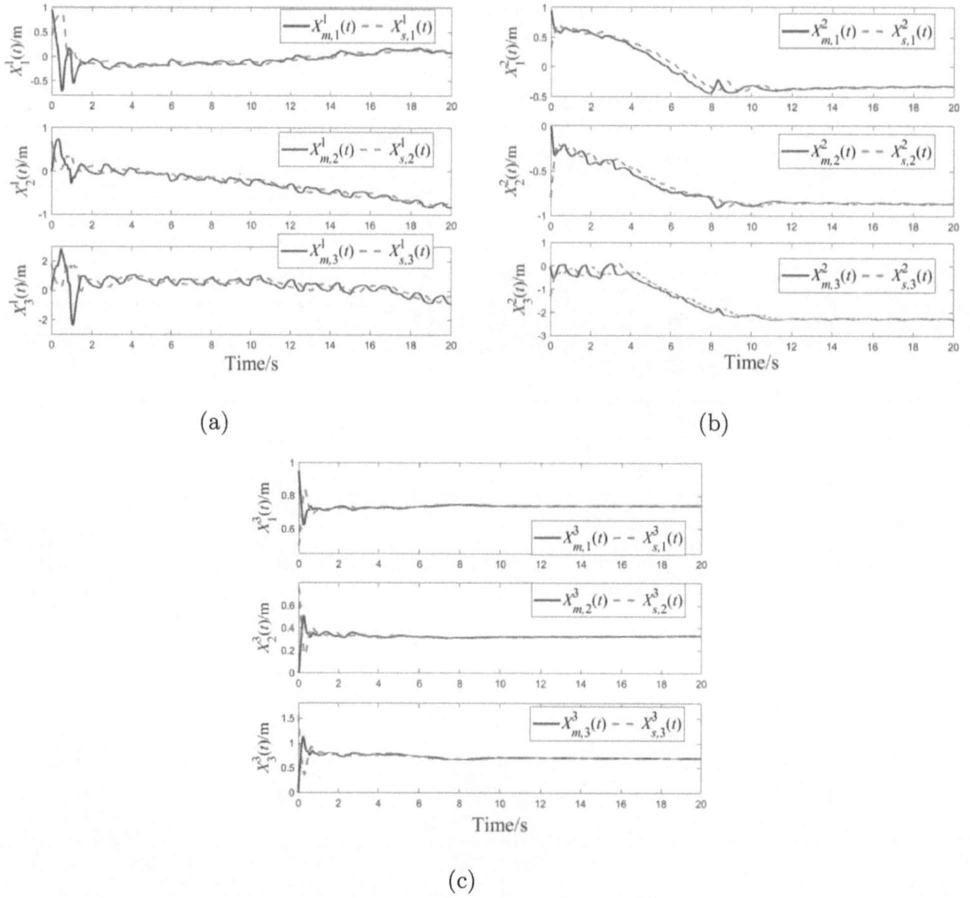

Figure 6.8 The position tracking between the master and slave robots under the Liu's control method [13]. Position tracking between (a) the first pair of master-slave robots, (b) the second pair of master-slave robots and (c) the third pair of master-slave robots.

6.6 CONCLUSION

In order to improve the performance of the traditional unconstrained coordinated control method, this chapter proposes a new unconstrained coordinated control strategy based on the motion synchronization of the slave robot for the multilateral tele-coordination control system. By introducing the prescribed performance control between the slave robots into the traditional unconstrained coordinated teleoperation control framework, and combining the auxiliary switched filtering method, a new neural network adaptive controller is designed, which ensures the cooperative performance of slave robots in the presence of operator differences in manipulation and external uncertainties (different communication delays, model uncertainty). While ensuring good coordination and cooperation performance, the use of force sensors is reduced, and hence the economic and technical burden of algorithm application is reduced. The chapter provides a new tele-coordination control strategy for engineers.

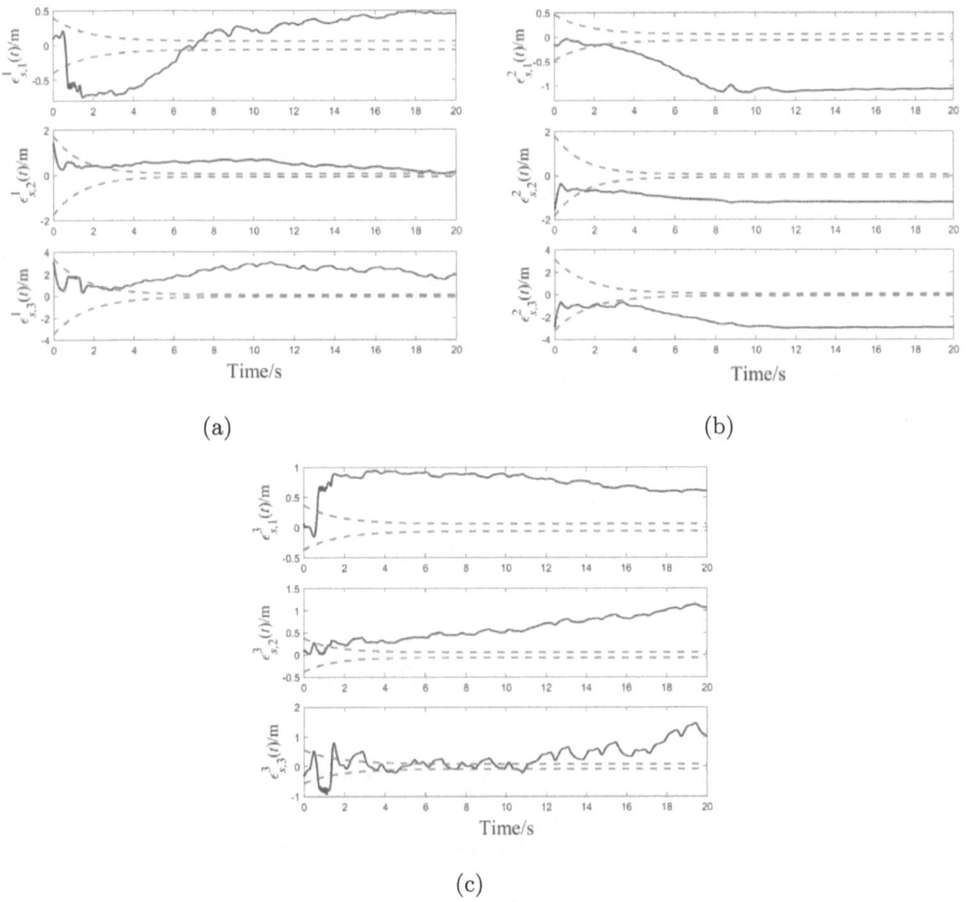

Figure 6.9 The synchronization errors $\boldsymbol{\epsilon}_s^i = [\epsilon_{s,1}^i, \epsilon_{s,2}^i, \epsilon_{s,3}^i]^{\mathrm{T}}$ under the Liu's control method [13]. Synchronization error (a) $\boldsymbol{\epsilon}_s^1$, (b) $\boldsymbol{\epsilon}_s^2$ and (c) $\boldsymbol{\epsilon}_s^3$.

Bibliography

[1] D. Sun and J. K. Mills, "Adaptive synchronized control for coordination of multi-robot assembly tasks," *IEEE Transactions on Robotics and Automation*, vol. 18, no. 4, pp. 498–510, 2002.

[2] T. H. Lee and C. J. Harris, *Adaptive Neural Network Control of Robotic Manipulators.* World Scientific, 1998.

[3] C. P. Bechlioulis, Z. Doulgeri, and G. A. Rovithakis, "Neuro-adaptive force/position control with prescribed performance and guaranteed contact maintenance," *IEEE Transactions on Neural Networks*, vol. 21, no. 12, pp. 1857–1868, 2010.

[4] A. K. Kostarigka and G. A. Rovithakis, "Adaptive dynamic output feedback neural network control of uncertain mimo nonlinear systems with prescribed

performance," *IEEE Transactions on Neural Networks and Learning Systems*, vol. 23, no. 1, pp. 138–149, 2011.

[5] D.-H. Zhai and Y. Xia, "Adaptive fuzzy control of multilateral asymmetric teleoperation for coordinated multiple mobile manipulators," *IEEE Transactions on Fuzzy Systems*, vol. 24, no. 1, pp. 57–70, 2015.

[6] I. G. Polushin, S. N. Dashkovskiy, A. Takhmar, and R. V. Patel, "A small gain framework for networked cooperative force-reflecting teleoperation," *Automatica*, vol. 49, no. 2, pp. 338–348, 2013.

[7] C.-C. Hua and P. X. Liu, "Delay-dependent stability criteria of teleoperation systems with asymmetric time-varying delays," *IEEE Transactions on Robotics*, vol. 26, no. 5, pp. 925–932, 2010.

[8] I. Sarras, E. Nuno, and L. Basanez, "An adaptive controller for nonlinear teleoperators with variable time-delays," *Journal of the Franklin Institute*, vol. 351, no. 10, pp. 4817–4837, 2014.

[9] S. Islam, P. X. Liu, and A. El Saddik, "Nonlinear control for teleoperation systems with time varying delay," *Nonlinear Dynamics*, vol. 76, no. 2, pp. 931–954, 2014.

[10] Z. Li, X. Cao, and N. Ding, "Adaptive fuzzy control for synchronization of nonlinear teleoperators with stochastic time-varying communication delays," *IEEE Transactions on Fuzzy Systems*, vol. 19, no. 4, pp. 745–757, 2011.

[11] Y. Yang, C. Hua, and X. Guan, "Coordination control for bilateral teleoperation with kinematics and dynamics uncertainties," *Robotics and Computer-Integrated Manufacturing*, vol. 30, no. 2, pp. 180–188, 2014.

[12] Y.-C. Liu, "Control of bilateral teleoperation system for robotic manipulators with uncertain kinematics and dynamics," in *Proceedings of the 4th Annual IEEE International Conference on Cyber Technology in Automation, Control and Intelligent*. IEEE, 2014, pp. 521–526.

[13] Y.-C. Liu and M.-H. Khong, "Adaptive control for nonlinear teleoperators with uncertain kinematics and dynamics," *IEEE/ASME Transactions on Mechatronics*, vol. 20, no. 5, pp. 2550–2562, 2015.

[14] C. I. Aldana Lopez, *Consensus control in robot networks and cooperative teleoperation: An operational space approach*. PhD dissertation, Universitat Politecnica de Catalunya, 2015.

[15] Z. Li, Y. Xia, and F. Sun, "Adaptive fuzzy control for multilateral cooperative teleoperation of multiple robotic manipulators under random network-induced delays," *IEEE Transactions on Fuzzy Systems*, vol. 22, no. 2, pp. 437–450, 2013.

IV

Heterogeneous Teleoperation

CHAPTER 7

Adaptive semi-autonomous control of heterogeneous teleoperation systems

THIS chapter addresses the adaptive task-space bilateral teleoperation for heterogeneous master and slave robots to guarantee stability and tracking performance, where a novel semi-autonomous teleoperation framework is developed to ensure the safety and enhance the efficiency of the robot in remote site. The basic idea is to stabilize the tracking error in task space while enhancing the efficiency of complex teleoperation by using redundant slave robot with sub-task control. By replacing the derivatives of position errors with the filtered outputs of their switching filters in the coordinate torque design, and employing the multiple Lyapunov-Krasovskii functionals method, the teleoperation system is proven to be state-independent input-to-output stable.

7.1 INTRODUCTION

In Chapters 3 to 6, an auxiliary switched filter control method is proposed for multi-type control tasks of networked teleoperated robot systems in complex external environments. On this basis, in order to adapt to the teleoperation control of slave robots with redundant degrees of freedom, this chapter will study the semi-autonomous adaptive control of heterogeneous teleoperation systems to further improve the control stability and safety of robots.

The so-called heterogeneity means that the master and slave robots have a non-uniform structure. The semi-autonomous control refers to the design of additional sub-task controllers using the redundant degrees of freedom of the slave robot through the null space control theory to autonomously complete tasks such as singularity avoidance, joint space obstacle avoidance and joint displacement limit avoidance without operator intervention, and do not affect the motion tracking between the master and slave robots. However, the use of slave robots with redundant degrees of freedom can create difficulties in teleoperated control design. Although the research

DOI: 10.1201/9781003382058-7

on related issues has made some progress [1,2], it still faces the limitations of application. As two important research results of semi-autonomous control of teleoperation systems, the control design and analysis method [1] only consider the constant communication delay. Liu et al. also consider the time-varying communication delay [2], but in the case of non-passive operator/environmental forces, the proposed method can only be used for teleoperation systems with nominal dynamics. Therefore, it is necessary to study semi-autonomous control algorithms of teleoperation systems with a wider application range.

In this chapter, a new adaptive control based on auxiliary switched filtering is proposed for heterogeneous teleoperation systems with asymmetric time-varying communication delays, passive/non-passive external forces, dynamics uncertainties, and input uncertainties. By adopting the semi-autonomous teleoperation control architecture [1], and replacing the end-effector position error derivative [1–3] with the auxiliary switched filter output, a new semi-autonomous teleoperation control architecture is developed. Combined with the adaptive control technology, the application scope of the existing algorithms [1,2] is further expanded.

7.2 PROBLEM FORMULATION

7.2.1 Robot dynamics

It is assumed that the dynamics of the master and slave robots are expressed as follows:

$$\begin{cases} \boldsymbol{M}_m(\boldsymbol{q}_m)\ddot{\boldsymbol{q}}_m + \boldsymbol{C}_m(\boldsymbol{q}_m, \dot{\boldsymbol{q}}_m)\dot{\boldsymbol{q}}_m + \boldsymbol{g}_m(\boldsymbol{q}_m) + \boldsymbol{f}_m(\dot{\boldsymbol{q}}_m) + \\ \qquad\qquad \boldsymbol{F}_m\dot{\boldsymbol{q}}_m + \boldsymbol{D}_m = \boldsymbol{J}_m^{\mathrm{T}}(\boldsymbol{q}_m)\boldsymbol{f}_h + \boldsymbol{\tau}_m \\ \boldsymbol{M}_s(\boldsymbol{q}_s)\ddot{\boldsymbol{q}}_s + \boldsymbol{C}_s(\boldsymbol{q}_s, \dot{\boldsymbol{q}}_s)\dot{\boldsymbol{q}}_s + \boldsymbol{g}_s(\boldsymbol{q}_s) + \boldsymbol{f}_s(\dot{\boldsymbol{q}}_s) + \\ \qquad\qquad \boldsymbol{F}_s\dot{\boldsymbol{q}}_s + \boldsymbol{D}_s = -\boldsymbol{J}_s^{\mathrm{T}}(\boldsymbol{q}_s)\boldsymbol{f}_e + \boldsymbol{\tau}_s \end{cases} \tag{7.1}$$

where the subscripts m and s are respectively the master and the slave robots, the dimension information of each component is $\boldsymbol{q}_m \in \mathbb{R}^n$, $\boldsymbol{q}_s \in \mathbb{R}^k$, $\boldsymbol{M}_m(\boldsymbol{q}_m)$, $\boldsymbol{C}_m(\boldsymbol{q}_m, \dot{\boldsymbol{q}}_m) \in \mathbb{R}^{n\times n}$, $\boldsymbol{M}_s(\boldsymbol{q}_s)$, $\boldsymbol{C}_s(\boldsymbol{q}_s, \dot{\boldsymbol{q}}_s) \in \mathbb{R}^{k\times k}$, $\boldsymbol{g}_m(\boldsymbol{q}_m) \in \mathbb{R}^n$, $\boldsymbol{g}_s(\boldsymbol{q}_s) \in \mathbb{R}^k$, $\boldsymbol{f}_m(\dot{\boldsymbol{q}}_m) \in \mathbb{R}^n$, $\boldsymbol{f}_s(\dot{\boldsymbol{q}}_s) \in \mathbb{R}^k$, $\boldsymbol{F}_m \in \mathbb{R}^{n\times n}$, $\boldsymbol{F}_s \in \mathbb{R}^{k\times k}$, $\boldsymbol{D}_m \in \mathbb{R}^n$, $\boldsymbol{D}_s \in \mathbb{R}^k$, $\boldsymbol{\tau}_m \in \mathbb{R}^n$, $\boldsymbol{\tau}_s \in \mathbb{R}^k$, $\boldsymbol{J}_m(\boldsymbol{q}_m) \in \mathbb{R}^{n\times n}$, $\boldsymbol{J}_s(\boldsymbol{q}_s) \in \mathbb{R}^{n\times k}$, $\boldsymbol{f}_h \in \mathbb{R}^n$, $\boldsymbol{f}_e \in \mathbb{R}^n$. For any $j \in \{m, s\}$, $\boldsymbol{F}_j$ is the viscous friction coefficient matrix of robot joints, $\boldsymbol{f}_j(\dot{\boldsymbol{q}}_j)$ is the Coulomb friction force vector, $\boldsymbol{D}_j$ is a bounded external disturbance, and the other components are defined in the same as (2.11). To achieve the semi-autonomous control, the slave robot is assumed to be a redundant manipulator, while the master robot is assumed to be a non-redundant one for simplicity. Thus, $n < k$.

The generalized end-effector positions $\boldsymbol{X}_m$ and $\boldsymbol{X}_s$ are:

$$\boldsymbol{X}_j = \boldsymbol{h}_j(\boldsymbol{q}_j),\ \dot{\boldsymbol{X}}_j = \boldsymbol{J}_j(\boldsymbol{q}_j)\dot{\boldsymbol{q}}_j,\ \forall j \in \{m, s\}$$

where $\boldsymbol{h}_m(\boldsymbol{q}_m) \in \mathbb{R}^n$ and $\boldsymbol{h}_s(\boldsymbol{q}_s) \in \mathbb{R}^n$ denote the mappings between the joint space and the task space, $\boldsymbol{J}_m(\boldsymbol{q}_m) = \dfrac{\partial \boldsymbol{h}_m(\boldsymbol{q}_m)}{\partial \boldsymbol{q}_m} \in \mathbb{R}^{n\times n}$ and $\boldsymbol{J}_s(\boldsymbol{q}_s) = \dfrac{\partial \boldsymbol{h}_s(\boldsymbol{q}_s)}{\partial \boldsymbol{q}_s} \in \mathbb{R}^{n\times k}$ are the Jacobian matrices that are assumed to be known. In following, denote $\boldsymbol{J}_m^{-1}(\boldsymbol{q}_m) \in$

$\mathbb{R}^{n\times n}$ be the inverse matrix of $\boldsymbol{J}_m(\boldsymbol{q}_m)$, and let $\boldsymbol{J}_s^+(\boldsymbol{q}_s) \in \mathbb{R}^{k\times n}$ denote the pseudo-inverse of $\boldsymbol{J}_s(\boldsymbol{q}_s)$, which is defined by

$$\begin{cases} \boldsymbol{J}_s^+(\boldsymbol{q}_s) = \boldsymbol{J}_s^{\mathrm{T}}(\boldsymbol{q}_s)\left(\boldsymbol{J}_s(\boldsymbol{q}_s)\boldsymbol{J}_s^{\mathrm{T}}(\boldsymbol{q}_s)\right)^{-1} \\ \boldsymbol{J}_s(\boldsymbol{q}_s)\boldsymbol{J}_s^+(\boldsymbol{q}_s) = \boldsymbol{I} \end{cases}$$

where $\boldsymbol{I}$ is an identity matrix with appropriate dimension.

Similar to the previous work [4], the operator dynamics of the human operator and the environment are assumed to be linear time invariant (LTI) models in the task-space:

$$\begin{cases} \boldsymbol{f}_h = \boldsymbol{f}_h^* - M_h\ddot{\boldsymbol{X}}_m - B_h\dot{\boldsymbol{X}}_m - K_h\boldsymbol{X}_m \\ \boldsymbol{f}_e = \boldsymbol{f}_e^* + M_e\ddot{\boldsymbol{X}}_s + B_e\dot{\boldsymbol{X}}_s + K_e\boldsymbol{X}_s \end{cases} \tag{7.2}$$

where M_h, M_e, B_h, B_e, K_h, and K_e are the unknown non-negative constant scalars corresponding to the mass, damping and stiffness of human operator and environment, with $\boldsymbol{f}_h^*$ and $\boldsymbol{f}_e^*$ being respectively the human operator and the environment exogenous force that are subjected to:

Assumption 7.1. *$\boldsymbol{f}_h^*$ and $\boldsymbol{f}_e^*$ are assumed to be locally essentially bounded.*

Then system (7.1) can then be rewritten as

$$\begin{cases} \boldsymbol{\mathcal{M}}_m(\boldsymbol{q}_m)\ddot{\boldsymbol{q}}_m + \boldsymbol{\mathcal{C}}_m(\boldsymbol{q}_m,\dot{\boldsymbol{q}}_m)\dot{\boldsymbol{q}}_m + \boldsymbol{g}_m(\boldsymbol{q}_m) + \\ \qquad \bar{\boldsymbol{f}}_m(\boldsymbol{q}_m,\dot{\boldsymbol{q}}_m) + \boldsymbol{D}_m = \boldsymbol{J}_m^{\mathrm{T}}(\boldsymbol{q}_m)\boldsymbol{f}_h^* + \boldsymbol{\tau}_m \\ \boldsymbol{\mathcal{M}}_s(\boldsymbol{q}_s)\ddot{\boldsymbol{q}}_s + \boldsymbol{\mathcal{C}}_s(\boldsymbol{q}_s,\dot{\boldsymbol{q}}_s)\dot{\boldsymbol{q}}_s + \boldsymbol{g}_s(\boldsymbol{q}_s) + \\ \qquad \bar{\boldsymbol{f}}_s(\boldsymbol{q}_s,\dot{\boldsymbol{q}}_s) + \boldsymbol{D}_s = -\boldsymbol{J}_s^{\mathrm{T}}(\boldsymbol{q}_s)\boldsymbol{f}_e^* + \boldsymbol{\tau}_s \end{cases} \tag{7.3}$$

where

$$\begin{aligned} \boldsymbol{\mathcal{M}}_m(\boldsymbol{q}_m) &= \boldsymbol{M}_m(\boldsymbol{q}_m) + \boldsymbol{J}_m^{\mathrm{T}}(\boldsymbol{q}_m)M_h\boldsymbol{J}_m(\boldsymbol{q}_m) \\ \boldsymbol{\mathcal{M}}_s(\boldsymbol{q}_s) &= \boldsymbol{M}_s(\boldsymbol{q}_s) + \boldsymbol{J}_s^{\mathrm{T}}(\boldsymbol{q}_s)M_e\boldsymbol{J}_s(\boldsymbol{q}_s) \\ \boldsymbol{\mathcal{C}}_m(\boldsymbol{q}_m,\dot{\boldsymbol{q}}_m) &= \boldsymbol{C}_m(\boldsymbol{q}_m,\dot{\boldsymbol{q}}_m) + \boldsymbol{J}_m^{\mathrm{T}}(\boldsymbol{q}_m)B_h\boldsymbol{J}_m(\boldsymbol{q}_m) + \boldsymbol{J}_m^{\mathrm{T}}(\boldsymbol{q}_m)M_h\dot{\boldsymbol{J}}_m(\boldsymbol{q}_m) \\ \boldsymbol{\mathcal{C}}_s(\boldsymbol{q}_s,\dot{\boldsymbol{q}}_s) &= \boldsymbol{C}_s(\boldsymbol{q}_s,\dot{\boldsymbol{q}}_s) + \boldsymbol{J}_s^{\mathrm{T}}(\boldsymbol{q}_s)B_e\boldsymbol{J}_s(\boldsymbol{q}_s) + \boldsymbol{J}_s^{\mathrm{T}}(\boldsymbol{q}_s)M_e\dot{\boldsymbol{J}}_s(\boldsymbol{q}_s) \\ \bar{\boldsymbol{f}}_m(\boldsymbol{q}_m,\dot{\boldsymbol{q}}_m) &= \boldsymbol{f}_m(\dot{\boldsymbol{q}}_m) + \boldsymbol{F}_m\dot{\boldsymbol{q}}_m + \boldsymbol{J}_m^{\mathrm{T}}(\boldsymbol{q}_m)K_h\boldsymbol{h}_m(\boldsymbol{q}_m) \\ \bar{\boldsymbol{f}}_s(\boldsymbol{q}_s,\dot{\boldsymbol{q}}_s) &= \boldsymbol{f}_s(\dot{\boldsymbol{q}}_s) + \boldsymbol{F}_s\dot{\boldsymbol{q}}_s + \boldsymbol{J}_s^{\mathrm{T}}(\boldsymbol{q}_s)K_e\boldsymbol{h}_s(\boldsymbol{q}_s) \end{aligned}$$

Remark 7.1. *According to the structural properties for robotic systems [5], for any $\boldsymbol{x} \in \mathbb{R}^n$ and $\boldsymbol{y} \in \mathbb{R}^k$, it has*

$$\begin{aligned} \boldsymbol{x}^{\mathrm{T}}(\dot{\boldsymbol{\mathcal{M}}}_m(\boldsymbol{q}_m) - 2\boldsymbol{\mathcal{C}}_m(\boldsymbol{q}_m,\dot{\boldsymbol{q}}_m))\boldsymbol{x} &= -2\boldsymbol{x}^{\mathrm{T}}\boldsymbol{J}_m^{\mathrm{T}}(\boldsymbol{q}_m)B_h\boldsymbol{J}_m(\boldsymbol{q}_m)\boldsymbol{x} \\ \boldsymbol{y}^{\mathrm{T}}(\dot{\boldsymbol{\mathcal{M}}}_s(\boldsymbol{q}_s) - 2\boldsymbol{\mathcal{C}}_s(\boldsymbol{q}_s,\dot{\boldsymbol{q}}_s))\boldsymbol{y} &= -2\boldsymbol{y}^{\mathrm{T}}\boldsymbol{J}_s^{\mathrm{T}}(\boldsymbol{q}_s)B_e\boldsymbol{J}_s(\boldsymbol{q}_s)\boldsymbol{y} \end{aligned}$$

Moreover, $\boldsymbol{\mathcal{M}}_j(\boldsymbol{q}_j)$ satisfies the Property 2.1, where $j \in \{m, s\}$.

7.2.2 Input uncertainty

In this chapter, we will consider the non-linear output characteristics of the robot actuator with dead-zone in the control. In this case, the control torque is re-written as

$$\boldsymbol{\tau}_j = \boldsymbol{\Gamma}_j(\bar{\boldsymbol{\tau}}_j),\ j \in \{m, s\}$$

where $\boldsymbol{\tau}_m = [\tau_{m,1}, \tau_{m,2}, \cdots, \tau_{m,n}]^{\mathrm{T}} \in \mathbb{R}^n$ and $\boldsymbol{\tau}_s = [\tau_{s,1}, \tau_{s,2}, \cdots, \tau_{s,k}]^{\mathrm{T}} \in \mathbb{R}^k$ are the actual output for the actuator, $\bar{\boldsymbol{\tau}}_m = [\bar{\tau}_{m,1}, \bar{\tau}_{m,2}, \cdots, \bar{\tau}_{m,n}]^{\mathrm{T}} \in \mathbb{R}^n$ and $\bar{\boldsymbol{\tau}}_s = [\bar{\tau}_{s,1}, \bar{\tau}_{s,2}, \cdots, \bar{\tau}_{s,k}]^{\mathrm{T}} \in \mathbb{R}^k$ are the real outputs of the controller (i.e., the input of the actuator), $\boldsymbol{\Gamma}_j(\cdot)$ is the dead-zone function, $j \in \{m, s\}$. The ith element is defined as

$$\tau_{j,i} = \Gamma_j(\bar{\tau}_{j,i}) = \begin{cases} m_{jir}(\bar{\tau}_{j,i} - b_{jir}), & \bar{\tau}_{j,i} \geq b_{jir} \\ m_{jil}(\bar{\tau}_{j,i} - b_{jil}), & \bar{\tau}_{j,i} < b_{jil} \\ 0, & \text{others} \end{cases} \tag{7.4}$$

where m_{jir}, m_{jil}, b_{jir}, and b_{jil} are the dead-zone nonlinear parameters, as shown in Fig. 7.1.

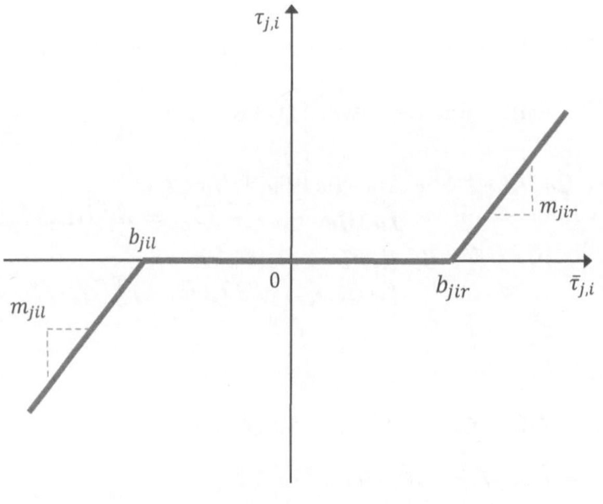

Figure 7.1 Dead-zone nonsmooth nonlinearity.

From a practical viewpoint, (7.4) can be rewritten as

$$\boldsymbol{\tau}_j = \boldsymbol{\Gamma}_j(\bar{\boldsymbol{\tau}}_j) = \boldsymbol{m}_j\bar{\boldsymbol{\tau}}_j + \boldsymbol{\Delta}_j$$

where $\boldsymbol{m}_m = \mathrm{diag}[m_{m,1}, m_{m,2}, \cdots, m_{m,n}]$, $\boldsymbol{m}_s = \mathrm{diag}[m_{s,1}, m_{s,2}, \cdots, m_{s,k}]$, with

$$m_{j,i} = \begin{cases} m_{jir}, & \bar{\tau}_{j,i} \geq b_{jir} \\ m_{jil}, & \bar{\tau}_{j,i} < b_{jir} \end{cases}$$

while $\boldsymbol{\Delta}_m = [\Delta_{m,1}, \Delta_{m,2}, \cdots, \Delta_{m,n}]^{\mathrm{T}}$ and $\boldsymbol{\Delta}_s = [\Delta_{s,1}, \Delta_{s,2}, \cdots, \Delta_{s,k}]^{\mathrm{T}}$, with

$$\Delta_{j,i} = \begin{cases} -m_{jir}b_{jir}, & \bar{\tau}_{j,i} \geq b_{jir} \\ -m_{jil}\bar{\boldsymbol{\tau}}_{ji}, & b_{jil} < \bar{\tau}_{j,i} < b_{jir} \\ -m_{jil}b_{jil}, & \bar{\tau}_{j,i} \leq b_{jil} \end{cases}$$

The key features of the input dead-zone satisfy:

Assumption 7.2. [6] *The dead-zone parameters m_{jir}, m_{jil}, b_{jir} and b_{jil} are unknown, but their signs are known, i.e., $m_{jir} > 0$, $m_{jil} > 0$, $b_{jir} > 0$, $b_{jil} < 0$.*

Assumption 7.3. [6] *There exist known constants $b_{r1} > 0$, $b_{r2} > 0$, $b_{l1} < 0$, $b_{l2} < 0$, $\delta_1 > 0$ and $\delta_2 > 0$, such that $m_{jir} \in [\delta_1, \delta_2]$, $m_{jil} \in [\delta_1, \delta_2]$, $b_{jir} \in [b_{r1}, b_{r2}]$, $b_{jil} \in [b_{l1}, b_{l2}]$.*

7.2.3 Control objectives

To measure the synchronization performance of master and slave robots, define the tracking errors as

$$\begin{cases} \boldsymbol{e}_m(t) = \boldsymbol{X}_m(t) - \boldsymbol{X}_s(t - d_s(t)) \\ \boldsymbol{e}_s(t) = \boldsymbol{X}_s(t) - \boldsymbol{X}_m(t - d_m(t)) \end{cases} \tag{7.5}$$

where $d_m(t)$ and $d_s(t)$ are the time-varying delays in the communication channels. And for the sake of simplifying expression, let

$$\boldsymbol{e}_{vm}(t) = \dot{\boldsymbol{X}}_m(t) - \boldsymbol{X}_{vds}(t, d_s(t)) = \dot{\boldsymbol{X}}_m(t) - \dot{\boldsymbol{X}}_s(\theta)|_{\theta=t-d_s(t)}$$
$$\boldsymbol{e}_{vs}(t) = \dot{\boldsymbol{X}}_s(t) - \boldsymbol{X}_{vdm}(t, d_m(t)) = \dot{\boldsymbol{X}}_s(t) - \dot{\boldsymbol{X}}_m(\theta)|_{\theta=t-d_m(t)}$$

In what follows, for all $j \in \{m, s\}$, the time delays are assumed to satisfy:

Assumption 7.4. *There exist bounded constant positive scalars $\bar{d}$ and $\tilde{d}$, such that $0 \leq d_j(t) \leq \bar{d}$ and $0 \leq |\dot{d}_j(t)| \leq \tilde{d}$.*

Therefore, the teleoperation control task in this chapter is defined as follows

(1) Task space teleoperation: In the presence of asymmetric time-varying communication delay, dynamic uncertainty and input uncertainty, the motion synchronization of master and slave robots in task space should be guaranteed

(2) Semi-autonomous control: Use the redundant degrees of freedom of the slave robot to realize the semi-autonomous control of the additional subtasks of the slave robot.

7.3 TASK SPACE TELEOPERATION

This section investigates a novel switched control framework for the task space teleoperation.

Firstly, define the auxiliary variables:

$$\begin{cases} \boldsymbol{\eta}_m = \dot{\boldsymbol{q}}_m + \boldsymbol{J}_m^{-1}(\boldsymbol{q}_m)\lambda_m \boldsymbol{e}_m \\ \boldsymbol{\eta}_s = \dot{\boldsymbol{q}}_s + \boldsymbol{J}_s^{+}(\boldsymbol{q}_s)\lambda_s \boldsymbol{e}_s - (\boldsymbol{I} - \boldsymbol{J}_s^{+}(\boldsymbol{q}_s)\boldsymbol{J}_s(\boldsymbol{q}_s))\boldsymbol{\phi}_s \end{cases} \tag{7.6}$$

where for all $j \in \{m, s\}$, $\lambda_j > 0$, $\boldsymbol{\phi}_s \in \mathbb{R}^k$ is the negative gradient of an appropriately defined function. Given that $\boldsymbol{J}_s(\boldsymbol{q}_s)[\boldsymbol{I} - \boldsymbol{J}_s^{+}(\boldsymbol{q}_s)\boldsymbol{J}_s(\boldsymbol{q}_s)] = \boldsymbol{0}$, the operator

$\boldsymbol{I} - \boldsymbol{J}_s^+(\boldsymbol{q}_s)\boldsymbol{J}_s(\boldsymbol{q}_s)$ projects vector $\boldsymbol{\phi}_s$ to the null space of the mapping $\boldsymbol{J}_s(\boldsymbol{q}_s)$. Hence, $\boldsymbol{\phi}_s$ can produce only a joint self-motion of the structure but no task space motion. In what follows, an appropriate $\boldsymbol{\phi}_s$ will be constructed to complete the sub-task control.

Substituting (7.6) into (7.1),

$$\begin{cases} \boldsymbol{\mathcal{M}}_m(\boldsymbol{q}_m)\dot{\boldsymbol{\eta}}_m + \boldsymbol{\mathcal{C}}_m(\boldsymbol{q}_m, \dot{\boldsymbol{q}}_m)\boldsymbol{\eta}_m = \boldsymbol{G}'_m + \boldsymbol{J}_m^{\mathrm{T}}(\boldsymbol{q}_m)\boldsymbol{f}_h^* + \\ \qquad \boldsymbol{m}_m\bar{\boldsymbol{\tau}}_m + \boldsymbol{\mathcal{M}}_m(\boldsymbol{q}_m)\boldsymbol{J}_m^{-1}(\boldsymbol{q}_m)\lambda_m\dot{d}_s\boldsymbol{X}_{vds}(t, d_s(t)) \\ \boldsymbol{\mathcal{M}}_s(\boldsymbol{q}_s)\dot{\boldsymbol{\eta}}_s + \boldsymbol{\mathcal{C}}_s(\boldsymbol{q}_s, \dot{\boldsymbol{q}}_s)\boldsymbol{\eta}_s = \boldsymbol{G}'_s - \boldsymbol{J}_s^{\mathrm{T}}(\boldsymbol{q}_s)\boldsymbol{f}_e^* + \\ \qquad \boldsymbol{m}_s\bar{\boldsymbol{\tau}}_s + \boldsymbol{\mathcal{M}}_s(\boldsymbol{q}_s)\boldsymbol{J}_s^{+}(\boldsymbol{q}_s)\lambda_s\dot{d}_m\boldsymbol{X}_{vdm}(t, d_m(t)) \end{cases} \tag{7.7}$$

where for $j \in \{m, s\}$,

$$\boldsymbol{G}'_j = \boldsymbol{\mathcal{M}}_j(\boldsymbol{q}_j)\boldsymbol{a}_j + \boldsymbol{\mathcal{C}}_j(\boldsymbol{q}_j, \dot{\boldsymbol{q}}_j)\boldsymbol{v}_j - \boldsymbol{g}_j(\boldsymbol{q}_j) - \bar{\boldsymbol{f}}_j(\boldsymbol{q}_j, \dot{\boldsymbol{q}}_j) - \boldsymbol{D}_j + \boldsymbol{\Delta}_j$$

with

$$\begin{aligned} \boldsymbol{a}_m &= \dot{\boldsymbol{J}}_m^{-1}(\boldsymbol{q}_m)\lambda_m\boldsymbol{e}_m + \boldsymbol{J}_m^{-1}(\boldsymbol{q}_m)\lambda_m\boldsymbol{e}_{vm} \\ \boldsymbol{v}_m &= \boldsymbol{J}_m^{-1}(\boldsymbol{q}_m)\lambda_m\boldsymbol{e}_m \\ \boldsymbol{a}_s &= \dot{\boldsymbol{J}}_s^{+}(\boldsymbol{q}_s)\lambda_s\boldsymbol{e}_s + \boldsymbol{J}_s^{+}(\boldsymbol{q}_s)\lambda_s\boldsymbol{e}_{vs} - \frac{\mathrm{d}}{\mathrm{d}t}\left((\boldsymbol{I} - \boldsymbol{J}_s^{+}(\boldsymbol{q}_s)\boldsymbol{J}_s(\boldsymbol{q}_s))\boldsymbol{\phi}_s\right) \\ \boldsymbol{v}_s &= \boldsymbol{J}_s^{+}(\boldsymbol{q}_s)\lambda_s\boldsymbol{e}_s - (\boldsymbol{I} - \boldsymbol{J}_s^{+}(\boldsymbol{q}_s)\boldsymbol{J}_s(\boldsymbol{q}_s))\boldsymbol{\phi}_s \end{aligned}$$

Denote

$$\begin{aligned} \boldsymbol{G}'_{j1} &= \boldsymbol{\mathcal{M}}_j(\boldsymbol{q}_j)\boldsymbol{a}_j + \boldsymbol{\mathcal{C}}_j(\boldsymbol{q}_j, \dot{\boldsymbol{q}}_j)\boldsymbol{v}_j - \boldsymbol{g}_j(\boldsymbol{q}_j) - \boldsymbol{F}_j\dot{\boldsymbol{q}}_j - \boldsymbol{J}_j^{\mathrm{T}}(\boldsymbol{q}_j)K_h\boldsymbol{h}_j(\boldsymbol{q}_j) \\ \boldsymbol{G}'_{j2} &= -\boldsymbol{f}_j(\dot{\boldsymbol{q}}_j) - \boldsymbol{D}_j + \boldsymbol{\Delta}_j \end{aligned}$$

Then $\boldsymbol{G}'_j = \boldsymbol{G}'_{j1} + \boldsymbol{G}'_{j2}$. Obviously, $\boldsymbol{G}'_{j2}$ is bounded. Without loss of generality, it is assumed that $|\boldsymbol{G}'_{j2}| \leq \omega_j$, where ω_j is an unknown positive constant. Using the parametric linearizability property in Property 2.5, $\boldsymbol{G}'_{j1} = \boldsymbol{Y}_{j0}(\boldsymbol{q}_j, \dot{\boldsymbol{q}}_j, \boldsymbol{a}_j, \boldsymbol{v}_j)\boldsymbol{\Theta}_j$, where $\boldsymbol{Y}_{j0}(\boldsymbol{q}_j, \dot{\boldsymbol{q}}_j, \boldsymbol{a}_j, \boldsymbol{v}_j) \in \mathbb{R}^{n_j \times o_j}$ is a matrix of known functions called regressor, $\boldsymbol{\Theta}_j \in \mathbb{R}^{o_j}$ is a vector of unknown parameters, $n_m = n$, $n_s = k$.

For all $j \in \{m, s\}$, $\iota \in \{m, s\}$ and $\iota \neq j$, to compensate the last term of (7.7), i.e., $\boldsymbol{\mathcal{M}}_j(\boldsymbol{q}_j)\boldsymbol{J}_j^{-1}(\boldsymbol{q}_j)\lambda_j\dot{d}_\iota\boldsymbol{X}_{vd\iota}(t, d_\iota(t))$, the adaptive technique based on linearly parameterizable of $\boldsymbol{\mathcal{M}}_j(\boldsymbol{q}_j)$ will also be used. For convenience, let $\bar{\boldsymbol{X}}_j(t) = \boldsymbol{J}_j^{-1}(\boldsymbol{q}_j)\boldsymbol{X}_{vd\iota}(t, d_\iota(t))$. Then

$$\boldsymbol{\mathcal{M}}_j(\boldsymbol{q}_j)\bar{\boldsymbol{X}}_j = \boldsymbol{Y}_j(\boldsymbol{q}_j, \bar{\boldsymbol{X}}_j)\boldsymbol{\theta}_j$$

where $\boldsymbol{Y}_j(\boldsymbol{q}_j, \bar{\boldsymbol{X}}_j) \in \mathbb{R}^{n_j \times p_j}$ is the matrix of known functions, $\boldsymbol{\theta}_j \in \mathbb{R}^{p_j}$ is the vector of unknown parameters.

Then let

$$\bar{\boldsymbol{\eta}}_j = [\bar{\eta}_{j,1}, \bar{\eta}_{j,2}, \cdots, \bar{\eta}_{j,n}]^{\mathrm{T}} = \boldsymbol{J}_j(\boldsymbol{q}_j)\boldsymbol{\eta}_j = \dot{\boldsymbol{X}}_j + \lambda_j\boldsymbol{e}_j,\ j \in \{m, s\}$$

Design the control inputs as

$$\begin{cases} \bar{\boldsymbol{\tau}}_m = -K_{m1}\boldsymbol{\eta}_m - K_{m2}\boldsymbol{J}_m^{\mathrm{T}}(\boldsymbol{q}_m)\bar{\boldsymbol{\eta}}_m - K_{m3}\boldsymbol{J}_m^{\mathrm{T}}(\boldsymbol{q}_m)\boldsymbol{y}_m - \dfrac{|\hat{\omega}_m|}{\delta_1}\mathrm{sgn}(\boldsymbol{\eta}_m) + \tilde{\boldsymbol{\tau}}_m \\ \bar{\boldsymbol{\tau}}_s = -K_{s1}\boldsymbol{\eta}_s - K_{s2}\boldsymbol{J}_s^{\mathrm{T}}(\boldsymbol{q}_s)\bar{\boldsymbol{\eta}}_s - K_{s3}\boldsymbol{J}_s^{\mathrm{T}}(\boldsymbol{q}_s)\boldsymbol{y}_s - \dfrac{|\hat{\omega}_s|}{\delta_1}\mathrm{sgn}(\boldsymbol{\eta}_s) + \tilde{\boldsymbol{\tau}}_s \end{cases} \tag{7.8}$$

where for all $j \in \{m, s\}$,

$$\tilde{\boldsymbol{\tau}}_j = -\frac{1}{\delta_1}\boldsymbol{Y}_{j0}(\boldsymbol{q}_j, \dot{\boldsymbol{q}}_j, \boldsymbol{a}_j, \boldsymbol{v}_j)\hat{\boldsymbol{\Theta}}_j\mathrm{sgn}(\boldsymbol{\eta}_j^{\mathrm{T}}\boldsymbol{Y}_{j0}(\boldsymbol{q}_j, \dot{\boldsymbol{q}}_j, \boldsymbol{a}_j, \boldsymbol{v}_j)\hat{\boldsymbol{\Theta}}_j) - \frac{K_{j4}}{\delta_1}\boldsymbol{Y}_j(\boldsymbol{q}_j, \bar{\boldsymbol{X}}_j)\boldsymbol{Y}_j^{\mathrm{T}}(\boldsymbol{q}_j, \bar{\boldsymbol{X}}_j)\boldsymbol{\eta}_j$$

and the adaptive laws are updated by

$$\begin{cases} \dot{\hat{\boldsymbol{\Theta}}}_j = \psi_{j1}\boldsymbol{Y}_{j0}^{\mathrm{T}}(\boldsymbol{q}_j, \dot{\boldsymbol{q}}_j, \boldsymbol{a}_j, \boldsymbol{v}_j)\boldsymbol{\eta}_j - \psi_{j2}(\hat{\boldsymbol{\Theta}}_j - \boldsymbol{\Theta}_j^*) \\ \dot{\hat{\omega}}_j = |\boldsymbol{\eta}_j| - \psi_{j3}(\hat{\omega}_j - \omega_j^*) \\ \dot{\hat{\boldsymbol{\theta}}}_j = -\lambda_j\tilde{d}\psi_{j4}\boldsymbol{Y}_j^{\mathrm{T}}(\boldsymbol{q}_j, \bar{\boldsymbol{X}}_j)\boldsymbol{\eta}_j\mathrm{sgn}\left(\boldsymbol{\eta}_j^{\mathrm{T}}\boldsymbol{Y}_j(\boldsymbol{q}_j, \bar{\boldsymbol{X}}_j)\hat{\boldsymbol{\theta}}_j\right) - \psi_{j5}\left(\hat{\boldsymbol{\theta}}_j - \boldsymbol{\theta}_j^*\right) \end{cases} \tag{7.9}$$

where K_{j1}, K_{j2}, K_{j3}, K_{j4}, ψ_{j1}, ψ_{j2}, $\cdots$, ψ_{j5} are the positive constant adaptive gains, $\boldsymbol{\Theta}_j^*$, ω_j^* and $\boldsymbol{\theta}_j^*$ are the nominal values of the parameters $\boldsymbol{\Theta}_j$, ω_j and $\boldsymbol{\theta}_j$.

In (7.8), $\boldsymbol{y}_m$ and $\boldsymbol{y}_s$ are the outputs for auxiliary switched filter based on dynamic compensation idea. Inspired by the work of previous chapters, for any $j \in \{m, s\}$, the switched error filter $\boldsymbol{y}_j = [y_{j,1}, y_{j,2}, \cdots, y_{j,n}]^{\mathrm{T}}$ is designed as

$$\dot{y}_{j,i} = \begin{cases} -P_{j1}s_{j,i} - e_{vj,i} - P_{j2}|\boldsymbol{X}_{vd\iota}(t, d_\iota(t))|\mathrm{sgn}(s_{j,i}) - P_{j3}\bar{\eta}_{j,i}, & \kappa_{j,i} = 1 \\ -P_{j1}s_{j,i} + e_{vj,i} - P_{j2}|\boldsymbol{X}_{vd\iota}(t, d_\iota(t))|\mathrm{sgn}(s_{j,i}) - P_{j3}\bar{\eta}_{j,i}, & \kappa_{j,i} = -1 \end{cases} \tag{7.10}$$

where $\iota \in \{m, s\}$ and $\iota \neq j$, P_{j1}, P_{j2}, $P_{j3} > 0$ are constant scalars,

$$\boldsymbol{s}_j = [s_{j,1}, s_{j,2}, \cdots, s_{j,n}]^{\mathrm{T}} = \boldsymbol{y}_j + \boldsymbol{\mathcal{K}}_j\boldsymbol{e}_j$$

with $\boldsymbol{\mathcal{K}}_j = \mathrm{diag}[\kappa_{j,1}, \kappa_{j,2}, \cdots, \kappa_{j,n}]$ being the switching rule and given by

$$\kappa_{j,i}(t) = \begin{cases} 1, & y_{j,i}(t)e_{j,i}(t) > 0 \text{ or } y_{j,i}(t)e_{j,i}(t) = 0,\ y_{j,i}(t^-)e_{j,i}(t^-) \leq 0 \\ -1, & y_{j,i}(t)e_{j,i}(t) < 0 \text{ or } y_{j,i}(t)e_{j,i}(t) = 0,\ y_{j,i}(t^-)e_{j,i}(t^-) > 0 \end{cases} \tag{7.11}$$

Due to the introduction of the switching rule $\boldsymbol{\mathcal{K}}_j$, the auxiliary compensation filter (7.10) is a switched system. In order to use the switched control theory to analyze this system, it is still necessary to model the system represented by (7.10) as a standard switched dynamic system. Similar to Remark 3.2, define discrete mode set $\mathcal{S} = \{1, 2, \cdots, 2^n\}$. Then there exists $\bar{r}_j : \boldsymbol{\mathcal{K}}_j(t) \to \mathcal{S}$ such that

$$\bar{r}_j(\boldsymbol{\mathcal{K}}_j(t)) = \ell, \text{ iff } \boldsymbol{\mathcal{K}}_j(t) \text{ is in the } \ell\text{th mode}$$

where $\ell \in \mathcal{S}$. To simplify the expression, let $r_j(t) = \bar{r}_j(\boldsymbol{\mathcal{K}}_j(t))$. Then, (7.10) can be regarded as a standard switched dynamic system with $r_j(t)$ as the switching signal, namely

$$\begin{aligned}\dot{\boldsymbol{y}}_j &= -P_{j1}[\boldsymbol{y}_j + \boldsymbol{\mathcal{K}}_j(t)\boldsymbol{e}_j] - P_{j2}\text{sgn}(\boldsymbol{s}_j)|\boldsymbol{X}_{vd\iota}(t, d_\iota(t))| - \boldsymbol{\mathcal{K}}_j(t)\boldsymbol{e}_{vj} - P_{j3}\bar{\boldsymbol{\eta}}_j \\ &:= \boldsymbol{h}(\boldsymbol{y}_j, \boldsymbol{e}_j, \boldsymbol{e}_{vj}, \boldsymbol{X}_{vd\iota}(t, d_\iota(t)), \dot{\boldsymbol{X}}_j, r_j(t)) \end{aligned} \tag{7.12}$$

Remark 7.2. *Let the time series $\{t_\iota^j\}_{\iota \geq 0}$ represent the switching moment of $r_j(t)$, then when $t \in [t_\iota^j, t_{\iota+1}^j)$, from the definition, it has*

$$\dot{\boldsymbol{s}}_j = -P_{j1}\boldsymbol{s}_j - P_{j2}\text{sgn}(\boldsymbol{s}_j)|\boldsymbol{X}_{vd\iota}(t, d_\iota(t))| + \boldsymbol{\mathcal{K}}_j(t)\dot{d}_\iota\boldsymbol{X}_{vd\iota}(t, d_\iota(t)) - P_{j3}\bar{\boldsymbol{\eta}}_j \tag{7.13}$$

Under the control framework proposed in this chapter, a complete closed-loop teleoperation system consists of robot dynamics (7.1), controller (7.8), adaptive update law (7.9) and the auxiliary switched filter (7.12). Due to the introduction of the subsystem (7.12) , the complete closed-loop system will also be a switched dynamic system.

In the rest of the chapter, let $\boldsymbol{x}_m = [\boldsymbol{\eta}_m^{\mathrm{T}}, \boldsymbol{q}_m^{\mathrm{T}}, \tilde{\boldsymbol{\Theta}}_m^{\mathrm{T}}, \tilde{\omega}_m, \tilde{\boldsymbol{\theta}}_m^{\mathrm{T}}, \boldsymbol{y}_m^{\mathrm{T}}, \boldsymbol{e}_m^{\mathrm{T}}]^{\mathrm{T}} \in \mathbb{R}^{4n+o_m+p_m+1}$, $\boldsymbol{x}_s = [\boldsymbol{\eta}_s^{\mathrm{T}}, \boldsymbol{q}_s^{\mathrm{T}}, \tilde{\boldsymbol{\Theta}}_s^{\mathrm{T}}, \tilde{\omega}_s, \tilde{\boldsymbol{\theta}}_s^{\mathrm{T}}, \boldsymbol{y}_s^{\mathrm{T}}, \boldsymbol{e}_s^{\mathrm{T}}]^{\mathrm{T}} \in \mathbb{R}^{2k+2n+o_2+p_s+1}$, $\boldsymbol{z}_m = [\boldsymbol{\eta}_m^{\mathrm{T}}, \tilde{\boldsymbol{\Theta}}_m^{\mathrm{T}}, \tilde{\omega}_m, \tilde{\boldsymbol{\theta}}_m^{\mathrm{T}}, \boldsymbol{y}_m^{\mathrm{T}}, \boldsymbol{e}_m^{\mathrm{T}}]^{\mathrm{T}} \in \mathbb{R}^{3n+o_m+p_m+1}$, $\boldsymbol{z}_s = [\boldsymbol{\eta}_s^{\mathrm{T}}, \tilde{\boldsymbol{\Theta}}_s^{\mathrm{T}}, \tilde{\omega}_s, \tilde{\boldsymbol{\theta}}_s^{\mathrm{T}}, \boldsymbol{y}_s^{\mathrm{T}}, \boldsymbol{e}_s^{\mathrm{T}}]^{\mathrm{T}} \in \mathbb{R}^{2n+o_s+p_s+k+1}$. And further, $\boldsymbol{x}_{mt} = \{\boldsymbol{x}_m(t+\theta)|\theta \in [-\bar{d}, 0]\} \in C([-\bar{d}, 0]; \mathbb{R}^{4n+o_m+p_m+1})$, $\boldsymbol{x}_{st} = \{\boldsymbol{x}_s(t+\theta)|\theta \in [-\bar{d}, 0]\} \in C([-\bar{d}, 0]; \mathbb{R}^{2k+2n+o_2+p_s+1})$, $\boldsymbol{z}_{mt} = \{\boldsymbol{z}_m(t+\theta)|\theta \in [-\bar{d}, 0]\} \in C([-\bar{d}, 0]; \mathbb{R}^{3n+o_m+p_m+1})$, $\boldsymbol{z}_{st} = \{\boldsymbol{z}_s(t+\theta)|\theta \in [-\bar{d}, 0]\} \in C([-\bar{d}, 0]; \mathbb{R}^{2n+o_s+p_s+k+1})$.

7.4 STABILITY ANALYSIS

This section gives the stability analysis of the closed-loop system. Since the existences of exogenous forces $\boldsymbol{f}_h^*$ and $\boldsymbol{f}_e^*$, the stability will be performed in the framework of state-independent input-to-output stability. Similar to the previous chapters, based on the multiple Lyapunov-Krasovskii functionals method, the following conclusions are established.

Theorem 7.1. *For any $j \in \{m, s\}$, if there exist positive constant gains K_{j1}, K_{j2}, K_{j3}, K_{j4}, P_{j1}, P_{j2}, P_{j3}, ψ_{j1}, ψ_{j2}, ψ_{j3}, ψ_{j4}, ψ_{j5}, such that*

$$\begin{cases} K_{j1} > \dfrac{1}{\delta_1}\vartheta_j \\ K_{j2} > \max\{\dfrac{1}{\delta_1}(1 + \dfrac{\delta_2 K_{j3}}{2} + \dfrac{1}{2}P_{j3} - B_{he_j}), 0\} \\ K_{j4} \geq \dfrac{\varepsilon_j \lambda_j \tilde{d}}{2} \\ P_{j1} > \dfrac{\delta_2 K_{j3}}{2} + \vartheta_j + \dfrac{1}{2}P_{j3} \\ P_{j2} \geq \tilde{d} \\ \psi_{j2} > 2\psi_{j1}\vartheta_j \\ \psi_{j3} > 2\vartheta_j \\ \psi_{j5} > 2\psi_{j4}(\vartheta_j + \dfrac{\lambda_j \tilde{d}}{2\varepsilon_{j1}}) \end{cases} \tag{7.14}$$

where $\vartheta_j > 0$, $\varepsilon_j = \varepsilon_{j1} + \varepsilon_{j2}$, $\varepsilon_{j1} > 0$, $\varepsilon_{j2} > 0$, *then the teleoperation system (7.3) with the controller (7.8) is SIIOS with* $\boldsymbol{z}_j$ *being the output.*

Proof. For the complete closed-loop systems, taking the following Lyapunov-Krasovskii functionals

$$V_j(\boldsymbol{x}_{jt}, r_j(t)) = \frac{1}{2}\boldsymbol{\eta}_j^{\mathrm{T}}\boldsymbol{\mathcal{M}}_j\boldsymbol{\eta}_j + \frac{1}{2\psi_{j1}}\tilde{\boldsymbol{\Theta}}_j^{\mathrm{T}}\tilde{\boldsymbol{\Theta}}_j + \frac{1}{2}(\hat{\omega}_j - \omega_j)^2 + \frac{1}{2\psi_{j4}}\tilde{\boldsymbol{\theta}}_j^{\mathrm{T}}\tilde{\boldsymbol{\theta}}_j + \frac{1}{2}\boldsymbol{s}_j^{\mathrm{T}}\boldsymbol{s}_j + \frac{\vartheta_j}{2}\int_{-\bar{d}}^{0}\boldsymbol{z}_{jt}^{\mathrm{T}}(\tau)\left(\frac{-\tau}{\bar{d}} + \frac{2(\tau+\bar{d})}{\bar{d}}\right)\boldsymbol{z}_{jt}(\tau)\mathrm{d}\tau$$

where $\vartheta_j > 0$.

Considering that

$$\frac{1}{2}\int_{-\bar{d}}^{0}\boldsymbol{z}_{jt}^{\mathrm{T}}(\tau)\left(\frac{-\tau}{\bar{d}} + \frac{2(\tau+\bar{d})}{\bar{d}}\right)\boldsymbol{z}_{jt}(\tau)\mathrm{d}\tau \le \int_{-\bar{d}}^{0}\boldsymbol{z}_{jt}^{\mathrm{T}}(\tau)\boldsymbol{z}_{jt}(\tau)\mathrm{d}\tau$$

one can verify

$$\begin{cases} \bar{\rho}_{j1}|\boldsymbol{z}_j(t)|^2 \le V_j(\boldsymbol{x}_{jt}, p) \le \bar{\rho}_{j2}\|\boldsymbol{z}_{jt}\|_{M_2}^2 \\ V_j(\boldsymbol{x}_{jt_{k+1}^j}, r_j(t_{k+1}^j)) \le V_j(\boldsymbol{x}_{jt_{k+1}^j}, r_j(t_k^j)) \end{cases} \tag{7.15}$$

where $p \in \mathcal{S}$, $k \ge 0$, $\bar{\rho}_{j2} = \max\left\{\frac{\rho_{j2}}{2}, \frac{1}{2\psi_{j1}}, \frac{1}{2\psi_{j4}}, 1, \vartheta_j\right\}$, $\bar{\rho}_{j1} = \min\left\{\frac{\rho_{j1}}{2}, \frac{1}{2\psi_{j1}}, \frac{1}{2\psi_{j4}}, \frac{1}{2}\right\}$.

To use Lemma 2.4, $V_j(\boldsymbol{x}_{jt}, r_j(t))$ should also satisfy (2.25) of Lemma 2.4. In fact, for any fixed $r_j(t) = p \in \mathcal{S}$, the upper right-hand derivative of $V_j(\boldsymbol{x}_{jt}, r_j(t))$ can be written as

$$D^+V_j(\boldsymbol{x}_{jt}, p) = \boldsymbol{\eta}_j^{\mathrm{T}}\boldsymbol{\mathcal{M}}_j\dot{\boldsymbol{\eta}}_j + \frac{1}{2}\boldsymbol{\eta}_j^{\mathrm{T}}\dot{\boldsymbol{\mathcal{M}}}_j\boldsymbol{\eta}_j + \frac{1}{\psi_{j1}}\tilde{\boldsymbol{\Theta}}_j^{\mathrm{T}}\dot{\tilde{\boldsymbol{\Theta}}}_j + (\hat{\omega}_j - \omega_j)\dot{\hat{\omega}}_j + \frac{1}{\psi_{j4}}\tilde{\boldsymbol{\theta}}_j^{\mathrm{T}}\dot{\tilde{\boldsymbol{\theta}}}_j + \boldsymbol{s}_j^{\mathrm{T}}\dot{\boldsymbol{s}}_j + \vartheta_j\boldsymbol{z}_j^{\mathrm{T}}\boldsymbol{z}_j - \vartheta_j\boldsymbol{z}_{jt}^{\mathrm{T}}(-\bar{d})\boldsymbol{z}_{jt}(-\bar{d}) - \frac{\vartheta_j}{2\bar{d}}\int_{-\bar{d}}^{0}\boldsymbol{z}_{jt}^{\mathrm{T}}(\tau)\boldsymbol{z}_{jt}(\tau)\mathrm{d}\tau$$

For the sake of simplifying expression, let $\boldsymbol{f}_{he_m} = \boldsymbol{f}_h^*$, $\boldsymbol{f}_{he_s} = \boldsymbol{f}_e^*$, $B_{he_m} = B_h$, $B_{he_s} = B_e$, $\mathfrak{J}_m(\boldsymbol{q}_m) = \boldsymbol{J}_m^{-1}(\boldsymbol{q}_m)$, $\mathfrak{J}_s(\boldsymbol{q}_s) = \boldsymbol{J}_s^{+}(\boldsymbol{q}_s)$. Along with the complete closed-loop systems, it holds the following results.

On the one hand, for all $j \in \{m, s\}$, $\iota \in \{m, s\}$ and $\iota \ne j$, according to Remark 7.1,

$$\begin{aligned} &\boldsymbol{\eta}_j^{\mathrm{T}}\boldsymbol{\mathcal{M}}_j\dot{\boldsymbol{\eta}}_j + \frac{1}{2}\boldsymbol{\eta}_j^{\mathrm{T}}\dot{\boldsymbol{\mathcal{M}}}_j\boldsymbol{\eta}_j \\ &\le -\boldsymbol{\eta}_j^{\mathrm{T}}\boldsymbol{m}_jK_{j1}\boldsymbol{\eta}_j - \boldsymbol{\eta}_j^{\mathrm{T}}\boldsymbol{m}_jK_{j2}\boldsymbol{J}_j^{\mathrm{T}}(\boldsymbol{q}_j)\bar{\boldsymbol{\eta}}_j + \boldsymbol{\eta}_j^{\mathrm{T}}\boldsymbol{J}_j^{\mathrm{T}}(\boldsymbol{q}_j)\boldsymbol{f}_{he_j} - \frac{m_j}{\delta_1}|\hat{\omega}_j||\boldsymbol{\eta}_j| - \\ &\quad \boldsymbol{\eta}_j^{\mathrm{T}}\boldsymbol{J}_j^{\mathrm{T}}(\boldsymbol{q}_j)B_{he_j}\boldsymbol{J}_j(\boldsymbol{q}_j)\boldsymbol{\eta}_j + \boldsymbol{\eta}_j^{\mathrm{T}}\boldsymbol{G}_j' - \frac{m_j}{\delta_1}|\boldsymbol{\eta}_j^{\mathrm{T}}\hat{\boldsymbol{G}}_{j1}'| + \lambda_j\dot{d}_\iota\boldsymbol{\eta}_j^{\mathrm{T}}\boldsymbol{\mathcal{M}}_j\bar{X}_j - \\ &\quad m_jK_{j3}\boldsymbol{\eta}_j^{\mathrm{T}}\boldsymbol{J}_j^{\mathrm{T}}(\boldsymbol{q}_j)\boldsymbol{y}_j - \frac{K_{j4}m_j}{\delta_1}\boldsymbol{\eta}_j^{\mathrm{T}}\boldsymbol{Y}_j(\boldsymbol{q}_j, \bar{\boldsymbol{X}}_j)\boldsymbol{Y}_j^{\mathrm{T}}(\boldsymbol{q}_j, \bar{\boldsymbol{X}}_j)\boldsymbol{\eta}_j \end{aligned}$$

$$
\begin{aligned}
&\le -\delta_1 \boldsymbol{\eta}_j^{\mathrm{T}} K_{j1} \boldsymbol{\eta}_j - \delta_1 \bar{\boldsymbol{\eta}}_j^{\mathrm{T}} K_{j2} \bar{\boldsymbol{\eta}}_j + \boldsymbol{\eta}_j^{\mathrm{T}} \boldsymbol{J}_j^{\mathrm{T}}(\boldsymbol{q}_j) \boldsymbol{J}_j(\boldsymbol{q}_j) \boldsymbol{\eta}_j + \frac{1}{4} \boldsymbol{f}_{he_j}^{\mathrm{T}} \boldsymbol{f}_{he_j} - \\
&\quad |\hat{\omega}_j||\boldsymbol{\eta}_j| - \boldsymbol{\eta}_j^{\mathrm{T}} \boldsymbol{J}_j^{\mathrm{T}}(\boldsymbol{q}_j) B_{he_j} \boldsymbol{J}_j(\boldsymbol{q}_j) \boldsymbol{\eta}_j + \boldsymbol{\eta}_j^{\mathrm{T}} \boldsymbol{G}_j' - |\boldsymbol{\eta}_j^{\mathrm{T}} \hat{\boldsymbol{G}}_{j1}'| + \\
&\quad \lambda_j \dot{d}_\iota \boldsymbol{\eta}_j^{\mathrm{T}} \boldsymbol{Y}_j(\boldsymbol{q}_j, \bar{\boldsymbol{X}}_j) \boldsymbol{\theta}_j - K_{j4} \boldsymbol{\eta}_j^{\mathrm{T}} \boldsymbol{Y}_j(\boldsymbol{q}_j, \bar{\boldsymbol{X}}_j) \boldsymbol{Y}_j^{\mathrm{T}}(\boldsymbol{q}_j, \bar{\boldsymbol{X}}_j) \boldsymbol{\eta}_j + \\
&\quad \frac{\delta_2 K_{j3}}{2} \bar{\boldsymbol{\eta}}_j^{\mathrm{T}} \bar{\boldsymbol{\eta}}_j + \frac{\delta_2 K_{j3}}{2} \boldsymbol{y}_j^{\mathrm{T}} \boldsymbol{y}_j \\
&\le -\boldsymbol{\eta}_j^{\mathrm{T}} \delta_1 K_{j1} \boldsymbol{\eta}_j - \bar{\boldsymbol{\eta}}_j^{\mathrm{T}} \left(\delta_1 K_{j2} + B_{he_j} - \frac{\delta_2 K_{j3}}{2} - 1 \right) \bar{\boldsymbol{\eta}}_j + \\
&\quad \boldsymbol{\eta}_j^{\mathrm{T}} \boldsymbol{G}_j' - |\boldsymbol{\eta}_j^{\mathrm{T}} \hat{\boldsymbol{G}}_{j1}'| - |\hat{\omega}_j||\boldsymbol{\eta}_j| + \lambda_j \dot{d}_\iota \boldsymbol{\eta}_j^{\mathrm{T}} \boldsymbol{Y}_j(\boldsymbol{q}_j, \bar{\boldsymbol{X}}_j) \boldsymbol{\theta}_j - \\
&\quad K_{j4} \boldsymbol{\eta}_j^{\mathrm{T}} \boldsymbol{Y}_j(\boldsymbol{q}_j, \bar{\boldsymbol{X}}_j) \boldsymbol{Y}_j^{\mathrm{T}}(\boldsymbol{q}_j, \bar{\boldsymbol{X}}_j) \boldsymbol{\eta}_j + \\
&\quad \frac{1}{4} \boldsymbol{f}_{he_j}^{\mathrm{T}} \boldsymbol{f}_{he_j} + \frac{\delta_2 K_{j3}}{2} \boldsymbol{y}_j^{\mathrm{T}} \boldsymbol{y}_j
\end{aligned}
$$

On the other hand,

$$
\begin{aligned}
&\boldsymbol{\eta}_j^{\mathrm{T}} \boldsymbol{G}_j' - |\boldsymbol{\eta}_j^{\mathrm{T}} \hat{\boldsymbol{G}}_{j1}'| + \tilde{\boldsymbol{\Theta}}_j^{\mathrm{T}} \psi_{j1}^{-1} \dot{\hat{\boldsymbol{\Theta}}}_j \\
&\quad = -|\boldsymbol{\eta}_j^{\mathrm{T}} \hat{\boldsymbol{G}}_{j1}'| - \tilde{\boldsymbol{\Theta}}_j^{\mathrm{T}} \psi_{j1}^{-1} \dot{\hat{\boldsymbol{\Theta}}}_j + \boldsymbol{\eta}_j^{\mathrm{T}} (\boldsymbol{G}_{j1}' + \boldsymbol{G}_{j2}') \\
&\quad \le -\frac{1}{\psi_{j1}} \tilde{\boldsymbol{\Theta}}_j^{\mathrm{T}} \left(\psi_{j1} \boldsymbol{Y}_{j0}^{\mathrm{T}}(\boldsymbol{q}_j, \dot{\boldsymbol{q}}_j, \boldsymbol{a}_j, \boldsymbol{v}_j) \boldsymbol{\eta}_j - \psi_{j2}(\hat{\boldsymbol{\Theta}}_j - \boldsymbol{\Theta}_j^*) \right) + \\
&\quad\quad \boldsymbol{\eta}_j^{\mathrm{T}} (\tilde{\boldsymbol{G}}_{j1}' + \boldsymbol{G}_{j2}') \\
&\quad \le \omega_j |\boldsymbol{\eta}_j| - \frac{\psi_{j2}}{2\psi_{j1}} |\tilde{\boldsymbol{\Theta}}_j|^2 + \frac{\psi_{j2}}{2\psi_{j1}} |\bar{\boldsymbol{\Theta}}_j|^2
\end{aligned}
$$

and

$$
\begin{aligned}
(\hat{\omega}_j - \omega_j)\dot{\hat{\omega}}_j &= (\hat{\omega}_j - \omega_j)|\boldsymbol{\eta}_j| - \psi_{j3}(\hat{\omega}_j - \omega_j)(\hat{\omega}_j - \omega_j^*) \\
&\le (\hat{\omega}_j - \omega_j)|\boldsymbol{\eta}_j| - \frac{\psi_{j3}}{2} \tilde{\omega}_j^2 + \frac{\psi_{j3}}{2} \bar{\omega}_j^2
\end{aligned}
$$

where $\bar{\boldsymbol{\Theta}}_j = \boldsymbol{\Theta}_j - \boldsymbol{\Theta}_j^*$, $\bar{\omega}_j = \omega_j - \omega_j^*$.

Since $|\dot{d}_\iota| \le \tilde{d}$, let $K_{j4} \ge \dfrac{\varepsilon_j \lambda_j \tilde{d}}{2}$, with$\varepsilon_j = \varepsilon_{j1} + \varepsilon_{j2}$, $\varepsilon_{j1} > 0$, $\varepsilon_{j2} > 0$, it also holds

$$
\begin{aligned}
&\lambda_j \dot{d}_\iota \boldsymbol{\eta}_j^{\mathrm{T}} \boldsymbol{Y}_j(\boldsymbol{q}_j, \bar{\boldsymbol{X}}_j) \boldsymbol{\theta}_j - K_{j4} |\boldsymbol{Y}_j^{\mathrm{T}}(\boldsymbol{q}_j, \bar{\boldsymbol{X}}_j) \boldsymbol{\eta}_j|^2 + \frac{1}{\psi_{j4}} \tilde{\boldsymbol{\theta}}_j^{\mathrm{T}} \dot{\hat{\boldsymbol{\theta}}}_j \\
&\quad \le \lambda_j \tilde{d} |\boldsymbol{\eta}_j^{\mathrm{T}} \boldsymbol{Y}_j(\boldsymbol{q}_j, \bar{\boldsymbol{X}}_j) \tilde{\boldsymbol{\theta}}_j| + \lambda_j \dot{d}_\iota \boldsymbol{\eta}_j^{\mathrm{T}} \boldsymbol{Y}_j(\boldsymbol{q}_j, \bar{\boldsymbol{X}}_j) \hat{\boldsymbol{\theta}}_j - \\
&\quad\quad K_{j4} \boldsymbol{\eta}_j^{\mathrm{T}} \boldsymbol{Y}_j(\boldsymbol{q}_j, \bar{\boldsymbol{X}}_j) \boldsymbol{Y}_j^{\mathrm{T}}(\boldsymbol{q}_j, \bar{\boldsymbol{X}}_j) \boldsymbol{\eta}_j + \frac{\psi_{j5}}{\psi_{j4}} \tilde{\boldsymbol{\theta}}_j^{\mathrm{T}} \left(\hat{\boldsymbol{\theta}}_j - \boldsymbol{\theta}_j^* \right) - \\
&\quad\quad \frac{1}{\psi_{j4}} \tilde{\boldsymbol{\theta}}_j^{\mathrm{T}} \left(-\lambda_j \tilde{d} \psi_{j4} \boldsymbol{Y}_j^{\mathrm{T}}(\boldsymbol{q}_j, \bar{\boldsymbol{X}}_j) \boldsymbol{\eta}_j \mathrm{sgn}(\boldsymbol{\eta}_j^{\mathrm{T}} \boldsymbol{Y}_j(\boldsymbol{q}_j, \bar{\boldsymbol{X}}_j) \hat{\boldsymbol{\theta}}_j) \right) \\
&\quad \le \lambda_j \tilde{d} |\boldsymbol{\eta}_j^{\mathrm{T}} \boldsymbol{Y}_j(\boldsymbol{q}_j, \bar{\boldsymbol{X}}_j) \tilde{\boldsymbol{\theta}}_j| + \lambda_j \tilde{d} |\boldsymbol{\eta}_j^{\mathrm{T}} \boldsymbol{Y}_j(\boldsymbol{q}_j, \bar{\boldsymbol{X}}_j) \boldsymbol{\theta}_j| + \\
&\quad\quad \frac{\psi_{j5}}{\psi_{j4}} \tilde{\boldsymbol{\theta}}_j^{\mathrm{T}} \left(\hat{\boldsymbol{\theta}}_j - \boldsymbol{\theta}_j^* \right) - K_{j4} \boldsymbol{\eta}_j^{\mathrm{T}} \boldsymbol{Y}_j(\boldsymbol{q}_j, \bar{\boldsymbol{X}}_j) \boldsymbol{Y}_j^{\mathrm{T}}(\boldsymbol{q}_j, \bar{\boldsymbol{X}}_j) \boldsymbol{\eta}_j
\end{aligned}
$$

$$\le -\frac{\psi_{j5}}{\psi_{j4}}\tilde{\boldsymbol{\theta}}_j^{\mathrm{T}}(\boldsymbol{\theta}_j - \hat{\boldsymbol{\theta}}_j) + \frac{\psi_{j5}}{\psi_{j4}}\tilde{\boldsymbol{\theta}}_j^{\mathrm{T}}\bar{\boldsymbol{\theta}}_j + \frac{\lambda_j \tilde{d}}{2\varepsilon_{j1}}|\tilde{\boldsymbol{\theta}}_j|^2 + \frac{\lambda_j \tilde{d}}{2\varepsilon_{j2}}|\boldsymbol{\theta}_j|^2$$
$$\le -\left(\frac{\psi_{j5}}{2\psi_{j4}} - \frac{\lambda_j \tilde{d}}{2\varepsilon_{j1}}\right)|\tilde{\boldsymbol{\theta}}_j|^2 + \frac{\psi_{j5}}{2\psi_{j4}}|\bar{\boldsymbol{\theta}}_j|^2 + \frac{\lambda_j \tilde{d}}{2\varepsilon_{j2}}|\boldsymbol{\theta}_j|^2$$

where $\bar{\boldsymbol{\theta}}_j = \boldsymbol{\theta}_j - \boldsymbol{\theta}_j^*$.

Given that$(|\boldsymbol{y}_j|^2 + |\boldsymbol{e}_j|^2) \le |\boldsymbol{s}_j|^2 \le 2(|\boldsymbol{y}_j|^2 + |\boldsymbol{e}_j|^2)$, then for those $P_{j2} \ge \tilde{d}$,

$$\boldsymbol{s}_j^{\mathrm{T}}\dot{\boldsymbol{s}}_j \le -P_{j1}|\boldsymbol{s}_j|^2 - P_{j2}|\boldsymbol{s}_j||\boldsymbol{X}_{vd\iota}(t, d_\iota(t))| +$$
$$\boldsymbol{s}_j^{\mathrm{T}}\boldsymbol{\mathcal{K}}_j(t)\dot{d}_\iota(t)\boldsymbol{X}_{vd\iota}(t, d_\iota(t)) - P_{j3}\boldsymbol{s}_j^{\mathrm{T}}\bar{\boldsymbol{\eta}}_j$$
$$\le -\left(P_{j1} - \frac{1}{2}P_{j3}\right)|\boldsymbol{s}_j|^2 - P_{j2}|\boldsymbol{s}_j||\boldsymbol{X}_{vd\iota}(t, d_\iota(t))| +$$
$$\tilde{d}|\boldsymbol{s}_j||\boldsymbol{X}_{vd\iota}(t, d_\iota(t))| + \frac{P_{j3}}{2}|\bar{\boldsymbol{\eta}}_j|^2$$
$$\le -\left(P_{j1} - \frac{1}{2}P_{j3}\right)|\boldsymbol{s}_j|^2 + \frac{P_{j3}}{2}|\bar{\boldsymbol{\eta}}_j|^2$$
$$\le -\left(P_{j1} - \frac{1}{2}P_{j3}\right)|\boldsymbol{y}_j|^2 - (P_{j1} - \frac{1}{2}P_{j3})|\boldsymbol{e}_j|^2 + \frac{P_{j3}}{2}|\bar{\boldsymbol{\eta}}_j|^2$$

Hence, if $K_{j2} \ge \max\left\{\frac{1}{\delta_1}\left(1 + \frac{\delta_2 K_{j3}}{2} + \frac{P_{j3}}{2} - B_{he_j}\right), 0\right\}$,

$$D^+V_j(\boldsymbol{x}_{jt}, p) \le -\delta_1 K_{j1}|\boldsymbol{\eta}_j|^2 - \frac{\psi_{j2}}{2\psi_{j1}}|\tilde{\boldsymbol{\Theta}}_j|^2 - \left(P_{j1} - \frac{1}{2}P_{j3} - \frac{\delta_2 K_{j3}}{2}\right)|\boldsymbol{y}_j|^2 -$$
$$\frac{\psi_{j3}}{2}\tilde{\omega}_j^2 - \left(P_{j1} - \frac{1}{2}P_{j3}\right)|\boldsymbol{e}_j|^2 - \left(\frac{\psi_{j5}}{2\psi_{j4}} - \frac{\lambda_j \tilde{d}}{2\varepsilon_{j1}}\right)|\tilde{\boldsymbol{\theta}}_j|^2 -$$
$$\frac{\vartheta_j}{2\bar{d}}\int_{-\bar{d}}^{0}\boldsymbol{z}_{jt}^{\mathrm{T}}(\tau)\boldsymbol{z}_{jt}(\tau)\mathrm{d}\tau + \frac{\psi_{j2}}{2\psi_{j1}}|\bar{\boldsymbol{\Theta}}_j|^2 + \frac{\psi_{j3}}{2}\bar{\omega}_j^2 +$$
$$\frac{\psi_{j5}}{2\psi_{j4}}|\bar{\boldsymbol{\theta}}_j|^2 + \frac{\lambda_j \tilde{d}}{2\varepsilon_{j2}}|\boldsymbol{\theta}_j|^2 + \frac{1}{4}\boldsymbol{f}_{he_j}^{\mathrm{T}}\boldsymbol{f}_{he_j} + \vartheta_j\boldsymbol{z}_j^{\mathrm{T}}\boldsymbol{z}_j$$
$$\le -(\delta_1 K_{j1} - \vartheta_j)|\boldsymbol{\eta}_j|^2 - \left(\frac{\psi_{j2}}{2\psi_{j1}} - \vartheta_j\right)|\tilde{\boldsymbol{\Theta}}_j|^2 - \left(\frac{\psi_{j3}}{2} - \vartheta_j\right)\tilde{\omega}_j^2 -$$
$$\left(P_{j1} - \frac{1}{2}P_{j3} - \frac{\delta_2 K_{j3}}{2} - \vartheta_j\right)|\boldsymbol{y}_j|^2 - \left(P_{j1} - \frac{1}{2}P_{j3} - \vartheta_j\right)|\boldsymbol{e}_j|^2 -$$
$$\left(\frac{\psi_{j5}}{2\psi_{j4}} - \vartheta_j - \frac{\lambda_j \tilde{d}}{2\varepsilon_{j1}}\right)|\tilde{\boldsymbol{\theta}}_j|^2 - \frac{\vartheta_j}{2\bar{d}}\int_{-\bar{d}}^{0}\boldsymbol{z}_{jt}^{\mathrm{T}}(\tau)\boldsymbol{z}_{jt}(\tau)\mathrm{d}\tau +$$
$$\frac{1}{4}\|\boldsymbol{f}_{he_j[0,t)}\|_\infty^2 + \frac{\psi_{j2}}{2\psi_{j1}}|\bar{\boldsymbol{\Theta}}_j|^2 + \frac{\psi_{j3}}{2}\bar{\omega}_j^2 +$$
$$\frac{\psi_{j5}}{2\psi_{j4}}|\bar{\boldsymbol{\theta}}_j|^2 + \frac{\lambda_j \tilde{d}}{2\varepsilon_{j2}}|\boldsymbol{\theta}_j|^2$$
$$\le -\mu_1|\boldsymbol{z}_j|^2 - \frac{\vartheta_j}{2\bar{d}}\int_{-\bar{d}}^{0}\boldsymbol{z}_{jt}^{\mathrm{T}}(\tau)\boldsymbol{z}_{jt}(\tau)\mathrm{d}\tau + \frac{1}{4}\|\boldsymbol{f}_{he_j[0,t)}\|_\infty^2 +$$

$$\frac{\psi_{j2}}{2\psi_{j1}}|\bar{\Theta}_j|^2+\frac{\psi_{j3}}{2}\bar{\omega}_j^2+\left(\frac{\psi_{j5}}{2\psi_{j4}}+\frac{\lambda_j\tilde{d}}{2\varepsilon_{j2}}\right)|\bar{\boldsymbol{\theta}}_j|^2$$

$$\le -\mu_2\|\boldsymbol{z}_{jt}\|_{M_2}^2+\frac{1}{4}\|\boldsymbol{f}_{he_j[0,t)}\|_\infty^2+\frac{\psi_{j2}}{2\psi_{j1}}|\bar{\Theta}_j|^2+\frac{\psi_{j3}}{2}\bar{\omega}_j^2+$$

$$\frac{\psi_{j5}}{2\psi_{j4}}|\bar{\boldsymbol{\theta}}_j|^2+\frac{\lambda_j\tilde{d}}{2\varepsilon_{j2}}|\boldsymbol{\theta}_j|^2 \tag{7.16}$$

where $\mu_1=\min\left\{\delta_1K_{j1}-\vartheta_j,\frac{\psi_{j2}}{2\psi_{j1}}-\vartheta_j,\frac{\psi_{j3}}{2}-\vartheta_j,P_{j1}-\frac{1}{2}P_{j3}-\frac{\delta_2K_{j3}}{2}-\vartheta_j,\frac{\psi_{j5}}{2\psi_{j4}}-\vartheta_j-\frac{\lambda_j\tilde{d}}{2\varepsilon_{j1}}\right\}$, $\mu_2=\min\left\{\mu_1,\frac{\vartheta_j}{2\bar{d}}\right\}$.

Furthermore, when $t\in[t_k^j,t_{k+1}^j)$, $k\ge 0$,

$$D^+V_j(\boldsymbol{x}_{jt},r_j(t))\le -\frac{\mu_2}{\bar{\rho}_{j2}}V_j(\boldsymbol{x}_{jt_j^k},r_j(t_k^j))+\frac{1}{4}\|\boldsymbol{f}_{he_j[0,\infty)}\|_\infty^2+$$

$$\frac{\psi_{j2}}{2\psi_{j1}}|\bar{\Theta}_j|^2+\frac{\psi_{j3}}{2}\bar{\omega}_j^2+\frac{\psi_{j5}}{2\psi_{j4}}|\bar{\boldsymbol{\theta}}_j|^2+\frac{\lambda_j\tilde{d}}{2\varepsilon_{j2}}|\boldsymbol{\theta}_j|^2$$

which means that $V_j(\boldsymbol{x}_{jt},r_j(t))$ satisfies (2.25) of Lemma 2.4. From Lemma 2.4, one can complete the proof. □

Remark 7.3. *By Theorem 7.1, if* $\lim_{t\to\infty}\boldsymbol{f}_e^*(t)=0$*, and taking a very small* ϑ_s*, the tracking error* $\mathbf{e}_s(t)$ *and the estimated errors* $\tilde{\Theta}_s(t)$ *and* $\tilde{\omega}_s(t)$ *will converge to a value very close to zero, which means that the slave robot can track the movement of the master robot very well. At this time,* $\boldsymbol{\eta}_s$ *will also converge close to zero, which is necessary to realize the additional subtask control of the slave robot in the next section. It should be noted that, the above conclusion does not depend on the characteristics of the master robot, and allows the slave robot to have contact with the environment. In the previous work [1], although the asymptotic convergence characteristics are obtained, it requires the master and slave robots to be in free motion, or the master robot exerts passive external force and the slave robot is in free motion.*

Remark 7.4. *The proposed adaptive semi-autonomous control scheme can handle the asymmetric time-varying delays, passive/non-passive external forces, model uncertainties and external disturbances of teleoperation systems in a unified framework. Compared with the existing important results [1, 2], it has the following advantages and disadvantages.*

(1) The method proposed in this chapter extends the Liu's conclusions [1] to the case of time-varying delays, and the design method is different from another work by Liu et al. [2]. In another work [2], the controller needs to use the real-time rate of change of time-delay, and in the scheme of this chapter, only the upper bound of the rate of change of time-delay is needed, which provides engineers with another implementation scheme.

(2) *Since both passive and non-passive external forces may exist in the teleoperation system, the research on non-passive external forces has received more and more attention [7–9]. Liu et al. also consider the case of non-passive external forces, but only for teleoperation systems with nominal dynamics, which limits the practical use of the algorithms [1, 2]. Compared with the methods [1, 2], the algorithm proposed in this chapter can deal with unknown model uncertainty and external disturbance by using adaptive control technology.*

(3) *The algorithms [1, 2] need to utilize precise operator/environmental dynamics parameters, namely k_h and k_e. However, considering the inherent complexity of operator/environmental dynamics modeling, the actual k_h and k_e are difficult to obtain accurately. In this chapter, adaptive techniques are used to estimate and compensate for unknown operator/environmental dynamics parameters, further expanding the application range of the algorithm.*

(4) *The algorithm proposed in this chapter also has some shortcomings. For example, it needs to use the upper bound information of the rate of change of the delay. Although a large upper bound of the rate of change can be selected for controller design, it is still impossible to completely cover all cases. Further research is required in the future.*

7.5 SEMI-AUTONOMOUS CONTROL OF SLAVE ROBOT

This section addresses the semi-autonomous control for the redundant slave robot. The main method for the sub-task control is similar to the Liu's work [1].

On the one hand, based on the definition of $\boldsymbol{\eta}_s$, the last term $(\boldsymbol{I}-\boldsymbol{J}_s^+(\boldsymbol{q}_s)\boldsymbol{J}_s(\boldsymbol{q}_s))\boldsymbol{\phi}_s$ can be seen as the desired velocity in the null space of $\boldsymbol{J}_s(\boldsymbol{q}_s)$, which only produces self-motion that changes its own configuration without affecting the trajectory of the end effector. Thus, additional subtask control can be accomplished by designing a suitable $\boldsymbol{\phi}_s$. On the other hand, for the redundant robot, the sub-task tracking error has been defined as Hsu's work [10]: $\boldsymbol{e}_{sst} = (\boldsymbol{I}-\boldsymbol{J}_s^+(\boldsymbol{q}_s)\boldsymbol{J}_s(\boldsymbol{q}_s))(\dot{\boldsymbol{q}}_s-\boldsymbol{\phi}_s)$. In fact, from the definition of $\boldsymbol{\eta}_s$, it is easy to verify:

$$\begin{aligned}(\boldsymbol{I}-\boldsymbol{J}_s^+(\boldsymbol{q}_s)\boldsymbol{J}_s(\boldsymbol{q}_s))\,\boldsymbol{\eta}_s &= (\boldsymbol{I}-\boldsymbol{J}_s^+(\boldsymbol{q}_s)\boldsymbol{J}_s(\boldsymbol{q}_s))\,\dot{\boldsymbol{q}}_s-(\boldsymbol{I}-\boldsymbol{J}_s^+(\boldsymbol{q}_s)\boldsymbol{J}_s(\boldsymbol{q}_s))\,\boldsymbol{\phi}_s\\ &= \boldsymbol{e}_{sst}\end{aligned}$$

According to Theorem 7.1, $\boldsymbol{\eta}_s$ is bounded. Since $\boldsymbol{I}-\boldsymbol{J}_s^+(\boldsymbol{q}_s)\boldsymbol{J}_s(\boldsymbol{q}_s)$ is also bounded at this time, it means that the subtask error $\boldsymbol{e}_{sst}$ will remain bounded. Obviously, $\boldsymbol{\phi}_s$ does not affect the subtask tracking error. Therefore, different subtasks can be controlled by introducing different auxiliary functions $\boldsymbol{\phi}_s$.

As discussed in the previous work [1], this chapter uses the gradient projection method to perform the semi-autonomous control task. In this framework, the auxiliary function $\boldsymbol{\phi}_s$ is given by

$$\boldsymbol{\phi}_s = -\frac{\partial}{\partial \boldsymbol{q}_s}\bar{\phi}(\boldsymbol{q}_s) \tag{7.17}$$

For different sub-task control objectives, an appropriate $\bar{\phi}(\boldsymbol{q}_s)$ can be constructed. For example, as shown in Li's work [11], for the joint limitation avoidance, it can be designed by

$$\bar{\phi}(\boldsymbol{q}_s) = -\sum_{i=1}^{k} \frac{1}{4} \frac{(q_{si}^{\max} - q_{si}^{\min})^2}{(q_{si} - q_{si}^{\max})(q_{si} - q_{si}^{\min})}$$

where $q_{si}^{\max}$ and $q_{si}^{\min}$ are the upper bound and lower bound of allowable values for the ith slave joint $q_{s,i}$. For other tasks, e.g., singularity avoidance, collision avoidance, it is also possible to design a specific $\bar{\phi}(\boldsymbol{q}_s)$. More details can be seen in Liu's work [1].

7.6 SIMULATION

In the simulation, the teleoperation system consists of two heterogeneous robots, as shown in Fig. 7.2. Among them, the master robot is a 2-DOF serial link manipulator, and the slave robot is a 3-DOF planar manipulator with redundant degrees of freedom, their dynamic models are shown in (7.1). For details, please refer to the existing works [12, 13].

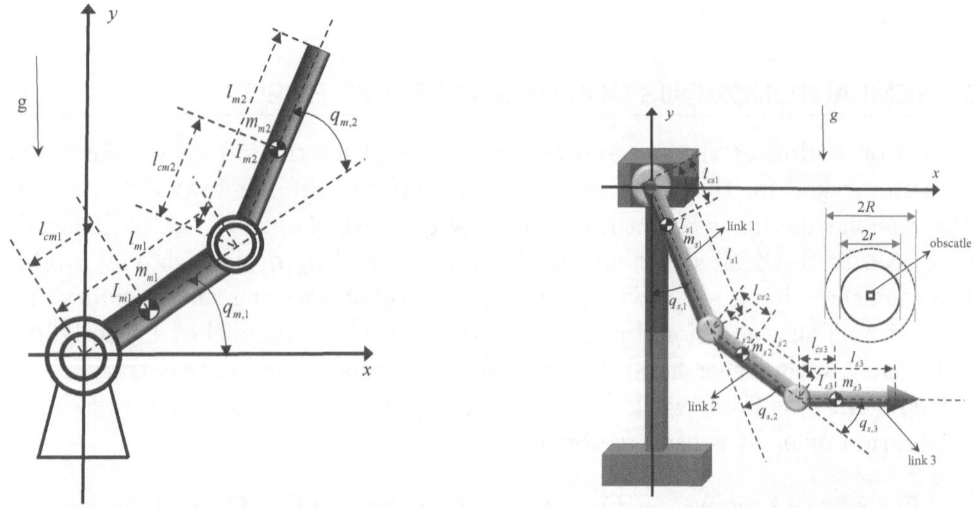

Figure 7.2 Schematic diagram of master and slave robots. (a) 2-DOF serial link master robotic manipulator, (b) 3-DOF planar slave robotic manipulator.

In simulation, let $l_{m1} = l_{m2} = 0.5$ m, $l_{s1} = 0.5$ m, $l_{s2} = l_{s3} = 0.4$ m, other physical parameters are the same as the ones in the references [12, 13]. The viscous friction coefficients $\boldsymbol{F}_m$ and $\boldsymbol{F}_s$ are respectively set as $\boldsymbol{F}_m = \mathrm{diag}[F_{m,1}, F_{m,2}] = \mathrm{diag}[0.2, 0.6]$ N · s and $\boldsymbol{F}_s = \mathrm{diag}[F_{s,1}, F_{s,2}, F_{s,3}] = \mathrm{diag}[1.3, 1.4, 1.1]$ N · s, Coulomb friction is $\boldsymbol{f}_m(\dot{\boldsymbol{q}}_m) = [0.2\mathrm{sgn}(\dot{q}_{m,1}), 0.1\mathrm{sgn}(\dot{q}_{m,2})]^{\mathrm{T}}$ N · m^2 and $\boldsymbol{f}_s(\dot{\boldsymbol{q}}_s) = [1\mathrm{sgn}(\dot{q}_{s,1}), 2\mathrm{sgn}(\dot{q}_{s,2}), 0.8\mathrm{sgn}(\dot{q}_{s,3})]^{\mathrm{T}}$ N · m^2. The bounded external disturbances are $\boldsymbol{D}_m = [0.2\sin(t),\ 0.3\cos(2t)]^{\mathrm{T}}$ N · m^2 and $\boldsymbol{D}_s = [0.8\sin(3t), 0.6\cos(5t), 1\cos(2t)]^{\mathrm{T}}$ N · m^2. The nonlinear parameters of the actuator dead zone are selected as: $m_{mir} = m_{mil} = m_{skr} = m_{skl} = 1$, $b_{mir} = b_{skr} = 1$, $b_{mil} = b_{skl} = -1$,

$i = 1, 2$, $k = 1, 2, 3$. Take $\delta_1 = \delta_2 = 1$. The operator/environment dynamics parameters are set to: $M_h = 0$, $B_h = 0$, $K_h = 1$, $M_e = 0$, $B_e = 0$, $K_e = 10$. In addition, according to the linear parameterization property of the dynamics, we have

$$\begin{aligned}
\boldsymbol{\Theta}_m &= [m_{m1}l_{cm1}^2 + m_{m2}l_{cm2}^2 + I_{m1} + I_{m2}, m_{m2}l_{cm2}^2 + I_{m2}, m_{m2},\\
&\qquad m_{m2}l_{cm2}, I_{m2}, M_h, F_{m1}, F_{m2}, K_h, m_{m1}l_{cm2}, B_h]^{\mathrm{T}}\\
\boldsymbol{\Theta}_s &= [m_{s1}l_{cs1}^2 + m_{s2}l_{cs2}^2 + m_{s3}l_{cs3}^2 + I_{s1} + I_{s2} + I_{s3}, m_{s2}l_{cs2}^2 + m_{s3}l_{cs3}^2 + I_{s2} + I_{s3},\\
&\qquad m_{s3}l_{cs3}^2 + I_{s3}, m_{s3}, m_{s2} + m_{s3}, m_{s1} + m_{s3}, m_{s3}l_{cs3}, m_{s2}l_{cs2}^2 + I_{s1} + I_{s2},\\
&\qquad F_{s1}, F_{s2}, F_{s3}, m_{s2}l_{cs2}, m_{s1}l_{cs1}, m_{s2}]^{\mathrm{T}}\\
\boldsymbol{\theta}_m &= [m_{m1}l_{cm1}^2 + m_{m2}l_{cm2}^2 + I_{m1} + I_{m2}, m_{m2}l_{cm2}^2 + I_{m2}, m_{m2}, m_{m2}l_{cm2}^2, I_{m2}]^{\mathrm{T}}\\
\boldsymbol{\theta}_s &= [m_{s1}l_{cs1}^2 + m_{s2}l_{cs2}^2 + m_{s3}l_{cs3}^2 + I_{s1} + I_{s2} + I_{s3}, m_{s2}l_{cs2}^2 + m_{s3}l_{cs3}^2 + I_{s2} + I_{s3},\\
&\qquad m_{s3}l_{cs3}^2 + I_{s3}, m_{s3}, m_{s2} + m_{s3}, m_{s1} + m_{s3}, m_{s3}l_{cs3}, m_{s2}l_{cs2}^2 + I_{s2} + I_{s3}]^{\mathrm{T}}
\end{aligned}$$

The initial parameter values of the closed-loop teleoperation system are set to: for any $t \in [-\bar{d}, 0]$, $\boldsymbol{q}_m(t) = [-\frac{\pi}{2}, \frac{\pi}{2}]^{\mathrm{T}}$, $\boldsymbol{q}_s(t) = [\frac{\pi}{3}, \frac{2\pi}{3}, -\frac{\pi}{4}]^{\mathrm{T}}$, $\dot{\boldsymbol{q}}_m(t) = \ddot{\boldsymbol{q}}_m = [0, 0]^{\mathrm{T}}$, $\dot{\boldsymbol{q}}_s(t) = \ddot{q}_s = [0, 0, 0]^{\mathrm{T}}$, $\boldsymbol{y}_m(t) = [-20, 10]$, $\boldsymbol{y}_s(t) = [20, -40]$, $\dot{\boldsymbol{y}}_m(t) = \dot{\boldsymbol{y}}_s(t) = [0, 0]^{\mathrm{T}}$, $\hat{\omega}_m(t) = 3$, $\hat{\omega}_s(t) = 2$, $\hat{\boldsymbol{\theta}}_m(t) = [0.4, 0.2, 0.6, 0.4, 0.2]^{\mathrm{T}}$, $\hat{\boldsymbol{\theta}}_s(t) = [0.1, 0.1, 0.1, 0.1, 0.1, 0.1, 0.2, 0.4]^{\mathrm{T}}$, $\hat{\boldsymbol{\Theta}}_m(t) = \mathbf{0}$ and $\hat{\boldsymbol{\Theta}}_s(t) = \mathbf{0}$, where $\mathbf{0}$ is a zero vector of suitable dimension. The nominal parameters are assumed to be: $\boldsymbol{\Theta}_j^* = 0.8\boldsymbol{\Theta}_j$, $\boldsymbol{\theta}_j^* = 0.8\boldsymbol{\theta}_j$, and $\omega_j^* = 0.8\omega_j$, $j \in \{m, s\}$. The exogenous forces exerted on the master and slave robots are shown in Fig. 7.3.

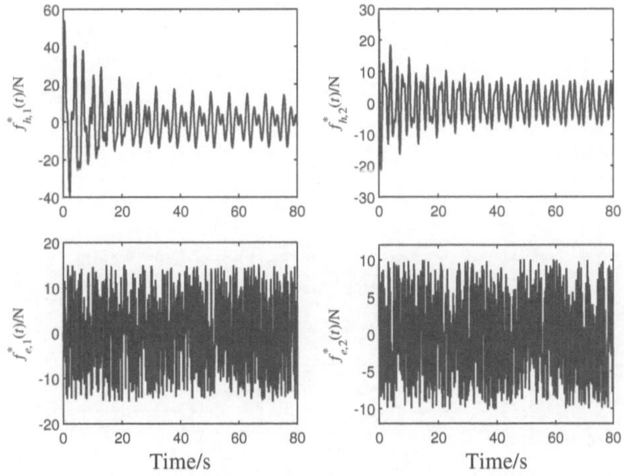

Figure 7.3 Exogenous forces $\boldsymbol{f}_h^*$ and $\boldsymbol{f}_e^*$.

7.6.1 Stability verification with practical communication delay

In order to better simulate the actual communication network, the actual Internet communication time delay between Beijing and Shenzhen measured in advance is used for simulation, and the measurement results are shown in Fig. 7.4. Under the above simulation settings, according to Theorem 7.1, the simulation results are shown in the Figs. 7.5–7.9.

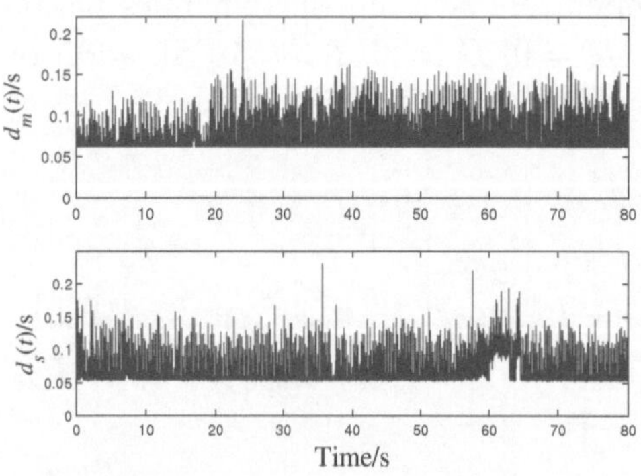

Figure 7.4 Actual Internet communication time delays $d_m(t)$ and $d_s(t)$.

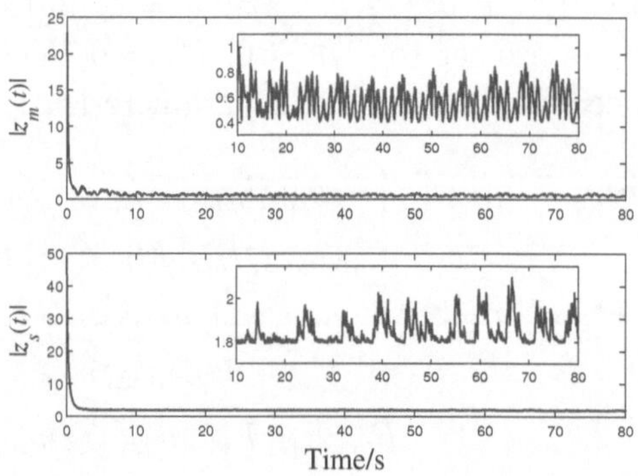

Figure 7.5 Response curves of $|z_m(t)|$ and $|z_s(t)|$.

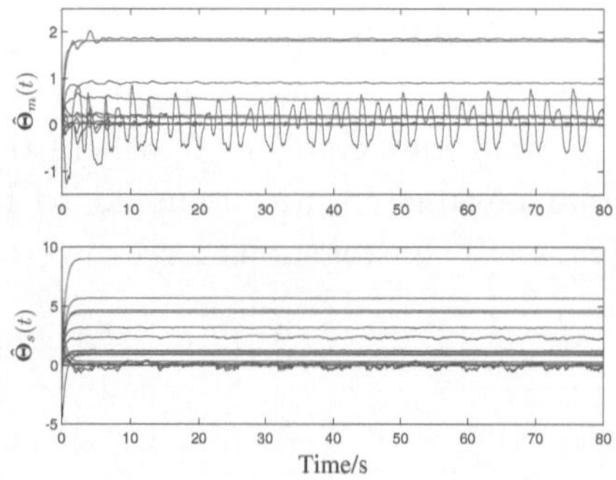

(a)

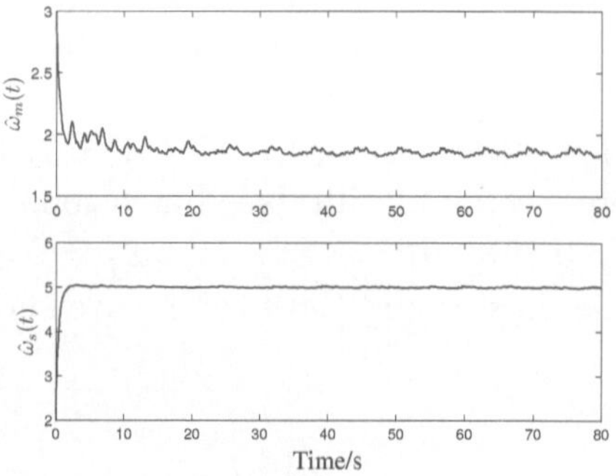

(b)

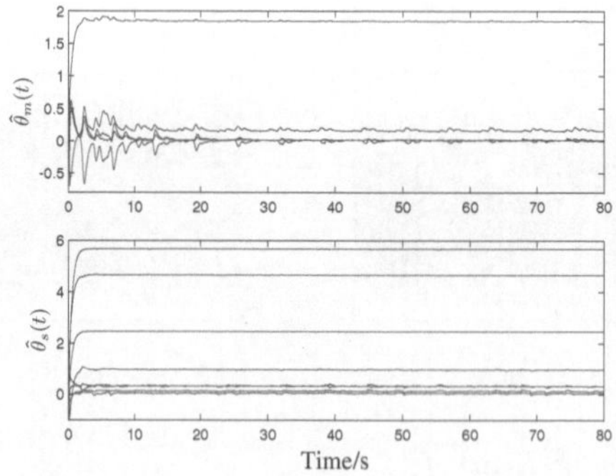

(c)

Figure 7.6 Adaptive parameter estimation without subtask control. (a) $\hat{\boldsymbol{\Theta}}_m(t)$ and $\hat{\boldsymbol{\Theta}}_s(t)$, (b) $\hat{\omega}_m(t)$ and $\hat{\omega}_s(t)$, (c)$\hat{\boldsymbol{\theta}}_m(t)$ and $\hat{\boldsymbol{\theta}}_s(t)$.

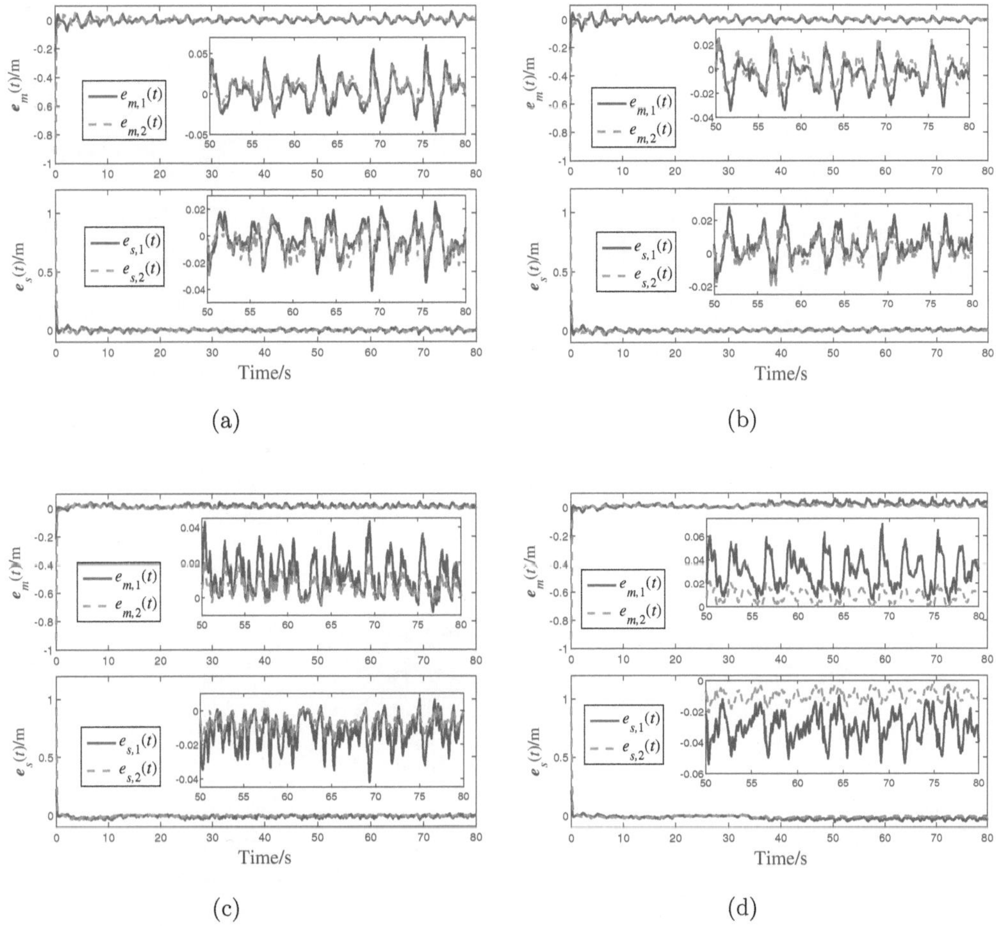

Figure 7.7 Position tracking errors. (a) Without the use of subtask control. (b) With subtask control to avoid configuration singularity. (c) With subtask control to limit the angle of joints. (d) With the use of collision avoidance control.

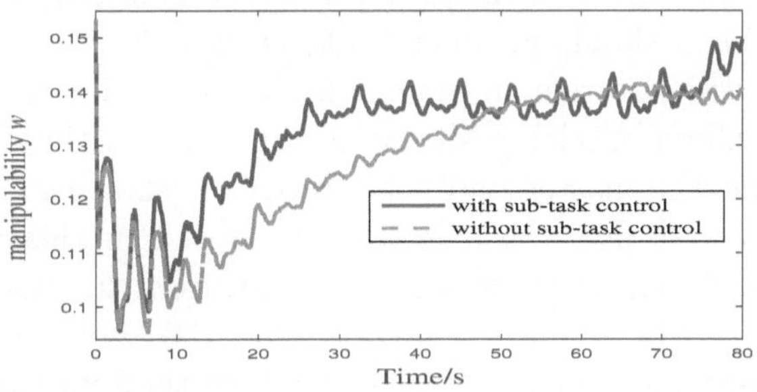

Figure 7.8 Measure of manipulability.

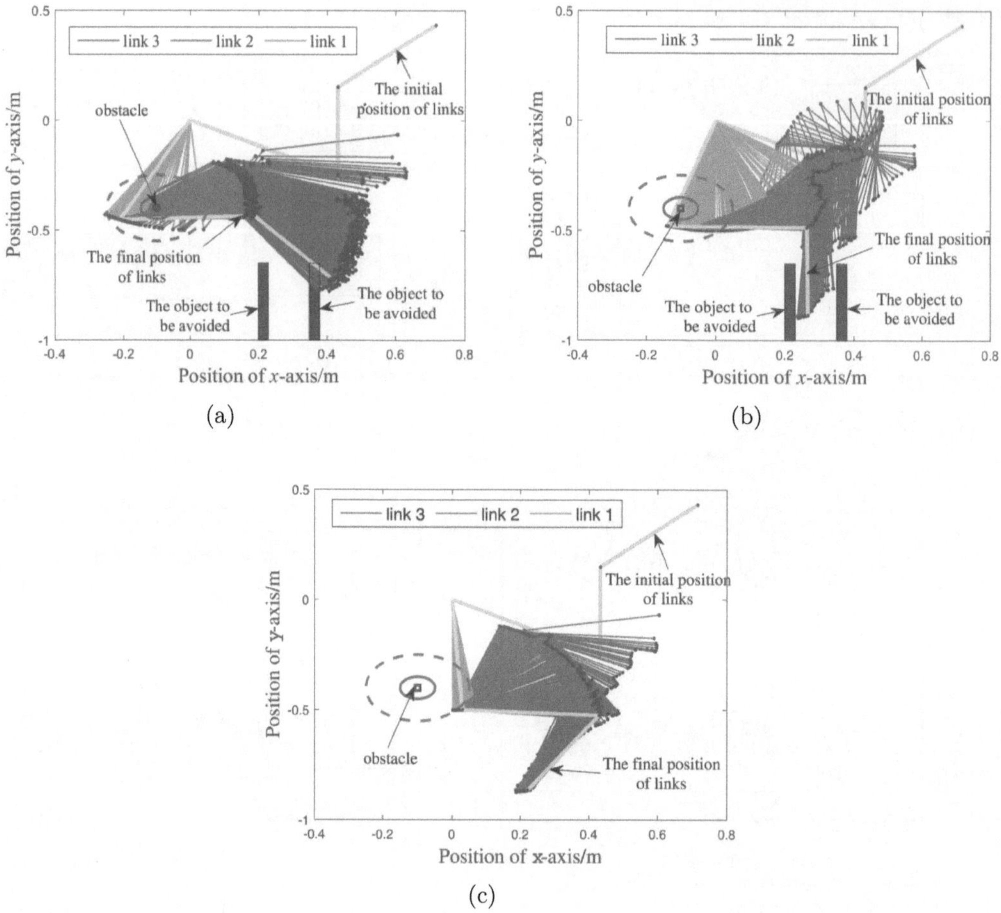

Figure 7.9 Configurations of the slave robot. (a) Without the use of subtask control. (b) With subtask control to limit the angle of joints. (c) With the use of collision avoidance control.

Fig. 7.5 shows the output $\boldsymbol{z}_m(t)$ ($\boldsymbol{z}_s(t)$) of the complete closed-loop system, their upper bounds depend on the size of the input, which is consistent with the theoretical analysis of Theorem 7.1. It should be noted that, by the definitions of $\boldsymbol{z}_m(t)$ and $\boldsymbol{z}_s(t)$, all their components will also remain bounded, and the estimation of the unknown parameters of the dynamic model is shown in Fig. 7.6. At this time, the position tracking error between the master and slave robots is shown in Fig. 7.7(a). From which, it can be observed that the algorithm proposed in this chapter has good tracking performance. When no subtask is executed, from the boundedness of $\boldsymbol{z}_s(t)$ shown by Fig. 7.5, $\boldsymbol{\eta}_s$ is also bounded. According to the above theoretical analysis, it can be used to realize additional subtask control, so the semi-autonomous control performance of the algorithm can be verified.

The first set of semi-autonomous control simulations will verify whether the algorithm can improve the manipulability of the slave robot to avoid singularity.

According to Yoshikawa's work [14], $w = \sqrt{|J_s(q_s)J_s^{\mathrm{T}}(q_s)|}$ is used to measure the manipulability of the robot at position q_s. Therefore, after adding the singularity avoidance subtask control, the manipulability coefficient w is shown in Fig. 7.8, from which it can be observed that the singularity avoidance subtask controller increases the manipulability by reconfiguring the robot joints.

The second subtask control concerns the joint limitation that may occur when the slave robot is manipulated within a small workspace. If there is no subtask control, the slave robot can move to and collide with the objects in the environment as shown in Fig. 7.9(a), where the two blue boxes denote objects in the environment that the slave robot has to be avoided via reconfiguring its joint angles. To avoid hitting objects in the environment, the slave robot needs to use its redundancy to reconfigure its joint angles. In the simulation, the passable area between the two objects is parallel the y-axis, which requires the joint angle $q_s = [q_{s,1}, q_{s,2}, q_{s,3}]^{\mathrm{T}}$ satisfies $q_{s,1}+q_{s,2}+q_{s,3} = 0$. Therefore, $\bar{\phi}(q_s)$ in (7.17) is set to $\bar{\phi}(q_s) = (q_{s,1} + q_{s,2} + q_{s,3})^2$. At this time, with the proposed control, the real-time joint angle configuration of the robot is shown in Fig. 7.9(b). It can be seen that the control algorithm successfully realizes the sub-task control through the narrow space by reconfiguring the joint angles.

In the third set of simulation, the collision avoidance subtask control is also performed. In this scenario, the collision avoidance algorithm is taken as the same as the existing ones [1,2]. In the simulation, the obstacle that the slave robot needs to avoid is assumed to be located at $X_0 = [-0.1, -0.4]^{\mathrm{T}}$m, and the collision distance and the safe distance are given as $R = 0.15$m and $r = 0.05$m, which are respectively shown as the dashed circle and solid circle in Fig. 7.9. The compared simulation results are then shown in Fig. 7.9(a and c). The two figures show the schematic diagrams of the slave robot joint configuration without adding and adding the obstacle avoidance subtask controller. It can be seen that the proposed obstacle avoidance algorithm can help the slave robot to use its redundancy to avoid the obstacle.

Note that, for all of the subtask control, the human operator only need to focus on manipulating the position of end-effector, and the slave robot tracks the master position in task space while automatically regulating its configuration via subtask controller to achieve the subtask. In above three sets of subtask controls, their corresponding position tracking errors in task space are shown in Fig. 7.7(b–d).

Remark 7.5. *Since the rate of change of the transmission delay of the actual communication channel is difficult to obtain, we choose $\tilde{d} = 4$ in the simulation to cover a wider time delay situation as much as possible. Although this pre-selected $\tilde{d}$ cannot cover all the cases in Fig. 7.4, the algorithm still obtains satisfactory control performance, as shown in Figs. 7.5–7.9.*

7.6.2 Simulations with large artificial communication delays

The time delays in Fig. 7.4 are very small. To better show the effectiveness of the proposed algorithm, in this set of simulation, we take two large communication delays, which are given in Fig. 7.10. In this case, under the same simulation setup as Subsection 7.6.1, the simulation results are shown in Figs. 7.11–7.13. From which, the satisfactory control performance is still obtained. Note that, given that the subtask

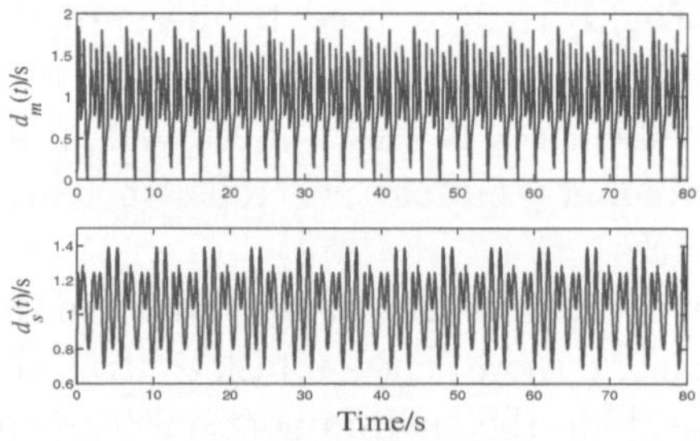

Figure 7.10 Artificial communication delays $d_m(t)$ and $d_s(t)$.

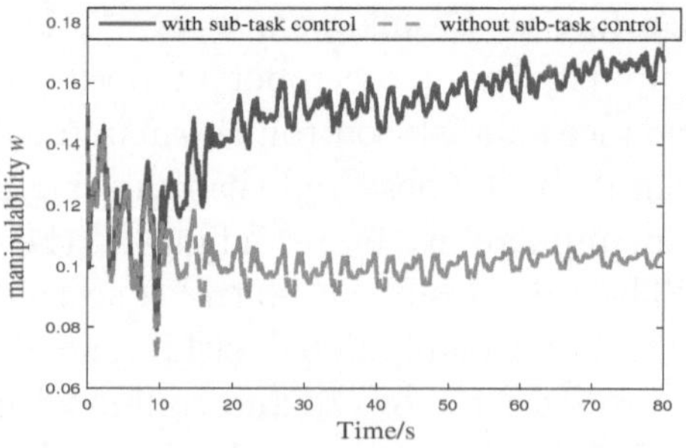

Figure 7.11 Measure of manipulability under the large artificial delays.

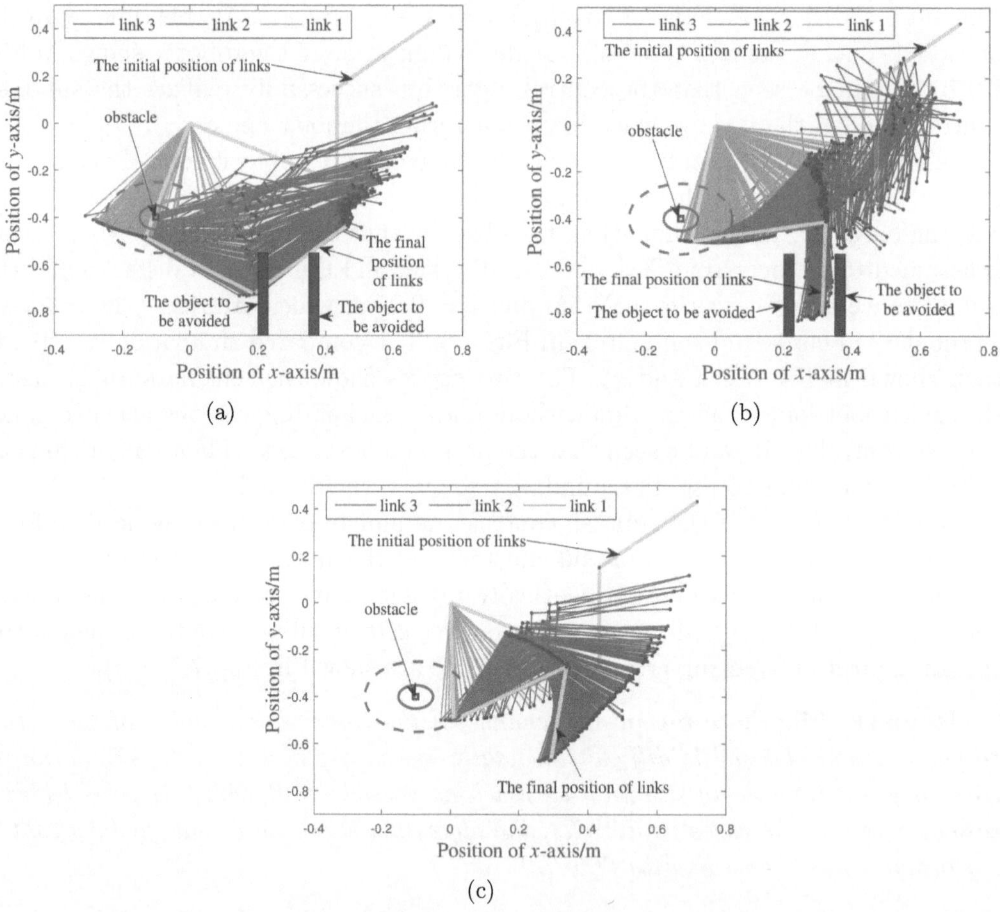

Figure 7.12 Configurations of the slave robot under the large artificial delays. (a) Without the use of subtask control. (b) With subtask control to limit the angle of joints. (c) With the use of collision avoidance control.

controllers can produce only a joint self-motion of the structure but no task space motion, to save the page space, only the tracking errors without using any subtask control are given in Fig. 7.13.

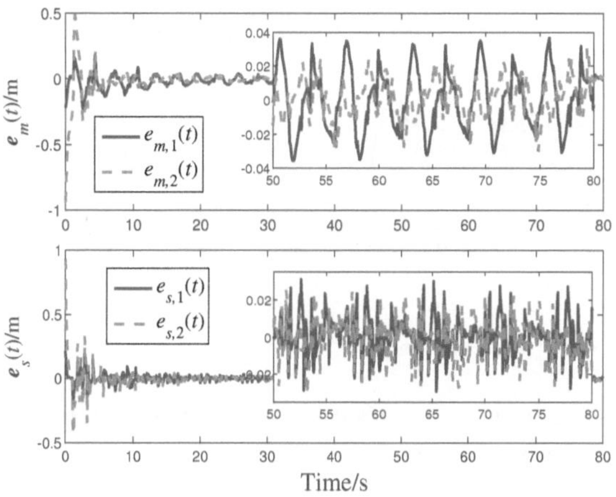

Figure 7.13 Position tracking errors without the use of subtask control under the large artificial delays.

7.6.3 Comparison with the existing work

Liu et al. provide a semi-autonomous control method for teleoperation systems with time-varying communication delays [2]. However, in the presence of non-passive external forces, the Liu's algorithm [2] can only be guaranteed to be effective for the teleoperation system whose dynamic model is accurately known, which limits its practicability. In this section, the comparison study between this chapter and the Liu's work [2] is presented. For the simulation setup, first, the slave robot is assumed to run in free motion, while the external force exerted on master robot from human operator is shown as Fig. 7.3 with $M_h = B_h = K_h \equiv 0$. Second, the external disturbances have the same values as the ones in Subsection 7.6.1, while the time delays are given as Fig. 7.4. Since the Liu's method does not consider the model uncertainty, the exact dynamic model of the robot in the simulation is assumed to be known for the simulation of Liu's algorithm. For comparison, the algorithm proposed in this chapter still assumes that the dynamic model is unknown, and uses adaptive control technology. When taking the same subtask controller as Subsection 7.6.1, the simulation results are as Figs. 7.14–7.17.

From the simulation results, although the control of Liu's method [2] uses accurate dynamic information, the algorithm proposed in this chapter still obtains better tracking performance after using the adaptive control technique. Further, as shown in Fig. 7.15(c), the adaptive control algorithm proposed in this chapter can also better improve the manipulability of the slave robot. Comparing Fig. 7.16(c and d), it can be found that the adaptive algorithm of this chapter obtains a smaller value of

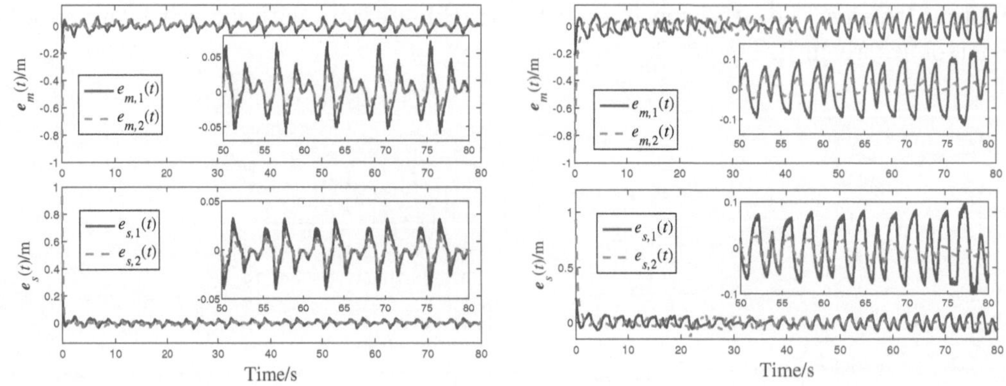

(a) (b)

Figure 7.14 Comparisons between this chapter and Liu's method [2] without the use of subtask control. (a) Tracking performance of this chapter. (b) Tracking performance of Liu's method [2].

$|q_{s,1} + q_{s,2} + q_{s,3}|$ in the final configuration of the robot joint angle, so it has better joint limitation subtask control performance.

7.6.4 Comparison on tracking response for different time delays

In the previous method [2], its control gains should satisfy (17), i.e., $2\beta_k(1 - \lambda\bar{T}) > \max\{k_v d_s^l + 1, k_v d_m^l + 1\}$. For different communication delays, the auxiliary gain λ should be $\lambda < \frac{1}{\bar{T}}$, which means that a larger $\bar{T}$[1] will cause a smaller λ. However, according to the discussion [15], if λ is lowered, the position difference between the master and slave robots contributes less to the control signal, resulting in the increase of the settling time for the position tracking response. Therefore, the tracking performance of the Liu's algorithm [2] is sensitive to the communication delays. A poor tracking performance may be obtained in the case of large communication delays. Compared with the method [2], as shown in (7.14), this chapter does not impose such a constraint. For a verification purpose, let the time delays be $d_m(t) = 0.3 + 0.6\cos^2(t)$ and $d_s(t) = 0.1 + 0.8\sin^2(t)$, then $\bar{T} = 1.8$, which is larger than the one in Subsection 7.6.3. In this case, a smaller $\lambda = 0.5$[2] is taken for the Liu's controller [2]. With the same simulation setups, the tracking performance is shown in Figs. 7.18 and 7.19. Compared with Subsection 7.6.3, the position tracking shown in Fig. 7.19 has a lower response speed.

[1] $\bar{T}$ has the same meaning as $\bar{d}$ in this chapter, representing the upper bound of delay.

[2] In Subsection 7.6.3, it is set to $\lambda = 1.5$.

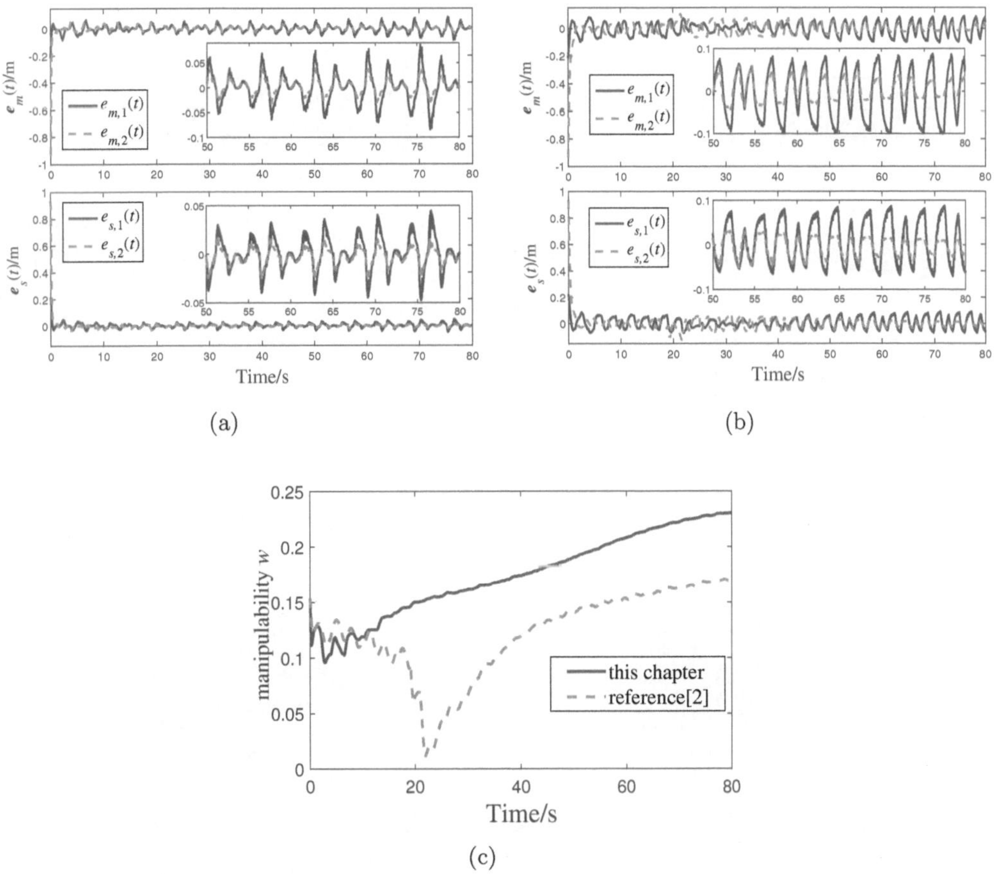

Figure 7.15 Comparisons between this chapter and Liu's method [2] with the use of subtask control to avoid configuration singularity. (a) Tracking performance of this chapter. (b) Tracking performance of Liu's method [2]. (c) Measure of manipulability.

7.6.5 Discussions

For the above simulation results, four important points need to be mentioned here.

(1) The error convergence rate in Fig. 7.13 is slower than that in Fig. 7.7. This is mainly due to the fact that a large $\bar{d}$ in Subsection 7.6.2 leads to a smaller parameter μ_2 in (7.16). From Lemma B.1 of our previous work [16], this will eventually reduce the convergence speed. In the actual implementation of the algorithm, μ_2 can be selected according to the largest $\bar{d}$ in the communication channel to obtain a satisfactory convergence speed for all time-delay situations.

(2) Although the performance of the algorithm proposed in this chapter is also related to time delay, the comparison of Subsection 7.6.3 and Subsection 7.6.4 shows that its sensitivity to time delay is weaker than that of Liu's algorithm [2].

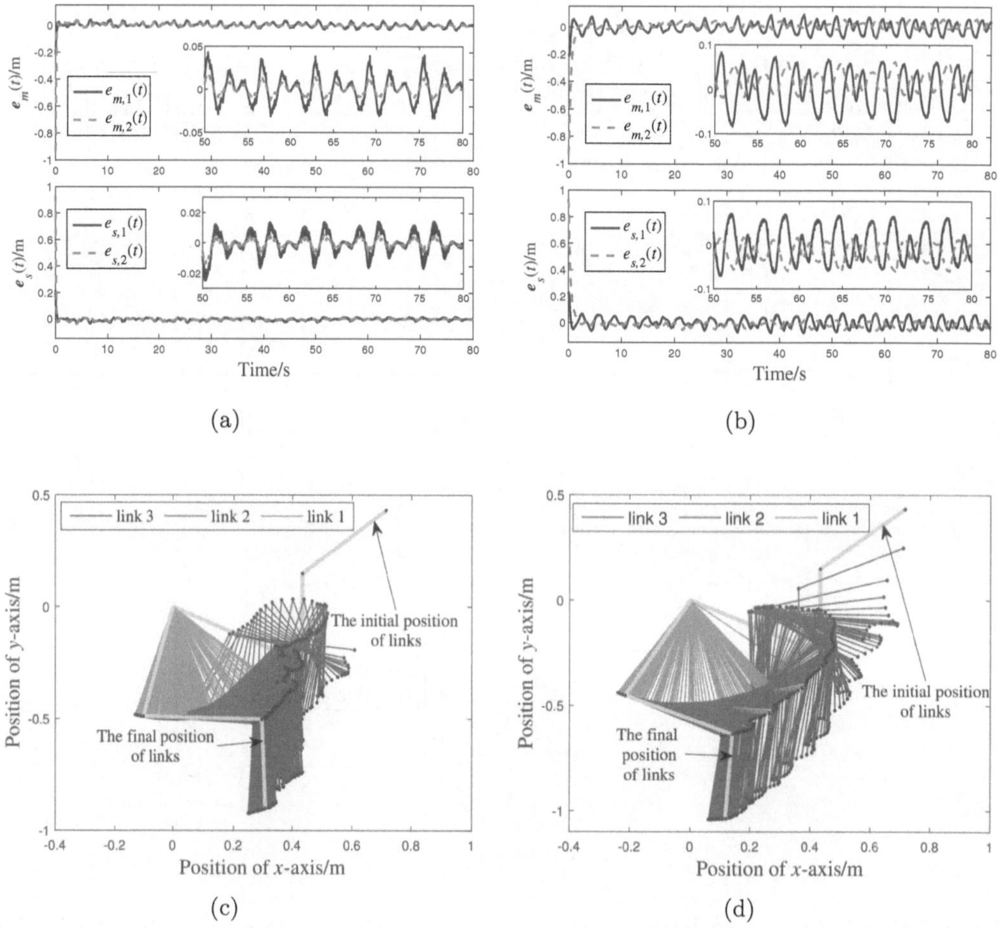

Figure 7.16 Comparisons between this chapter and Liu's method [2] with subtask control to limit the angle of joints. (a) Tracking performance of this chapter. (b) Tracking performance of Liu's method [2]. (c) Configuration of the slave robot of this chapter. (d) Configuration of the slave robot of Liu's method [2].

(3) From the simulation results in Subsection 7.6.3, although the real time $\dot{d}_m$ and $\dot{d}_s$ are used in Liu's control design [2], a better performance is obtained by this chapter. As discussed in Remark 7.4, given that the rate of change of the delay is very difficult to measure or estimate in practice, the proposed delay-dependent control scheme of this chapter provides another option to the engineer.

(4) In Subsection 7.6.3 and Subsection 7.6.4, although the exact dynamics are used in the simulations of Liu's method [2], a better performance is still obtained by the proposed adaptive control of this chapter. In addition, the proposed control also has a faster position tracking response than Liu's method [2] in the case of large communication delays.

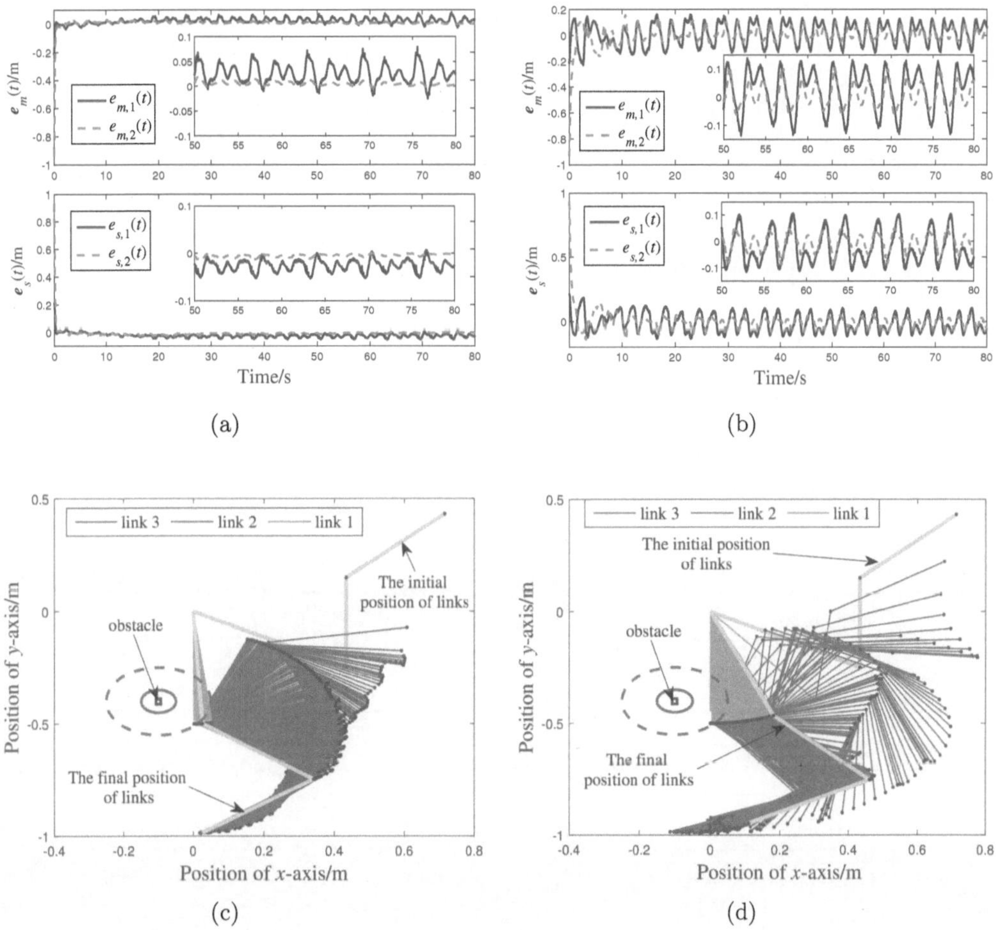

Figure 7.17 Comparisons between this chapter and Liu's method [2] with the use of collision avoidance control. (a) Tracking performance of this chapter. (b) Tracking performance of Liu's method [2]. (c) Configuration of the slave robot of this chapter. (d) Configuration of the slave robot of Liu's method [2].

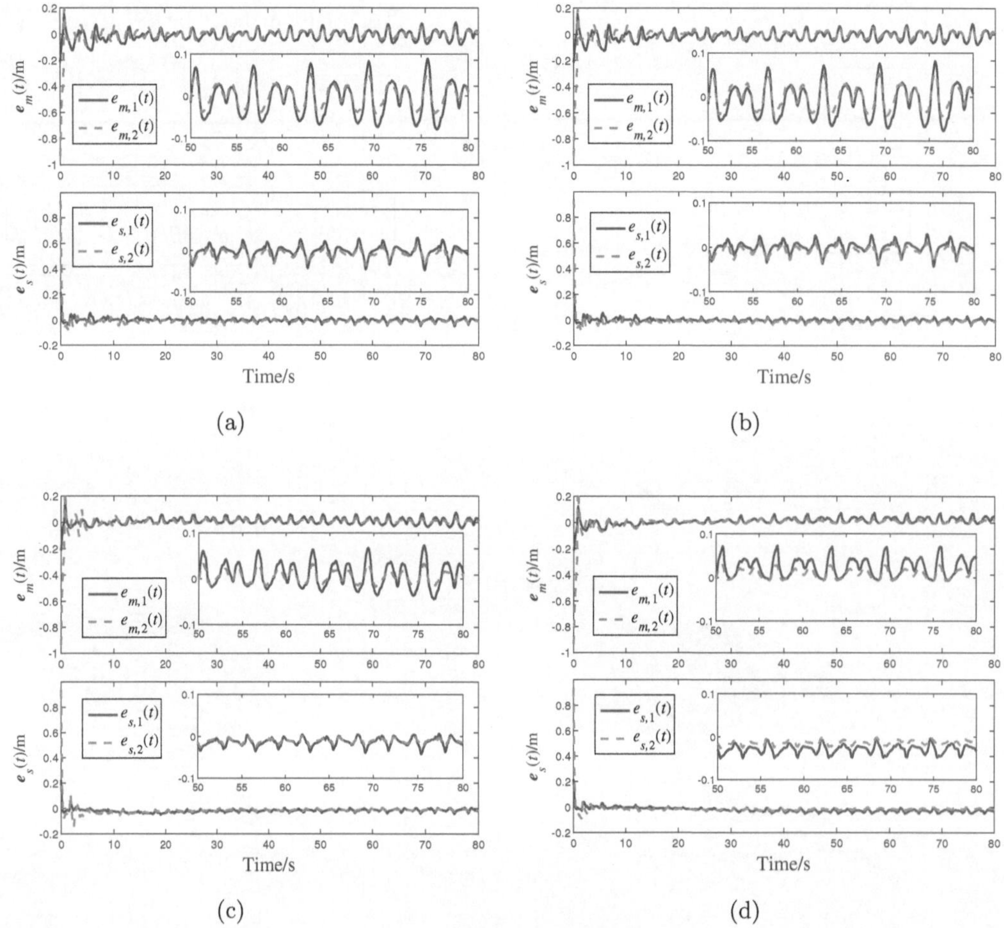

Figure 7.18 Tracking errors under the control of this chapter. (a) Without using any subtask control. (b) With the use of subtask control to avoid configuration singularity. (c) With subtask control to limit the angle of joints. (d) With the use of collision avoidance subtask control.

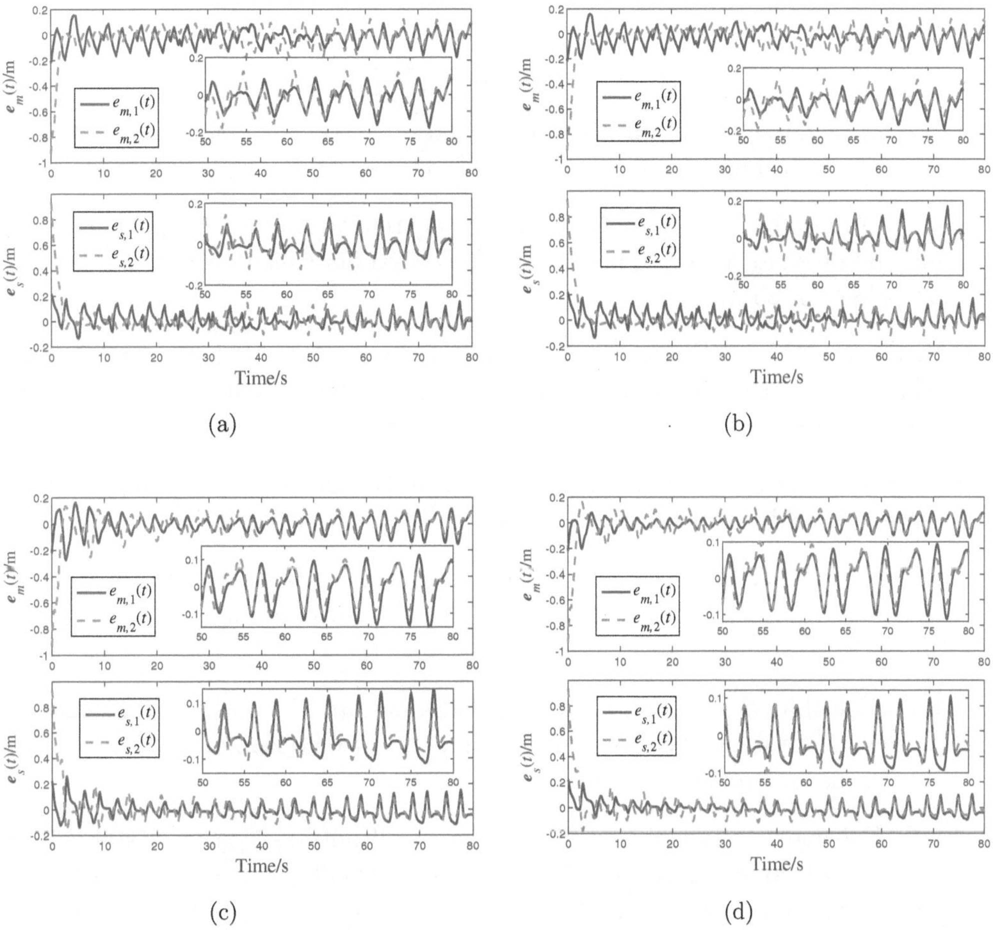

Figure 7.19 Tracking errors under the control of Liu's method [2]. (a) Without using any subtask control. (b) With the use of subtask control to avoid configuration singularity. (c) With subtask control to limit the angle of joints. (d) With the use of collision avoidance subtask control.

7.7 CONCLUSION

In this chapter, a new semi-autonomous adaptive control method based on the auxiliary switched filter control scheme is proposed for the heterogeneous bilateral teleoperation system with asymmetric time-varying communication delay, passive/non-passive external force, dynamic uncertainty and input uncertainty. While realizing the synchronization of the master and slave robot motion, the redundant joints of the slave robot can be used to realize additional subtask control such as singularity avoidance, joint-limit avoidance and obstacle avoidance. Compared with the existing semi-autonomous control methods of heterogeneous teleoperation systems, the algorithm proposed in this chapter has a wider application range, better control effect, and improves the manipulability of the teleoperated robots in complex dynamic environments. the effectiveness of the algorithm is verified by simulation results.

Bibliography

[1] Y.-C. Liu and N. Chopra, "Control of semi-autonomous teleoperation system with time delays," *Automatica*, vol. 49, no. 6, pp. 1553–1565, 2013.

[2] Y.-C. Liu, "Task-space bilateral teleoperation systems for heterogeneous robots with time-varying delays," *Robotica*, vol. 33, no. 10, pp. 2065–2082, 2015.

[3] Y.-C. Liu and N. Chopra, "Controlled synchronization of heterogeneous robotic manipulators in the task space," *IEEE Transactions on Robotics*, vol. 28, no. 1, pp. 268–275, 2011.

[4] R. Moreau, M. T. Pham, M. Tavakoli, M. Le, and T. Redarce, "Sliding-mode bilateral teleoperation control design for master–slave pneumatic servo systems," *Control Engineering Practice*, vol. 20, no. 6, pp. 584–597, 2012.

[5] T. H. Lee and C. J. Harris, *Adaptive Neural Network Control of Robotic Manipulators.* World Scientific, 1998.

[6] X.-S. Wang, C.-Y. Su, and H. Hong, "Robust adaptive control of a class of nonlinear systems with unknown dead-zone," *Automatica*, vol. 40, no. 3, pp. 407–413, 2004.

[7] I. G. Polushin, S. N. Dashkovskiy, A. Takhmar, and R. V. Patel, "A small gain framework for networked cooperative force-reflecting teleoperation," *Automatica*, vol. 49, no. 2, pp. 338–348, 2013.

[8] I. G. Polushin, A. Tayebi, and H. J. Marquez, "Control schemes for stable teleoperation with communication delay based on ios small gain theorem," *Automatica*, vol. 42, no. 6, pp. 905–915, 2006.

[9] I. Sarras, E. Nuno, and L. Basanez, "An adaptive controller for nonlinear teleoperators with variable time-delays," *Journal of the Franklin Institute*, vol. 351, no. 10, pp. 4817–4837, 2014.

[10] P. Hsu, J. Mauser, and S. Sastry, "Dynamic control of redundant manipulators," *Journal of Robotic Systems*, vol. 6, no. 2, pp. 133–148, 1989.

[11] L. Li, W. A. Gruver, Q. Zhang, and W. Chen, "Real-time control of redundant robots subject to multiple criteria," in *Proceedings of the IEEE International Conference on Robotics and Automation.* IEEE, 1998, pp. 115–120.

[12] R. Kelly, V. S. Davila, and J. A. L. Perez, *Control of Robot Manipulators in Joint Space.* Springer Science & Business Media, 2005.

[13] L. Sciavicco and B. Siciliano, *Modelling and Control of Robot Manipulators.* Springer Science & Business Media, 2001.

[14] T. Yoshikawa, "Analysys and control of robot manipulators with redundancy. robotic research," *Robotics Research: The First International Syposium*, pp. 735–747, 1984.

[15] F. Hashemzadeh, I. Hassanzadeh, and M. Tavakoli, "Teleoperation in the presence of varying time delays and sandwich linearity in actuators," *Automatica*, vol. 49, no. 9, pp. 2813–2821, 2013.

[16] D.-H. Zhai and Y. Xia, "Adaptive fuzzy control of multilateral asymmetric teleoperation for coordinated multiple mobile manipulators," *IEEE Transactions on Fuzzy Systems*, vol. 24, no. 1, pp. 57–70, 2015.

V

Finite-time Teleoperation

CHAPTER 8

Adaptive finite-time teleoperation control

TO achieve faster master-slave robot motion synchronization, this chapter addresses the adaptive finite-time control problem of nonlinear teleoperation system in the presence of asymmetric time-varying delays. To achieve the finite-time control objective, the nonsmooth control technique is employed, where the nonsmooth switched filter is developed. The proposed finite-time controller provides faster convergence speed than the adaptive controllers in the previous chapters.

8.1 INTRODUCTION

In Chapters 3 and 4, the bounded synchronization control of telerobot is realized. However, from a practical point of view, in the actual teleoperation system, people always hope that the master robot and the slave robot can achieve synchronization faster, which involves the problem of limited time control design for the teleoperation system. However, at present, there are few research results on the finite-time stability analysis and controller design of teleoperation system in complex network and external environment. The main reasons are: the time-varying delay of network transmission and the complexity of robot environment force make the control difficult. Yang et al. propose a fuzzy adaptive controller based on nonsingular terminal sliding mode technology for nonlinear teleoperation system, which realizes finite-time synchronous control [1]. However, it still has some limitations, such as only dealing with constant communication delay, and assuming that the master and slave robots are in free motion. In 2015, Yang et al. further combined the barrier Lyapunov function method to realize the predefined performance finite-time control of teleoperation system [2]. However, it still faces the same limitations as the previous algorithm [1]. Although there are many finite-time control schemes for nonlinear dynamic systems [3–10], considering that the related algorithms are not aimed at teleoperation systems and do not consider the communication time delay, they are difficult to use in teleoperation.

This chapter addresses the adaptive finite-time control problem of nonlinear teleoperation system in the presence of asymmetric time-varying delays. To achieve the

DOI: 10.1201/9781003382058-8

finite-time control objective, the nonsmooth control technique is employed, where the nonsmooth switched filter is investigated.

8.2 PROBLEM FORMULATION

The dynamics of the teleoperation system considered in this chapter is consistent with the description of (3.1). For the convenience, it is repeated as follows:

$$\begin{aligned} \boldsymbol{M}_m(\boldsymbol{q}_m)\ddot{\boldsymbol{q}}_m + \boldsymbol{C}_m(\boldsymbol{q}_m, \dot{\boldsymbol{q}}_m)\dot{\boldsymbol{q}}_m + \boldsymbol{g}_m(\boldsymbol{q}_m) &= \boldsymbol{f}_h + \boldsymbol{\tau}_m \\ \boldsymbol{M}_s(\boldsymbol{q}_s)\ddot{\boldsymbol{q}}_s + \boldsymbol{C}_s(\boldsymbol{q}_s, \dot{\boldsymbol{q}}_s)\dot{\boldsymbol{q}}_s + \boldsymbol{g}_s(\boldsymbol{q}_s) &= \boldsymbol{\tau}_s - \boldsymbol{f}_e \end{aligned}$$

When implementing finite-time control for the teleoperated robotic system (3.1), the following definitions are further made. First, let $\boldsymbol{X}_m \in \mathbb{R}^n$ and $\boldsymbol{X}_s \in \mathbb{R}^n$ represent the coordinates of the master and slave robot end effector poses in the task space respectively, which are defined as $\boldsymbol{X}_j = \boldsymbol{h}_j(\boldsymbol{q}_j)$, $\dot{\boldsymbol{X}}_j = \boldsymbol{J}_j(\boldsymbol{q}_j)\dot{\boldsymbol{q}}_j$, where $j \in \{m, s\}$, $\boldsymbol{h}_m(\boldsymbol{q}_m) \in \mathbb{R}^n$ and $\boldsymbol{h}_s(\boldsymbol{q}_s) \in \mathbb{R}^n$ are the mappings from joint space to task space, $\boldsymbol{J}_m(\boldsymbol{q}_m) = \dfrac{\partial \boldsymbol{h}_m(\boldsymbol{q}_m)}{\partial \boldsymbol{q}_m} \in \mathbb{R}^{n\times n}$ and $\boldsymbol{J}_s(\boldsymbol{q}_s) = \dfrac{\partial \boldsymbol{h}_s(\boldsymbol{q}_s)}{\partial \boldsymbol{q}_s} \in \mathbb{R}^{n\times n}$ are the Jacobian matrices. In this chapter, we further assume that the Jacobian matrix satisfies:

Assumption 8.1. *Jacobian matrices $\boldsymbol{J}_m(\boldsymbol{q}_m)$ and $\boldsymbol{J}_s(\boldsymbol{q}_s)$ are known and not singular.*

Similar to the previous works [11,12], this chapter models the operator and environment dynamics as follows:

$$\begin{cases} \boldsymbol{f}_h = \boldsymbol{J}_m^{\mathrm{T}}(\boldsymbol{q}_m)\left(\boldsymbol{f}_h^* - \boldsymbol{M}_h\ddot{\boldsymbol{X}}_m - \boldsymbol{B}_h\dot{\boldsymbol{X}}_m - \boldsymbol{K}_h\boldsymbol{X}_m\right) \\ \boldsymbol{f}_e = \boldsymbol{J}_s^{\mathrm{T}}(\boldsymbol{q}_s)\left(\boldsymbol{f}_e^* + \boldsymbol{M}_e\ddot{\boldsymbol{X}}_s + \boldsymbol{B}_e\dot{\boldsymbol{X}}_s + \boldsymbol{K}_e\boldsymbol{X}_s\right) \end{cases} \tag{8.1}$$

where $\boldsymbol{M}_h$, $\boldsymbol{M}_e$, $\boldsymbol{B}_h$, $\boldsymbol{B}_e$, $\boldsymbol{K}_h$ and $\boldsymbol{K}_e$ are unknown non-negative constant value matrices, representing the mass, damping and stiffness coefficients, respectively. $\boldsymbol{f}_h^*$ and $\boldsymbol{f}_e^*$ are the exogenous forces of the operator and the environment on the robot, respectively.

Assumption 8.2. *[11] The operator exogenous force $\boldsymbol{f}_h^*$ and the environment exogenous force $\boldsymbol{f}_e^*$ are locally essential bounded.*

Denote $\boldsymbol{M}_{he_m} = \boldsymbol{M}_h$, $\boldsymbol{M}_{he_s} = \boldsymbol{M}_e$, $\boldsymbol{B}_{he_m} = \boldsymbol{B}_h$, $\boldsymbol{B}_{he_s} = \boldsymbol{B}_e$, $\boldsymbol{K}_{he_m} = \boldsymbol{K}_h$, $\boldsymbol{K}_{he_s} = \boldsymbol{K}_e$. Then (3.1) can be written as

$$\begin{cases} \boldsymbol{\mathcal{M}}_m(\boldsymbol{q}_m)\ddot{\boldsymbol{q}}_m + \boldsymbol{\mathcal{C}}_m(\boldsymbol{q}_m, \dot{\boldsymbol{q}}_m)\dot{\boldsymbol{q}}_m + \boldsymbol{g}_m(\boldsymbol{q}_m) \\ \qquad = \boldsymbol{J}_m^{\mathrm{T}}(\boldsymbol{q}_m)\boldsymbol{f}_h^* - \boldsymbol{J}_m^{\mathrm{T}}(\boldsymbol{q}_m)\boldsymbol{K}_h\boldsymbol{X}_m + \boldsymbol{\tau}_m \\ \boldsymbol{\mathcal{M}}_s(\boldsymbol{q}_s)\ddot{\boldsymbol{q}}_s + \boldsymbol{\mathcal{C}}_s(\boldsymbol{q}_s, \dot{\boldsymbol{q}}_s)\dot{\boldsymbol{q}}_s + \boldsymbol{g}_s(\boldsymbol{q}_s) = -\boldsymbol{J}_s^{\mathrm{T}}(\boldsymbol{q}_s)\boldsymbol{f}_e^* - \boldsymbol{J}_s^{\mathrm{T}}(\boldsymbol{q}_s)\boldsymbol{K}_e\boldsymbol{X}_s + \boldsymbol{\tau}_s \end{cases} \tag{8.2}$$

where for any $j \in \{m, s\}$,

$$\begin{aligned} \boldsymbol{\mathcal{M}}_j(\boldsymbol{q}_j) &= \boldsymbol{M}_j(\boldsymbol{q}_j) + \boldsymbol{J}_j^{\mathrm{T}}(\boldsymbol{q}_j)\boldsymbol{M}_{he_j}\boldsymbol{J}_j(\boldsymbol{q}_j) \\ \boldsymbol{\mathcal{C}}_j(\boldsymbol{q}_j, \dot{\boldsymbol{q}}_j) &= \boldsymbol{C}_j(\boldsymbol{q}_j, \dot{\boldsymbol{q}}_j) + \boldsymbol{J}_j^{\mathrm{T}}(\boldsymbol{q}_j)\boldsymbol{B}_{he_j}\boldsymbol{J}_j(\boldsymbol{q}_j) + \boldsymbol{J}_j^{\mathrm{T}}(\boldsymbol{q}_j)\boldsymbol{M}_{he_j}\dot{\boldsymbol{J}}_j(\boldsymbol{q}_j) \end{aligned}$$

For dynamics (8.2), from Property 2.1~Property 2.5, the following properties hold [13]:

Property 8.1. *The inertia matrix* $\mathcal{M}_j(\boldsymbol{q}_j)$ *is symmetric, positive definite and bounded; that is, there are* $\rho_{j1} > 0$ *and* $\rho_{j2} > 0$*, such that* $\rho_{j1}\boldsymbol{I} \leq \mathcal{M}_j(\boldsymbol{q}_j) \leq \rho_{j2}\boldsymbol{I}$*, where* $\boldsymbol{I}$ *is of suitable dimension identity matrix.*

Property 8.2. *For any* $\boldsymbol{x} \in \mathbb{R}^n$*,*

$$\boldsymbol{x}^{\mathrm{T}}\left(\dot{\mathcal{M}}_j(\boldsymbol{q}_j) - 2\mathcal{C}_j(\boldsymbol{q}_j, \dot{\boldsymbol{q}}_j)\right)\boldsymbol{x} = -2\boldsymbol{x}^{\mathrm{T}}\boldsymbol{J}_j^{\mathrm{T}}(\boldsymbol{q}_j)\boldsymbol{B}_{he_j}\boldsymbol{J}_j(\boldsymbol{q}_j)\boldsymbol{x}$$

where $\boldsymbol{B}_{he_m} = \boldsymbol{B}_h$*,* $\boldsymbol{B}_{he_s} = \boldsymbol{B}_e$*.*

Property 8.3. *The dynamics (8.2) can be linearly parameterized, i.e.*

$$\mathcal{M}_j(\boldsymbol{q}_j)\boldsymbol{x} + \mathcal{C}_j(\boldsymbol{q}_j, \dot{\boldsymbol{q}}_j)\boldsymbol{y} + \boldsymbol{g}_j(\boldsymbol{q}_j) = \boldsymbol{Y}_j(\boldsymbol{q}_j, \dot{\boldsymbol{q}}_j, \boldsymbol{x}, \boldsymbol{y})\boldsymbol{\Theta}_j$$

where $\boldsymbol{Y}_j(\boldsymbol{q}_j, \dot{\boldsymbol{q}}_j, \boldsymbol{x}, \boldsymbol{y}) \in \mathbb{R}^{n\times l}$ *is the known regression matrix,* $\boldsymbol{\Theta}_j \in \mathbb{R}^l$ *is the unknown parameter vector.*

In this chapter, the control objective of the finite-time control task is: in the presence of variable communication time delay, model uncertainty and non-passive external force, through the action of the controller, the tracking errors (3.2) of the master and slave robots can converge to a certain bound in finite time.

8.3 FINITE-TIME CONTROLLER

To achieve the control goal, combined with non-smooth control theory, this section will present the design of the finite-time controller based on the switched filter-based adaptive control framework proposed in Chapter 3.

First, for any $j \in \{m, s\}$, define the auxiliary control variable

$$\boldsymbol{\eta}_j(t) = \dot{\boldsymbol{q}}_j(t) + \lambda_j\boldsymbol{e}_j(t) \tag{8.3}$$

where λ_j is a positive constant.

To simplify the expression, define

$$\boldsymbol{e}_{vj}(t) = \dot{\boldsymbol{q}}_j(t) - \boldsymbol{q}_{vd\iota}(t, d_\iota(t))$$

where $j \in \{m, s\}$, $\iota \in \{m, s\}$, $\iota \neq j$; $\boldsymbol{q}_{vdm}(t, d_m(t)) = \dot{\boldsymbol{q}}_m(\theta)|_{\theta=t-d_m(t)}$; $\boldsymbol{q}_{vds}(t, d_s(t)) = \dot{\boldsymbol{q}}_s(\theta)|_{\theta=t-d_s(t)}$. Denote $\boldsymbol{f}^*_{he_m}(t) = \boldsymbol{f}^*_h(t)$, $\boldsymbol{f}^*_{he_s}(t) = -\boldsymbol{f}^*_e(t)$.

Substituting (8.3) into (8.2),

$$\begin{aligned}&\mathcal{M}_j(\boldsymbol{q}_j)\dot{\boldsymbol{\eta}}_j + \mathcal{C}_j(\boldsymbol{q}_j, \dot{\boldsymbol{q}}_j)\boldsymbol{\eta}_j \\ &\quad= \boldsymbol{Y}_j(t)\boldsymbol{\Theta}_j + \boldsymbol{J}_j^{\mathrm{T}}(\boldsymbol{q}_j)\boldsymbol{f}^*_{he_j} + \boldsymbol{\tau}_j + \mathcal{M}_j(\boldsymbol{q}_j)\lambda_j\dot{d}_\iota(t)\boldsymbol{q}_{vd\iota}(t, d_\iota(t))\end{aligned} \tag{8.4}$$

where $\iota \in \{m, s\}$, $\iota \neq j$; $\boldsymbol{Y}_j(t)\boldsymbol{\Theta}_j := \boldsymbol{Y}_j(\boldsymbol{q}_j, \dot{\boldsymbol{q}}_j, \boldsymbol{e}_j, \boldsymbol{e}_{vj})\boldsymbol{\Theta}_j$ satisfies

$$\boldsymbol{Y}_j(t)\boldsymbol{\Theta}_j = \mathcal{M}_j(\boldsymbol{q}_j)\lambda_j\boldsymbol{e}_{vj} + \mathcal{C}_j(\boldsymbol{q}_j, \dot{\boldsymbol{q}}_j)\lambda_j\boldsymbol{e}_j - \boldsymbol{g}_j(\boldsymbol{q}_j) - \boldsymbol{J}_j^{\mathrm{T}}(\boldsymbol{q}_j)\boldsymbol{K}_{he_j}\boldsymbol{X}_j$$

$\boldsymbol{\Theta}_j \in \mathbb{R}^{o_{j1}}$. From Property 8.3, the last term of (8.4) can be written as

$$\boldsymbol{Y}_{j1}(t)\boldsymbol{\theta}_j := \boldsymbol{Y}_{j1}(\boldsymbol{q}_j, \boldsymbol{q}_{vd\iota}(t, d_\iota(t)))\boldsymbol{\theta}_j = \mathcal{M}_j(\boldsymbol{q}_j)\boldsymbol{q}_{vd\iota}(t, d_\iota(t))$$

where $\boldsymbol{\theta}_j \in \mathbb{R}^{o_{j2}}$.

Then,

$$\boldsymbol{\mathcal{M}}_j(\boldsymbol{q}_j)\dot{\boldsymbol{\eta}}_j + \boldsymbol{\mathcal{C}}_j(\boldsymbol{q}_j, \dot{\boldsymbol{q}}_j)\boldsymbol{\eta}_j = \boldsymbol{Y}_j(t)\boldsymbol{\Theta}_j + \lambda_j \dot{d}_\iota(t)\boldsymbol{Y}_{j1}(t)\boldsymbol{\theta}_j + \boldsymbol{J}_j^{\mathrm{T}}(\boldsymbol{q}_j)\boldsymbol{f}_{he_j}^* + \boldsymbol{\tau}_j \tag{8.5}$$

Design the controller as

$$\begin{aligned}\boldsymbol{\tau}_j(t) = &-\boldsymbol{Y}_j(t)\hat{\boldsymbol{\Theta}}_j(t) - K_{j1}\boldsymbol{\eta}_j(t) - K_{j2}\|\boldsymbol{\eta}_{jt}\|_{M_2}^{1+\sigma}\mathrm{sgn}(\boldsymbol{\eta}_j(t)) - \\ &K_{j3}\boldsymbol{J}_j^{\mathrm{T}}(\boldsymbol{q}_j(t))\boldsymbol{J}_j(\boldsymbol{q}_j(t))\boldsymbol{\eta}_j(t) - K_{j4}\boldsymbol{y}_j(t) - K_{j5}\boldsymbol{Y}_{j1}(t)\boldsymbol{Y}_{j1}^{\mathrm{T}}(t)\boldsymbol{\eta}_j(t)\end{aligned} \tag{8.6}$$

where $j \in \{m, s\}$, for any $i = 1, 2, 3, 4$, $K_{ji} > 0$ is a constant; $\boldsymbol{y}_j(t)$ is the state of auxiliary switched filter; $\sigma \in (0, 1)$ is a constant; $\hat{\boldsymbol{\Theta}}_j(t)$ and $\hat{\boldsymbol{\theta}}_j(t)$ are the estimations of the unknown parameter vectors $\boldsymbol{\Theta}_j$ and $\boldsymbol{\theta}_j$, with the update laws

$$\begin{cases} \dot{\hat{\boldsymbol{\Theta}}}_j(t) = \psi_{j1}\boldsymbol{Y}_j^{\mathrm{T}}(t)\boldsymbol{\eta}_j(t) - \psi_{j2}\hat{\boldsymbol{\Theta}}_j(t) \\ \dot{\hat{\boldsymbol{\theta}}}_j(t) = -\lambda_j \tilde{d}\psi_{j3}\boldsymbol{Y}_{j1}^{\mathrm{T}}(t)\boldsymbol{\eta}_j(t)\mathrm{sgn}\left(\boldsymbol{\eta}_j^{\mathrm{T}}(t)\boldsymbol{Y}_{j1}(t)\hat{\boldsymbol{\theta}}_j(t)\right) - \psi_{j4}\hat{\boldsymbol{\theta}}_j(t) \end{cases} \tag{8.7}$$

where $\psi_{j1}, \psi_{j2}, \cdots, \psi_{j4} > 0$.

Let $\tilde{\boldsymbol{\Theta}}_j = \boldsymbol{\Theta}_j - \hat{\boldsymbol{\Theta}}_j$, from (8.5) and (8.6),

$$\begin{aligned}&\boldsymbol{\mathcal{M}}_j(\boldsymbol{q}_j(t))\dot{\boldsymbol{\eta}}_j(t) + \boldsymbol{\mathcal{C}}_j(\boldsymbol{q}_j(t), \dot{\boldsymbol{q}}_j(t))\boldsymbol{\eta}_j(t) \\ &\quad = \boldsymbol{Y}_j(t)\tilde{\boldsymbol{\Theta}}_j + \lambda_j\dot{d}_\iota(t)\boldsymbol{Y}_{j1}(t)\boldsymbol{\theta}_j + \boldsymbol{J}_j^{\mathrm{T}}(\boldsymbol{q}_j(t))\boldsymbol{f}_{he_j}^*(t) - K_{j1}\boldsymbol{\eta}_j(t) - \\ &\qquad K_{j2}\|\boldsymbol{\eta}_{jt}\|_{M_2}^{1+\sigma}\mathrm{sgn}(\boldsymbol{\eta}_j(t)) - K_{j3}\boldsymbol{J}_j^{\mathrm{T}}(\boldsymbol{q}_j(t))\boldsymbol{J}_j(\boldsymbol{q}_j(t))\boldsymbol{\eta}_j(t) - \\ &\qquad K_{j4}\boldsymbol{y}_j(t) - K_{j5}\boldsymbol{Y}_{j1}(t)\boldsymbol{Y}_{j1}^{\mathrm{T}}(t)\boldsymbol{\eta}_j(t)\end{aligned} \tag{8.8}$$

In (8.8), the auxiliary switched filter $\boldsymbol{y}_j$ is still designed based on the idea of dynamic compensation, and its design is similar to Section 3.3. Firstly, define the switching rule $\boldsymbol{\mathcal{K}}_j(t) = \mathrm{diag}\{\kappa_{j,1}(t), \kappa_{j,2}(t), \cdots, \kappa_{j,n}(t)\}$ as

$$\kappa_{j,i}(t) = \begin{cases} 1, & y_{j,i}(t)e_{j,i}(t) > 0 \text{ or } y_{j,i}(t)e_{j,i}(t) = 0,\ y_{j,i}(t^-)e_{j,i}(t^-) \le 0 \\ -1, & y_{j,i}(t)e_{j,i}(t) < 0 \text{ or } y_{j,i}(t)e_{j,i}(t) = 0,\ y_{j,i}(t^-)e_{j,i}(t^-) > 0 \end{cases} \tag{8.9}$$

Let $\boldsymbol{s}_j(t) = \boldsymbol{y}_j(t) + \boldsymbol{\mathcal{K}}_j(t)\boldsymbol{e}_j(t)$, $\bar{\boldsymbol{y}}_j(t) = [\boldsymbol{y}_j^{\mathrm{T}}(t), \boldsymbol{e}_j^{\mathrm{T}}(t)]^{\mathrm{T}}$, $\bar{\boldsymbol{y}}_{jt}(\tau) := \{\bar{\boldsymbol{y}}_j(t+\tau),\ \tau \in [-\bar{d}, 0]\} \in C([-\bar{d}, 0]; \mathbb{R}^{2n})$, for any $i = 1, 2, \cdots, n$ and $j \in \{m, s\}$, $\boldsymbol{y}_j(t)$ is designed as

$$\dot{y}_{j,i}(t) = \begin{cases} -P_{j1}s_{j,i}(t) - P_{j2}\|\bar{\boldsymbol{y}}_{jt}\|_{M_2}^{1+\sigma}\mathrm{sgn}(s_{j,i}(t)) - e_{vj,i}(t) - P_{j3}\eta_{j,i}(t) + u_{j,i}(t), \\ \qquad \kappa_{j,i}(t) = 1 \\ -P_{j1}s_{j,i}(t) - P_{j2}\|\bar{\boldsymbol{y}}_{jt}|_{M_2}^{1+\sigma}\mathrm{sgn}(s_{j,i}(t)) + e_{vj,i}(t) - P_{j3}\eta_{j,i}(t) + u_{j,i}(t), \\ \qquad \kappa_{j,i}(t) = -1 \end{cases} \tag{8.10}$$

where

$$u_{j,i}(t) = \begin{cases} -P_{j4}\dfrac{s_{j,i}(t)}{|s_{j,i}(t)|^2}|q_{vd\iota,i}(t, d_\iota(t))|^2, & s_{j,i}(t) \neq 0 \\ 0, & s_{j,i}(t) = 0 \end{cases}$$

$\iota \in \{m, s\}$, $\iota \neq j$; $P_{jk} > 0$ $(k = 1, 2, 3, 4)$ is a constant; $y_{j,i}(t)$, $s_{j,i}(t)$ and $\eta_{j,i}(t)$ are the ith element of $\boldsymbol{y}_j(t)$, $\boldsymbol{s}_j(t)$ and $\boldsymbol{\eta}_j(t)$, respectively. Similar to Remark 3.2, $\boldsymbol{\mathcal{K}}_j(t)$ has 2^n modes, and denote $\mathcal{S} = \{1, 2, 3, \cdots, 2^n\}$; then build the one-to-one mapping $\bar{r}_j : \boldsymbol{\mathcal{K}}_j(t) \to \mathcal{S}$, model $\boldsymbol{\mathcal{K}}_j(t)$ to be a standard switching signal $r_j(t) := \bar{r}_j(\boldsymbol{\mathcal{K}}_j(t))$. Finally, the auxiliary switched filter (8.10) can be modeled as a standard switched dynamic system, i.e.,

$$\begin{aligned}\dot{\boldsymbol{y}}_j(t) &= -P_{j1}\left(\boldsymbol{y}_j(t) + \boldsymbol{\mathcal{K}}_j(t)\boldsymbol{g}(\boldsymbol{e}_j(t))\right) - P_{j2}\|\bar{\boldsymbol{y}}_{jt}\|_{M_2}^{1+\sigma}\operatorname{sgn}(\boldsymbol{s}_j(t)) - \\ &\quad \boldsymbol{\mathcal{K}}_j(t)\boldsymbol{e}_{vj}(t) - P_{j3}\boldsymbol{\eta}_j(t) + \boldsymbol{u}_j(t) \\ &:= \boldsymbol{h}_1(\boldsymbol{y}_j(t), \boldsymbol{e}_j(t), \boldsymbol{e}_{vj}(t), \boldsymbol{q}_{vd\iota}(t, d_\iota(t)), \dot{\boldsymbol{q}}_j(t), r_j(t))\end{aligned} \tag{8.11}$$

where $\boldsymbol{u}_j(t) = [u_{j,1}(t), u_{j,2}(t), \cdots, u_{j,n}(t)]^{\mathrm{T}}$.

Similarly, let $\{t_k^j\}_{k\geq 0}$ denote the switching sequences of $r_j(t)$. Then for any $t \in [t_k^j, t_{k+1}^j)$, by using the definition of $\boldsymbol{s}_j(t)$,

$$\begin{aligned}\dot{\boldsymbol{s}}_j(t) &= -P_{j1}\boldsymbol{s}_j(t) - P_{j2}\|\bar{\boldsymbol{y}}_{jt}\|_{M_2}^{1+\sigma}\operatorname{sgn}(\boldsymbol{s}_j(t)) - P_{j3}\boldsymbol{\eta}_j(t) + \\ &\quad \boldsymbol{u}_j(t) + \boldsymbol{\mathcal{K}}_j(t)\dot{d}_\iota(t)\boldsymbol{q}_{vd\iota}(t, d_\iota(t))\end{aligned} \tag{8.12}$$

With the designed finite-time controller, the complete closed-loop teleoperation system consists of subsystems (8.5)~(8.7) and (8.11), which can be regarded as a switched nonlinear time-delay system.

8.4 STABILITY ANALYSIS

For the complete closed-loop teleoperation system, the state and output are denoted as $\boldsymbol{x}_j = [\boldsymbol{\eta}_j^{\mathrm{T}}, \boldsymbol{q}_j^{\mathrm{T}}, \tilde{\boldsymbol{\Theta}}_j^{\mathrm{T}}, \tilde{\boldsymbol{\theta}}_j^{\mathrm{T}}, \boldsymbol{y}_j^{\mathrm{T}}, \boldsymbol{e}_j^{\mathrm{T}}]^{\mathrm{T}} \in \mathbb{R}^{4n+o_{j1}+o_{j2}}$, $\boldsymbol{z}_j = [\boldsymbol{\eta}_j^{\mathrm{T}}, \tilde{\boldsymbol{\Theta}}_j^{\mathrm{T}}, \tilde{\boldsymbol{\theta}}_j^{\mathrm{T}}, \boldsymbol{y}_j^{\mathrm{T}}, \boldsymbol{e}_j^{\mathrm{T}}]^{\mathrm{T}} \in \mathbb{R}^{3n+o_{j1}+o_{j2}}$, where $j \in \{m, s\}$. Also, let $\boldsymbol{x}_{jt} = \{\boldsymbol{x}_j(t+\tau) : \tau \in [-\bar{d}, 0]\} \in C([-\bar{d}, 0]; \mathbb{R}^{4n+o_{j1}+o_{j2}})$, $\boldsymbol{z}_{jt} = \{\boldsymbol{z}_j(t+\tau) : \tau \in [-\bar{d}, 0]\} \in C([-\bar{d}, 0]; \mathbb{R}^{3n+o_{j1}+o_{j2}})$. Then, we have the following conclusion.

Theorem 8.1. *If for any $j \in \{m, s\}$ and $i = 1, 2, 3, 4$, the control gains K_{ji}, K_{j5}, P_{ji}, ψ_{ji} and λ_j satisfy*

$$\begin{cases} K_{j1} > \dfrac{K_{j4} + P_{j3} + 1}{2} + \vartheta_j + K_{j2}\varepsilon_{j0} + \varepsilon_{j1} + \varepsilon_{j2} + P_{j2}\varepsilon_{j3} \\ K_{j3} \geq \dfrac{\varepsilon_j}{2} \\ K_{j5} \geq \dfrac{\lambda_j \tilde{d}\varepsilon_{jt1}}{2} + \dfrac{\lambda_j \tilde{d}\varepsilon_{jt2}}{2} \\ \dfrac{\psi_{j2}}{\psi_{j1}} > 2(\vartheta_j + K_{j2}\varepsilon_{j0} + \varepsilon_{j1} + \varepsilon_{j2} + P_{j2}\varepsilon_{j3}) \\ \dfrac{\psi_{j4}}{\psi_{j3}} > 2(\vartheta_j + \dfrac{\lambda_j \tilde{d}}{2\varepsilon_{jt1}} + K_{j2}\varepsilon_{j0} + \varepsilon_{j1} + \varepsilon_{j2} + P_{j2}\varepsilon_{j3}) \\ P_{j1} > \dfrac{P_{j3} + K_{j4} + \tilde{d}}{2} + \vartheta_j + K_{j2}\varepsilon_{j0} + \varepsilon_{j1} + \varepsilon_{j2} + P_{j2}\varepsilon_{j3} \\ P_{j4} \geq \dfrac{\tilde{d}}{2} \\ \vartheta_j > 2\bar{d}(K_{j2}\varepsilon_{j0} + \varepsilon_{j1} + \varepsilon_{j2} + P_{j2}\varepsilon_{j3}) \end{cases} \tag{8.13}$$

where ϑ_j, ε_j, ε_{j0}, ε_{j1}, ε_{j2}, ε_{j3}, ε_{jt1}, ε_{jt2} are some auxiliary positive constants, then the complete closed-loop teleoperation system is FTSIIOpS.

Proof. For closed-loop teleoperation system, the Lyapunov-Krasovskii functional is taken as

$$V_j(\boldsymbol{x}_{jt}, r_j(t)) = \frac{1}{2}\boldsymbol{\eta}_j^{\mathrm{T}}\boldsymbol{\mathcal{M}}_j\boldsymbol{\eta}_j + \frac{1}{2\psi_{j1}}\tilde{\boldsymbol{\Theta}}_j^{\mathrm{T}}\tilde{\boldsymbol{\Theta}}_j + \frac{1}{2\psi_{j3}}\tilde{\boldsymbol{\theta}}_j^{\mathrm{T}}\tilde{\boldsymbol{\theta}}_j + \frac{\vartheta_j}{2}\int_{-\bar{d}}^{0}\boldsymbol{z}_{jt}^{\mathrm{T}}(\tau)\left[\frac{-\tau}{\bar{d}} + \frac{2(\tau+\bar{d})}{\bar{d}}\right]\boldsymbol{z}_{jt}(\tau)\mathrm{d}\tau + \frac{1}{2}\boldsymbol{s}_j^{\mathrm{T}}\boldsymbol{s}_j$$

From the definitions of $\boldsymbol{\mathcal{K}}_j(t)$, $\boldsymbol{s}_j(t)$, and $\bar{\boldsymbol{y}}_j(t)$, it has $|\bar{\boldsymbol{y}}_j(t)|^2 \le |\boldsymbol{s}_j(t)|^2 \le 2|\bar{\boldsymbol{y}}_j(t)|^2$. Further, since

$$\frac{1}{2}\int_{-\bar{d}}^{0}\boldsymbol{z}_{jt}^{\mathrm{T}}(\tau)\left[\frac{-\tau}{\bar{d}} + \frac{2(\tau+\bar{d})}{\bar{d}}\right]\boldsymbol{z}_{jt}(\tau)\mathrm{d}\tau \le \int_{-\bar{d}}^{0}\boldsymbol{z}_{jt}^{\mathrm{T}}(\tau)\boldsymbol{z}_{jt}(\tau)\mathrm{d}\tau$$

then for any $k \ge 0$,

$$\begin{cases} \bar{\rho}_{j1}|\boldsymbol{z}_j(t)|^2 \le V_j(\boldsymbol{x}_{jt}, r_j(t)) \le \bar{\rho}_{j2}\|\boldsymbol{z}_{jt}\|_{M_2}^2 \\ V_j(\boldsymbol{x}_{jt_{k+1}^j}, r_j(t_{k+1}^j)) \le V_j(\boldsymbol{x}_{jt_{k+1}^j}, r_j(t_k^j)) \end{cases}$$

where $\bar{\rho}_{j2} = \max\left\{\frac{\rho_{j2}}{2}, \frac{1}{2\psi_{j1}}, \frac{1}{2\psi_{j3}}, 1, \vartheta_j\right\}$; $\bar{\rho}_{j1} = \min\left\{\frac{\rho_{j1}}{2}, \frac{1}{2\psi_{j1}}, \frac{1}{2\psi_{j3}}, \frac{1}{2}\right\}$.

For any fixed $r_j(t) = p \in \mathcal{S}$, the time derivative of $V_j(x_{jt}, p)$ can be calculated by

$$\begin{aligned} &D^+V_j(\boldsymbol{x}_{jt}, p) \\ &= \boldsymbol{\eta}_j^{\mathrm{T}}\boldsymbol{\mathcal{M}}_j\dot{\boldsymbol{\eta}}_j + \frac{1}{2}\boldsymbol{\eta}_j^{\mathrm{T}}\dot{\boldsymbol{\mathcal{M}}}_j\boldsymbol{\eta}_j + \frac{1}{\psi_{j1}}\tilde{\boldsymbol{\Theta}}_j^{\mathrm{T}}\dot{\tilde{\boldsymbol{\Theta}}}_j + \frac{1}{\psi_{j3}}\tilde{\boldsymbol{\theta}}_j^{\mathrm{T}}\dot{\tilde{\boldsymbol{\theta}}}_j + \boldsymbol{s}_j^{\mathrm{T}}\dot{\boldsymbol{s}}_j + \\ &\quad \vartheta_j\boldsymbol{z}_j^{\mathrm{T}}\boldsymbol{z}_j - \frac{\vartheta_j}{2}\boldsymbol{z}_{jt}^{\mathrm{T}}(-\bar{d})\boldsymbol{z}_{jt}(-\bar{d}) - \frac{\vartheta_j}{2\bar{d}}\int_{-\bar{d}}^{0}\boldsymbol{z}_{jt}^{\mathrm{T}}(\tau)\boldsymbol{z}_{jt}(\tau)\mathrm{d}\tau \end{aligned} \tag{8.14}$$

Specifically, from (8.8) and (8.11),

$$\begin{aligned} &\boldsymbol{\eta}_j^{\mathrm{T}}\boldsymbol{\mathcal{M}}_j\dot{\boldsymbol{\eta}}_j + \frac{1}{2}\boldsymbol{\eta}_j^{\mathrm{T}}\dot{\boldsymbol{\mathcal{M}}}_j\boldsymbol{\eta}_j \\ &\le \boldsymbol{\eta}_j^{\mathrm{T}}\left(\mathbf{Y}_j\tilde{\boldsymbol{\Theta}}_j + \lambda_j\dot{d}_\iota(t)\mathbf{Y}_{j1}\boldsymbol{\theta}_j\right) + \boldsymbol{\eta}_j^{\mathrm{T}}\boldsymbol{J}_j^{\mathrm{T}}(\boldsymbol{q}_j)\boldsymbol{f}_{he_j}^* - K_{j1}|\boldsymbol{\eta}_j|^2 - \\ &\quad K_{j2}\|\boldsymbol{\eta}_{jt}\|_{M_2}^{1+\sigma}|\boldsymbol{\eta}_j| - K_{j3}|\boldsymbol{J}_j(\boldsymbol{q}_j)\boldsymbol{\eta}_j|^2 - K_{j4}\boldsymbol{\eta}_j^{\mathrm{T}}\boldsymbol{y}_j - K_{j5}\boldsymbol{\eta}_j^{\mathrm{T}}\mathbf{Y}_{j1}\mathbf{Y}_{j1}^{\mathrm{T}}\boldsymbol{\eta}_j \\ &\le \boldsymbol{\eta}_j^{\mathrm{T}}\mathbf{Y}_j\tilde{\boldsymbol{\Theta}}_j + \frac{\lambda_j\tilde{d}}{2\varepsilon_{jt1}}|\tilde{\boldsymbol{\theta}}_j|^2 + \lambda_j\tilde{d}|\boldsymbol{\eta}_j^{\mathrm{T}}\mathbf{Y}_{j1}\hat{\boldsymbol{\theta}}_j| - \left(K_{j1} - \frac{K_{j4}+1}{2}\right)|\boldsymbol{\eta}_j|^2 - \\ &\quad K_{j2}\|\boldsymbol{\eta}_{jt}\|_{M_2}^{1+\sigma}|\boldsymbol{\eta}_j| - \left(K_{j3} - \frac{\varepsilon_j}{2}\right)|\boldsymbol{J}_j(\boldsymbol{q}_j)\boldsymbol{\eta}_j|^2 + \frac{K_{j4}}{2}|\boldsymbol{y}_j|^2 + \frac{1}{2\varepsilon_j}|\boldsymbol{f}_{he_j}^*|^2 - \\ &\quad \left(K_{j5} - \frac{\lambda_j\tilde{d}\varepsilon_{jt1}}{2}\right)\boldsymbol{\eta}_j^{\mathrm{T}}\mathbf{Y}_{j1}\mathbf{Y}_{j1}^{\mathrm{T}}\boldsymbol{\eta}_j \\ &= \boldsymbol{\eta}_j^{\mathrm{T}}\mathbf{Y}_j\tilde{\boldsymbol{\Theta}}_j + \frac{\lambda_j\tilde{d}}{2\varepsilon_{jt1}}|\tilde{\boldsymbol{\theta}}_j|^2 + \lambda_j\tilde{d}|\boldsymbol{\eta}_j^{\mathrm{T}}\mathbf{Y}_{j1}\hat{\boldsymbol{\theta}}_j| - \left(K_{j1} - \frac{K_{j4}+1}{2}\right)|\boldsymbol{\eta}_j|^2 - \end{aligned}$$

$$K_{j2}\|\boldsymbol{\eta}_{jt}\|_{M_2}^{1+\sigma}(|\boldsymbol{\eta}_j|+\varepsilon_{j0})+K_{j2}\|\boldsymbol{\eta}_{jt}\|_{M_2}^{1+\sigma}\varepsilon_{j0}-\left(K_{j3}-\frac{\varepsilon_j}{2}\right)|\boldsymbol{J}_j(\boldsymbol{q}_j)\boldsymbol{\eta}_j|^2+$$
$$\frac{K_{j4}}{2}|\boldsymbol{y}_j|^2+\frac{1}{2\varepsilon_j}|\boldsymbol{f}_{he_j}^*|^2-\left(K_{j5}-\frac{\lambda_j\tilde{d}\varepsilon_{jt1}}{2}\right)\boldsymbol{\eta}_j^{\mathrm{T}}\mathbf{Y}_{j1}\mathbf{Y}_{j1}^{\mathrm{T}}\boldsymbol{\eta}_j$$
$$\leq \boldsymbol{\eta}_j^{\mathrm{T}}\mathbf{Y}_j\tilde{\boldsymbol{\Theta}}_j+\frac{\lambda_j\tilde{d}}{2\varepsilon_{jt1}}|\tilde{\boldsymbol{\theta}}_j|^2+\lambda_j\tilde{d}|\boldsymbol{\eta}_j^{\mathrm{T}}\mathbf{Y}_{j1}\hat{\boldsymbol{\theta}}_j|-\left(K_{j1}-\frac{K_{j4}+1}{2}\right)|\boldsymbol{\eta}_j|^2-$$
$$K_{j2}\varepsilon_{j0}\|\boldsymbol{\eta}_{jt}\|_{M_2}^{1+\sigma}+K_{j2}\varepsilon_{j0}\|\boldsymbol{z}_{jt}\|_{M_2}^{1+\sigma}-\left(K_{j3}-\frac{\varepsilon_j}{2}\right)|\boldsymbol{J}_j(\boldsymbol{q}_j)\boldsymbol{\eta}_j|^2+$$
$$\frac{K_{j4}}{2}|\boldsymbol{y}_j|^2+\frac{1}{2\varepsilon_j}|\boldsymbol{f}_{he_j}^*|^2-\left(K_{j5}-\frac{\lambda_j\tilde{d}\varepsilon_{jt1}}{2}\right)\boldsymbol{\eta}_j^{\mathrm{T}}\mathbf{Y}_{j1}\mathbf{Y}_{j1}^{\mathrm{T}}\boldsymbol{\eta}_j \tag{8.15}$$

where ε_j, ε_{j0} and ε_{jt1} are some positive constants.

In addition, there exist positive constants ε_{j1}, ε_{j2} and ε_{jt2} such that

$$\frac{1}{\psi_{j1}}\tilde{\boldsymbol{\Theta}}_j^{\mathrm{T}}\dot{\tilde{\boldsymbol{\Theta}}}_j+\frac{1}{\psi_{j3}}\tilde{\boldsymbol{\theta}}_j^{\mathrm{T}}\dot{\tilde{\boldsymbol{\theta}}}_j$$
$$=-\frac{1}{\psi_{j1}}\tilde{\boldsymbol{\Theta}}_j^{\mathrm{T}}\dot{\hat{\boldsymbol{\Theta}}}_j-\frac{1}{\psi_{j3}}\tilde{\boldsymbol{\theta}}_j^{\mathrm{T}}\dot{\hat{\boldsymbol{\theta}}}_j$$
$$=-\frac{1}{\psi_{j1}}\tilde{\boldsymbol{\Theta}}_j^{\mathrm{T}}(\psi_{j1}\mathbf{Y}_j^{\mathrm{T}}\boldsymbol{\eta}_j-\psi_{j2}\hat{\boldsymbol{\Theta}}_j)-$$
$$\frac{1}{\psi_{j3}}\tilde{\boldsymbol{\theta}}_j^{\mathrm{T}}(-\lambda_j\tilde{d}\psi_{j3}\mathbf{Y}_{j1}^{\mathrm{T}}\boldsymbol{\eta}_j\mathrm{sgn}(\boldsymbol{\eta}_j^{\mathrm{T}}\mathbf{Y}_{j1}\hat{\boldsymbol{\theta}}_j)-\psi_{j4}\hat{\boldsymbol{\theta}}_j)$$
$$\leq-\tilde{\boldsymbol{\Theta}}_j^{\mathrm{T}}\mathbf{Y}_j^{\mathrm{T}}\boldsymbol{\eta}_j-\lambda_j\tilde{d}|\hat{\boldsymbol{\theta}}_j^{\mathrm{T}}\mathbf{Y}_{j1}^{\mathrm{T}}\boldsymbol{\eta}_j|+\lambda_j\tilde{d}|\boldsymbol{\theta}_j^{\mathrm{T}}\mathbf{Y}_{j1}^{\mathrm{T}}\boldsymbol{\eta}_j|-$$
$$\frac{\psi_{j2}}{\psi_{j1}}\tilde{\boldsymbol{\Theta}}_j^{\mathrm{T}}(\tilde{\boldsymbol{\Theta}}_j-\boldsymbol{\Theta}_j)-\frac{\psi_{j4}}{\psi_{j3}}\tilde{\boldsymbol{\theta}}_j^{\mathrm{T}}(\tilde{\boldsymbol{\theta}}_j-\boldsymbol{\theta}_j)$$
$$\leq-\tilde{\boldsymbol{\Theta}}_j^{\mathrm{T}}\mathbf{Y}_j^{\mathrm{T}}\boldsymbol{\eta}_j-\lambda_j\tilde{d}|\hat{\boldsymbol{\theta}}_j^{\mathrm{T}}\mathbf{Y}_{j1}^{\mathrm{T}}\boldsymbol{\eta}_j|-\frac{\psi_{j2}}{2\psi_{j1}}|\tilde{\boldsymbol{\Theta}}_j|^2-\frac{\psi_{j4}}{2\psi_{j3}}|\tilde{\boldsymbol{\theta}}_j|^2+\frac{\psi_{j2}}{2\psi_{j1}}|\boldsymbol{\Theta}_j|^2+$$
$$(\frac{\psi_{j4}}{2\psi_{j3}}+\frac{\lambda_j\tilde{d}}{2\varepsilon_{jt2}})|\boldsymbol{\theta}_j|^2-\varepsilon_{j1}\|\tilde{\boldsymbol{\Theta}}_{jt}\|_{M_2}^{1+\sigma}+\varepsilon_{j1}\|\tilde{\boldsymbol{\Theta}}_{jt}\|_{M_2}^{1+\sigma}-$$
$$\varepsilon_{j2}\|\tilde{\boldsymbol{\theta}}_{jt}\|_{M_2}^{1+\sigma}+\varepsilon_{j2}\|\tilde{\boldsymbol{\theta}}_{jt}\|_{M_2}^{1+\sigma}+\frac{\lambda_j\tilde{d}\varepsilon_{jt2}}{2}\boldsymbol{\eta}_j^{\mathrm{T}}\mathbf{Y}_{j1}\mathbf{Y}_{j1}^{\mathrm{T}}\boldsymbol{\eta}_j$$
$$\leq-\tilde{\boldsymbol{\Theta}}_j^{\mathrm{T}}\mathbf{Y}_j^{\mathrm{T}}\boldsymbol{\eta}_j-\lambda_j\tilde{d}|\hat{\boldsymbol{\theta}}_j^{\mathrm{T}}\mathbf{Y}_{j1}^{\mathrm{T}}\boldsymbol{\eta}_j|-\frac{\psi_{j2}}{2\psi_{j1}}|\tilde{\boldsymbol{\Theta}}_j|^2-\frac{\psi_{j4}}{2\psi_{j3}}|\tilde{\boldsymbol{\theta}}_j|^2+\frac{\psi_{j2}}{2\psi_{j1}}|\boldsymbol{\Theta}_j|^2+$$
$$(\frac{\psi_{j4}}{2\psi_{j3}}+\frac{\lambda_j\tilde{d}}{2\varepsilon_{jt2}})|\boldsymbol{\theta}_j|^2-\varepsilon_{j1}\|\tilde{\boldsymbol{\Theta}}_{jt}\|_{M_2}^{1+\sigma}+\varepsilon_{j1}\|\boldsymbol{z}_{jt}\|_{M_2}^{1+\sigma}-$$
$$\varepsilon_{j2}\|\tilde{\boldsymbol{\theta}}_{jt}\|_{M_2}^{1+\sigma}+\varepsilon_{j2}\|\boldsymbol{z}_{jt}\|_{M_2}^{1+\sigma}+\frac{\lambda_j\tilde{d}\varepsilon_{jt2}}{2}\boldsymbol{\eta}_j^{\mathrm{T}}\mathbf{Y}_{j1}\mathbf{Y}_{j1}^{\mathrm{T}}\boldsymbol{\eta}_j \tag{8.16}$$

Similarly,

$$\boldsymbol{s}_j^{\mathrm{T}}\dot{\boldsymbol{s}}_j\leq-(P_{j1}-\frac{P_{j3}}{2})|\boldsymbol{s}_j|^2-P_{j2}\|\bar{\boldsymbol{y}}_{jt}\|_{M_2}^{1+\sigma}|\boldsymbol{s}_j|+\frac{P_{j3}}{2}|\boldsymbol{\eta}_j|^2+$$
$$\boldsymbol{s}_j^{\mathrm{T}}\boldsymbol{u}_j+\boldsymbol{s}_j^{\mathrm{T}}\boldsymbol{\mathcal{K}}_j(t)\dot{d}_\iota(t)\boldsymbol{q}_{vd\iota}(t,d_\iota(t))$$

$$\le -(P_{j1} - \frac{P_{j3}}{2})|\boldsymbol{s}_j|^2 - P_{j2}\varepsilon_{j3}\|\bar{\boldsymbol{y}}_{jt}\|_{M_2}^{1+\sigma} + P_{j2}\varepsilon_{j3}\|\boldsymbol{z}_{jt}\|_{M_2}^{1+\sigma} + \frac{P_{j3}}{2}|\boldsymbol{\eta}_j|^2 + \boldsymbol{s}_j^{\mathrm{T}}\boldsymbol{u}_j + \boldsymbol{s}_j^{\mathrm{T}}\boldsymbol{\mathcal{K}}_j(t)\dot{d}_\iota(t)\boldsymbol{q}_{vd\iota}(t, d_\iota(t))$$

where $\varepsilon_{j3} > 0$.

Considering that for any $i = 1, 2, \cdots, n$, when $s_{j,i}(t) \neq 0$,

$$\kappa_{j,i}(t)\dot{d}_\iota(t)s_{j,i}(t)q_{vd\iota,i}(t, d_\iota(t)) + s_{j,i}(t)u_{j,i}(t) \le \frac{\tilde{d}}{2}|s_{j,i}(t)|^2 + \frac{\tilde{d}}{2}|q_{vd\iota,i}(t - d_\iota(t))|^2 - P_{j4}|q_{vd\iota,i}(t, d_\iota(t))|^2$$

and when $s_{j,i}(t) = 0$,

$$\kappa_{j,i}(t)\dot{d}_\iota(t)s_{j,i}(t)q_{vd\iota,i}(t, d_\iota(t)) + s_{j,i}(t)u_{j,i}(t) = 0$$

thus for $P_{j4} \ge \frac{\tilde{d}}{2}$, for any $s_{j,i}(t) \in \mathbb{R}$,

$$\kappa_{j,i}(t)\dot{d}_\iota(t)s_{j,i}(t)q_{vd\iota,i}(t, d_\iota(t)) + s_{j,i}(t)u_{j,i}(t) \le \frac{\tilde{d}}{2}|s_{j,i}(t)|^2$$

and further

$$\begin{aligned}
&\boldsymbol{\mathcal{K}}_j(t)\dot{d}_\iota(t)\boldsymbol{s}_j^{\mathrm{T}}(t)\boldsymbol{q}_{vd\iota}(t, d_\iota(t)) + \boldsymbol{s}_j^{\mathrm{T}}(t)\boldsymbol{u}_j(t) \\
&= \sum_{i=1}^{n}[\kappa_{j,i}(t)\dot{d}_\iota(t)s_{j,i}(t)q_{vd\iota,i}(t, d_\iota(t)) + s_{j,i}(t)u_{j,i}(t)] \\
&\le \frac{\tilde{d}}{2}|\boldsymbol{s}_j(t)|^2
\end{aligned}$$

Finally,

$$\begin{aligned}
\boldsymbol{s}_j^{\mathrm{T}}\dot{\boldsymbol{s}}_j &\le -(P_{j1} - \frac{P_{j3} + \tilde{d}}{2})|\boldsymbol{s}_j|^2 - P_{j2}\varepsilon_{j3}\|\bar{\boldsymbol{y}}_{jt}\|_{M_2}^{1+\sigma} + P_{j2}\varepsilon_{j3}\|\boldsymbol{z}_{jt}\|_{M_2}^{1+\sigma} + \frac{P_{j3}}{2}|\boldsymbol{\eta}_j|^2 \\
&\le -\left(P_{j1} - \frac{P_{j3} + \tilde{d}}{2}\right)\left(|\boldsymbol{y}_j|^2 + |\boldsymbol{e}_j|^2\right) - P_{j2}\varepsilon_{j3}\|\bar{\boldsymbol{y}}_{jt}\|_{M_2}^{1+\sigma} + \\
&\quad P_{j2}\varepsilon_{j3}\|\boldsymbol{z}_{jt}\|_{M_2}^{1+\sigma} + \frac{P_{j3}}{2}|\boldsymbol{\eta}_j|^2
\end{aligned} \tag{8.17}$$

Substitute (8.15)~(8.17) into (8.14), and let $K_{j3} \ge \frac{\varepsilon_j}{2}$, $K_{j5} \ge \frac{\lambda_j\tilde{d}\varepsilon_{jt1}}{2} + \frac{\lambda_j\tilde{d}\varepsilon_{jt2}}{2}$, then

$$\begin{aligned}
D^+V_j(\boldsymbol{x}_{jt}, p) \le &-\left(K_{j1} - \frac{K_{j4} + P_{j3} + 1}{2}\right)|\boldsymbol{\eta}_j|^2 - \frac{\psi_{j2}}{2\psi_{j1}}|\tilde{\boldsymbol{\Theta}}_j|^2 - \\
&\left(\frac{\psi_{j4}}{2\psi_{j3}} - \frac{\lambda_j\tilde{d}}{2\varepsilon_{jt1}}\right)|\tilde{\boldsymbol{\theta}}_j|^2 - \left(P_{j1} - \frac{P_{j3} + K_{j4} + \tilde{d}}{2}\right)|\boldsymbol{y}_j|^2 - \\
&\left(P_{j1} - \frac{P_{j3} + \tilde{d}}{2}\right)|\boldsymbol{e}_j|^2 - K_{j2}\varepsilon_{j0}\|\boldsymbol{\eta}_{jt}\|_{M_2}^{1+\sigma} - \varepsilon_{j1}\|\tilde{\boldsymbol{\Theta}}_{jt}\|_{M_2}^{1+\sigma} -
\end{aligned}$$

$$
\begin{aligned}
&\varepsilon_{j2}\|\tilde{\boldsymbol{\theta}}_{jt}\|_{M_2}^{1+\sigma} - P_{j2}\varepsilon_{j3}\|\bar{\boldsymbol{y}}_{jt}\|_{M_2}^{1+\sigma} + \frac{\psi_{j2}}{2\psi_{j1}}|\boldsymbol{\Theta}_j|^2 + \left(\frac{\psi_{j4}}{2\psi_{j3}} + \frac{\lambda_j \tilde{d}}{2\varepsilon_{jt2}}\right)|\boldsymbol{\theta}_j|^2 + \\
&\frac{1}{2\varepsilon_j}|\boldsymbol{f}_{he_j}^*|^2 + (K_{j2}\varepsilon_{j0} + \varepsilon_{j1} + \varepsilon_{j2} + P_{j2}\varepsilon_{j3})\|\boldsymbol{z}_{jt}\|_{M_2}^{1+\sigma} + \\
&\vartheta_j \boldsymbol{z}_j^{\mathrm{T}}\boldsymbol{z}_j - \frac{\vartheta_j}{2\bar{d}}\int_{-\bar{d}}^{0}\boldsymbol{z}_{jt}^{\mathrm{T}}(\tau)\boldsymbol{z}_{jt}(\tau)\mathrm{d}\tau \\
\le\; & -\mu_{j1}\|\boldsymbol{z}_{jt}\|_{M_2}^2 - K_{j2}\varepsilon_{j0}\|\boldsymbol{\eta}_{jt}\|_{M_2}^{1+\sigma} - \varepsilon_{j1}\|\tilde{\boldsymbol{\Theta}}_{jt}\|_{M_2}^{1+\sigma} - \varepsilon_{j2}\|\tilde{\boldsymbol{\theta}}_{jt}\|_{M_2}^{1+\sigma} - \\
&P_{j2}\varepsilon_{j3}\|\bar{\boldsymbol{y}}_{jt}\|_{M_2}^{1+\sigma} + (K_{j2}\varepsilon_{j0} + \varepsilon_{j1} + \varepsilon_{j2} + P_{j2}\varepsilon_{j3})\|\boldsymbol{z}_{jt}\|_{M_2}^2 + \\
&\frac{\psi_{j2}}{2\psi_{j1}}|\boldsymbol{\Theta}_j|^2 + \left(\frac{\psi_{j4}}{2\psi_{j3}} + \frac{\lambda_j \tilde{d}}{2\varepsilon_{jt1}}\right)|\boldsymbol{\theta}_j|^2 + \frac{1}{2\varepsilon_j}|\boldsymbol{f}_{he_j}^*|^2 + K_{j2}\varepsilon_{j0} + \\
&\varepsilon_{j1} + \varepsilon_{j2} + P_{j2}\varepsilon_{j3}
\end{aligned} \tag{8.18}
$$

where $\mu_{j1} = \min\left\{K_{j1} - \dfrac{K_{j4} + P_{j3} + 1}{2} - \vartheta_j, \dfrac{\psi_{j2}}{2\psi_{j1}} - \vartheta_j, \dfrac{\psi_{j4}}{2\psi_{j3}} - \vartheta_j - \dfrac{\lambda_j \tilde{d}}{2\varepsilon_{jt1}}, P_{j1} - \dfrac{P_{j3} + K_{j4} + \tilde{d}}{2} - \vartheta_j, \dfrac{\vartheta_j}{2\bar{d}}\right\}$.

Further, let $\mu_{j2} = \min\{K_{j2}\varepsilon_{j0}, \varepsilon_{j1}, \varepsilon_{j2}, P_{j2}\varepsilon_{j3}\}$, $\tilde{\mu}_{j1} = \mu_{j1} - K_{j2}\varepsilon_{j0} - \varepsilon_{j1} - \varepsilon_{j2} - P_{j2}\varepsilon_{j3}$, $\bar{\mu}_{j1} = (1 - \epsilon_j)\tilde{\mu}_{j1}$, $\xi_j = K_{j2}\varepsilon_{j0} + \varepsilon_{j1} + \varepsilon_{j2} + P_{j2}\varepsilon_{j3}$, where $\epsilon_j \in (0, 1)$. Without loss of generality, it can be seen from (8.18) that when

$$
\begin{aligned}
\|\boldsymbol{z}_{jt}\|_{M_2} \ge \frac{1}{\sqrt{\epsilon_j \tilde{\mu}_{j1}}}&\left(\sqrt{\frac{\psi_{j2}}{2\psi_{j1}}}|\boldsymbol{\Theta}_j| + \sqrt{\left(\frac{\psi_{j4}}{2\psi_{j3}} + \frac{\lambda_j \tilde{d}}{2\varepsilon_{jt2}}\right)}|\boldsymbol{\theta}_j| + \right. \\
&\left.\sqrt{\frac{1}{2\varepsilon_j}}\|\boldsymbol{f}_{he_j[0,\infty)}^*\|_\infty\right) + \xi_j
\end{aligned}
$$

it holds

$$
\begin{aligned}
D^+V_j(\boldsymbol{x}_{jt}, p) &\le -\bar{\mu}_{j1}\|\boldsymbol{z}_{jt}\|_{M_2}^2 - \mu_{j2}\|\boldsymbol{z}_{jt}\|_{M_2}^{1+\sigma} \\
&\le -\frac{\bar{\mu}_{j1}}{\bar{\rho}_{j2}}V_j(\boldsymbol{x}_{jt}, p) - \frac{\mu_{j2}}{(\bar{\rho}_{j2})^{\frac{1+\sigma}{2}}}(V_j(\boldsymbol{x}_{jt}, p))^{\frac{1+\sigma}{2}}
\end{aligned} \tag{8.19}
$$

From Lemma 2.6,

$$
|\boldsymbol{z}_j(t)| \le \beta_j(\|\boldsymbol{z}_{j0}\|, t) + \gamma_{j1}(\|\boldsymbol{f}_{he_j[0,\infty)}^*\|_\infty) + \gamma_{j2}(|\boldsymbol{\Theta}_j|) + \gamma_{j3}(|\boldsymbol{\theta}_j|) + \bar{\xi}_j
$$

where $\gamma_{j1}(s) = \dfrac{\bar{\rho}_{j1}^{-1}\bar{\rho}_{j2}}{\sqrt{\epsilon_j \tilde{\mu}_{j1}}}\sqrt{\dfrac{1}{2\varepsilon_j}}s$; $\gamma_{j2}(s) = \dfrac{\bar{\rho}_{j1}^{-1}\bar{\rho}_{j2}}{\sqrt{\epsilon_j \tilde{\mu}_{j1}}}\sqrt{\dfrac{\psi_{j2}}{2\psi_{j1}}}s$; $\gamma_{j3}(s) = \dfrac{\bar{\rho}_{j1}^{-1}\bar{\rho}_{j2}}{\sqrt{\epsilon_j \tilde{\mu}_{j1}}} \times \sqrt{(\dfrac{\psi_{j4}}{2\psi_{j3}} + \dfrac{\lambda_j \tilde{d}}{2\varepsilon_{jt2}})}s$; $\bar{\xi}_j = \bar{\rho}_{j1}^{-1}\bar{\rho}_{j2}\xi_j$; $\beta_j(\cdot,\cdot) \in \mathcal{GKL}$, the settling time of $\mathcal{GKL}$-function $\beta_j(\cdot,\cdot)$ satisifes

$$
T_{j0}(\boldsymbol{x}_{j0}) \le \frac{\bar{\rho}_{j2}}{\bar{\mu}_{j1}(1 - \sigma)}\ln\frac{\bar{\mu}_{j1}(\bar{\rho}_{j2})^{\frac{1-\sigma}{2}}\|\boldsymbol{z}_0\|_\infty^{2(1-\sigma)} + \mu_{j2}}{\mu_{j2}}
$$

□

Remark 8.1. *From the definition of the finite-time state-independent input-to-output stability and the proof of Theorem 8.1, it can be inferred that the complete output signal* $|\boldsymbol{z}_m(t)|$ *and* $|\boldsymbol{z}_s(t)|$ *can always remain bounded. Moreover, if the external forces* $\boldsymbol{f}_h^*(t) \to 0$ *and* $\boldsymbol{f}_e^*(t) \to 0$, $|\boldsymbol{z}_m(t)|$ *and* $|\boldsymbol{z}_s(t)|$ *converge into the sets* $\gamma_{m2}(|\boldsymbol{\Theta}_m|) + \gamma_{m3}(|\boldsymbol{\theta}_m|) + \bar{\xi}_m$ *and* $\gamma_{s2}(|\boldsymbol{\Theta}_s|) + \gamma_{s3}(|\boldsymbol{\theta}_s|) + \bar{\xi}_s$ *in finite time* $T_{m0}(x_{m0})$ *and* $T_{s0}(x_{s0})$*, respectively. More precisely, from Lemma2.6,* $|\boldsymbol{z}_j(t)|$ *will eventually converge into* $\frac{1}{\sqrt{\epsilon_j \bar{\mu}_{j1}}}\Big(\sqrt{\frac{\psi_{j2}}{2\psi_{j1}}}|\boldsymbol{\Theta}_j| + \sqrt{(\frac{\psi_{j4}}{2\psi_{j3}} + \frac{\lambda_j \tilde{d}}{2\varepsilon_{jt2}})}|\boldsymbol{\theta}_j| + \sqrt{\frac{1}{2\varepsilon_j}}\|\boldsymbol{f}^*_{he_j[0,\infty)}\|_\infty\Big) + \xi_j$.

Remark 8.2. *For the designed control algorithm, on the one hand, the conditions to be satisfied for the stability of the closed-loop system are shown in (8.13), which can be solved according to Remark 8.4 given later; on the other hand, in order to reducethe control chattering caused by* $\boldsymbol{y}_j$ *and* $\text{sgn}(\boldsymbol{\eta}_j)$ *in (8.6), it is suggested that the gains* K_{j2}*,* P_{j2} *and* K_{j4} *take some small values.*

Remark 8.3. *Compared with the existing results, the finite-time control proposed in this chapter has the following advantages:*

(1) The adaptive control problem of teleoperation systems with time-varying delay based on linear parameterization has been studied in previous works [14]. However, according to the existing scheme [14], accurate time-delay rate of change is essential for controller design, and an auxiliary variable communication method is proposed to measure accurate time delay information. However, when this method is applied in practice, it still faces some challenges due to network-induced uncertainties (quantization error, packet loss, clock asynchrony, etc.). In this chapter, if $K_{j2} = P_{j2} \equiv 0$*, according to the proof of Theorem 8.1, when only the upper bound of the time delay and its rate of change is known, the stability condition that guarantees the bounded tracking error can be obtained. Since it does not require precise time-delay rate-of-change information, the method proposed in this chapter is more robust.*

(2) The finite-time control problem has been studied in many fields, including multi-agent systems [7], space vehicle systems [5], robotic systems [6], etc. Although Yang et al. propose a fuzzy adaptive finite-time control scheme of teleoperation systems, it only considers the constant communication time delay [1]. None of the existing research considers the finite-time control of the teleoperation system with time-varying delay.

(3) In addition to the time delay constraint, the finite-time control scheme proposed in this chapter also overcomes the following two shortcomings of Yang's method [1]:

① The actual robot system inevitably needs to be in contact with the operator and the environment, and the stability conclusion of Yang's method [1] can only guarantee the finite time convergence of the closed-loop system when the master and slave robots move freely.

②*From Theorem 2 of Yang's work* [1]*, for any* $j \in \{m, s\}$*, the control gain* ξ_j *should satisfy* $\xi_j \geq \|\tilde{\theta}_j \varsigma_j(X_j)\| + \|\tilde{\omega}_j\|$*, which strictly depends on parameter estimation errors* $\tilde{\theta}_j$ *and* $\tilde{\omega}_j$*. Although Theorem 1 of Yang's work* [1] *proves that* $\tilde{\theta}_j$ *and* $\tilde{\omega}_j$ *are bounded, it is impossible to give their theoretical bounds. In this regard, a feasible way is to take a sufficiently large* ξ_j*. However, when* ξ_j *is sufficiently large, the* $\xi_j \mathrm{sgn}(s_j)$ *introduced into the control torque (21) in Yang's method* [1] *will bring serious chattering and actuator saturation to the control, making the control algorithm difficult to implement.*

8.5 SIMULATION

In order to verify the effectiveness of the algorithm, similar to Section 3.6, a two-degree-of-freedom serial manipulator is selected as the simulation object, and the robot dynamics model and parameters are given in Nuno's work [15].

For any $j \in \{m, s\}$, denote $\boldsymbol{q}_j = [q_{j,1}, q_{j,2}]^{\mathrm{T}}$. Assume that the position of the robot end effector in task space is

$$\boldsymbol{X}_j = \begin{bmatrix} l_{j1}\cos(q_{j,1}) + l_{j2}\cos(q_{j,1} + q_{j,2}) \\ l_{j1}\sin(q_{j,1}) + l_{j2}\sin(q_{j,1} + q_{j,2}) \end{bmatrix}$$

Further, the Jacobian matrix can be written as

$$\boldsymbol{J}_j(\boldsymbol{q}_j) = \begin{bmatrix} -l_{j1}\sin(q_{j,1}) - l_{j2}\sin(q_{j,1} + q_{j,2}) & -l_{j2}\sin(q_{j,1} + q_{j,2}) \\ l_{j1}\cos(q_{j,1}) + l_{j2}\cos(q_{j,1} + q_{j,2}) & l_{j2}\cos(q_{j,1} + q_{j,2}) \end{bmatrix}$$

So when the robot dynamics is expressed in a linear parameterized form, its unknown parameter vector is $\boldsymbol{\Theta}_j = [m_{j1}, m_{j2}, \boldsymbol{M}_{he_j}, \boldsymbol{B}_{he_j}, \boldsymbol{K}_{he_j}]^{\mathrm{T}}$ and $\boldsymbol{\theta}_j = [m_{j1}, m_{j2}, \boldsymbol{M}_{he_j}]^{\mathrm{T}}$. From the parameters given in Nuno's work [15], $\rho_{m2} = 0.966\ 9$ and $\rho_{s2} = 0.665\ 6$. In the simulation, the dynamic parameters of the operator and the environment are set as $\boldsymbol{M}_h = 0$, $\boldsymbol{B}_h = 2$, $\boldsymbol{K}_h = 10$, $\boldsymbol{M}_e = 0$, $\boldsymbol{B}_e = 5$, $\boldsymbol{K}_e = 20$. Assuming exogenous forces from the operator and the environment are shown in Fig. 8.1, where $f^*_{e,1}(t)$ is set to $20 \times (2 \times \mathrm{rand} - 1)$ when $t \in [0, 5\mathrm{s}]$ and $10 \times (2 \times \mathrm{rand} - 1)$ when $t \in [5\mathrm{s}, 12\mathrm{s}]$, and 0 at other times; $f^*_{e,2}(t)$ is set to $30 \times (2 \times \mathrm{rand} - 1)$ when $t \in [0, 8\mathrm{s}]$, and 0 at other times; rand is the random number generation function in MATLAB. The initial parameters of the master and slave robots are set as: for any $t \in [-\bar{d}, 0]$, $\boldsymbol{q}_m(t) = [0, 0]^{\mathrm{T}}$, $\boldsymbol{q}_s(t) = [-\pi, \pi]^{\mathrm{T}}$, $\dot{\boldsymbol{q}}_m(t) = \ddot{\boldsymbol{q}}_m = [0, 0]^{\mathrm{T}}$, $\dot{\boldsymbol{q}}_s(t) = \ddot{\boldsymbol{q}}_s = [0, 0]^{\mathrm{T}}$, $\boldsymbol{y}_m(t) = [-50, 50]^{\mathrm{T}}$, $\boldsymbol{y}_s(t) = [50, -50]^{\mathrm{T}}$, $\dot{\boldsymbol{y}}_m(t) = \dot{\boldsymbol{y}}_s(t) = [0, 0]^{\mathrm{T}}$, $\hat{\boldsymbol{\theta}}_m(t) = 50\boldsymbol{\theta}_m$, $\hat{\boldsymbol{\theta}}_s(t) = 50\boldsymbol{\theta}_s$, $\hat{\boldsymbol{\Theta}}_m(t) = \hat{\boldsymbol{\Theta}}_s(t) = [0, 0, 0, 0, 0]^{\mathrm{T}}$.

Assume that the upper bound of the time delay and its rate of change satisfy $\bar{d} = 2$ and $\tilde{d} = 2$. For any $j \in \{m, s\}$, set the auxiliary control gain to $\sigma = 0.5$, $\lambda_j = 2$, $\varepsilon_{jt1} = \varepsilon_{jt2} = 2.5$, $\vartheta_j = 6.4$. Under the given initial parameters, it is possible to set $\mu_{j2} = 0.25$ and $\bar{\mu}_{j1} = 0.05$. Take $\epsilon_j = 0.5$, then $\tilde{\mu}_{j1} = 0.1$. According to Theorem 8.1, the remaining (auxiliary) control gains can be taken as: $\varepsilon_j = 5$, $\varepsilon_{j0} = 50$, $\varepsilon_{j1} = \varepsilon_{j2} = 0.25$, $\varepsilon_{j3} = 50$, $K_{j1} = 12$, $K_{j2} = 0.01$, $K_{j3} = 5$, $K_{j4} = 1$, $K_{j5} = 10$, $P_{j1} = 12$, $P_{j2} = 0.01$, $P_{j3} = 5$, $P_{j4} = 2$, $\psi_{j1} = \psi_{j3} = 1$, $\psi_{j2} = 16$, $\psi_{j3} = 18$.

Remark 8.4. *After getting* $\tilde{\mu}_{j1}$ *and* μ_{j2}*, (auxiliary) control gain satisfying (8.13) can be obtained according to the following steps:*

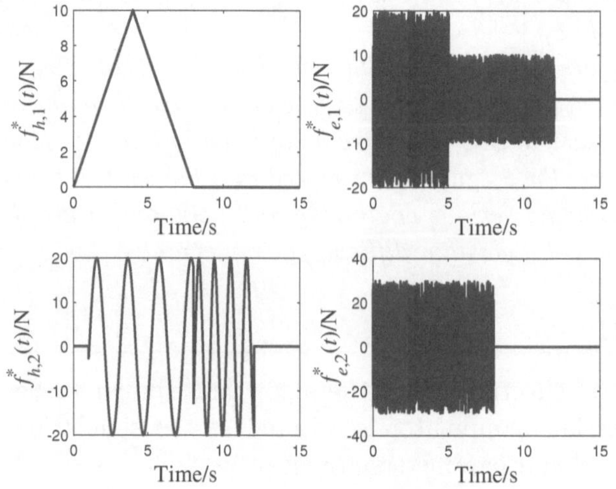

Figure 8.1 Operator exogenous force $\boldsymbol{f}_h^*$ and environmental exogenous force $\boldsymbol{f}_e^*$.

(1) Since K_{j4} and P_{j3} are not constrained, they can take any suitable value.

(2) From the given λ_j, $\tilde{d}$, ε_{jt1} and ε_{jt2}, calculate P_{j4} and K_{j5}.

(3) From ϑ_j and $\bar{d}$, calculate $K_{j2}\varepsilon_{j0} + \varepsilon_{j1} + \varepsilon_{j2} + P_{j2}\varepsilon_{j3}$. To avoid excessive control chattering caused by $\|\boldsymbol{\eta}_{jt}\|_{M_2}^{1+\sigma}\mathrm{sgn}(\boldsymbol{\eta}_j)$ and $\boldsymbol{y}_j$ in (8.6), for the control gains K_{j2} and P_{j2}, some small values can be taken. After that, calculate ε_{j0}, ε_{j1}, ε_{j2} and ε_{j3}.

(4) Based on the control gain obtained above, it can be solved to get K_{j1}, P_{j1}, $\dfrac{\psi_{j4}}{\psi_{j3}}$ and $\dfrac{\psi_{j2}}{\psi_{j1}}$; further, by assigning any bounded positive value to $\psi_{j3}(\psi_{j1})$, it can be solved to get $\psi_{j4}(\psi_{j2})$.

(5) For any fixed ε_j, calculate K_{j3}.

8.5.1 Stability verification

The first set of simulations is still used to verify the closed-loop stability of the proposed algorithm. First, according to the above simulation settings, let the communication delay be $d_m(t) = 1.8\sin^2(t)$ and $d_s(t) = 0.8\sin^2(2t)$. Then, under the excitation of exogenous force shown in Fig. 8.1, the response trajectories of the closed-loop teleoperation system are shown in Figs. 8.2–8.5. Among them, Fig. 8.2 gives the repose curves of the complete closed-loop master and slave systems. One can find that the stable value of $|\boldsymbol{z}_m(t)|$ ($|\boldsymbol{z}_s(t)|$) is less than 33.410 4 (59.906 3). Moreover, the cost time that $|\boldsymbol{z}_m(t)|$ ($|\boldsymbol{z}_s(t)|$) enters set $\{\boldsymbol{z}_m(t)||\boldsymbol{z}_m(t)| \leq 33.410\,4\}$ ($\{\boldsymbol{z}_s(t)||\boldsymbol{z}_s(t)| \leq 59.906\,3\}$) is fairly less than 3 s. All of those are in accordance with the analysis in Remark 8.1. From Fig. 8.2, all the position tracking errors and the parameter estimation errors are bounded. In fact, Fig. 8.3 shows the joint position trajectories of the master and slave robots, where it can be found that the proposed algorithm has a good tracking performance. Moreover, as shown in Fig. 8.4 the joint velocities of the master and slave robots can remain bounded. Finally, the control inputs are given in Fig. 8.5. Note that the chattering of $\boldsymbol{\tau}_j$ is mainly caused by two factors: the random exogenous force $\boldsymbol{f}_e^*(t)$ in Fig. 8.1, and the terms $-K_{j2}\|\boldsymbol{\eta}_{jt}\|_{M_2}^{1+\sigma}\mathrm{sgn}(\boldsymbol{\eta}_j)$ and $-K_{j4}\boldsymbol{y}_j$ also increase the bad performance.

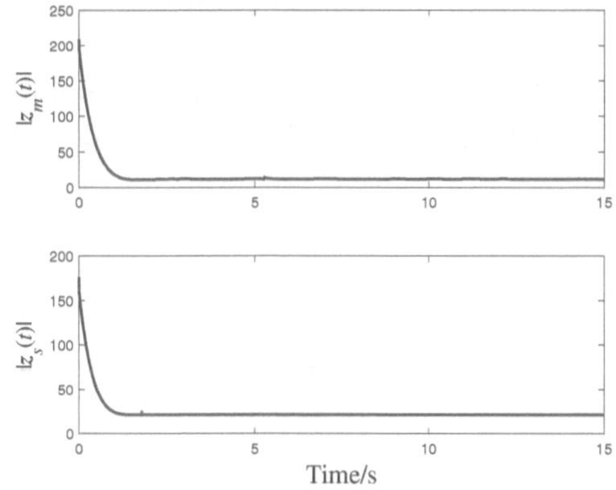

Figure 8.2 The response curves of $|\boldsymbol{z}_m(t)|$ and $|\boldsymbol{z}_s(t)|$.

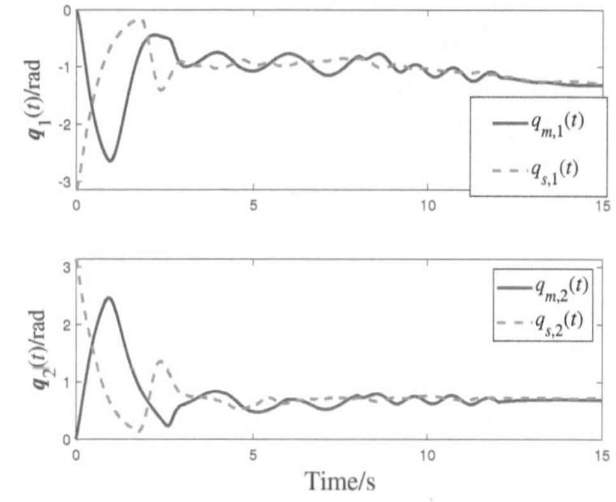

Figure 8.3 The joint positions of the master and the slave robots.

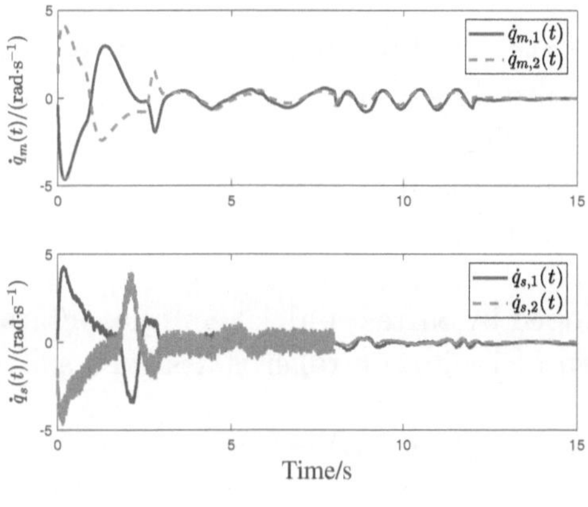

Figure 8.4 The joint velocities of the master and the slave robots.

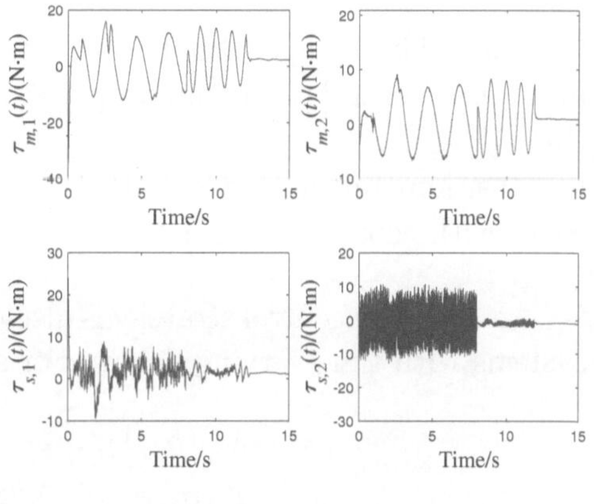

Figure 8.5 The control inputs for the master and the slave robots.

8.5.2 Comparison study

To illustrate the superiority of the algorithm, three sets of simulation comparisons are provided here.

Firstly, for the teleoperation system (3.1) with constant communication delays, a famous adaptive controller design is given in Nuno's work [15], that is,

$$\begin{aligned}\boldsymbol{\tau}_j &= -\hat{\boldsymbol{M}}_j(\boldsymbol{q}_j)\lambda_j\dot{\boldsymbol{e}}_j - \hat{\boldsymbol{C}}_j(\boldsymbol{q}_j,\dot{\boldsymbol{q}}_j)\lambda_j\boldsymbol{e}_j + \hat{\boldsymbol{g}}_j(\boldsymbol{q}_j) - \bar{K}_{j1}\boldsymbol{\eta}_j - \bar{K}_{j2}\dot{\boldsymbol{e}}_j \\ &= -\boldsymbol{Y}_j(\boldsymbol{q}_j,\dot{\boldsymbol{q}}_j,\boldsymbol{e}_j,\dot{\boldsymbol{e}}_j)\boldsymbol{\Theta}_j - \bar{K}_{j1}\boldsymbol{\eta}_j - \bar{K}_{j2}\dot{\boldsymbol{e}}_j\end{aligned}$$

In the first set of comparative simulation, consistent with Nuno's work [15], the communication delay is set as a constant delay, and the feedforward and feedback delays are both 0.4s. With Nuno's adaptive controller [15], the simulation results are shown in Figs. 8.6 and 8.7. From which, it can be seen that the finite-time adaptive control algorithm proposed in this chapter has a faster master-slave synchronization response speed than the ordinary asymptotic control algorithm proposed by Nuno et al. [15].

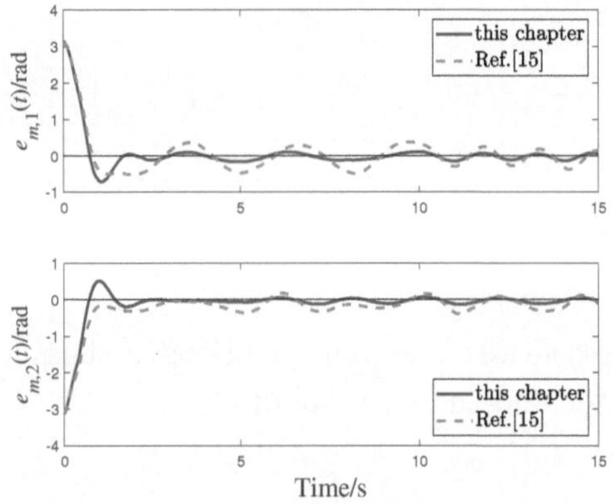

Figure 8.6 The comparison between the finite-time control algorithm proposed in this chapter and Nuno's algorithm [15] on the position tracking error of the mater robot with constant time delay.

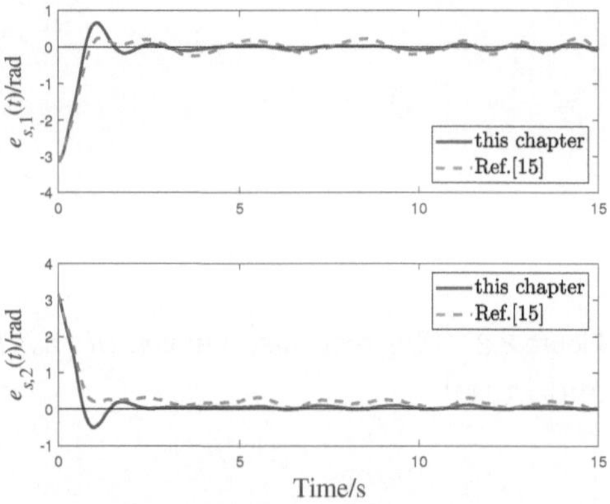

Figure 8.7 The comparison between the finite-time control algorithm proposed in this chapter and Nuno's algorithm [15] on the position tracking error of the slave robot with constant time delay.

An adaptive control strategy is also developed by Sarras et al. [14] for teleoperation systems with time-varying delays. For system (3.1), its controller is designed as

$$\begin{aligned}\boldsymbol{\tau}_j &= -\hat{\boldsymbol{M}}_j(\boldsymbol{q}_j)\lambda_j\dot{\boldsymbol{e}}_j - \hat{\boldsymbol{C}}_j(\boldsymbol{q}_j,\dot{\boldsymbol{q}}_j)\lambda_j\boldsymbol{e}_j + \hat{\boldsymbol{g}}_j(\boldsymbol{q}_j) - K_j\boldsymbol{\eta}_j \\ &= -\boldsymbol{Y}_j(\boldsymbol{q}_j,\dot{\boldsymbol{q}}_j,\boldsymbol{e}_j,\dot{\boldsymbol{e}}_j)\boldsymbol{\Theta}_j - K_j\boldsymbol{\eta}_j\end{aligned} \tag{8.20}$$

where $\dot{\boldsymbol{e}}_j = \dot{\boldsymbol{q}}_j - (1-\dot{d}_\iota)\boldsymbol{q}_{vd\iota}(t, d_\iota(t))$, $j \in \{m, s\}$, $\iota \in \{m, s\}$, $\iota \neq j$.

The second group of comparative studies will be carried out between the Sarras's algorithm [14] and the finite-time algorithm proposed in this chapter. Assume that the communication delays between the master and slave robots are $d_m(t) = 2.8\sin^2(t)$ and $d_s(t) = 1.8\sin^2(2t)$. Moreover, in order to ensure the fairness of the comparison, it is assumed that the robot actuators have the same output capability under the two algorithms, and the output upper bound is set to 10N·m. Then according to Sarras's work [14], let $\lambda_m = \lambda_s = 0.2$, and the values of other control gains are consistent with

Sarras's work [14]. Under the excitation of the exogenous force shown in Fig. 8.1, the tracking response of the closed-loop system is shown in Figs. 8.8 and 8.9. From the comparison results, it can be found that the finite-time control algorithm proposed in this chapter has a faster convergence speed.

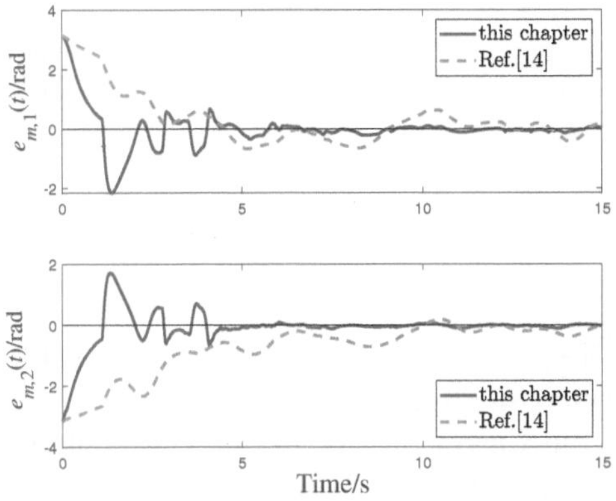

Figure 8.8 The comparison between the finite-time control algorithm proposed in this chapter and Sarras's algorithm [14] on the position tracking error of the master robot with time-varying delay.

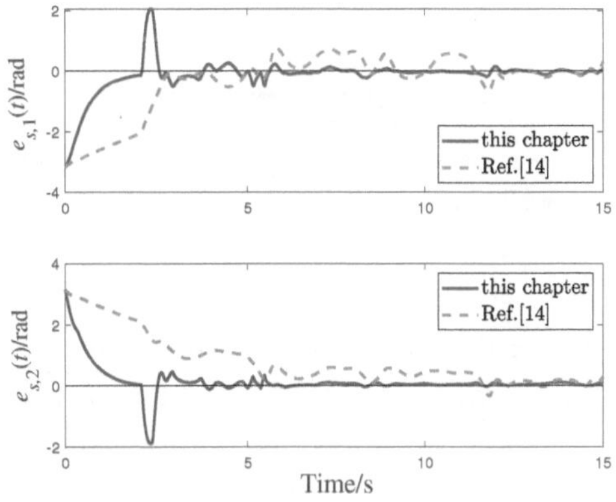

Figure 8.9 The comparison between the finite-time control algorithm proposed in this chapter and Sarras's algorithm [14] on the position tracking error of the slave robot with time-varying delay.

Under the same simulation settings, the FTSIIOpS algorithm proposed in this chapter is also compared with the SIIOS method in Chapter 3. The simulation results are shown in Figs. 8.10 and 8.11. Obviously, the finite-time control algorithm proposed in this chapter obtains a faster convergence speed.

8.6 CONCLUSION

We have examined the adaptive finite-time control problem for a class of nonlinear teleoperation system in the presence of time-varying delay as well as parameter uncertainties. To achieve finite-time joint position tracking control, a novel adaptive finite-time coordination framework based on subsystem decomposition has been investigated. By introducing the investigated error filtering into our controller design, the complete closed-loop teleoperation system has been modeled into a special class of switched system, which are composed by two interconnected subsystems. With the help of the obtained finite-time stability conclusion for switched delayed system, the sufficient FISIIOpS criteria for the complete closed-loop systems are obtained. As the conclusion, the master or slave joint position error can converge into a deterministic domain in finite time with the bounded velocities. Finally, the simulation verifies the effectiveness.

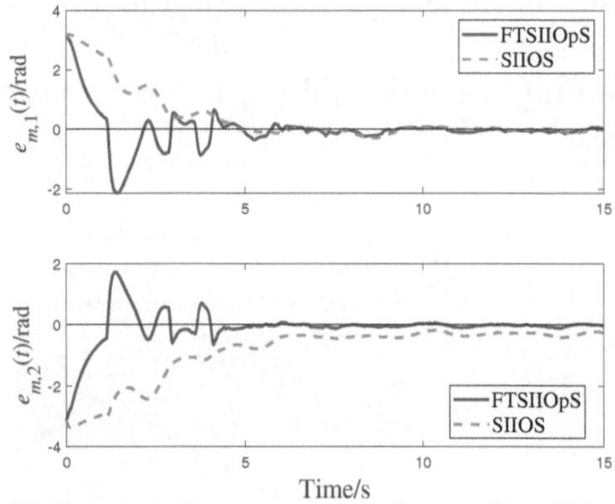

Figure 8.10 The comparison of the position tracking error of the master robot with time-varying delay under the control of the FTSIIOpS algorithm and the SIIOS algorithm.

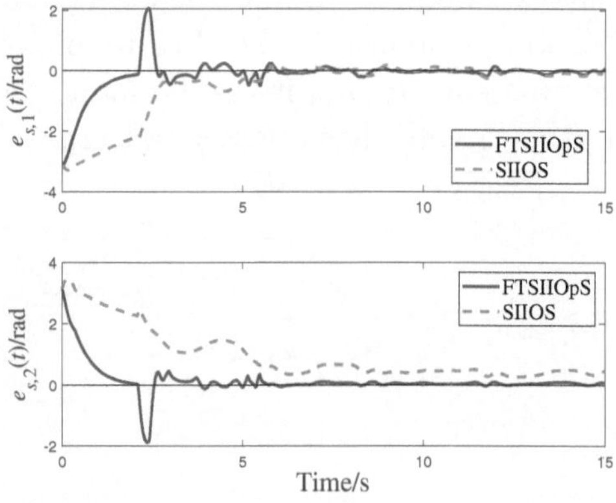

Figure 8.11 The comparison of the position tracking error of the slave robot with time-varying delay under the control of the FTSIIOpS algorithm and the SIIOS algorithm.

Bibliography

[1] Y. Yang, C. Hua, and X. Guan, "Adaptive fuzzy finite-time coordination control for networked nonlinear bilateral teleoperation system," *IEEE Transactions on Fuzzy Systems*, vol. 22, no. 3, pp. 631–641, 2013.

[2] Y. Yang, C. Hua, and X. Guan, "Finite time control design for bilateral teleoperation system with position synchronization error constrained," *IEEE Transactions on Cybernetics*, vol. 46, no. 3, pp. 609–619, 2015.

[3] Z. Man, A. P. Paplinski, and H. R. Wu, "A robust mimo terminal sliding mode control scheme for rigid robotic manipulators," *IEEE Transactions on Automatic Control*, vol. 39, no. 12, pp. 2464–2469, 1994.

[4] Y. Hong, Z.-P. Jiang, and G. Feng, "Finite-time input-to-state stability and applications to finite-time control design," *SIAM Journal on Control and Optimization*, vol. 48, no. 7, pp. 4395–4418, 2010.

[5] A.-M. Zou, K. D. Kumar, Z.-G. Hou, and X. Liu, "Finite-time attitude tracking control for spacecraft using terminal sliding mode and Chebyshev neural network," *IEEE Transactions on Systems, Man, and Cybernetics, Part B (Cybernetics)*, vol. 41, no. 4, pp. 950–963, 2011.

[6] S. Yu, X. Yu, B. Shirinzadeh, and Z. Man, "Continuous finite-time control for robotic manipulators with terminal sliding mode," *Automatica*, vol. 41, no. 11, pp. 1957–1964, 2005.

[7] Y. Cao, W. Ren, and Z. Meng, "Decentralized finite-time sliding mode estimators and their applications in decentralized finite-time formation tracking," *Systems & Control Letters*, vol. 59, no. 9, pp. 522–529, 2010.

[8] L. Wang, T. Chai, and L. Zhai, "Neural-network-based terminal sliding-mode control of robotic manipulators including actuator dynamics," *IEEE Transactions on Industrial Electronics*, vol. 56, no. 9, pp. 3296–3304, 2009.

[9] Y. Kang and P. Zhao, "Finite-time stability and input-to-state stability of stochastic nonlinear systems," in *Proceedings of the 31st Chinese Control Conference*. IEEE, 2012, pp. 1529–1534.

[10] S. Li and X. Wang, "Finite-time consensus and collision avoidance control algorithms for multiple auvs," *Automatica*, vol. 49, no. 11, pp. 3359–3367, 2013.

[11] P. Malysz and S. Sirouspour, "Nonlinear and filtered force/position mappings in bilateral teleoperation with application to enhanced stiffness discrimination," *IEEE Transactions on Robotics*, vol. 25, no. 5, pp. 1134–1149, 2009.

[12] T. Tsuji, P. G. Morasso, K. Goto, and K. Ito, "Human hand impedance characteristics during maintained posture," *Biological Cybernetics*, vol. 72, no. 6, pp. 475–485, 1995.

[13] R. Kelly, V. S. Davila, and J. A. L. Perez, *Control of Robot Manipulators in Joint Space*. Springer Science & Business Media, 2005.

[14] I. Sarras, E. Nuno, and L. Basanez, "An adaptive controller for nonlinear teleoperators with variable time-delays," *Journal of the Franklin Institute*, vol. 351, no. 10, pp. 4817–4837, 2014.

[15] E. Nuno, R. Ortega, and L. Basanez, "An adaptive controller for nonlinear teleoperators," *Automatica*, vol. 46, no. 1, pp. 155–159, 2010.

CHAPTER 9

Finite-time adaptive anti-windup teleoperation control

ALTHOUGH the finite-time controller designed based on non-smooth control technology can provide faster convergence speed, the introduction of non-smooth feedback terms will also lead to large control gain, which is easy to cause actuator output saturation and affect the control performance. On the basis of Chapter 8, this chapter further considers the input saturation constraint and designs an anti-saturation finite-time controller. A switched control framework is developed for nonlinear teleoperation system to unify the study of finite-time control, actuator saturation, model uncertainties, and asymmetric time-varying delays.

9.1 INTRODUCTION

To achieve faster convergence performance, many finite-time control techniques have been investigated for robotic systems [1]. Man et al. investigate a terminal sliding mode (TSM) control scheme [2], while an improved TSM controller is developed in 2005 [3], and further a neural-network-based nonsingular TSM control scheme is proposed for robotic manipulators including actuator dynamics [4]. Since they do not consider any time delays, their schemes cannot be directly applied to teleoperation system. Recently, motivated by Wang's work [4], Yang et al. investigate a novel nonsingular fast terminal sliding mode (NFTSM) controller for the networked bilateral teleoperation system [5]. It is further applied to the prescribed performance finite-time control of teleoperation system [6]. However, in the previous works [5,6], they do not address the input saturation and, the communication delays are assumed to be constant, which unavoidably renders the application limitations. In fact, the varying time delays are widely existed in teleoperation systems, which can severely affect the closed-loop control performance. In addition, due to the introduction of the nonlinear (nonsmooth) error terms, i.e., $\boldsymbol{e}^{\smile\sigma_1}$ and $\boldsymbol{e}^{\smile\sigma_2}$ where $\sigma_1 \in (0,1)$, $\sigma_2 > 1$, in the previous finite-time controllers [5,6], the finite-time control system is more likely to reach

DOI: 10.1201/9781003382058-9

the input saturation. Both of those largely affect their practical implementation. It is, therefore, the purpose of this chapter is to shorten such a gap by providing a rather general framework.

Motivated by the above discussions, this chapter aims to develop a novel finite-time control design for nonlinear teleoperation systems that is applicable to a wider area. The investigated design can handle the finite-time control, actuator saturation, asymmetric time-varying delays and model uncertainties in a unified framework, where the finite-time synchronization between the master and slave robots is obtained, whether the robots contact with the human operator/environment or not. Specifically, to cope with the actuator saturation, a modified anti-windup compensator is investigated, which is different from the existing ones [7,8]. By jointing with the non-smooth generalized switched filter technique, a novel finite-time anti-windup synchronous algorithm is proposed. Under the developed framework, the complete closed-loop teleoperation system consists of the master/slave robot dynamics with the specified finite-time control torque, the anti-windup compensator, and the switched filter. Similar to our previous work [9], the state-independent input-to-output stability approach is utilized to cope with the possible non-passive external forces applied by the human operator and/or the environment. Based on the multiple Lyapunov-Krasovskii functionals method, the complete closed-loop teleoperation system is SIIOpS and on this basis, it is proved to be finite-time SIIOpS.

9.2 PROBLEM FORMULATION

This chapter considers the finite-time adaptive control problem of a teleoperation system with input saturation constraints. Assuming that the slave robot has n DOF, the dynamic model is described as follows:

$$\begin{cases} \boldsymbol{M}_m(\boldsymbol{q}_m)\ddot{\boldsymbol{q}}_m + \boldsymbol{C}_m(\boldsymbol{q}_m, \dot{\boldsymbol{q}}_m)\dot{\boldsymbol{q}}_m + \boldsymbol{g}_m(\boldsymbol{q}_m) + \boldsymbol{f}_m(\boldsymbol{q}_m, \dot{\boldsymbol{q}}_m) = \boldsymbol{J}_m^{\mathrm{T}}(\boldsymbol{q}_m)\boldsymbol{f}_h + \boldsymbol{\tau}_m \\ \boldsymbol{M}_s(\boldsymbol{q}_s)\ddot{\boldsymbol{q}}_s + \boldsymbol{C}_s(\boldsymbol{q}_s, \dot{\boldsymbol{q}}_s)\dot{\boldsymbol{q}}_s + \boldsymbol{g}_s(\boldsymbol{q}_s) + \boldsymbol{f}_s(\boldsymbol{q}_s, \dot{\boldsymbol{q}}_s) = -\boldsymbol{J}_s^{\mathrm{T}}(\boldsymbol{q}_s)\boldsymbol{f}_e + \boldsymbol{\tau}_s \end{cases} \tag{9.1}$$

where $\boldsymbol{f}_m(\boldsymbol{q}_m, \dot{\boldsymbol{q}}_m)$ and $\boldsymbol{f}_s(\boldsymbol{q}_s, \dot{\boldsymbol{q}}_s)$ are the total disturbances of master and slave robots, respectively. The definitions of other components are the same as those of (4.1).

In this chapter, the dynamic model of the operator and the environment is defined as

$$\begin{cases} \boldsymbol{f}_h = \boldsymbol{f}_h^* - \boldsymbol{\Phi}_h(\boldsymbol{q}_m, \dot{\boldsymbol{q}}_m, \ddot{\boldsymbol{q}}_m) \\ \boldsymbol{f}_e = \boldsymbol{f}_e^* + \boldsymbol{\Phi}_e(\boldsymbol{q}_s, \dot{\boldsymbol{q}}_s, \ddot{\boldsymbol{q}}_s) \end{cases} \tag{9.2}$$

where $\boldsymbol{f}_h^*$ and $\boldsymbol{f}_e^*$ are the exogenous forces of the operator and the environment, respectively. $\boldsymbol{\Phi}_h(\boldsymbol{q}_m, \dot{\boldsymbol{q}}_m, \ddot{\boldsymbol{q}}_m)$ and $\boldsymbol{\Phi}_e(\boldsymbol{q}_s, \dot{\boldsymbol{q}}_s, \ddot{\boldsymbol{q}}_s)$ are inertial/viscoelastic coefficients of operator and environment, respectively, which are passive linear or nonlinear functions. In addition,

Assumption 9.1. *Exogenous forces $\boldsymbol{f}_h^*$ and $\boldsymbol{f}_e^*$ are locally essential bounded.*

Assumption 9.2. *$\boldsymbol{\Phi}_h(\boldsymbol{q}_m, \dot{\boldsymbol{q}}_m, \ddot{\boldsymbol{q}}_m)$ and $\boldsymbol{\Phi}_e(\boldsymbol{q}_s, \dot{\boldsymbol{q}}_s, \ddot{\boldsymbol{q}}_s)$ are some continuous functions.*

Remark 9.1. *The assumptions for the operator and environment dynamics model in (9.2) are based on the Haddadi's work [10]. First, according to some famous works [10–13], the time-domain dynamics models of operators can be classified into the following three categories:*

(1) *Nonlinear model:* $\boldsymbol{\Phi}_h(\boldsymbol{q}_m, \dot{\boldsymbol{q}}_m, \ddot{\boldsymbol{q}}_m) = \boldsymbol{I}_h(\boldsymbol{X}_m)\ddot{\boldsymbol{X}}_m + \boldsymbol{H}_h(\boldsymbol{X}_m, \dot{\boldsymbol{X}}_m)$*, where both* $\boldsymbol{I}_h(\boldsymbol{X}_m)$ *and* $\boldsymbol{H}_h(\boldsymbol{X}_m, \dot{\boldsymbol{X}}_m)$ *are continuously differentiable functions with respect to their own variables.*

(2) *Nonlinear viscoelastic function plus linear damping model:* $\boldsymbol{\Phi}_h(\boldsymbol{q}_m, \dot{\boldsymbol{q}}_m, \ddot{\boldsymbol{q}}_m) = \bar{\boldsymbol{\Phi}}_h(\boldsymbol{X}_m, \dot{\boldsymbol{X}}_m) + \boldsymbol{B}_h(t)\dot{\boldsymbol{X}}_m$.

(3) *Linear spring mass damper model:* $\boldsymbol{\Phi}_h(\boldsymbol{q}_m, \dot{\boldsymbol{q}}_m, \ddot{\boldsymbol{q}}_m) = \boldsymbol{M}_h(t)\ddot{\boldsymbol{X}}_m + \boldsymbol{B}_h(t)\dot{\boldsymbol{X}}_m + \boldsymbol{K}_h(t)(\boldsymbol{X}_m - \boldsymbol{X}_{m0})$*, where* $\boldsymbol{X}_{m0}$ *is the contact between the operator and the robot point.*

For different application scenarios, the dynamic model of the operator is different, and it usually has attitude dependence, time dependence and operator dependence. Therefore, the operator dynamics model is very complex. This chapter adopts the general model shown in eq.(9.2) to represent the operator's dynamics model. Similarly, the kinetic model of the environment and the kinetic model of the operator have similar characteristics and are therefore also represented by a general model of the form (9.2).

Remark 9.2. *In (9.2), by setting different* $\boldsymbol{f}_h^*$ *and* $\boldsymbol{f}_e^*$*, the operator and the environmental forces can be modeled as passive or non-passive models*[1]*. In the remainder of this chapter, the passive or non-passive operator/environmental forces that the proposed algorithm can handle all satisfy the defined form of (9.2).*

Since it is difficult to obtain an accurate dynamic model in practical engineering applications, this chapter assumes that

$$\begin{aligned}\boldsymbol{M}_j(\cdot) &= \boldsymbol{M}_{jo}(\cdot) + \Delta\boldsymbol{M}_j(\cdot)\\ \boldsymbol{C}_j(\cdot,\cdot) &= \boldsymbol{C}_{jo}(\cdot,\cdot) + \Delta\boldsymbol{C}_j(\cdot,\cdot)\\ \boldsymbol{g}_j(\cdot) &= \boldsymbol{g}_{jo}(\cdot) + \Delta\boldsymbol{g}_j(\cdot)\end{aligned}$$

where $\boldsymbol{M}_{jo}(\cdot)$, $\boldsymbol{C}_{jo}(\cdot,\cdot)$ and $\boldsymbol{g}_{jo}(\cdot)$ denote the nominal values of $\boldsymbol{M}_j(\cdot)$, $\boldsymbol{C}_j(\cdot,\cdot)$ and $\boldsymbol{g}_j(\cdot)$, while $\Delta\boldsymbol{M}_j(\cdot)$, $\Delta\boldsymbol{C}_j(\cdot,\cdot)$ and $\Delta\boldsymbol{g}_j(\cdot)$ are the corresponding uncertainties.

Then, system (9.1) can be rewritten as

$$\begin{cases}\boldsymbol{M}_{mo}(\boldsymbol{q}_m)\ddot{\boldsymbol{q}}_m + \boldsymbol{C}_{mo}(\boldsymbol{q}_m, \dot{\boldsymbol{q}}_m)\dot{\boldsymbol{q}}_m + \boldsymbol{g}_{mo}(\boldsymbol{q}_m) - \\ \qquad \boldsymbol{p}_m(\boldsymbol{q}_m, \dot{\boldsymbol{q}}_m, \ddot{\boldsymbol{q}}_m) = \boldsymbol{J}_m^{\mathrm{T}}(\boldsymbol{q}_m)\boldsymbol{f}_h^* + \boldsymbol{\tau}_m \\ \boldsymbol{M}_{so}(\boldsymbol{q}_s)\ddot{\boldsymbol{q}}_s + \boldsymbol{C}_{so}(\boldsymbol{q}_s, \dot{\boldsymbol{q}}_s)\dot{\boldsymbol{q}}_s + \boldsymbol{g}_{so}(\boldsymbol{q}_s) - \\ \qquad \boldsymbol{p}_s(\boldsymbol{q}_s, \dot{\boldsymbol{q}}_s, \ddot{\boldsymbol{q}}_s) = -\boldsymbol{J}_s^{\mathrm{T}}(\boldsymbol{q}_s)\boldsymbol{f}_e^* + \boldsymbol{\tau}_s\end{cases} \tag{9.3}$$

---

[1]From Nuno's work [14], an operator (environment) force is called passive, if for any $t \geq 0$, there exists $k_1 \geq 0$ ($k_2 \geq 0$) such that the mapping from velocity to force satisfies $-\int_0^t \boldsymbol{f}_h^{\mathrm{T}}\dot{\boldsymbol{X}}_m \mathrm{d}t \geq -k_1$ ($\int_0^t \boldsymbol{f}_e^{\mathrm{T}}\dot{\boldsymbol{X}}_s \mathrm{d}t \geq -k_2$)

where for any $j \in \{m, s\}$,

$$\boldsymbol{p}_j(\boldsymbol{q}_j, \dot{\boldsymbol{q}}_j, \ddot{\boldsymbol{q}}_j) = -\boldsymbol{f}_j(\boldsymbol{q}_j, \dot{\boldsymbol{q}}_j) - \Delta \boldsymbol{M}_j(\boldsymbol{q}_j)\ddot{\boldsymbol{q}}_j - \Delta \boldsymbol{C}_j(\boldsymbol{q}_j, \dot{\boldsymbol{q}}_j)\dot{\boldsymbol{q}}_j - \Delta \boldsymbol{g}_j(\boldsymbol{q}_j) + \boldsymbol{J}_j^{\mathrm{T}}(\boldsymbol{q}_j)\boldsymbol{\Phi}_{he_j}(\boldsymbol{q}_j, \dot{\boldsymbol{q}}_j, \ddot{\boldsymbol{q}}_j)$$

where $\boldsymbol{\Phi}_{he_m}(\cdot,\cdot,\cdot) = \boldsymbol{\Phi}_h(\cdot,\cdot,\cdot)$; $\boldsymbol{\Phi}_{he_s}(\cdot,\cdot,\cdot) = \boldsymbol{\Phi}_e(\cdot,\cdot,\cdot)$.

The nominal dynamic model of the robot satisfies the Properties 2.1 and 2.4, and according to Ge's work [15], it further has

Property 9.1. *For a manipulator with a rotating joint, the gravity vector $\boldsymbol{g}_{jo}(\boldsymbol{q}_j)$ is bounded; that is, there is a constant positive scalar g_i such that the ith gravity component $g_{jo,i}(\boldsymbol{q}_j)$ satisfies $|g_{jo,i}(\boldsymbol{q}_j)| \le g_i$, where $i = 1, 2, \cdots, n$.*

Let $\boldsymbol{f}_j(\boldsymbol{q}_j, \dot{\boldsymbol{q}}_j) = \boldsymbol{f}_{cj}(\boldsymbol{q}_j, \dot{\boldsymbol{q}}_j) + \boldsymbol{f}_{pj}(\boldsymbol{q}_j, \dot{\boldsymbol{q}}_j)$, where $\boldsymbol{f}_{cj}(\boldsymbol{q}_j, \dot{\boldsymbol{q}}_j)$ and $\boldsymbol{f}_{pj}(\boldsymbol{q}_j, \dot{\boldsymbol{q}}_j)$ are the continuous and discontinuous parts of the total disturbance $\boldsymbol{f}_j(\boldsymbol{q}_j, \dot{\boldsymbol{q}}_j)$, respectively. $\boldsymbol{f}_{pj}(\boldsymbol{q}_j, \dot{\boldsymbol{q}}_j)$ satisfies

Assumption 9.3. *There is an unknown positive constant D_j such that $|\boldsymbol{f}_{pj}(\boldsymbol{q}_j, \dot{\boldsymbol{q}}_j)| \le D_j$.*

Denote

$$\boldsymbol{p}_{cj}(\boldsymbol{q}_j, \dot{\boldsymbol{q}}_j, \ddot{\boldsymbol{q}}_j) = -\boldsymbol{f}_{cj}(\boldsymbol{q}_j, \dot{\boldsymbol{q}}_j) - \Delta \boldsymbol{M}_j(\boldsymbol{q}_j)\ddot{\boldsymbol{q}}_j - \Delta \boldsymbol{C}_j(\boldsymbol{q}_j, \dot{\boldsymbol{q}}_j)\dot{\boldsymbol{q}}_j - \Delta \boldsymbol{g}_j(\boldsymbol{q}_j) + \boldsymbol{J}_j^{\mathrm{T}}(\boldsymbol{q}_j)\boldsymbol{\Phi}_{he_j}(\boldsymbol{q}_j, \dot{\boldsymbol{q}}_j, \ddot{\boldsymbol{q}}_j)$$

Then

$$\boldsymbol{p}_j(\boldsymbol{q}_j, \dot{\boldsymbol{q}}_j, \ddot{\boldsymbol{q}}_j) = -\boldsymbol{f}_{pj}(\boldsymbol{q}_j, \dot{\boldsymbol{q}}_j) + \boldsymbol{p}_{cj}(\boldsymbol{q}_j, \dot{\boldsymbol{q}}_j, \ddot{\boldsymbol{q}}_j)$$

Since the exact model of $\boldsymbol{p}_j(\boldsymbol{q}_j, \dot{\boldsymbol{q}}_j, \ddot{\boldsymbol{q}}_j)$ is difficult to obtain and may not satisfy the linear parameterization conditions, in the follow-up research in this chapter, the fuzzy logic system will be used to estimate and approximate it. According to fuzzy logic system, for given unknown nonlinear function $\boldsymbol{p}_{cj}(\boldsymbol{q}_j, \dot{\boldsymbol{q}}_j, \ddot{\boldsymbol{q}}_j) + \boldsymbol{J}_j^{\mathrm{T}}(\boldsymbol{q}_j)\boldsymbol{J}_j(\boldsymbol{q}_j)\dot{\boldsymbol{q}}_j$, there is an optimized approximation parameter $\boldsymbol{\Theta}_j$ such that

$$\boldsymbol{p}_{cj}(\boldsymbol{q}_j, \dot{\boldsymbol{q}}_j, \ddot{\boldsymbol{q}}_j) + \boldsymbol{J}_j^{\mathrm{T}}(\boldsymbol{q}_j)\boldsymbol{J}_j(\boldsymbol{q}_j)\dot{\boldsymbol{q}}_j = \boldsymbol{\Theta}_j^{\mathrm{T}}\varphi(\boldsymbol{Z}_j) + \boldsymbol{o}_j(\boldsymbol{Z}_j)$$

where $\boldsymbol{Z}_j = [\boldsymbol{q}_j^{\mathrm{T}}, \dot{\boldsymbol{q}}_j^{\mathrm{T}}, \ddot{\boldsymbol{q}}_j^{\mathrm{T}}]^{\mathrm{T}}$; $\boldsymbol{o}_j(\boldsymbol{Z}_j)$ is the bounded approximation error, satisfying $|\boldsymbol{o}_j(\boldsymbol{Z}_j)| \le o_{jN} < \infty$.

Remark 9.3. *In the above formula, $\boldsymbol{J}_j^{\mathrm{T}}(\boldsymbol{q}_j)\boldsymbol{J}_j(\boldsymbol{q}_j)\dot{\boldsymbol{q}}_j$ is the residual term in the subsequent derivation process of this chapter (see eq.(9.16), which is $\dot{\boldsymbol{q}}_j^{\mathrm{T}}\boldsymbol{J}_j^{\mathrm{T}}(\boldsymbol{q}_j)\boldsymbol{f}_{he_j}^* \le \dot{\boldsymbol{q}}_j^{\mathrm{T}}\boldsymbol{J}_j^{\mathrm{T}}(\boldsymbol{q}_j)\boldsymbol{J}_j(\boldsymbol{q}_j)\dot{\boldsymbol{q}}_j + \frac{1}{4}|\boldsymbol{f}_{he_j}^*|^2$). In addition, as discussed by the previous work [16], the Jacobian matrix in practical systems also faces many uncertainties. To adapt to this situation, this chapter proposes an adaptive Jacobian controller based on fuzzy logic systems.*

System (9.1) also faces the input saturation constraint, assuming the control output $\boldsymbol{\tau}_j = [\tau_{j,1}, \tau_{j,2}, cdots, \tau_{j,n}]^{\mathrm{T}}$ satisfies

$$\tau_{j,i} = \begin{cases} \tau_{j,i}, & |\tau_{j,i}| \leq M_i \\ M_i \mathrm{sgn}(\tau_{j,i}), & |\tau_{j,i}| > M_i \end{cases}, \quad i = 1, 2, \cdots, n$$

where M_i is assumed to satisfy $g_i < M_i < \infty$.

For all $j \in \{m, s\}$, define the position tracking error $\boldsymbol{e}_j(t)$ and velocity error $\boldsymbol{e}_{vj}(t)$ as

$$\begin{cases} \boldsymbol{e}_j(t) = \boldsymbol{q}_j(t) - \boldsymbol{q}_\iota(t - d_\iota(t)) \\ \boldsymbol{e}_{vj}(t) = \dot{\boldsymbol{q}}_j(t) - \boldsymbol{q}_{vd\iota}(t, d_\iota(t)) \end{cases}$$

where $\iota \in \{m, s\}$, $\iota \neq j$, $\boldsymbol{q}_{vd\iota}(t, d_\iota(t)) = \dot{\boldsymbol{q}}_\iota(\theta)|_{\theta = t - d_\iota(t)}$, $d_\iota(t)$ is the corresponding communication delay, which satisfies:

Assumption 9.4. *There exist bounded constant positive scalars $\bar{d}$ and $\tilde{d}$ such that $0 \leq d_\iota(t) \leq \bar{d}$, $0 \leq |\dot{d}_\iota(t)| \leq \tilde{d}$.*

In this chapter, the control objective is: in the presence of actuator saturation, asymmetric time-varying communication delays, and uncertain kinematics/dynamics, design a non-smooth controller such that no matter the robot is in contact with the operator/environment or not , the position error $\boldsymbol{e}_j(t)$ can converge to a specified region within a limited time.

9.3 ANTI-WINDUP FINITE-TIME CONTROLLER

To deal with input saturation constraints, this section uses an anti-saturation compensation control strategy for robot control.

For the master and slave robots, design the following control torque $\boldsymbol{\tau}_j = [\tau_{j,1}, \tau_{j,2}, \cdots, \tau_{j,n}]^{\mathrm{T}}$:

$$\tau_{j,i} = \begin{cases} \bar{\tau}_{j,i}, & |\bar{\tau}_{j,i}| \leq M_i \\ M_i \mathrm{sgn}(\bar{\tau}_{j,i}), & |\bar{\tau}_{j,i}| > M_i \end{cases}, \quad i = 1, 2, \cdots, n \tag{9.4}$$

where $j \in \{m, s\}$, $\bar{\boldsymbol{\tau}}_j = [\bar{\tau}_{j,1}, \bar{\tau}_{j,2}, \cdots, \bar{\tau}_{j,n}]^{\mathrm{T}}$ is

$$\begin{aligned} \bar{\boldsymbol{\tau}}_j = {} & \boldsymbol{G}_{jo}(\boldsymbol{q}_j) - \hat{\boldsymbol{\Theta}}_j^T \boldsymbol{\varphi}(\boldsymbol{Z}_j) - \hat{\omega}_j \dot{\boldsymbol{q}}_j - K_{j0} \dot{\boldsymbol{q}}_j^{\smile\sigma_1} - K_{j1} \dot{\boldsymbol{q}}_j - K_{j2} \phi_{j2}(\boldsymbol{e}_j) - \\ & K_{j3} \boldsymbol{e}_j^{\smile\sigma_1} - \boldsymbol{K}_{j4} \boldsymbol{y}_j - K_{j5} \hat{\boldsymbol{\tau}}_j - \frac{1}{2} |\boldsymbol{\varphi}(\boldsymbol{Z}_j)|^2 \dot{\boldsymbol{q}}_j \end{aligned} \tag{9.5}$$

where $\sigma_1 = \dfrac{\gamma_1}{\gamma_2}$, γ_1 and γ_2 are constant positive odd scalars, with $\gamma_1 < \gamma_2$; $\phi_{j2}(\boldsymbol{e}_j) = \boldsymbol{e}_j + \lambda_j \boldsymbol{e}_j^{\smile\sigma_2}$, $\lambda_j > 0$; $\sigma_2 = \dfrac{\gamma_4}{\gamma_3}$, γ_3 and γ_4 are also constant positive odd scalars, with $\gamma_3 < \gamma_4$; $\boldsymbol{y}_j$ is the filter output that will be subsequently defined; K_{j0}, K_{j1}, K_{j2}, K_{j3} and K_{j5} are some constant positive scalars, $\boldsymbol{K}_{j4}$ is $n \times 2n$ non-zero constant matrix;

$\hat{\omega}_j = \dfrac{1}{2\varepsilon_{j1}}\hat{\upsilon}_j$, $\varepsilon_{j1} > 0$, $\upsilon_j = o_{jN}^2 + D_j^2$, $\hat{\boldsymbol{\Theta}}_j$ and $\hat{\upsilon}_j$ are respectively the estimations of $\boldsymbol{\Theta}_j$ and υ_j, with the adaptive update laws given by

$$\begin{cases} \dot{\hat{\boldsymbol{\Theta}}}_j = \Lambda_{j1}\boldsymbol{\varphi}(\boldsymbol{Z}_j)\dot{\boldsymbol{q}}_j^{\mathrm{T}} - \Lambda_{j2}(\hat{\boldsymbol{\Theta}}_j - \boldsymbol{\Theta}_j^*) \\ \dot{\hat{\omega}}_j = \Lambda_{j3}|\dot{\boldsymbol{q}}_j|^2 - \Lambda_{j4}(\hat{\omega}_j - \omega_j^*) \end{cases} \tag{9.6}$$

where $\boldsymbol{\Theta}_j^*$ and ω_j^* are respectively the nominal values of $\boldsymbol{\Theta}_j$ and ω_j; Λ_{j1}, Λ_{j2}, Λ_{j3} and Λ_{j4} are some positive constants. $\hat{\boldsymbol{\tau}}_j$ is the state of the anti-windup compensator. To deal with the actuator saturation, the dynamics of anti-windup compensator $\hat{\boldsymbol{\tau}}_j$ is given by

$$\dot{\hat{\boldsymbol{\tau}}}_j = \begin{cases} -K_{j6}\dfrac{\dot{\boldsymbol{q}}_j^{\mathrm{T}}\Delta\boldsymbol{\tau}_j}{|\hat{\boldsymbol{\tau}}_j|^2}\hat{\boldsymbol{\tau}}_j + K_{j7}\hat{\boldsymbol{\tau}}_j + 2K_{j8}|\boldsymbol{e}_j|\mathrm{sgn}(\hat{\boldsymbol{\tau}}_j), & \dot{\boldsymbol{q}}_j^{\mathrm{T}}\Delta\boldsymbol{\tau}_j \geq \dfrac{K_{j7}+K_{j8}}{K_{j6}}|\hat{\boldsymbol{\tau}}_j|^2 \\ -K_{j6}\dfrac{\dot{\boldsymbol{q}}_j^{\mathrm{T}}\Delta\boldsymbol{\tau}_j}{|\hat{\boldsymbol{\tau}}_j|^2}\hat{\boldsymbol{\tau}}_j - K_{j7}\hat{\boldsymbol{\tau}}_j + 2K_{j8}|\boldsymbol{e}_j|\mathrm{sgn}(\hat{\boldsymbol{\tau}}_j), & 0 < \dot{\boldsymbol{q}}_j^{\mathrm{T}}\Delta\boldsymbol{\tau}_j < \dfrac{K_{j7}+K_{j8}}{K_{j6}}|\hat{\boldsymbol{\tau}}_j|^2 \\ 2K_{j8}|\boldsymbol{e}_j|\mathrm{sgn}(\hat{\boldsymbol{\tau}}_j), & \dot{\boldsymbol{q}}_j^{\mathrm{T}}\Delta\boldsymbol{\tau}_j \leq 0,\ |\hat{\boldsymbol{\tau}}_j| < \delta \\ -K_{j9}\hat{\boldsymbol{\tau}}_j, & \dot{\boldsymbol{q}}_j^{\mathrm{T}}\Delta\boldsymbol{\tau}_j \leq 0,\ |\hat{\boldsymbol{\tau}}_j| \geq \delta \end{cases} \tag{9.7}$$

where $\Delta\boldsymbol{\tau}_j = \boldsymbol{\tau}_j - \bar{\boldsymbol{\tau}}_j$; δ is a positive design parameter which should be chosen as an appropriate value; K_{j6}, K_{j7}, K_{j8} and K_{j9} are some positive constants.

Remark 9.4. *The anti-windup compensator (9.7) is different from the existing one* [7]*. In Chen's work* [7]*, when $|\hat{\boldsymbol{\tau}}_j| < \delta$, the update law should be $\dot{\hat{\boldsymbol{\tau}}}_j = 0$; n this case, once $\hat{\boldsymbol{\tau}}_j$ enters the set $\{\hat{\boldsymbol{\tau}}_j : |\hat{\boldsymbol{\tau}}_j| < \delta\}$, the designed compensator will not work when the actuator saturation is occurred. However in practical teleoperation implementation, when $|\hat{\boldsymbol{\tau}}_j| < \delta$, there still exists the possibility to occur the actuator saturation. For example, if the motion of a master robot which is manipulated by a human operator is suddenly changed and becomes very fast, the actuator saturation will happen when a slave robot cannot follow the motion in real time. Another example is the saturation case caused by contacting with obstacle. If there exists an obstacle in the operation space, the actuator saturation may also occur when the slave robot encounters obstruction. Thus, the anti-windup compensator* [7] *may not work well due to $\dot{\hat{\boldsymbol{\tau}}}_j = 0$. To overcome this shortcoming, this chapter adopts a new antiwindup compensator for teleoperation systems of the form (9.7).*

In (9.5), $\boldsymbol{y}_j$ is the state output of a non-smooth auxiliary switched filter designed based on the idea of dynamic compensation. The specific design is as follows:

For the sake of simplifying expression, denote

$$\boldsymbol{E}_j = \mathrm{diag}\left[e_{j,1}^{\frac{\gamma_4-\gamma_3}{\gamma_3}}, e_{j,2}^{\frac{\gamma_4-\gamma_3}{\gamma_3}}, \cdots, e_{j,n}^{\frac{\gamma_4-\gamma_3}{\gamma_3}}\right]$$

and

$$\bar{\phi}_{j2}(\boldsymbol{e}_j, \boldsymbol{e}_{vj}) = \phi_{j2}(\boldsymbol{e}_j) - \dot{d}_\iota(t)\boldsymbol{E}_j\boldsymbol{q}_{vd\iota}(t, d_\iota(t)) = \boldsymbol{e}_{vj} + \lambda_j\boldsymbol{E}_j\boldsymbol{e}_{vj}$$

Also, let $\bar{\boldsymbol{e}}_j = [(\phi_{j2}(\boldsymbol{e}_j))^{\mathrm{T}}, \hat{\boldsymbol{\tau}}_j]^{\mathrm{T}} \in \mathbb{R}^{2n}$, $\bar{\boldsymbol{e}}_{vj} = [(\bar{\phi}_{j2}(\boldsymbol{e}_j, \boldsymbol{e}_{vj}))^{\mathrm{T}}, \dot{\hat{\boldsymbol{\tau}}}_j^{\mathrm{T}}]^{\mathrm{T}} \in \mathbb{R}^{2n}$. Then, $\boldsymbol{y}_j = [y_{j,1}, y_{j,2}, \cdots, y_{j,(2n)}]^{\mathrm{T}}$ is designed as

$$\dot{y}_{j,i} = \begin{cases} -P_{j1}s_{j,i} - P_{j2}s_{j,i}^{\smile\sigma_1} - \bar{e}_{vj,i} + u_{j,i}, & \kappa_{j,i}(t) = 1 \\ -P_{j1}s_{j,i} - P_{j2}s_{j,i}^{\smile\sigma_1} + \bar{e}_{vj,i} + u_{j,i}, & \kappa_{j,i}(t) = -1 \end{cases} \tag{9.8}$$

where $i = 1, 2, \cdots, 2n$, $\boldsymbol{s}_j = [s_{j,1}, s_{j,2}, \cdots, s_{j,(2n)}]^{\mathrm{T}} = \boldsymbol{y}_j + \boldsymbol{\mathcal{K}}_j(t)\bar{\boldsymbol{e}}_j$; $\bar{e}_{vj,i}(t)$ is the i-th element of $\bar{\boldsymbol{e}}_{vj}(t)$; $u_{j,i}$ is given by

$$u_{j,i} = \begin{cases} -P_{j3}\mathrm{sgn}(s_{j,i})|e_{j,i}^{\frac{\gamma_4-\gamma_3}{\gamma_3}} q_{vd\iota,i}(t, d_\iota(t))|^2, & i = 1, 2, \cdots, n \\ 0, & i = n+1, n+2, \cdots, 2n \end{cases}$$

where P_{j1}, P_{j2} and P_{j3} are some positive constants.

To guarantee the stability of the closed-loop teleoperation system with a modified anti-windup compensator, similar to previous chapters, the switching rule $\boldsymbol{\mathcal{K}}_j(t) = \mathrm{diag}[\kappa_{j,1}(t), \kappa_{j,2}(t), \cdots, \kappa_{j,(2n)}(t)] \in \mathbb{R}^{2n\times 2n}$ is given by

$$\kappa_{j,i}(t) = \begin{cases} 1, & y_{j,i}(t)\bar{e}_{j,i}(t) > 0 \text{ or } y_{j,i}(t)\bar{e}_{j,i}(t) = 0,\ y_{j,i}(t^-)\bar{e}_{j,i}(t^-) \le 0 \\ -1, & y_{j,i}(t)\bar{e}_{j,i}(t) < 0 \text{ or } y_{j,i}(t)\bar{e}_{j,i}(t) = 0,\ y_{j,i}(t^-)\bar{e}_{j,i}(t^-) > 0 \end{cases} \tag{9.9}$$

where $\bar{e}_{j,i}(t)$ is the ith element of $\bar{\boldsymbol{e}}_j(t)$.

Rewrite (9.8) into the vector form as

$$\dot{\boldsymbol{y}}_j = -P_{j1}(\boldsymbol{y}_j + \boldsymbol{\mathcal{K}}_j\bar{\boldsymbol{e}}_j) - P_{j2}(\boldsymbol{y}_j + \boldsymbol{\mathcal{K}}_j\bar{\boldsymbol{e}}_j)^{\smile\sigma_1} - \boldsymbol{\mathcal{K}}_j\bar{\boldsymbol{e}}_{vj} + \boldsymbol{u}_j \tag{9.10}$$

where $\boldsymbol{u}_j = [u_{j,1}, u_{j,2}, \cdots, u_{j,(2n)}]^{\mathrm{T}}$; $\bar{\boldsymbol{e}}_j = [\bar{e}_{j,1}, \bar{e}_{j,2}, \cdots, \bar{e}_{j,(2n)}]^{\mathrm{T}}$; $\bar{\boldsymbol{e}}_{vj} = [\bar{e}_{vj,1}, \bar{e}_{vj,2}, \cdots, \bar{e}_{vj,(2n)}]^{\mathrm{T}}$.

From the definition, $\boldsymbol{\mathcal{K}}_j(t)$ has 2^{2n} possible values, i.e.,

$$\begin{gathered} \mathrm{diag}[-1, -1, -1, \cdots, -1] \in \mathbb{R}^{2n\times 2n} \\ \mathrm{diag}[1, -1, -1, \cdots, -1] \in \mathbb{R}^{2n\times 2n} \\ \mathrm{diag}[-1, 1, -1, \cdots, -1] \in \mathbb{R}^{2n\times 2n} \\ \mathrm{diag}[1, 1, -1, \cdots, -1] \in \mathbb{R}^{2n\times 2n} \\ \vdots \\ \mathrm{diag}[-1, 1, 1, \cdots, 1] \in \mathbb{R}^{2n\times 2n} \\ \mathrm{diag}[1, 1, 1, \cdots, 1] \in \mathbb{R}^{2n\times 2n} \end{gathered}$$

Number the 2^{2n} states in turn as 1, 2, $\cdots$, 2^{2n}. Denote the index set be $\mathcal{S} = \{1, 2, \cdots, 2^{2n}\}$. Define a mapping $\bar{r}_j : \boldsymbol{\mathcal{K}}_j(t) \to \mathcal{S}$ as

$$\bar{r}_j(\boldsymbol{\mathcal{K}}_j(t)) = i, \ \textit{iff}\ \boldsymbol{\mathcal{K}}_j(t) \textit{is in the ith mode}$$

where $i \in \mathcal{S}$. For the sake of simplifying the expression, let $r_j(t) = \bar{r}_j(\boldsymbol{\mathcal{K}}_j(t))$. Then (9.10) can be written as a norm state-dependent switched system:

$$\dot{\boldsymbol{y}}_j(t) = h(\boldsymbol{y}_j(t), \bar{\boldsymbol{e}}_j(t), \bar{\boldsymbol{e}}_{vj}(t), \boldsymbol{q}_{vd\iota}(t, d_\iota(t)), r_j(t)) \tag{9.11}$$

where $r_j(t)$ is a norm state-dependent switching signal.

Remark 9.5. *Let time sequence* $\{t^j_k\}_{k\geq 1}$ *denote the switching time of* $r_j(t)$*, with* $t^j_0 = t_0 = 0$*. Then for any* $t \in [t^j_k, t^j_{k+1})$*,*

$$\dot{\boldsymbol{s}}_j(t) = -P_{j1}\boldsymbol{s}_j(t) - P_{j2}(\boldsymbol{s}_j(t))^{\smile\sigma_1} + \bar{\boldsymbol{u}}_j(t) \tag{9.12}$$

where

$$\bar{\boldsymbol{u}}_j(t) = \boldsymbol{u}_j(t) + [(\boldsymbol{\mathcal{K}}_j(t)\dot{d}_\iota(t)\boldsymbol{E}_j(t)\boldsymbol{q}_{vd\iota}(t, d_\iota(t)))^{\mathrm{T}}, \boldsymbol{0}^{\mathrm{T}}]^{\mathrm{T}}$$

From the definitions of $\boldsymbol{s}_j(t)$ *and* $\boldsymbol{\mathcal{K}}_j(t)$*, it is easy to verify that*

$$|\boldsymbol{y}_j(t)|^2 + |\bar{\boldsymbol{e}}_j(t)|^2 \leq |\boldsymbol{s}_j(t)|^2 \leq 2(|\boldsymbol{y}_j(t)|^2 + |\bar{\boldsymbol{e}}_j(t)|^2) \tag{9.13}$$

For the sake of simplifying notation, when $t \in [t^j_k, t^j_{k+1})$*,* $k \geq 0$*, it is assumed that* $r_j(t) = \mathfrak{p}_{jk} \in \mathcal{S}$*.*

Under the proposed control design scheme, the closed-loop teleoperation system is composed of the robot dynamics (9.3), the generalized control torque (9.4), the anti-windup compensator (9.7) and the auxiliary switched filter dynamic system (9.11).

9.4 STABILITY ANALYSIS

This section presents the stability analysis of the resulting closed-loop teleoperation system. Similar to the previous chapters, the analysis is still based on the finite-time state-independent input-to-output stability framework, and adopts the multiple Lyapunov-Krasovskii functionals analysis method.

Firstly, let $\boldsymbol{x}_j = [\boldsymbol{q}_j^{\mathrm{T}}, \dot{\boldsymbol{q}}_j^{\mathrm{T}}, \|\tilde{\boldsymbol{\Theta}}_j\|_F, \tilde{\omega}_j, \boldsymbol{y}_j^{\mathrm{T}}, \boldsymbol{e}_j^{\mathrm{T}}, \hat{\boldsymbol{\tau}}_j^{\mathrm{T}}]^{\mathrm{T}} \in \mathbb{R}^{6n+2}$, $\boldsymbol{z}_j = [\dot{\boldsymbol{q}}_j^{\mathrm{T}}, \|\tilde{\boldsymbol{\Theta}}_j\|_F, \tilde{\omega}_j, \boldsymbol{y}_j^{\mathrm{T}}, \phi_{j2}(\boldsymbol{e}_j)^{\mathrm{T}}, \hat{\boldsymbol{\tau}}_j^{\mathrm{T}}]^{\mathrm{T}} \in \mathbb{R}^{5n+2}$, $\bar{\boldsymbol{z}}_j = [\dot{\boldsymbol{q}}_j^{\mathrm{T}}, \boldsymbol{y}_j^{\mathrm{T}}, \phi_{j2}(\boldsymbol{e}_j)^{\mathrm{T}}, \hat{\boldsymbol{\tau}}_j]^{\mathrm{T}} \in \mathbb{R}^{5n}$. denote the new state and outputs: $\boldsymbol{x}_{jt} = \{\boldsymbol{x}_j(t+\theta), \theta \in [-\bar{d}, 0]\} \in C([-\bar{d}, 0]; \mathbb{R}^{6n+2})$, $\boldsymbol{z}_{jt} = \{\boldsymbol{z}_j(t+\theta), \theta \in [-\bar{d}, 0]\} \in C([-\bar{d}, 0]; \mathbb{R}^{5n+2})$, $\bar{\boldsymbol{z}}_{jt} = \{\bar{\boldsymbol{z}}_j(t+\theta), \theta \in [-\bar{d}, 0]\} \in C([-\bar{d}, 0]; \mathbb{R}^{5n})$. Define the auxiliary variables as

$$\begin{aligned}\Upsilon_{j0} = \min\Bigg\{ & K_{j1} - \frac{K_{j2}}{2} - \frac{\lambda_{\max}(\boldsymbol{K}_{j4}\boldsymbol{K}_{j4}^{\mathrm{T}})}{2} - \frac{K_{j5}}{2} - \varepsilon_{j1} - \vartheta_j, \\ & P_{j1} - \frac{1}{2} - \vartheta_j,\ P_{j1} - \frac{K_{j5}}{2} - K_{j7} - K_{j8} - \vartheta_j, \\ & P_{j1} - \frac{K_{j2}}{2} - nK_{j8} - \vartheta_j,\ \frac{\vartheta_j}{2\bar{d}} \Bigg\} \\ \Upsilon_{j1} = \min\Bigg\{ & K_{j0} - \frac{K_{j3}}{1+\sigma_1}, P_{j2}(1-\varepsilon_{j2}) \Bigg\}\end{aligned}$$

By using multiple Lyapunov-Krasovskii functionals analysis method, we have

Theorem 9.1. *With the controller (9.4)~(9.7) and (9.10), when Assumption 9.3 and Assumption 9.4 are satisfied, if the constant positive scalar* ϑ_j *and the (auxiliary) positive control gains* K_{j0}*,* K_{j1}*,* K_{j2}*,* K_{j3}*,* K_{j5}*,* K_{j6}*,* K_{j7}*,* K_{j8}*,* K_{j9}*,* Λ_{j1}*,* Λ_{j2}*,* Λ_{j3}*,*

Λ_{j4}, P_{j1}, P_{j2}, P_{j3}, σ_1, σ_2, λ_j, ε_{j1}, ε_{j2} and non-zero constant matrix $\boldsymbol{K}_{j4} \in \mathbb{R}^{n\times 2n}$ satisfy

$$\begin{cases} \Upsilon_{j0} > \Upsilon_{j1} > 0 \\ \Lambda_{j2} > 2\Lambda_{j1}\vartheta_j \\ \Lambda_{j4} > 2\Lambda_{j3}\vartheta_j \\ K_{j6} \geq 1 \\ P_{j2} \geq \dfrac{K_{j3}\sigma_1}{(1+\sigma_1)\varepsilon_{j2}} \\ P_{j3} \geq \bar{d} \end{cases} \tag{9.14}$$

then for the initial state $|\hat{\boldsymbol{\tau}}_j(0)| > \delta$, and for any other initial conditions and any bounded exogenous forces $\boldsymbol{f}_h^*$ and $\boldsymbol{f}_e^*$, the teleoperation system (9.3) is:

(1) SIIOpS with $\boldsymbol{z}_m$ and $\boldsymbol{z}_s$ being the outputs. And further, the velocity $\dot{\boldsymbol{q}}_j$, position error $\boldsymbol{e}_j$ and, adaptive (fuzzy) estimation errors $\tilde{\boldsymbol{\Theta}}_j$ and $\tilde{\omega}_j$ are bounded.

(2) FTSIIOpS with $\bar{\boldsymbol{z}}_m$ and $\bar{\boldsymbol{z}}_s$ being the outputs. The position error $\boldsymbol{e}_j$ will converge to the region $\{\boldsymbol{e}_j \in \mathbb{R}^n \| |\phi_{j2}(\boldsymbol{e}_j)| \leq \Phi_j\}$ in finite time, where $\Phi_j = \sqrt{\dfrac{\bar{\alpha}_{j2}}{\bar{\alpha}_{j1}}}\sqrt{\dfrac{\bar{\Xi}_j + \frac{1}{4}\|\boldsymbol{f}^*_{he_j[0,t)}\|^2_\infty}{\epsilon_j(\Upsilon_{j0}-\Upsilon_{j1})}}$, $\epsilon_j \in (0,1)$, $\bar{\Xi}_j = \dfrac{1}{2}\sup\|\tilde{\boldsymbol{\Theta}}_j\|_F^2 + \sup|\tilde{\omega}_j| + \Upsilon_{j1}$, $\bar{\alpha}_{j1} = \min\{\frac{\rho_{j1}}{2}, \frac{1}{2}\}$, $\bar{\alpha}_{j2} = \max\{\frac{\rho_{j2}}{2}, 1, \vartheta_j\}$.

Proof. First, the proof of SIIOpS is given. Take the following Lyapunov-Krasovskii functional: $V_j(\boldsymbol{x}_{jt}, r_j(t)) = V_1 + V_2 + V_3 + V_4$, where

$$\begin{aligned} V_1 &= \frac{1}{2}\dot{\boldsymbol{q}}_j^{\mathrm{T}}\boldsymbol{M}_{jo}(\boldsymbol{q}_j)\dot{\boldsymbol{q}}_j + \frac{1}{2}\hat{\boldsymbol{\tau}}_j^{\mathrm{T}}\hat{\boldsymbol{\tau}}_j \\ V_2 &= \frac{1}{2}\mathrm{Tr}(\tilde{\boldsymbol{\Theta}}_j^{\mathrm{T}}\Lambda_{j1}^{-1}\tilde{\boldsymbol{\Theta}}_j) + \frac{1}{2}\tilde{\omega}_j\Lambda_{j3}^{-1}\tilde{\omega}_j \\ V_3 &= \frac{1}{2}\boldsymbol{s}_j^{\mathrm{T}}\boldsymbol{s}_j \\ V_4 &= \frac{\vartheta_j}{2}\int_{-\bar{d}}^{0}\boldsymbol{z}_{jt}^{\mathrm{T}}(\tau)\left[\frac{-\tau}{\bar{d}} + \frac{2(\tau+\bar{d})}{\bar{d}}\right]\boldsymbol{z}_{jt}(\tau)\mathrm{d}\tau \end{aligned}$$

with $\tilde{\boldsymbol{\Theta}}_j = \boldsymbol{\Theta}_j - \hat{\boldsymbol{\Theta}}_j$; $\tilde{\omega}_j = \omega_j - \hat{\omega}_j$.

If and only if $y_{j,i}(t)\bar{e}_{j,i}(t) = 0$ and $y_{j,i}(t^-)\bar{e}_{j,i}(t^-) > 0$, $\kappa_{j,i}(t)$ switches from 1 to -1; if and only if $y_{j,i}(t)\bar{e}_{j,i}(t) = 0$ and $y_{j,i}(t^-)\bar{e}_{j,i}(t^-) < 0$, $\kappa_{j,i}(t)$ switch -1 to 1. Then for $t = t^j_{k+1}$, $k \geq 0$, from Property 2.1 and (9.13), by considering the fact that

$$\frac{1}{2}\int_{-\bar{d}}^{0}\boldsymbol{z}_{jt}^{\mathrm{T}}(\tau)\left[\frac{-\tau}{\bar{d}} + \frac{2(\tau+\bar{d})}{\bar{d}}\right]\boldsymbol{z}_{jt}(\tau)\mathrm{d}\tau \leq \int_{-\bar{d}}^{0}\boldsymbol{z}_{jt}^{\mathrm{T}}(\tau)\boldsymbol{z}_{jt}(\tau)\mathrm{d}\tau$$

it is easy to verify that

$$\begin{cases} \alpha_{j1}|\boldsymbol{z}_j|^2 \leq V_j(\boldsymbol{x}_{jt}, r_j(t)) \leq \alpha_{j2}\|\boldsymbol{z}_{jt}\|^2_{M_2} \\ V_j(\boldsymbol{x}_{jt^j_{k+1}}, r_j(t^j_{k+1})) \leq V_j(\boldsymbol{x}_{jt^j_{k+1}}, r_j(t^j_k)) \end{cases} \tag{9.15}$$

where $\alpha_{j1} = \min\left\{\frac{\rho_{j1}}{2}, \frac{1}{2\Lambda_{j1}}, \frac{1}{2\Lambda_{j3}}, \frac{1}{2}\right\}$; $\alpha_{j2} = \max\left\{\frac{\rho_{j1}}{2}, \frac{1}{2\Lambda_{j1}}, \frac{1}{2\Lambda_{j3}}, 1, \vartheta_j\right\}$.

For V_1, it has

$$
\begin{aligned}
D^+V_1 = {} & \dot{\boldsymbol{q}}_j^{\mathrm{T}}\Big(\boldsymbol{J}_j^{\mathrm{T}}(\boldsymbol{q}_j)\boldsymbol{f}_{he_j}^* - \boldsymbol{F}_{pj}(\boldsymbol{q}_j,\dot{\boldsymbol{q}}_j) + \boldsymbol{P}_{cj}(\boldsymbol{q}_j,\dot{\boldsymbol{q}}_j,\ddot{\boldsymbol{q}}_j) - \hat{\boldsymbol{\Theta}}_j^{\mathrm{T}}\boldsymbol{\varphi}(\boldsymbol{Z}_j) - \\
& \hat{\omega}_j\dot{\boldsymbol{q}}_j - K_{j1}\dot{\boldsymbol{q}}_j - K_{j2}\phi_{j2}(\boldsymbol{e}_j)\Big) + \dot{\boldsymbol{q}}_j^{\mathrm{T}}\Big(- K_{j0}\dot{\boldsymbol{q}}_j^{\smile\sigma_1} - K_{j3}\boldsymbol{e}_j^{\smile\sigma_1} - \\
& K_{j4}\boldsymbol{y}_j - K_{j5}\hat{\boldsymbol{\tau}}_j - \frac{1}{2}|\boldsymbol{\varphi}(\boldsymbol{Z}_j)|^2\dot{\boldsymbol{q}}_j + \Delta\boldsymbol{\tau}_j\Big) + \hat{\boldsymbol{\tau}}_j^{\mathrm{T}}\dot{\hat{\boldsymbol{\tau}}}_j \\
\le {} & \frac{1}{4}|\boldsymbol{f}_{he_j}^*|^2 + \dot{\boldsymbol{q}}_j^{\mathrm{T}}\tilde{\boldsymbol{\Theta}}_j^{\mathrm{T}}\boldsymbol{\varphi}(\boldsymbol{Z}_j) + \varepsilon_{j1} + \tilde{\omega}_j|\dot{\boldsymbol{q}}_j|^2 + \frac{K_{j2}}{2}|\phi_{j2}(\boldsymbol{e}_j)|^2 + \\
& \frac{1}{2}|\boldsymbol{y}_j|^2 - \frac{1}{2}|\boldsymbol{\varphi}(\boldsymbol{Z}_j)|^2|\dot{\boldsymbol{q}}_j|^2 + \dot{\boldsymbol{q}}_j^{\mathrm{T}}\Delta\boldsymbol{\tau}_j + \frac{K_{j5}}{2}|\hat{\boldsymbol{\tau}}_j|^2 + \hat{\boldsymbol{\tau}}_j^{\mathrm{T}}\dot{\hat{\boldsymbol{\tau}}}_j - \\
& \left(K_{j1} - \frac{K_{j2}}{2} - \frac{\lambda_{\max}(K_{j4}K_{j4}^{\mathrm{T}})}{2} - \frac{K_{j5}}{2}\right)|\dot{\boldsymbol{q}}_j|^2 - \\
& \left(K_{j0} - \frac{K_{j3}}{1+\sigma_1}\right)|\dot{\boldsymbol{q}}_j|^{1+\sigma_1} + \frac{K_{j3}\sigma_1}{1+\sigma_1}|\boldsymbol{e}_j|^{1+\sigma_1}
\end{aligned} \tag{9.16}
$$

where $\boldsymbol{f}_{he_m}^* = \boldsymbol{f}_h^*$; $\boldsymbol{f}_{he_s}^* = -\boldsymbol{f}_e^*$.

Given that if $K_{j6} \ge 1$,

(1) when $\dot{\boldsymbol{q}}_j^{\mathrm{T}}\Delta\boldsymbol{\tau}_j \ge \frac{K_{j7}+K_{j8}}{K_{j6}}|\hat{\boldsymbol{\tau}}_j|^2$, it has

$$
\begin{aligned}
\dot{\boldsymbol{q}}_j^{\mathrm{T}}\Delta\boldsymbol{\tau}_j + \hat{\boldsymbol{\tau}}_j^{\mathrm{T}}\dot{\hat{\boldsymbol{\tau}}}_j &= -(K_{j6}-1)\dot{\boldsymbol{q}}_j^{\mathrm{T}}\Delta\boldsymbol{\tau}_j + K_{j7}|\hat{\boldsymbol{\tau}}_j|^2 + 2K_{j8}\sum_{i=1}^{n}|\boldsymbol{e}_j||\hat{\tau}_{j,i}| \\
&\le (K_{j7}+K_{j8})|\hat{\boldsymbol{\tau}}_j|^2 + K_{j8}n|\boldsymbol{e}_j|^2
\end{aligned}
$$

(2) when $0 < \dot{\boldsymbol{q}}_j^{\mathrm{T}}\Delta\boldsymbol{\tau}_j < \frac{K_{j7}+K_{j8}}{K_{j6}}|\hat{\boldsymbol{\tau}}_j|^2$, it has

$$
\dot{\boldsymbol{q}}_j^{\mathrm{T}}\Delta\boldsymbol{\tau}_j + \hat{\boldsymbol{\tau}}_j^{\mathrm{T}}\dot{\hat{\boldsymbol{\tau}}}_j \le -(K_{j7}-K_{j8})|\hat{\boldsymbol{\tau}}_j|^2 + K_{j8}n|\boldsymbol{e}_j|^2
$$

(3) when $\dot{\boldsymbol{q}}_j^{\mathrm{T}}\Delta\boldsymbol{\tau}_j \le 0$ and $|\hat{\boldsymbol{\tau}}_j| < \delta$, it has

$$
\dot{\boldsymbol{q}}_j^{\mathrm{T}}\Delta\boldsymbol{\tau}_j + \hat{\boldsymbol{\tau}}_j^{\mathrm{T}}\dot{\hat{\boldsymbol{\tau}}}_j \le K_{j8}|\hat{\boldsymbol{\tau}}_j|^2 + K_{j8}n|\boldsymbol{e}_j|^2
$$

(4) when $\dot{\boldsymbol{q}}_j^{\mathrm{T}}\Delta\boldsymbol{\tau}_j \le 0$ and $|\hat{\boldsymbol{\tau}}_j| \ge \delta$, it has

$$
\dot{\boldsymbol{q}}_j^{\mathrm{T}}\Delta\boldsymbol{\tau}_j + \hat{\boldsymbol{\tau}}_j^{\mathrm{T}}\dot{\hat{\boldsymbol{\tau}}}_j \le 0
$$

Hence, by letting $K_{j6} \ge 1$, it has

$$
\dot{\boldsymbol{q}}_j^{\mathrm{T}}\Delta\boldsymbol{\tau}_j + \hat{\boldsymbol{\tau}}_j^{\mathrm{T}}\dot{\hat{\boldsymbol{\tau}}}_j \le (K_{j7}+K_{j8})|\hat{\boldsymbol{\tau}}_j|^2 + K_{j8}n|\boldsymbol{e}_j|^2
$$

And further,

$$\begin{aligned}D^{+}V_1 \leq & \frac{1}{4}|\boldsymbol{f}_{he_j}^{*}|^2 + \dot{\boldsymbol{q}}_j^{\mathrm{T}}\tilde{\boldsymbol{\Theta}}_j^{\mathrm{T}}\boldsymbol{\varphi}(Z_j) + \varepsilon_{j1} + \tilde{\omega}_j|\dot{\boldsymbol{q}}_j|^2 + \frac{K_{j2}}{2}|\phi_{j2}(\boldsymbol{e}_j)|^2 + \\ & \frac{1}{2}|\boldsymbol{y}_j|^2 + \dot{\boldsymbol{q}}_j^{\mathrm{T}}\Delta\boldsymbol{\tau}_j + \left(K_{j7} + K_{j8} + \frac{K_{j5}}{2}\right)|\hat{\boldsymbol{\tau}}_j|^2 - \\ & \left(K_{j1} - \frac{K_{j2}}{2} - \frac{\lambda_{\max}(\boldsymbol{K}_{j4}\boldsymbol{K}_{j4}^{\mathrm{T}})}{2} - \frac{K_{j5}}{2}\right)|\dot{\boldsymbol{q}}_j|^2 - \\ & \left(K_{j0} - \frac{K_{j3}}{1+\sigma_1}\right)|\dot{\boldsymbol{q}}_j|^{1+\sigma_1} + \frac{K_{j3}\sigma_1}{1+\sigma_1}|\boldsymbol{e}_j|^{1+\sigma_1} - \\ & \frac{1}{2}|\boldsymbol{\varphi}(\boldsymbol{Z}_j)|^2|\dot{\boldsymbol{q}}_j|^2 + K_{j8}n|\boldsymbol{e}_j|^2 \end{aligned} \tag{9.17}$$

Let $\bar{\boldsymbol{\Theta}}_j = \boldsymbol{\Theta}_j - \boldsymbol{\Theta}_j^{*}$, $\bar{\omega}_j = \omega_j - \omega_j^{*}$, for V_2, it has

$$\begin{aligned}D^{+}V_2 = & -\mathrm{Tr}\left(\tilde{\boldsymbol{\Theta}}_j^{\mathrm{T}}\boldsymbol{\varphi}(\boldsymbol{Z}_j)\dot{\boldsymbol{q}}_j^{\mathrm{T}} + \frac{\Lambda_{j2}}{\Lambda_{j1}}\tilde{\boldsymbol{\Theta}}_j^{\mathrm{T}}(\tilde{\boldsymbol{\Theta}}_j - \bar{\boldsymbol{\Theta}}_j)\right) - \\ & \tilde{\omega}_j|\dot{\boldsymbol{q}}_j|^2 - \frac{\Lambda_{ji4}}{\Lambda_{ji3}}\tilde{\omega}_j^{i\mathrm{T}}(\tilde{\omega}_j - \bar{\omega}_j) \\ \leq & -\mathrm{Tr}(\tilde{\boldsymbol{\Theta}}_j^{\mathrm{T}}\boldsymbol{\varphi}(\boldsymbol{Z}_j)\dot{\boldsymbol{q}}_j^{\mathrm{T}}) - \frac{\Lambda_{j2}}{2\Lambda_{j1}}\|\tilde{\boldsymbol{\Theta}}_j\|_F^2 + \frac{\Lambda_{j2}}{2\Lambda_{j1}}\|\bar{\boldsymbol{\Theta}}_j\|_F^2 + \\ & \tilde{\omega}_j|\dot{\boldsymbol{q}}_j|^2 - \frac{\Lambda_{j4}}{2\Lambda_{j3}}|\tilde{\omega}_j|^2 + \frac{\Lambda_{j4}}{2\Lambda_{j3}}|\bar{\omega}_j|^2 \end{aligned} \tag{9.18}$$

For any $t \in [t_k, t_{k+1})$, by taking $P_{j3} \geq \tilde{d}$, there then exists $\varepsilon_{j2} \in (0,1)$ such that

$$\begin{aligned}D^{+}V_3 = & \boldsymbol{s}_j^{\mathrm{T}}\dot{\boldsymbol{s}}_j \\ = & -P_{j1}|\boldsymbol{s}_j|^2 - P_{j2}|\boldsymbol{s}_j|^{1+\sigma_1} + \boldsymbol{s}_j^{\mathrm{T}}\bar{\boldsymbol{u}}_j \\ \leq & -P_{j1}\left(|\boldsymbol{y}_j|^2 + |\phi_{j2}(\boldsymbol{e}_j)|^2 + |\hat{\boldsymbol{\tau}}_j|^2\right) - P_{j2}\left(|\boldsymbol{y}_j|^2 + |\phi_{j2}(\boldsymbol{e}_j)|^2 + |\hat{\boldsymbol{\tau}}_j|^2\right)^{\frac{1+\sigma_1}{2}} + \\ & \sum_{i=1}^{n}\left(s_{j,i}u_{j,i} + s_{j,i}\kappa_{j,i}\dot{d}_\iota(t)e_{j,i}^{\frac{\gamma_4-\gamma_3}{\gamma_3}}q_{vd\iota,i}(t, d_\iota(t))\right) \\ \leq & -P_{j1}\left(|\boldsymbol{y}_j|^2 + |\phi_{j2}(\boldsymbol{e}_j)|^2 + |\hat{\boldsymbol{\tau}}_j|^2\right) - P_{j2}\varepsilon_{j2}|\phi_{j2}(\boldsymbol{e}_j)|^{\frac{1+\sigma_1}{2}} - \\ & P_{j2}(1-\varepsilon_{j2})\left(|\boldsymbol{y}_j|^2 + |\phi_{j2}(\boldsymbol{e}_j)|^2 + |\hat{\boldsymbol{\tau}}_j|^2\right)^{\frac{1+\sigma_1}{2}} \\ \leq & -P_{j1}\left(|\boldsymbol{y}_j|^2 + |\phi_{j2}(\boldsymbol{e}_j)|^2 + |\hat{\boldsymbol{\tau}}_j|^2\right) - P_{j2}\varepsilon_{j2}|\boldsymbol{e}_j|^{\frac{1+\sigma_1}{2}} - \\ & P_{j2}(1-\varepsilon_{j2})\left(|\boldsymbol{y}_j|^2 + |\phi_{j2}(\boldsymbol{e}_j)|^2 + |\hat{\boldsymbol{\tau}}_j|^2\right)^{\frac{1+\sigma_1}{2}} \end{aligned} \tag{9.19}$$

Finally, for V_4, it has

$$\begin{aligned}D^{+}V_4 = & \vartheta_j\boldsymbol{z}_j^{\mathrm{T}}\boldsymbol{z}_j - \frac{\vartheta_j}{2}\boldsymbol{z}_{jt}^{\mathrm{T}}(-\bar{d})\boldsymbol{z}_{jt}(-\bar{d}) - \frac{\vartheta_j}{2\bar{d}}\int_{-\bar{d}}^{0}\boldsymbol{z}_{jt}^{\mathrm{T}}(\tau)\boldsymbol{z}_{jt}(\tau)\mathrm{d}\tau \\ \leq & \vartheta_j\boldsymbol{z}_j^{\mathrm{T}}\boldsymbol{z}_j - \frac{\vartheta_j}{2\bar{d}}\int_{-\bar{d}}^{0}\boldsymbol{z}_{jt}^{\mathrm{T}}(\tau)\boldsymbol{z}_{jt}(\tau)\mathrm{d}\tau \end{aligned} \tag{9.20}$$

Jointing (9.17)∼(9.20), for any $t \in [t_k^j, t_{k+1}^j)$, $k \geq 0$, let $P_{j2} \geq \frac{K_{j3}\sigma_1}{(1+\sigma_1)\varepsilon_{j2}}$, one can obtain

$$\begin{aligned}
D^+V_j(\boldsymbol{x}_{jt}, \mathfrak{p}_{jk}) \leq & -\Big(K_{j1} - \frac{K_{j2}}{2} - \frac{\lambda_{\max}(\boldsymbol{K}_{j4}\boldsymbol{K}_{j4}^{\mathrm{T}})}{2} - \frac{K_{j5}}{2} - \vartheta_j\Big)|\dot{\boldsymbol{q}}_j|^2 - \\
& (P_{j1} - \frac{1}{2} - \vartheta_j)|\boldsymbol{y}_j|^2 - \Big(P_{j1} - \frac{K_{j2}}{2} - nK_{j8} - \vartheta_j\Big)|\phi_{j2}(\boldsymbol{e}_j)|^2 - \\
& \Big(P_{j1} - \frac{K_{j5}}{2} - K_{j7} - K_{j8} - \vartheta_j\Big)|\hat{\boldsymbol{\tau}}_j|^2 - \Big(\frac{\Lambda_{j2}}{2\Lambda_{j1}} - \vartheta_j\Big)\|\tilde{\boldsymbol{\Theta}}_j\|_F^2 - \\
& \Big(\frac{\Lambda_{j4}}{2\Lambda_{j3}} - \vartheta_j\Big)|\tilde{\omega}_j|^2 + \frac{1}{4}|\boldsymbol{f}_{he_j}^*|^2 + \Xi_j - \frac{\vartheta_j}{2\bar{d}}\int_{-\bar{d}}^{0} \boldsymbol{z}_{jt}^{\mathrm{T}}(\tau)\boldsymbol{z}_{jt}(\tau)\mathrm{d}\tau - \\
& \Big(P_{j2}\varepsilon_{j2} - \frac{K_{j3}\sigma_1}{1+\sigma_1}\Big)|\boldsymbol{e}_j|^{1+\sigma_1} - \Big(K_{j0} - \frac{K_{j3}}{1+\sigma_1}\Big)|\dot{\boldsymbol{q}}_j|^{1+\sigma_1} - \\
& \frac{1}{2}|\boldsymbol{\varphi}(Z_j)|^2|\dot{\boldsymbol{q}}_j|^2 - P_{j2}(1-\varepsilon_{j2})\Big(|\boldsymbol{y}_j|^2 + |\phi_{j2}(\boldsymbol{e}_j)|^2 + |\hat{\boldsymbol{\tau}}_{jo}|^2\Big)^{\frac{1+\sigma_1}{2}} \\
\leq & -\alpha_j\|\boldsymbol{z}_{jt}\|_{M_2}^2 + \frac{1}{4}|\boldsymbol{f}_{he_j}^*|^2 + \Xi_j
\end{aligned}$$

where $\alpha_j = \min\Big\{K_{j1} - \frac{K_{j2}}{2} - \frac{\lambda_{\max}(\boldsymbol{K}_{j4}\boldsymbol{K}_{j4}^{\mathrm{T}})}{2} - \frac{K_{j5}}{2} - \vartheta_j, P_{j1} - \frac{K_{j5}}{2} - K_{j7} - K_{j8} - \vartheta_j,\ P_{j1} - \frac{K_{j2}}{2} - nK_{j8} - \vartheta_j, P_{j1} - \frac{1}{2} - \vartheta_j,\ \frac{\Lambda_{j2}}{2\Lambda_{j1}} - \vartheta_j,\ \frac{\Lambda_{j4}}{2\Lambda_{j3}} - \vartheta_j,\ \frac{\vartheta_j}{2\bar{d}}\Big\}$,
$\Xi_j = \varepsilon_{j1} + \frac{\Lambda_{j2}}{2\Lambda_{j1}}\|\bar{\boldsymbol{\Theta}}_j\|_F^2 + \frac{\Lambda_{j4}}{2\Lambda_{j3}}|\bar{\omega}_j|^2$.

Following Lemma 2.4, it is easy to verify that the complete closed-loop master (slave) system is SIIOpS with the input $\boldsymbol{f}_h^*$ ($\boldsymbol{f}_e^*$) and the output $\boldsymbol{z}_m$ ($\boldsymbol{z}_s$). Furthermore, from Definition 2.4, the velocity $\dot{\boldsymbol{q}}_j$, the position error function $\phi_{j2}(\boldsymbol{e}_j)$ and, the adaptive (fuzzy) estimation error $\tilde{\boldsymbol{\Theta}}_j$ and $\tilde{\omega}_j$ will remain bounded for all time. Moreover, from the definition of $\phi_{j2}(\boldsymbol{e}_j)$ the position error $\boldsymbol{e}_j$ is also bounded.

Next, the finite time synchronization performance is given. Taking the Lyapunov-Krasovskii functional $\bar{V}_j(\boldsymbol{x}_{jt}, r_j(t)) = V_1 + V_3 + \bar{V}_4$, where

$$\bar{V}_4 = \frac{\vartheta_j}{2}\int_{-\bar{d}}^{0} \bar{\boldsymbol{z}}_{jt}^{\mathrm{T}}(\tau)\left[\frac{-\tau}{\bar{d}} + \frac{2(\tau+\bar{d})}{\bar{d}}\right]\bar{\boldsymbol{z}}_{jt}(\tau)\mathrm{d}\tau$$

Similar to (9.15), it has

$$\begin{cases} \bar{\alpha}_{j1}|\bar{\boldsymbol{z}}_j|^2 \leq \bar{V}_j(\boldsymbol{x}_{jt}, r_j(t)) \leq \bar{\alpha}_{j2}\|\bar{\boldsymbol{z}}_{jt}\|_{M_2}^2 \\ \bar{V}_j(\boldsymbol{x}_{jt_{k+1}^j}, r_j(t_{k+1}^j)) \leq \bar{V}_j(\boldsymbol{x}_{jt_{k+1}^j}, r_j(t_k^j)) \end{cases}$$

where $\bar{\alpha}_{j1} = \min\{\frac{\rho_{j1}}{2}, \frac{1}{2}\}$; $\bar{\alpha}_{j2} = \max\{\frac{\rho_{j2}}{2}, 1, \vartheta_j\}$. Moreover,

$$\begin{aligned}
D^+V_1 \leq & \frac{1}{4}|\boldsymbol{f}_{he_j}^*|^2 + \dot{\boldsymbol{q}}_j^{\mathrm{T}}\tilde{\boldsymbol{\Theta}}_j^{\mathrm{T}}\boldsymbol{\varphi}(\boldsymbol{Z}_j) + |\tilde{\omega}_j| + \frac{K_{j2}}{2}|\phi_{j2}(\boldsymbol{e}_j)|^2 + \frac{1}{2}|\boldsymbol{y}_j|^2 + \\
& K_{j8}n|\boldsymbol{e}_j|^2 - \left[K_{j1} - \frac{K_{j2}}{2} - \frac{\lambda_{\max}(\boldsymbol{K}_{j4}\boldsymbol{K}_{j4}^{\mathrm{T}})}{2} - \frac{K_{j5}}{2} - \varepsilon_{j1}\right]|\dot{\boldsymbol{q}}_j|^2 -
\end{aligned}$$

$$(K_{j0}-\frac{K_{j3}}{1+\sigma_1})|\dot{\boldsymbol{q}}_j|^{1+\sigma_1}+\left(\frac{K_{j5}}{2}+K_{j7}+K_{j8}\right)|\hat{\boldsymbol{\tau}}_j|^2+$$
$$\frac{K_{j3}\sigma_1}{1+\sigma_1}|\boldsymbol{e}_j|^{1+\sigma_1}-\frac{1}{2}|\boldsymbol{\varphi}(\boldsymbol{Z}_j)|^2|\dot{\boldsymbol{q}}_j|^2$$
$$\leq-\left[K_{j1}-\frac{K_{j2}}{2}-\frac{\lambda_{\max}(\boldsymbol{K}_{j4}\boldsymbol{K}_{j4}^{\mathrm{T}})}{2}-\frac{K_{j5}}{2}-\varepsilon_{j1}\right]|\dot{\boldsymbol{q}}_j|^2-\Bigg(K_{j0}-$$
$$\frac{K_{j3}}{1+\sigma_1}\Bigg)|\dot{\boldsymbol{q}}_j|^{1+\sigma_1}+\left(K_{j7}+K_{j8}+\frac{K_{j5}}{2}\right)|\hat{\boldsymbol{\tau}}_j|^2+\frac{K_{j3}\sigma_1}{1+\sigma_1}|\boldsymbol{e}_j|^{1+\sigma_1}+$$
$$K_{j8}n|\boldsymbol{e}_j|^2+\frac{K_{j2}}{2}|\phi_{j2}(\boldsymbol{e}_j)|^2+\frac{1}{2}\|\tilde{\boldsymbol{\Theta}}_j\|_F^2+|\tilde{\omega}_j|+\frac{1}{2}|\boldsymbol{y}_j|^2+\frac{1}{4}|\boldsymbol{f}_{he_j}^*|^2 \tag{9.21}$$

Combining (9.19)~(9.21), for any $t\in[t_k^j,t_{k+1}^j)$, $k\geq 0$, it holds

$$D^+\bar{V}_j(\boldsymbol{x}_{jt},\mathfrak{p}_{jk})$$
$$\leq-\left[K_{j1}-\frac{K_{j2}}{2}-\frac{\lambda_{\max}(\boldsymbol{K}_{j4}\boldsymbol{K}_{j4}^{\mathrm{T}})}{2}-\frac{K_{j5}}{2}-\varepsilon_{j1}-\vartheta_j\right]|\dot{\boldsymbol{q}}_j|^2-$$
$$\left(P_{j1}-\frac{1}{2}-\vartheta_j\right)|\boldsymbol{y}_j|^2-\left(P_{j1}-\frac{K_{j2}}{2}-nK_{j8}-\vartheta_j\right)|\phi_{j2}(\boldsymbol{e}_j)|^2-$$
$$\left(P_{j1}-\frac{K_{j5}}{2}-K_{j7}-K_{j8}-\vartheta_j\right)|\hat{\boldsymbol{\tau}}_j|^2-\frac{\vartheta_j}{2\bar{d}}\int_{-\bar{d}}^{0}\bar{\boldsymbol{z}}_{jt}^{\mathrm{T}}(\tau)\bar{\boldsymbol{z}}_{jt}(\tau)\mathrm{d}\tau-$$
$$\Upsilon_{j1}|\bar{\boldsymbol{z}}_j|^{1+\sigma_1}+\frac{1}{2}\|\tilde{\boldsymbol{\Theta}}_j\|_F^2+|\tilde{\omega}_j|+\frac{1}{4}|\boldsymbol{f}_{he_j}^*|^2$$
$$\leq-(\Upsilon_{j0}-\Upsilon_{j1})\|\bar{\boldsymbol{z}}_{jt}\|_{M_2}^2-\Upsilon_{j1}\|\bar{\boldsymbol{z}}_{jt}\|_{M_2}^{1+\sigma_1}+\frac{1}{2}\|\tilde{\boldsymbol{\Theta}}_j\|_F^2+$$
$$|\tilde{\omega}_j|+\frac{1}{4}|\boldsymbol{f}_{he_j}^*|^2+\Upsilon_{j1}$$

From the proof of SIIOpS, $\|\tilde{\boldsymbol{\Theta}}_j\|_F$ and $|\tilde{\omega}_j|$ are bounded. Denote $\bar{\Xi}_j=\frac{1}{2}\sup\|\tilde{\boldsymbol{\Theta}}_j\|_F^2+\sup|\tilde{\omega}_j|+\Upsilon_{j1}$, then it has $\bar{\Xi}_j<+\infty$ and further

$$D^+\bar{V}_j(\boldsymbol{x}_{jt},\mathfrak{p}_{jk})\leq-(\Upsilon_{j0}-\Upsilon_{j1})\|\bar{\boldsymbol{z}}_{jt}\|_{M_2}^2-\Upsilon_{j1}\|\bar{\boldsymbol{z}}_{jt}\|_{M_2}^{1+\sigma_1}+\frac{1}{4}|\boldsymbol{f}_{he_j}^*|^2+\bar{\Xi}_j$$

Under the hypotheses (9.14), it has $\Upsilon_{j0}>\Upsilon_{j1}>0$. Then according to Lemma 2.6, the complete closed-loop master (slave) system is also FTSIIOpS, with the output $\bar{\boldsymbol{z}}_m$ ($\bar{\boldsymbol{z}}_s$), and the input $\boldsymbol{f}_h^*$ ($\boldsymbol{f}_e^*$). In this scenario, the position error $\boldsymbol{e}_j$ will converge into the set $\{\boldsymbol{e}_j\in\mathbb{R}^n||\phi_{j2}(\boldsymbol{e}_j)|\leq\Phi_j\}$ in finite time, where $\Phi_j=\sqrt{\frac{\bar{\alpha}_{j2}}{\bar{\alpha}_{j1}}}\sqrt{\frac{\bar{\Xi}_j+\frac{1}{4}\|\boldsymbol{f}_{he_j[0,t)}^*\|_\infty^2}{\epsilon_j(\Upsilon_{j0}-\Upsilon_{j1})}}$, $\epsilon_j\in(0,1)$, with the setting time

$$T_{0j}\leq\frac{2}{(1-\epsilon_j)(\Upsilon_{j0}-\Upsilon_{j1})(1-\sigma_1)}\ln\left[\frac{(1-\epsilon_j)(\Upsilon_{j0}-\Upsilon_{j1})\bar{\alpha}_{j2}\|\bar{\boldsymbol{z}}_{j0}\|_\infty^{1-\sigma_1}+\Upsilon_{j1}}{\Upsilon_{j1}}\right]$$

The claim of the theorem is thus established. □

Remark 9.6. *For Theorem 9.1, the (auxiliary) control gain can be solved by the following steps:*

(1) Since they are independent of Theorem 9.1, the constant positive scalars ε_{j1}*,* $\varepsilon_{j2} \in (0,1)$*,* K_{j2}*,* K_{j3}*,* $\boldsymbol{K}_{j4}$*,* K_{j5}*,* K_{j7}*,* K_{j8}*,* K_{j9}*,* λ_j*,* Λ_{j1}*,* Λ_{j3}*, and positive odds* γ_1*,* γ_2*,* γ_3 *and* γ_4 *can be given first.*

(2) From $P_{j2} \geq \dfrac{K_{j3}\sigma_1}{(1+\sigma_1)\varepsilon_{j2}}$ *and* $\Upsilon_{j1} > 0$*, calculate* P_{j2} *and* K_{j0}*.*

(3) According to the characteristics of the communication network, we can get $\bar{d}$ *and* $\tilde{d}$*. From (9.14), calculate* P_{j3}*.*

(4) By solving $\Upsilon_{j0} > \Upsilon_{j1}$*, obtain* ϑ_j*,* K_{j1} *and* P_{j1}*.*

(5) After getting ϑ_j*,* Λ_{j1} *and* Λ_{j3}*, by (9.14), get* Λ_{j2} *and* Λ_{j4}*.*

(6) From $K_{j6} \geq 1$*, select* K_{j6}*.*

Remark 9.7. *From the design of anti-windup compensator (9.7), if there is no saturation,* $\hat{\boldsymbol{\tau}}_j$ *will stabilize to* δ*. At this time, a non-zero constant vector* $K_{j5}\hat{\boldsymbol{\tau}}_j$ *will contribute to the output of the auxiliary control torque* $\bar{\boldsymbol{\tau}}_j$*, thereby affecting the stability of state tracking error. On the one hand, by choosing a sufficiently small* δ*,* $K_{j5}\hat{\boldsymbol{\tau}}_j$ *will also have a sufficiently small value, thereby reducing the impact on the steady-state performance of closed-loop control. On the other hand, by introducing an auxiliary filter variable* $\boldsymbol{y}_j$ *in the control torque (9.5), the steady-state tracking errors that may be caused by* $K_{j5}\hat{\boldsymbol{\tau}}_j$ *can be further weakened. Mathematically, by (9.17) and (9.19), if* $P_{j1} > \dfrac{K_{j5}}{2} + K_{j7} + K_{j8}$*, no matter whether input saturation occurs or not, and regardless of the value of* $\hat{\boldsymbol{\tau}}_j$*, there will be* $-(P_{j1} - \dfrac{K_{j5}}{2} - K_{j7} - K_{j8})|\hat{\boldsymbol{\tau}}_j|^2 \leq 0$*.*

Remark 9.8. *In Chapter 4, for the teleoperation system under the input saturation constraint, a saturation control design scheme based on the auxiliary switched filter is proposed. In this chapter, the scheme is further extended to the finite time control of the teleoperation system, and a new anti-saturation compensation control is adopted. Compared with the method in Chapter 4, the new saturation control method proposed in this chapter has the following characteristics:*

(1) Faster master-slave synchronization performance.

(2) The auxiliary switched filter does not need to use the acceleration information, which can reduce the undesired control chattering.

9.5 SIMULATION STUDY

In this section, some simulation examples will be used to verify the effectiveness of the algorithm. It is assumed that the teleoperation system consists of a pair of two-degree-of-freedom serial manipulators with rotating joints, and the nominal dynamic model can be found in Nuno's work [17]. Simulation settings are as follows.

First, the settings of the nominal physical parameters of the robot are consistent with those in Nuno's work [17]. The model uncertainty of the robot is assumed as follows:

$$\begin{aligned}
\Delta \boldsymbol{M}_m(\boldsymbol{q}_m) &= 0.3\sin(2t)\boldsymbol{M}_{mo}(\boldsymbol{q}_m)\\
\Delta \boldsymbol{C}_m(\boldsymbol{q}_m, \dot{\boldsymbol{q}}_m) &= 0.2\sin(3t)\boldsymbol{C}_{mo}(\boldsymbol{q}_m, \dot{\boldsymbol{q}}_m)\\
\Delta \boldsymbol{g}_m(\boldsymbol{q}_m) &= 0.1\sin(4t)\boldsymbol{g}_{mo}(\boldsymbol{q}_m)\\
\Delta \boldsymbol{M}_s(\boldsymbol{q}_s) &= 0.5\cos(5t)\boldsymbol{M}_{so}(\boldsymbol{q}_s)\\
\Delta \boldsymbol{C}_s(\boldsymbol{q}_s, \dot{\boldsymbol{q}}_s) &= 0.3\cos(5t)\boldsymbol{C}_{so}(\boldsymbol{q}_s, \dot{\boldsymbol{q}}_s)\\
\Delta \boldsymbol{g}_s(\boldsymbol{q}_s) &= 0.1\cos(5t)\boldsymbol{g}_{so}(\boldsymbol{q}_s)
\end{aligned}$$

According to the simulation setup, it has $\max\{g_1, g_2\} = 19.341\,3$, where g_1, g_2 are defined in Property 9.1. The levels of actuator saturation for both the master and slave robots are set to be $M_1 = M_2 = 20$. The external forces imposed by the human operator as well as the environment are assumed to be $\boldsymbol{f}_h = \boldsymbol{f}_h^* - K_h\boldsymbol{X}_m - B_h\dot{\boldsymbol{X}}_m$ and $\boldsymbol{f}_e = \boldsymbol{f}_e^* + K_e\boldsymbol{X}_s + B_e\dot{\boldsymbol{X}}_s$, where in our simulation, $K_h = 30$, $K_e = 50$, $B_h = B_e = 1$, the exogenous forces from the human operator and the environment are given as Fig. 9.1. We also set the fractions as $\boldsymbol{f}_m(\dot{\boldsymbol{q}}_m) = 0.5\text{sgn}(\dot{\boldsymbol{q}}_m) + 0.5\dot{\boldsymbol{q}}_m$ and $\boldsymbol{f}_s(\dot{\boldsymbol{q}}_s) = 0.3\text{sgn}(\dot{\boldsymbol{q}}_s) + 0.3\dot{\boldsymbol{q}}_s$; and assume the bounded disturbances as $\boldsymbol{f}_{dm} = [0.8\sin(1.2t), 0.8\cos(0.2t)]^{\mathrm{T}}$ and $\boldsymbol{f}_{ds} = [1.6\cos(0.5t), 1.6\sin(1.5t)]^{\mathrm{T}}$. Thus, $\boldsymbol{f}_m(\boldsymbol{q}_m, \dot{\boldsymbol{q}}_m) = \boldsymbol{f}_m(\dot{\boldsymbol{q}}_m) + \boldsymbol{f}_{dm}$, $\boldsymbol{f}_s(\boldsymbol{q}_s, \dot{\boldsymbol{q}}_s) = \boldsymbol{f}_s(\dot{\boldsymbol{q}}_s) + \boldsymbol{f}_{ds}$. In addition, the time delays are set as Fig. 9.2. For any $t \in [-\bar{d}, 0]$, the initial state of the teleoperation system is assumed to be: $\boldsymbol{q}_m(t) = [0, 0]^{\mathrm{T}}$, $\boldsymbol{q}_s(t) = [\pi, -\pi]^{\mathrm{T}}$, $\dot{\boldsymbol{q}}_m(t) = \dot{\boldsymbol{q}}_s(t) = \ddot{\boldsymbol{q}}_m(t) = \ddot{\boldsymbol{q}}_s(t) = [0, 0]^{\mathrm{T}}$, $\hat{\boldsymbol{\tau}}_m(t) = [-1, 1]^{\mathrm{T}}$, $\hat{\boldsymbol{\tau}}_s(t) = [1, -1]^{\mathrm{T}}$, $\boldsymbol{y}_m(t) = [-3, 10, 15, -10]^{\mathrm{T}}$, $\boldsymbol{y}_s(t) = [-10, 10, -20, 10]^{\mathrm{T}}$, $\hat{\omega}_m(t) = \hat{\omega}_s(t) = 0$, $\hat{\boldsymbol{\Theta}}_m(t) = \hat{\boldsymbol{\Theta}}_2(t) = \boldsymbol{0}_{5\times 2}$, where $\boldsymbol{0}_{5\times 2}$ is a 5×2 dimensional matrix with all zero elements.

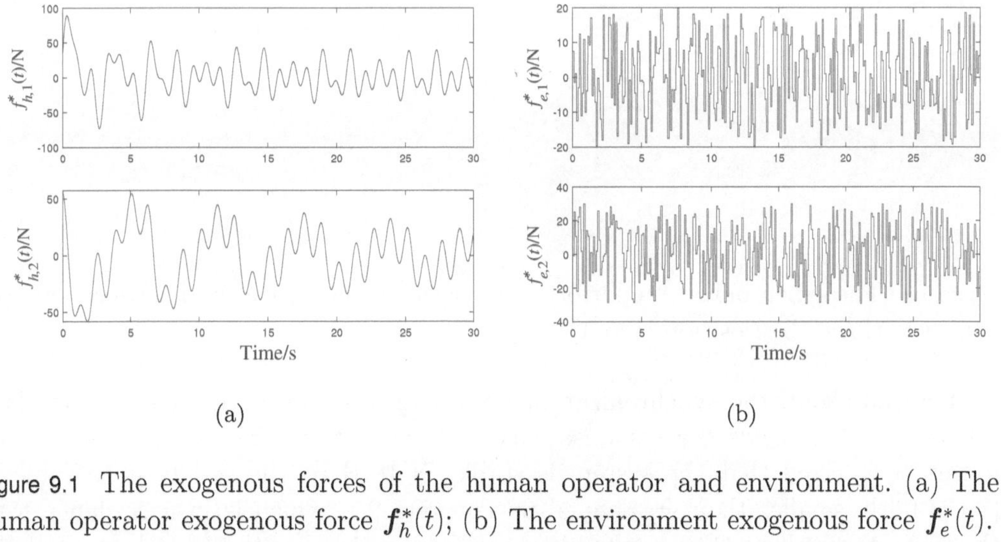

Figure 9.1 The exogenous forces of the human operator and environment. (a) The human operator exogenous force $\boldsymbol{f}_h^*(t)$; (b) The environment exogenous force $\boldsymbol{f}_e^*(t)$.

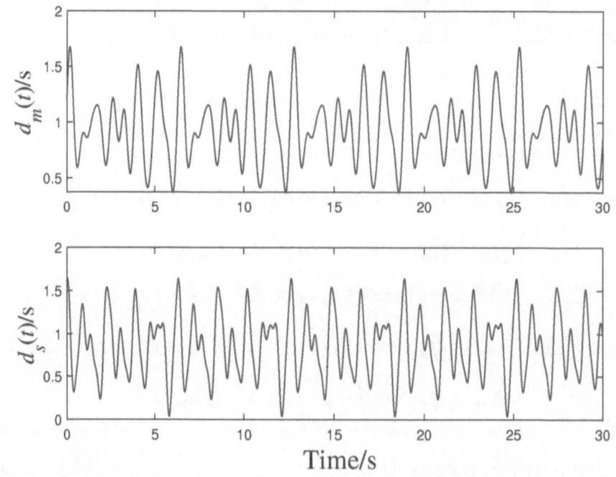

Figure 9.2 The communication delays $d_m(t)$ and $d_s(t)$.

9.5.1 Stability verification

In the first set of simulations, the stability of the closed-loop system will be verified first. Under the above simulation settings, according to Theorem 9.1, the simulation results are as follows Figs. 9.3–9.6.

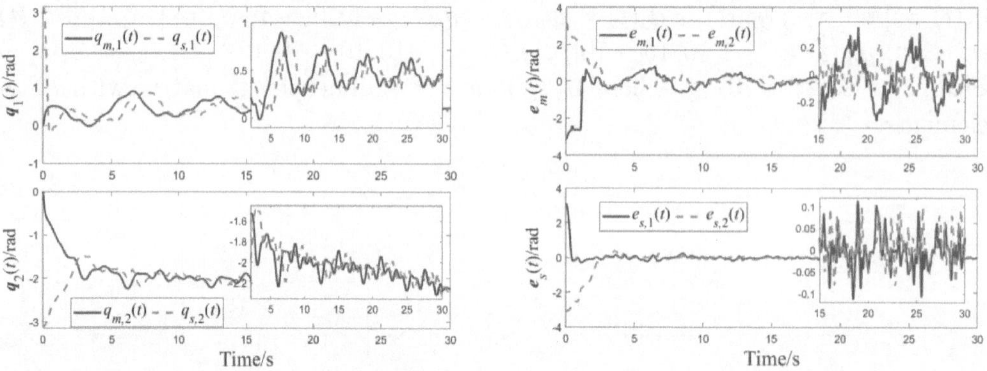

(a) (b)

Figure 9.3 The synchronization performance between the master and slave robots. (a) Joint positions. (b) Position errors.

Fig. 9.3 shows the synchronization performance between the master and slave robots. Fig. 9.4 shows the response curves of the outputs $\bar{z}_m$ and $\bar{z}_s$, from which it can be inferred that the system is stable. Note that, the steady-state value of $|\bar{z}_j|$ is fairly smaller than Φ_j defined in Theorem 9.1, which is in accordance with the theoretic analysis. The boundedness of $|\bar{z}_j|$ also implies the boundedness of the estimation errors, as shown in Fig. 9.5. The (fuzzy) adaptive estimated values $\hat{\boldsymbol{\Theta}}_j$ and

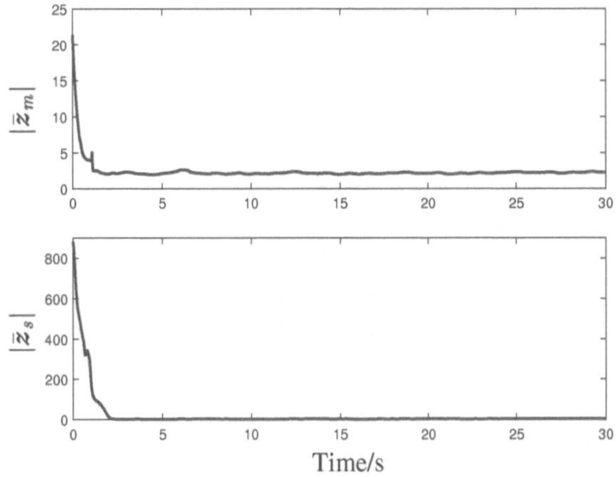

Figure 9.4 The response trajectories of the outputs $|\bar{z}_j|$.

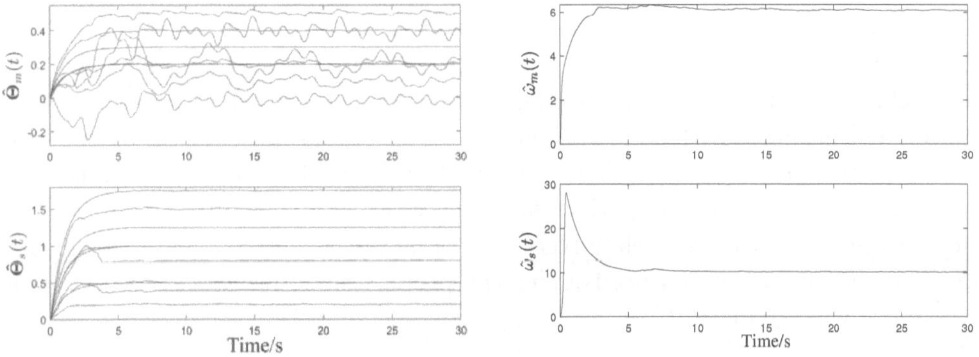

(a) (b)

Figure 9.5 The adaptive estimations. (a) $\hat{\Theta}_j$, (b) $\hat{\omega}_j$.

$\hat{\omega}_j$ are given in Fig. 9.5(a and b). Finally, the control inputs are shown in Fig. 9.6, where the peak value of the control torque $\boldsymbol{\tau}_j$ is less than or equal to 20 N·m. These simulation results show that the propose control algorithm can ensure the stability of the closed-loop control.

The other two sets of simulations will be used to observe the control performances for different control parameters..

On the one hand, from the theoretic analysis in Theorem 9.1, one can choose appropriate control gains such that α_j, $\Upsilon_{j0} - \Upsilon_{j1}$ and/or Υ_{j1} are sufficiently large, to reduce the size of the final region. To verify this comment, by choosing K_{j0}, K_{j1}, K_{j2}, K_{j3}, P_{j1} and ϑ_j to get the larger α_j and $\Upsilon_{j0} - \Upsilon_{j1}$, the second simulation is conducted, where the tracking performance is shown in Fig. 9.7. Compared with Fig. 9.3, one can find the steady tracking errors are reduced by taking larger α_j and $\Upsilon_{j0} - \Upsilon_{j1}$.

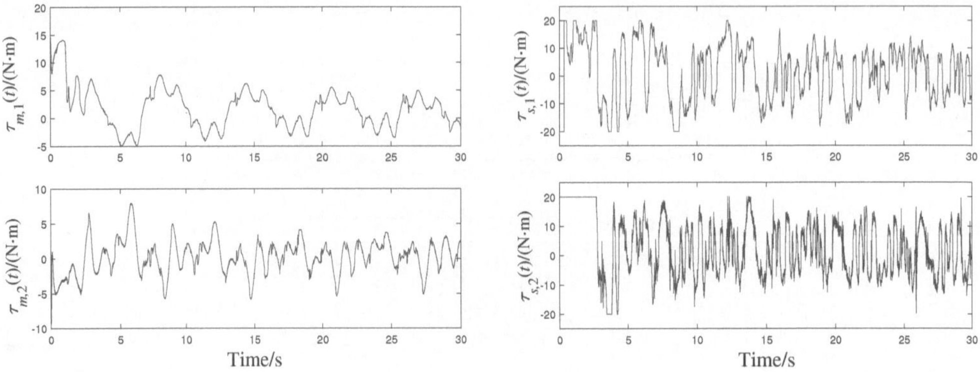

(a) (b)

Figure 9.6 The control inputs $\boldsymbol{\tau}_m(t)$ and $\boldsymbol{\tau}_s(t)$. (a) $\boldsymbol{\tau}_m(t)$, (b) $\boldsymbol{\tau}_s(t)$.

On the other hand, we also verify the discussion of Remark 9.7. From Remark 9.7, the possible steady-state errors caused by $K_{j5}\hat{\boldsymbol{\tau}}_j$ can be reduced by the introduction of filter variable $\boldsymbol{y}_j$ in the controller design. In the third set of simulation, let $\delta = 15^2$, the tracking performance in this scenario is shown in Fig. 9.8. As it is observed, the tracking performance with a small steady-state tracking error can still be guaranteed for a large δ. For comparison, if we remove the filter (9.10) from the design of the controller (9.5), the tracking performance is given in Fig. 9.9. From Figs. 9.7 to 9.9, one can infer that the possible adverse effect on steady- state control performance caused by $K_{j5}\hat{\boldsymbol{\tau}}_j$ can be reduced under the proposed control algorithm, which is in accordance with Remark 9.7.

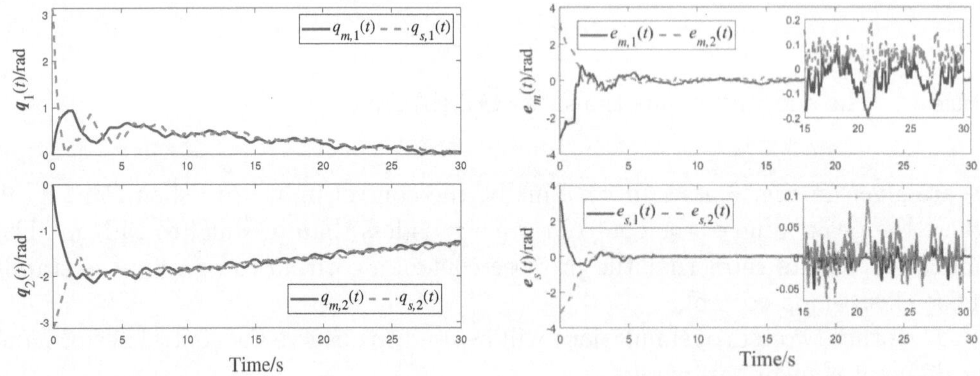

(a) (b)

Figure 9.7 The synchronization performance between the master and slave robots with larger α_j and $\Upsilon_{j0} - \Upsilon_{j1}$. (a) Joint positions. (b) Position errors.

[2]It is much larger than 0.01 in the above two sets of simulations.

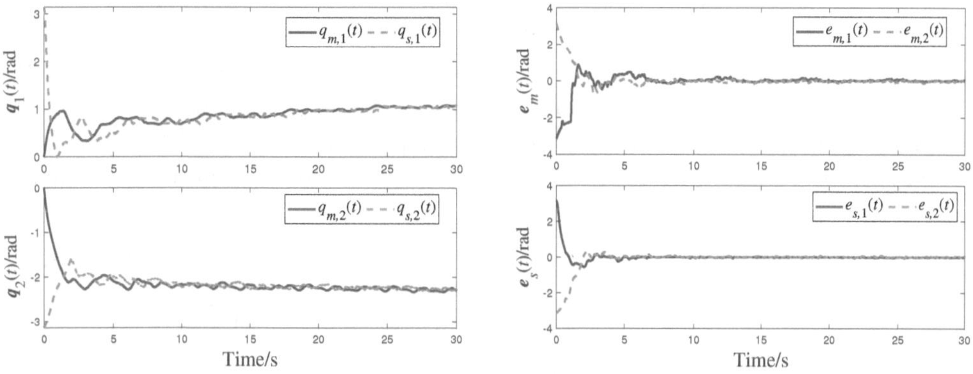

(a) (b)

Figure 9.8 The synchronization performance between the master and slave robots with larger δ. (a) Joint positions. (b) Position errors.

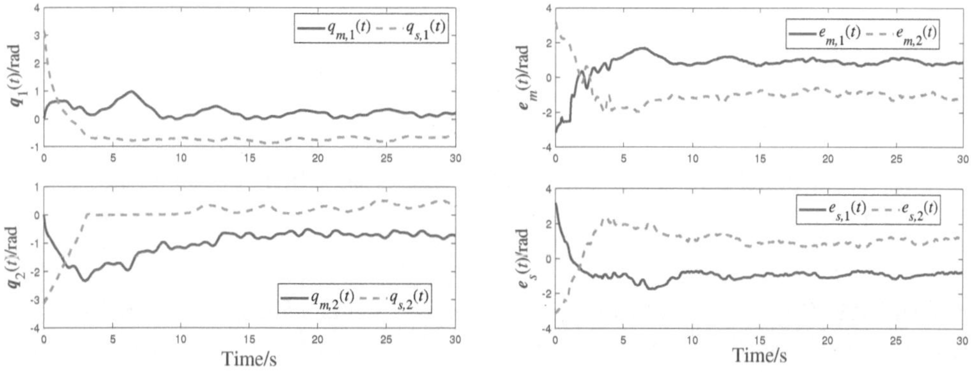

(a) (b)

Figure 9.9 The synchronization performance between the master and slave robots with larger δ when the filter term is removed from controller (9.5). (a) Joint positions. (b) Position errors.

9.5.2 Comparisons with the asymptotic control methods

Hashemzadeh et al. proposed an effective anti-saturation control method for teleoperation system, where the designed nonlinear proportional plus damping controller is given as [18]:

$$\boldsymbol{\tau}_j = \boldsymbol{G}_{jo}(\boldsymbol{q}_j) - P_j(\boldsymbol{e}_j) - K_j\dot{\boldsymbol{q}}_j,\ j \in \{m, s\}$$

where $P_j(\cdot)$ is a special nonlinear function. Compared with the method proposed in this chapter, the Hashemzadeh's control design can only guarantee the stable control of the teleoperation system with nominal dynamic model under the passive operator/environmental external force in theory. In this comparison study, the exogenous forces $\boldsymbol{f}_h^*$ and $\boldsymbol{f}_e^*$ on time interval $[0, 30\text{s}]$ are given in Fig. 9.1, while for $t > 30$ s, $\boldsymbol{f}_h^* = \boldsymbol{f}_e^* \equiv 0$. Then, the tracking performance of Hashemzadeh's method [18] is shown as Fig. 9.10. It is easy to see that the non-zero exogenous forces $\boldsymbol{f}_h^*$ and $\boldsymbol{f}_e^*$ on $[0, 30\text{s}]$ have a great influence on the tracking performance of Hashemzadeh's method [18]. Under the same simulation setup, the control performance of this chapter is given as Fig. 9.3. One can find that the proposed finite-time anti-windup algorithm of this chapter has a faster convergence speed than Hashemzadeh's method [18].

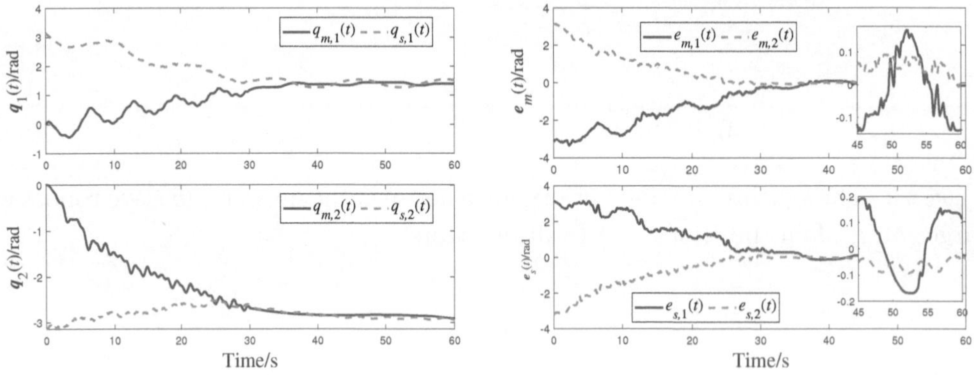

Figure 9.10 The synchronization performance under the Hashemzadeh's algorithm [18]. (a) Joint positions, (b) Position errors.

Remark 9.9. *The position tracking error in this chapter is bounded, while in Hashemzadeh's work [18], it is proved to be able to converge to zero asymptotically. However, there are two basic premises for Hashemzadeh's method [18] to obtain the asymptotic stability.*

① *Firstly, the zero convergence of Hashemzadeh's method [18] can only be obtained when both the master and slave robots run in free motion. In fact, for the non-zero external forces, it can still only guarantee the asymptotic boundedness.*

② *the model uncertainties and external disturbances are not considered by Hashemzadeh's method [18].*

In Hua's work [19], a novel output-feedback based adaptive control framework is developed for teleoperation system with time-varying delays and actuator saturation, where a saturated proportion plus saturated damping controller is designed by using the estimated velocity. In the second study, the comparison study between Hua's work [19] and this chapter is performed. Note that, in Hua's work [19], a fast terminal sliding-mode velocity observer is used to estimate the unknown velocities. To be more fair, the true velocity signal is used to replace the estimated one in Hua's control design [19] in this simulation. Under the same simulation setup, the tracking performance of Hua's method [19] is shown in Fig. 9.11. On the one hand, due to the introduction of adaptive law, the Hua's algorithm [19] has a better control performance than Hashemzadeh's method [18], by comparing Figs. 9.11 and 9.10. Note that, the smaller steady-state errors shown in Fig. 9.10 is mainly because of the external forces f_h^* and f_e^* are set to be zero after time $t > 30$ s. On the other hand, for this chapter and Hua's work [19] that have taken into account the model uncertainty, their control performances have been presented in Figs. 9.11 and 9.3, where the faster convergence speed is obtained in Fig. 9.3. The simulation results show that the finite-time control algorithm proposed in this chapter has better synchronization performance than the traditional control algorithm proposed in Hua's work [19].

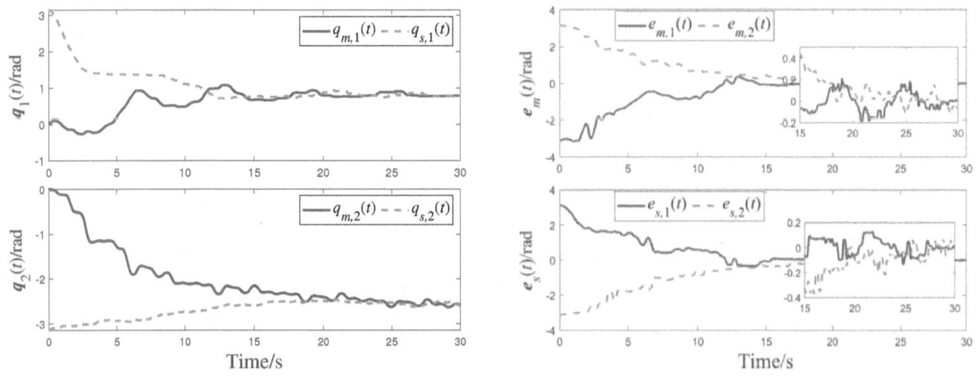

Figure 9.11 The synchronization performance under the Hua's algorithm [19]. (a) Joint positions, (b) Position errors.

In our previous works [9,20], the adaptive controllers based on switching control technique are investigated for teleoperation system with non-passive external forces and varying time delays, which correspond to Chapter 5 and Chapter 4 in this book, respectively. Noting that, in those works [9,20], only the asymptotic boundedness is obtained. In the third comparison study, we will verify whether the new finite-time control algorithm has faster synchronization performance. Under the same simulation setup, the control performance of the adaptive fuzzy controller [9] and the general adaptive controller [20] is shown in Figs. 9.12 and 9.13, respectively. Compared with Fig. 9.3, the proposed novel finite-time anti-windup algorithm still has a faster convergence speed than the previous works [9,20].

To better show the faster convergence speed of this chapter, the comparison results with the works [9,18–20] are unified to Fig. 9.14.

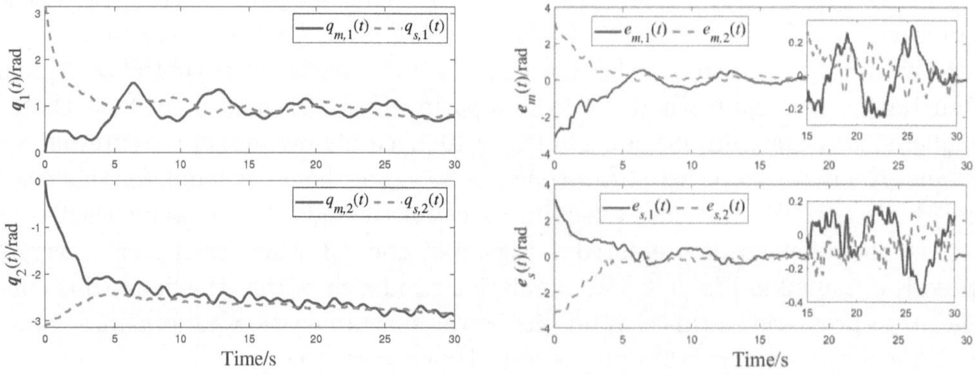

(a) (b)

Figure 9.12 The synchronization performance with our adaptive fuzzy controller [9]. (a) Joint positions, (b) Position errors.

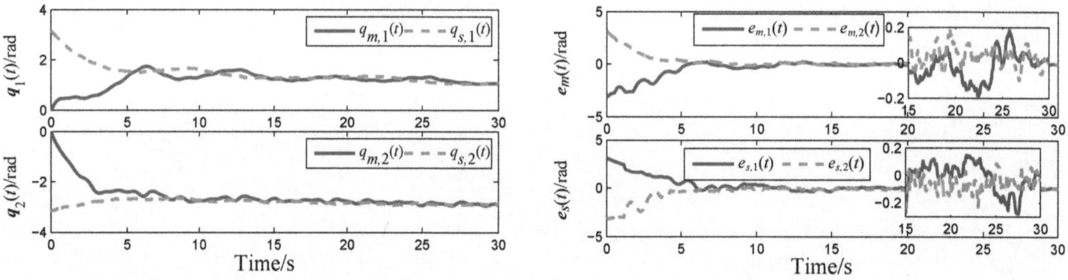

(a) (b)

Figure 9.13 The synchronization performance with our general adaptive controller [20]. (a) Joint positions, (b) Position errors.

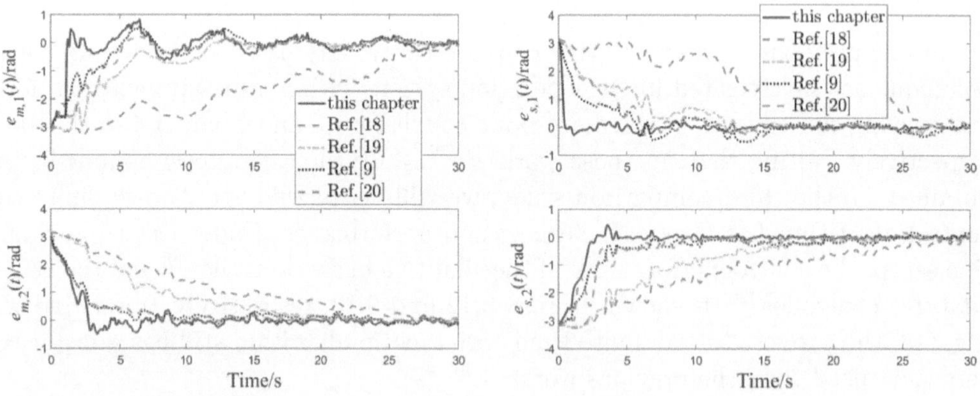

(a) (b)

Figure 9.14 The synchronization performance between this chapter and the works [9, 18–20] with the same simulation setup. (a) Tracking errors at master side, (b) Tracking errors at slave side.

9.5.3 Comparisons with the finite-time control methods

The finite-time control of teleoperation system has been investigated in Yang's works [5,6]. This subsection presents the comparisons between this chapter and those finite-time methods [5,6].

In Yang's work [5], an NFTSM-based finite-time controller is developed for teleoperation system with constant communication delays. To verify the effectiveness, it is assumed that the time delays are $d_m(t) = d_s(t) = 2$, while both the master and slave robots run in free motion. Then under the above simulation setup (the same actuator saturation, the same model uncertainties and disturbances), when we take the control parameters given in Yang's work [5], the simulation results are shown as Figs. 9.15 and 9.16, which give the the synchronization performance and control torque of Yang's work [5]. From Fig. 9.15, the satisfactory synchronization performance is obtained. Compared with Fig. 9.6, however, the serious control chattering is found in Fig. 9.16, which is fairly undesired in practical implementation.

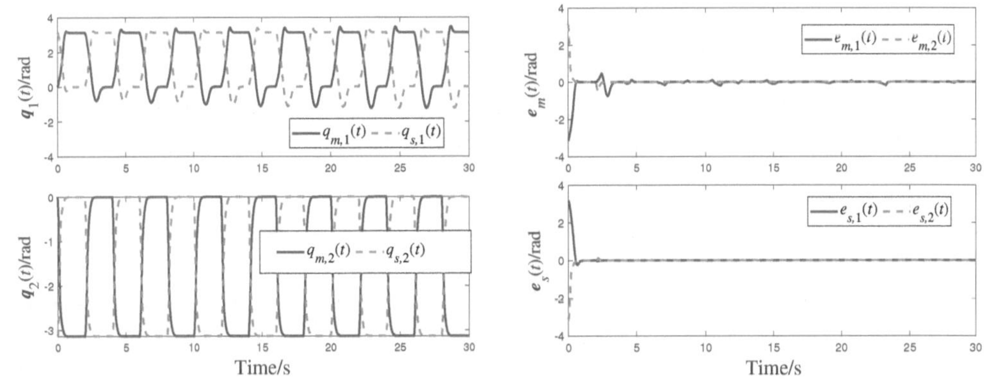

(a) (b)

Figure 9.15 The synchronization performance of Yang's method [5] when the robots run in free motion while the time delays are constant. (a) Joint positions. (b) Position errors.

Remark 9.10. *The main cause for the serious chattering of Fig. 9.16 is the utilization of sign function in Yang's controller* [5], *i.e.,* $\hat{\omega}_j \text{sgn}(s_j) + \xi_j \text{sgn}(s_j)$[3]*. Based on Theorem 2 of Yang's work* [5], ξ_j *should meet* $\xi_j \geq |\tilde{\theta}_j^{\mathrm{T}} \varsigma_j(X_j)| + |\omega_j - \hat{\omega}_j|$.

(1) *From Theorem 1 of Yang's work* [5], *one can only guarantee the boundedness of* $\tilde{\theta}_j$ *and* $\omega_j - \hat{\omega}_j$, *while their upper bounds cannot be obtained. To ensure the finite time convergence,* ξ_j *should take some large values. Hence the high chattering will be induced by* $\xi_j \text{sgn}(s_j)$.

(2) *In addition, the adaptive update law* $\dot{\hat{\omega}}_j = |s_j|$ *is given by eq.(23)* [5]*. Obviously,* $\hat{\omega}_j$ *is non-decreased. Before* $|s_j|$ *becomes zero,* $\hat{\omega}_j$ *may be with some very high*

[3]More details can be seen in eq.(21) [5], where $j \in \{m, s\}$, s_j is the sliding mode surface.

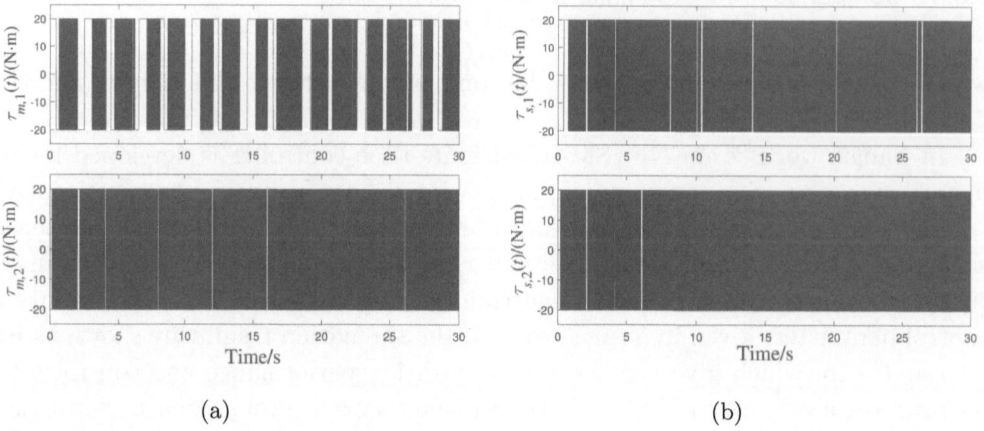

Figure 9.16 The control inputs of Yang's method [5] when the robots run in free motion while the time delays are constant. (a) $\boldsymbol{\tau}_m$, (b) $\boldsymbol{\tau}_s$.

values. In fact, in this set of simulation, the response curves of $\hat{\omega}_m$ *and* $\hat{\omega}_s$ *are given in Fig. 9.17. Obviously, both* $\hat{\omega}_m$ *and* $\hat{\omega}_s$ *have very high values. Thus,* $\hat{\omega}_j\mathrm{sgn}(s_j)$ *will also cause the control chattering.*

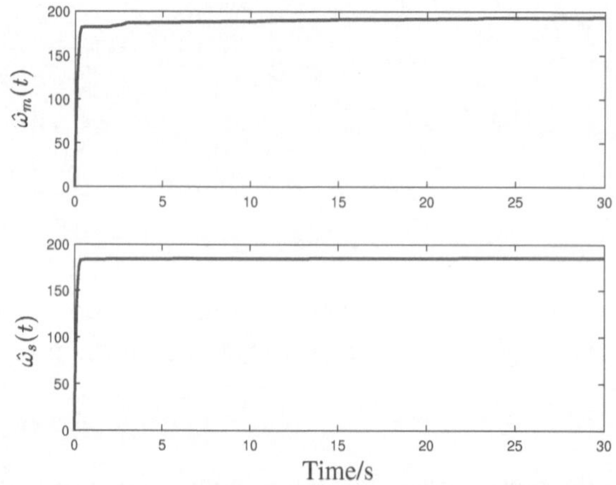

Figure 9.17 Response curves of $\hat{\omega}_m$ and $\hat{\omega}_s$ in Yang's method [5].

To meet the practical implementation, the effectiveness of Yang's algorithm [5] is also verified when the communication delays are time varying or the master/slave robot contacts with the human operator/environment. When the external forces are set to be the same as the ones in above simulations, under the same control setup, the tracking errors are shown in Fig. 9.18(a). When the varying time delays are given by Fig. 9.2, the tracking errors are shown in Fig. 9.18(b). On the one hand, comparing Figs. 9.18(a) and 9.15(b), one can find that the non-zero external forces have an adverse effect on the convergence speed. Therefore, to reduce the conservative, it is necessary to address the external forces for the control and stability analysis of teleoperation system. On the other hand, from Figs. 9.18(b) and 9.15(b), the synchronization performance of Yang's work [5] is also adversely affected by the varying time delays. Although Fig. 9.18(b) shows the closed-loop system is stable, from Fig. 9.3(b), the performance of Yang's work [5] in the time-varying delay scenario is worse than the control of this chapter that has considered the varying delays in the control design.

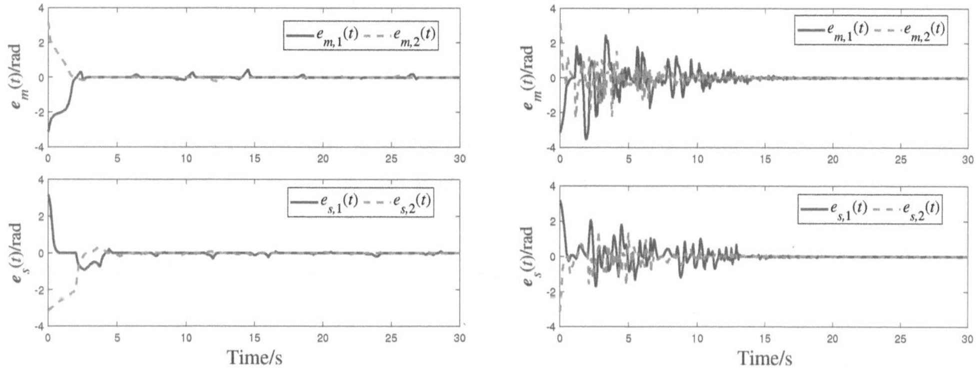

(a) (b)

Figure 9.18 The tracking errors of Yang's method [5] when the communication delays are time varying or the external forces are not zero. (a) The tracking errors with nonzero external forces. (b) The tracking errors with the varying time delays shown in Fig. 9.2.

In another of Yang's work [6], a terminal sliding mode based finite time control method is investigated for teleoperation system with position error constraint. Compared with the finite-time control of this chapter, the transient-state synchronization performance is considered in Yang's work [6] by introducing the position error constraint. However, the communication delays in Yang's work [6] have still been assumed to be constant and the actuator saturation is also not considered. Unfortunately, it is found that the control performance of Yang's meyhod [6] is adversely affected by the varying time delays and actuator saturation. In fact, from Yang's control scheme [6], if we do not consider the capacity limitation of the actuators, when the master and slave robots run in free motion and the communication delays are constant, the simulation results are given as Fig. 9.19, where the satisfactory performance is guaranteed.

For comparisons, when the actuator capacity is limited, we consider the following cases:

CASE 1: The master and slave robots contact with the human operator and the environment while the time delays are constant.

CASE 2: The master and slave robots contact with the human operator and the environment while the communication delays are time varying.

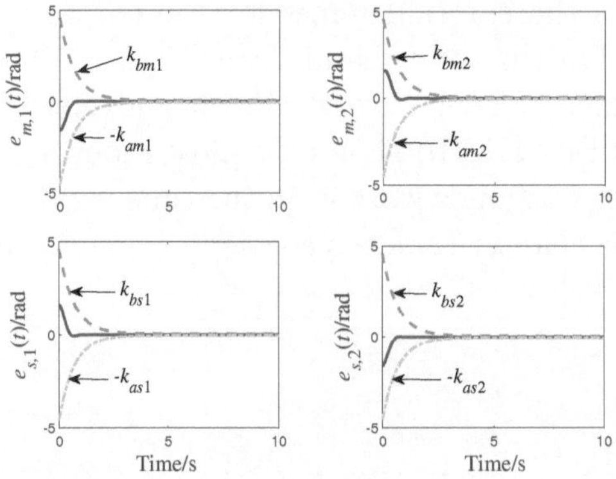

Figure 9.19 The synchronization performance of Yang's method [6] when both the master and slave robots run in free motion and the communication delays are constant.

CASE 3: The master and slave robots run in free motion while the communication delays are time varying.

Under the same control parameters, the comparison results are shown in Fig. 9.20. From Figs. 9.19 and 9.20, one can find that the Yang's control [6] has a poor performance in the actuator saturation case and/or varying time-delay case, which limits its practical application. The improvement of Yang's method [6] will be given in the following chapters of this book.

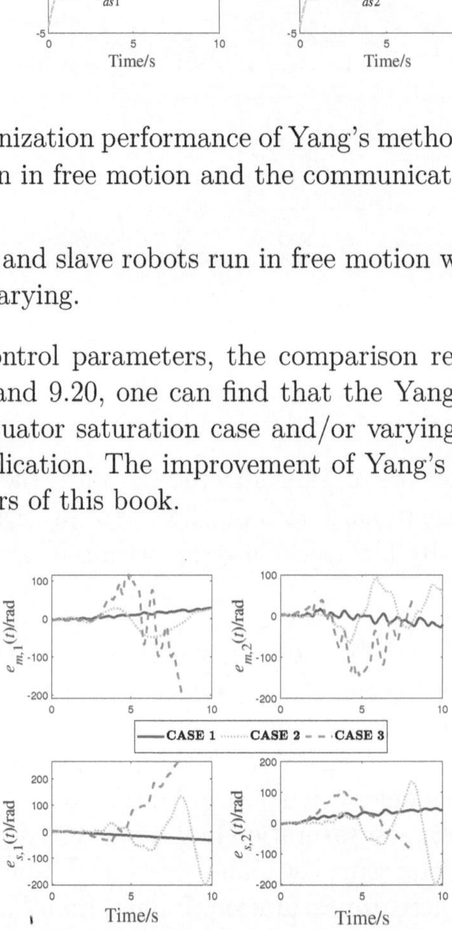

Figure 9.20 The tracking errors of **CASE 1**, **CASE 2** and **CASE 3** under the control of Yang's method [6].

9.6 CONCLUSION

In this chapter, the finite-time tracking control for nonlinear teleoperation systems is addressed, where the model uncertainties, actuator saturation, asymmetric time-varying delays, and passive/non-passive external forces are considered. In the developed control framework, the fuzzy logic system is used to approximate the unknown nonlinear robotic dynamics. To analyze and handle actuator saturation, the anti-windup control framework is adopted, and a modified anti-windup compensator is developed. To study the finite-time control, actuator saturation and asymmetric time-varying delays in a unified framework, a novel non-smooth generalized switched filter is also designed. By the introduction of the designed anti-windup compensator and generalized switched filter in the finite-time adaptive fuzzy control algorithm, the complete closed-loop teleoperation system is modeled as a special switched system, which consists of the master/slave robotic dynamics with the designed finite-time controller, the anti-windup compensator, and the non-smooth generalized switched filter. To cope with the non-zero exogenous forces, the state-independent input-to-output stability framework is used. By using the multiple Lypaunov-Krasovskii functionals method, the complete closed-loop system is SIIOpS and based on this, it is proved to be FTSIIOpS, where the asymptotic convergence of the adaptive estimation errors and the finite-time convergence of the position tracking errors are obtained. Simulation results of the proposed control method and the comparison studies have demonstrated its efficiency.

Bibliography

[1] L. Fang, T. Li, Z. Li, and R. Li, "Adaptive terminal sliding mode control for anti-synchronization of uncertain chaotic systems," *Nonlinear Dynamics*, vol. 74, no. 4, pp. 991–1002, 2013.

[2] Z. Man, A. P. Paplinski, and H. R. Wu, "A robust mimo terminal sliding mode control scheme for rigid robotic manipulators," *IEEE Transactions on Automatic Control*, vol. 39, no. 12, pp. 2464–2469, 1994.

[3] S. Yu, X. Yu, B. Shirinzadeh, and Z. Man, "Continuous finite-time control for robotic manipulators with terminal sliding mode," *Automatica*, vol. 41, no. 11, pp. 1957–1964, 2005.

[4] L. Wang, T. Chai, and L. Zhai, "Neural-network-based terminal sliding-mode control of robotic manipulators including actuator dynamics," *IEEE Transactions on Industrial Electronics*, vol. 56, no. 9, pp. 3296–3304, 2009.

[5] Y. Yang, C. Hua, and X. Guan, "Adaptive fuzzy finite-time coordination control for networked nonlinear bilateral teleoperation system," *IEEE Transactions on Fuzzy Systems*, vol. 22, no. 3, pp. 631–641, 2013.

[6] Y. Yang, C. Hua, and X. Guan, "Finite time control design for bilateral teleoperation system with position synchronization error constrained," *IEEE Transactions on Cybernetics*, vol. 46, no. 3, pp. 609–619, 2015.

[7] M. Chen, S. S. Ge, and B. Ren, "Adaptive tracking control of uncertain mimo nonlinear systems with input constraints," *Automatica*, vol. 47, no. 3, pp. 452–465, 2011.

[8] Y. Yang, C. Ge, H. Wang, X. Li, and C. Hua, "Adaptive neural network based prescribed performance control for teleoperation system under input saturation," *Journal of the Franklin Institute*, vol. 352, no. 5, pp. 1850–1866, 2015.

[9] D.-H. Zhai and Y. Xia, "Adaptive fuzzy control of multilateral asymmetric teleoperation for coordinated multiple mobile manipulators," *IEEE Transactions on Fuzzy Systems*, vol. 24, no. 1, pp. 57–70, 2015.

[10] A. Haddadi, K. Razi, and K. Hashtrudi-Zaad, "Operator dynamics consideration for less conservative coupled stability condition in bilateral teleoperation," *IEEE/ASME Transactions on Mechatronics*, vol. 20, no. 5, pp. 2463–2475, 2015.

[11] H. Gomi and R. Osu, "Task-dependent viscoelasticity of human multijoint arm and its spatial characteristics for interaction with environments," *Journal of Neuroscience*, vol. 18, no. 21, pp. 8965–8978, 1998.

[12] P. Malysz and S. Sirouspour, "Nonlinear and filtered force/position mappings in bilateral teleoperation with application to enhanced stiffness discrimination," *IEEE Transactions on Robotics*, vol. 25, no. 5, pp. 1134–1149, 2009.

[13] T. Tsuji, P. G. Morasso, K. Goto, and K. Ito, "Human hand impedance characteristics during maintained posture," *Biological Cybernetics*, vol. 72, no. 6, pp. 475–485, 1995.

[14] E. Nuno, L. Basanez, and R. Ortega, "Passivity-based control for bilateral teleoperation: a tutorial," *Automatica*, vol. 47, no. 3, pp. 485–495, 2011.

[15] T. H. Lee and C. J. Harris, *Adaptive Neural Network Control of Robotic Manipulators*. World Scientific, 1998.

[16] Y. Zhao and C. C. Cheah, "Neural network control of multifingered robot hands using visual feedback," *IEEE Transactions on Neural Networks*, vol. 20, no. 5, pp. 758–767, 2009.

[17] E. Nuno, R. Ortega, and L. Basanez, "An adaptive controller for nonlinear teleoperators," *Automatica*, vol. 46, no. 1, pp. 155–159, 2010.

[18] F. Hashemzadeh, I. Hassanzadeh, and M. Tavakoli, "Teleoperation in the presence of varying time delays and sandwich linearity in actuators," *Automatica*, vol. 49, no. 9, pp. 2813–2821, 2013.

[19] C. Hua, Y. Yang, and P. X. Liu, "Output-feedback adaptive control of networked teleoperation system with time-varying delay and bounded inputs," *IEEE/ASME Transactions on Mechatronics*, vol. 20, no. 5, pp. 2009–2020, 2014.

[20] D.-H. Zhai and Y. Xia, "Adaptive control for teleoperation system with varying time delays and input saturation constraints," *IEEE Transactions on Industrial Electronics*, vol. 63, no. 11, pp. 6921–6929, 2016.

VI

Prescribed-performance Teleoperation

CHAPTER 10

Prescribed performance task-space teleoperation control

SINCE the finite time control can only improve the convergence speed and do nothing to other transient state performance index like overshoot, this chapter provides a prescribed performance control framework for teleoperation system to guarantee the given transient-state and steady-state synchronization performances between the master and slave robots, which simultaneously handles the unknown kinematics/dynamics, asymmetric varying time delays, and prescribed performance control in a unified switched control framework.

10.1 INTRODUCTION

Networked teleoperated robot control is one of the hot research topics in the field of robotic system and control since the beginning of the 21st century [1,2]. In the past few decades, related control theory research emerges one after another, such as time-delay stability methods [3–13], model uncertainty methods [14–19], non-passive external forces [20–23], actuator saturation [24,25] et al. The above research obtains asymptotic or bounded steady-state synchronization performance while ensuring closed-loop stability. However, for many control systems, in addition to the steady-state performance requirements, transient performance such as overshoot and convergence time are also considered. For example, in a tele-surgery robot system, the robot is required to have a satisfactory control speed and higher precision to ensure the health status of the patient. Obviously, the above studies are not sufficient to deal with such control needs.

In order to satisfy the simultaneous control of the transient and steady-state performance of the system, researchers have also proposed many control methods [26–30]. Among them, a simple and effective method is predefined performance

DOI: 10.1201/9781003382058-10

control, including PPC method [29,31] and BLF strategy [30,32]. Under the predefined performance control strategy, by transforming the corresponding predefined performance requirements into the controller design, it can achieve the high-performance control requirements that the tracking error converges to any specified steady-state region, the convergence speed is not less than a predefined value, and the maximum overshoot is not greater than a pre-specified sufficiently small value. In recent years, predefined performance control strategies have also been used for time-delay nonlinear systems. Yang et al. studied the adaptive predefined performance control of teleoperation systems under constant communication time delay [33–35]. Among them, a prescribed performance controller is investigated for the teleoperation system [34]. The design problem of prescribed performance control under input saturation constraints is also considered [33]. Based on the BLF strategy, Yang et al. study the finite-time predefined performance control of teleoperation systems [35]. However, considering that the existing designs [33–35] are only for constant communication delay, and it is assumed that the master and slave robots are not affected by external forces (i.e., in the case of free motion), they still face limitations in practical use. Therefore, it is necessary to study the new predefined performance control of teleoperation system in complex network environment.

In view of the fact that in the actual teleoperation system, the contact between the robot and the operator/environment appears more on the end effector, task space control has gradually become one of the research hotspots in the teleoperation system control in recent years [4,11,36–38]. Compared with joint space control, task space control allows the master and slave robots to have different structures, which brings a lot of convenience to practical engineering applications. In the existing task space control of robots, a common assumption is that the kinematics and Jacobian matrices of the robot from joint space to task space are precisely known, which introduces conservatism to the control. In fact, the robot needs to be in contact with the environment in the actual working process, and its structural parameters may change with different tasks due to factors such as deformation [39]. In recent years, some researchers have introduced adaptive Jacobian control technology into task space control, which has solved this problem well, for example, the robotic systems [40,41] and the teleoperation systems [42], etc. However, since they do not consider the communication delay or only consider the constant time delay, it is difficult to directly generalize to the teleoperation system with time-varying communication delay.

Based on the PPC strategy, this chapter studies the task space prescribed performance control of teleoperation systems. Based on the auxiliary switched filter control method, a new prescribed performance controller is designed for teleoperation system. In the presence of time-varying communication delay and the robot is in contact with the operator/environment, the synchronous control of the prescribed performance of the teleoperation system is realized, which further expands the application scope of the existing theory.

10.2 PROBLEM FORMULATION

10.2.1 Dynamics of robots

In this chapter, the teleoperation system consists of a pair of robots with revolute joints, and the dynamic model is as follows:

$$\begin{cases} \boldsymbol{M}_m(\boldsymbol{q}_m)\ddot{\boldsymbol{q}}_m + \boldsymbol{C}_m(\boldsymbol{q}_m,\dot{\boldsymbol{q}}_m)\dot{\boldsymbol{q}}_m + \boldsymbol{g}_m(\boldsymbol{q}_m) + \boldsymbol{f}_m(\boldsymbol{q}_m,\dot{\boldsymbol{q}}_m) = \boldsymbol{J}_m^{\mathrm{T}}(\boldsymbol{q}_m)\boldsymbol{f}_h + \boldsymbol{\tau}_m \\ \boldsymbol{M}_s(\boldsymbol{q}_s)\ddot{\boldsymbol{q}}_s + \boldsymbol{C}_s(\boldsymbol{q}_s,\dot{\boldsymbol{q}}_s)\dot{\boldsymbol{q}}_s + \boldsymbol{g}_s(\boldsymbol{q}_s) + \boldsymbol{f}_s(\boldsymbol{q}_s,\dot{\boldsymbol{q}}_s) = -\boldsymbol{J}_s^{\mathrm{T}}(\boldsymbol{q}_s)\boldsymbol{f}_e + \boldsymbol{\tau}_s \end{cases} \tag{10.1}$$

where $j \in \{m,s\}$, $\boldsymbol{q}_j \in \mathbb{R}^{n_j}$, $\boldsymbol{M}_j(\boldsymbol{q}_j)$, $\boldsymbol{C}_j(\boldsymbol{q}_j,\dot{\boldsymbol{q}}_j) \in \mathbb{R}^{n_j \times n_j}$, $\boldsymbol{g}_j(\boldsymbol{q}_j) \in \mathbb{R}^{n_j}$, $\boldsymbol{f}_j(\boldsymbol{q}_j,\dot{\boldsymbol{q}}_j) \in \mathbb{R}^{n_j}$, $\boldsymbol{\tau}_j \in \mathbb{R}^{n_j}$, $\boldsymbol{J}_j(\boldsymbol{q}_j) \in \mathbb{R}^{n \times n_j}$, $\boldsymbol{f}_h \in \mathbb{R}^n$, $\boldsymbol{f}_e \in \mathbb{R}^n$; $\boldsymbol{f}_j(\boldsymbol{q}_j,\dot{\boldsymbol{q}}_j)$ is the robot's disturbance uncertainty and friction force vector; and the other parameters are defined in (2.11).

The generalized end-effector positions $\boldsymbol{X}_m \in \mathbb{R}^n$ and $\boldsymbol{X}_s \in \mathbb{R}^n$ of the master and slave robots in task space can be given, respectively, by

$$\boldsymbol{X}_j = \boldsymbol{h}_j(\boldsymbol{q}_j),\ \dot{\boldsymbol{X}}_j = \boldsymbol{J}_j(\boldsymbol{q}_j)\dot{\boldsymbol{q}}_j,\ j \in \{m,s\}$$

Similar to Haddadi's work [43], the human operator and environment dynamics are assumed to be

$$\begin{cases} \boldsymbol{f}_h = \boldsymbol{f}_h^*(t) - \boldsymbol{\phi}_h(\boldsymbol{q}_m,\dot{\boldsymbol{q}}_m,\ddot{\boldsymbol{q}}_m) \\ \boldsymbol{f}_e = \boldsymbol{f}_e^*(t) + \boldsymbol{\phi}_e(\boldsymbol{q}_s,\dot{\boldsymbol{q}}_s,\ddot{\boldsymbol{q}}_s) \end{cases} \tag{10.2}$$

where $\boldsymbol{\phi}_h(\boldsymbol{q}_m,\dot{\boldsymbol{q}}_m,\ddot{\boldsymbol{q}}_m)$ and $\boldsymbol{\phi}_e(\boldsymbol{q}_s,\dot{\boldsymbol{q}}_s,\ddot{\boldsymbol{q}}_s)$ represent, respectively, the inertial and viscoelastic properties of the human arm and the environment, which can be a linear or nonlinear passive dynamic function with respect to their arguments, $\boldsymbol{f}_h^*$ and $\boldsymbol{f}_e^*$ are the operator's and the environment's exogenous force. Due to the existences of $\boldsymbol{f}_h^*$ and $\boldsymbol{f}_e^*$, the operator/environment forces $\boldsymbol{f}_h$ and $\boldsymbol{f}_e$ may be non-passive. It is also assumed that:

Assumption 10.1. *The exogenous forces $\boldsymbol{f}_h^*$ and $\boldsymbol{f}_e^*$ are assumed to be locally essentially bounded.*

Assumption 10.2. *There exist non-negative constants k_{j1}, k_{j2}, k_{j3} and k_{j4}, $j \in \{m,s\}$, such that*

$$|\boldsymbol{\phi}_h(\boldsymbol{q}_m,\dot{\boldsymbol{q}}_m,\ddot{\boldsymbol{q}}_m)| \le k_{m1}|\ddot{\boldsymbol{q}}_m| + k_{m2}|\dot{\boldsymbol{q}}_m|^2 + k_{m3}|\dot{\boldsymbol{q}}_m| + k_{m4}$$
$$|\boldsymbol{\phi}_e(\boldsymbol{q}_s,\dot{\boldsymbol{q}}_s,\ddot{\boldsymbol{q}}_s)| \le k_{s1}|\ddot{\boldsymbol{q}}_s| + k_{s2}|\dot{\boldsymbol{q}}_s|^2 + k_{s3}|\dot{\boldsymbol{q}}_s| + k_{s4}$$

Remark 10.1. *Assumption 10.2 is reasonable. For example, in many previous works [19, 44], the passive functions are modeled by some linear impedance models, i.e., $\boldsymbol{\phi}_h(\boldsymbol{q}_m,\dot{\boldsymbol{q}}_m,\ddot{\boldsymbol{q}}_m) = \boldsymbol{M}_m\ddot{\boldsymbol{X}}_m + \boldsymbol{B}_m\dot{\boldsymbol{X}}_m + \boldsymbol{K}_m\boldsymbol{X}_m$ and $\boldsymbol{\phi}_e(\boldsymbol{q}_s,\dot{\boldsymbol{q}}_s,\ddot{\boldsymbol{q}}_s) = \boldsymbol{M}_s\ddot{\boldsymbol{X}}_s + \boldsymbol{B}_s\dot{\boldsymbol{X}}_s + \boldsymbol{K}_s\boldsymbol{X}_s$. Then for al $j \in \{m,s\}$, the matrices $\boldsymbol{M}_j$, $\boldsymbol{B}_j$ and $\boldsymbol{K}_j$ represent the hand/environment inertia, damping, and stiffness, respectively, which may be time varying but bounded. Given that the definitions of $\boldsymbol{h}_j(\boldsymbol{q}_j)$ and $\boldsymbol{J}_j(\boldsymbol{q}_j)$, without loss of generality, it is assumed that $|\dot{\boldsymbol{J}}_j(\boldsymbol{q}_j)\dot{\boldsymbol{q}}_j| \le k_{j0}|\dot{\boldsymbol{q}}_j|^2$, where $k_{j0} \ge 0$. Therefore one can obtain that $|\boldsymbol{\phi}_h(\boldsymbol{q}_m,\dot{\boldsymbol{q}}_m,\ddot{\boldsymbol{q}}_m)| \le k_{m1}|\ddot{\boldsymbol{q}}_m| + k_{m2}|\dot{\boldsymbol{q}}_m|^2 + k_{m3}|\dot{\boldsymbol{q}}_m| + k_{m4}$ and $|\boldsymbol{\phi}_e(\boldsymbol{q}_s,\dot{\boldsymbol{q}}_s,\ddot{\boldsymbol{q}}_s)| \le k_{s1}|\ddot{\boldsymbol{q}}_s| + k_{s2}|\dot{\boldsymbol{q}}_s|^2 + k_{s3}|\dot{\boldsymbol{q}}_s| + k_{s4}$. See the simulation example later for more details.*

In practices, it is difficult to obtain the exact dynamics of a robot. In what follows, let $\boldsymbol{M}_j(\cdot) = \boldsymbol{M}_{jo}(\cdot) + \Delta \boldsymbol{M}_j(\cdot)$, $\boldsymbol{C}_j(\cdot,\cdot) = \boldsymbol{C}_{jo}(\cdot,\cdot) + \Delta \boldsymbol{C}_j(\cdot,\cdot)$, $\boldsymbol{g}_j(\cdot) = \boldsymbol{g}_{jo}(\cdot) + \Delta \boldsymbol{g}_j(\cdot)$, where $\boldsymbol{M}_{jo}(\cdot)$, $\boldsymbol{C}_{jo}(\cdot,\cdot)$ and $\boldsymbol{g}_{jo}(\cdot)$ are the nominal values of $\boldsymbol{M}_j(\cdot)$, $\boldsymbol{C}_j(\cdot,\cdot)$ and $\boldsymbol{g}_j(\cdot)$, respectively, $\Delta \boldsymbol{M}_j(\cdot)$, $\Delta \boldsymbol{C}_j(\cdot,\cdot)$ and $\Delta \boldsymbol{g}_j(\cdot)$ represent the corresponding uncertainties. Then system (10.1) can be rewritten as

$$\begin{cases} \boldsymbol{M}_{mo}(\boldsymbol{q}_m)\ddot{\boldsymbol{q}}_m + \boldsymbol{C}_{mo}(\boldsymbol{q}_m, \dot{\boldsymbol{q}}_m)\dot{\boldsymbol{q}}_m + \boldsymbol{g}_{mo}(\boldsymbol{q}_m) + \boldsymbol{\tau}_{dm} = \boldsymbol{J}_m^{\mathrm{T}}(\boldsymbol{q}_m)\boldsymbol{f}_h^* + \boldsymbol{\tau}_m \\ \boldsymbol{M}_{so}(\boldsymbol{q}_s)\ddot{\boldsymbol{q}}_s + \boldsymbol{C}_{so}(\boldsymbol{q}_s, \dot{\boldsymbol{q}}_s)\dot{\boldsymbol{q}}_s + \boldsymbol{g}_{so}(\boldsymbol{q}_s) + \boldsymbol{\tau}_{ds} = -\boldsymbol{J}_s^{\mathrm{T}}(\boldsymbol{q}_s)\boldsymbol{f}_e^* + \boldsymbol{\tau}_s \end{cases} \tag{10.3}$$

where

$$\begin{aligned} \boldsymbol{\tau}_{dm} &= \Delta \boldsymbol{M}_m(\boldsymbol{q}_m)\ddot{\boldsymbol{q}}_m + \Delta \boldsymbol{C}_m(\boldsymbol{q}_m, \dot{\boldsymbol{q}}_m)\dot{\boldsymbol{q}}_m + \Delta \boldsymbol{g}_m(\boldsymbol{q}_m) + \\ &\qquad \boldsymbol{f}_m(\boldsymbol{q}_m, \dot{\boldsymbol{q}}_m) + \boldsymbol{J}_m^{\mathrm{T}}(\boldsymbol{q}_m)\boldsymbol{\phi}_h(\boldsymbol{q}_m, \dot{\boldsymbol{q}}_m, \ddot{\boldsymbol{q}}_m) \\ \boldsymbol{\tau}_{ds} &= \Delta \boldsymbol{M}_s(\boldsymbol{q}_s)\ddot{\boldsymbol{q}}_s + \Delta \boldsymbol{C}_s(\boldsymbol{q}_s, \dot{\boldsymbol{q}}_s)\dot{\boldsymbol{q}}_s + \Delta \boldsymbol{g}_s(\boldsymbol{q}_s) + \\ &\qquad \boldsymbol{f}_s(\boldsymbol{q}_s, \dot{\boldsymbol{q}}_s) + \boldsymbol{J}_s^{\mathrm{T}}(\boldsymbol{q}_s)\boldsymbol{\phi}_h(\boldsymbol{q}_s, \dot{\boldsymbol{q}}_s, \ddot{\boldsymbol{q}}_s) \end{aligned}$$

In this chapter, the unknown dynamics (may be non-parametric) is assumed to be:

Assumption 10.3. *For any $j \in \{m, s\}$, there exist non-negative constants k_{j5}, k_{j6}, k_{j7} and k_{j8}, such that*

$$\begin{aligned} &|\Delta \boldsymbol{M}_j(\boldsymbol{q}_j)\ddot{\boldsymbol{q}}_j + \Delta \boldsymbol{C}_j(\boldsymbol{q}_j, \dot{\boldsymbol{q}}_j)\dot{\boldsymbol{q}}_j + \Delta \boldsymbol{g}_j(\boldsymbol{q}_j) + \boldsymbol{f}_j(\boldsymbol{q}_j, \dot{\boldsymbol{q}}_j)| \\ &\qquad \le k_{j5}|\ddot{\boldsymbol{q}}_j| + k_{j6}|\dot{\boldsymbol{q}}_j|^2 + k_{j7}|\dot{\boldsymbol{q}}_j| + k_{j8} \end{aligned}$$

For all $j \in \{m, s\}$, the robotic system of the form (10.3) satisfies Property 2.1 and Property 2.4, and it also has

Property 10.1. [45] *For any $\boldsymbol{x} \in \mathbb{R}^{n_j}$, $\boldsymbol{Y}_j(\boldsymbol{q}_j, \boldsymbol{x})\boldsymbol{\theta}_j = \boldsymbol{J}_j(\boldsymbol{q}_j)\boldsymbol{x}$, where $\boldsymbol{Y}_j(\boldsymbol{q}_j, \boldsymbol{x}) \in \mathbb{R}^{n_j \times k_j}$ is the matrix of known functions called regressor, and $\boldsymbol{\theta}_j \in \mathbb{R}^{k_j}$ is the vector of unknown parameters.*

For the robots with revolute joints, without loss of generality, there exists $\bar{k}_{j0} \ge 0$ such that $|\boldsymbol{J}_j^{\mathrm{T}}(\boldsymbol{q}_j)\boldsymbol{x}| \le \bar{k}_{j0}|\boldsymbol{x}|$ for all $\boldsymbol{x} \in \mathbb{R}^n$. Therefore from Assumption 10.2 and Assumption 10.3, for all $j \in \{m, s\}$,

$$\begin{aligned} |\boldsymbol{\tau}_{dj}| \le (\bar{k}_{j0}k_{j1} + k_{j5})|\ddot{\boldsymbol{q}}_j| + (\bar{k}_{j0}k_{j2} + k_{j6})|\dot{\boldsymbol{q}}_j|^2 + \\ (\bar{k}_{j0}k_{j3} + k_{j7})|\dot{\boldsymbol{q}}_j| + (\bar{k}_{j0}k_{j4} + k_{j8}) \end{aligned}$$

For simplicity, denote $\omega_{j1} = \bar{k}_{j0}k_{j1} + k_{j5}$, $\omega_{j2} = \bar{k}_{j0}k_{j2} + k_{j6}$, $\omega_{j3} = \bar{k}_{j0}k_{j3} + k_{j7}$, $\omega_{j4} = \bar{k}_{j0}k_{j4} + k_{j8}$, which are unknown.

10.2.2 Control objectives

Denote $d_m(t)$ and $d_s(t)$ be the time delays from the master robot to the slave robot and from the slave robot to the master robot, respectively. Thus, the tracking errors can be defined as

$$\begin{cases} \boldsymbol{e}_m(t) = \boldsymbol{X}_m(t) - \boldsymbol{X}_s(t - d_s(t)) \\ \boldsymbol{e}_s(t) = \boldsymbol{X}_s(t) - \boldsymbol{X}_m(t - d_m(t)) \end{cases} \tag{10.4}$$

In the following text, the communication delays are assumed to be bounded and have bounded rates of change, i.e.,

Assumption 10.4. *There exist constants* $0 < \bar{d} < \infty$ *and* $0 \leq \tilde{d} < \infty$*, such that for all* $j \in \{m, s\}$*,* $0 \leq d_j(t) \leq \bar{d}$ *and* $|\dot{d}_j(t)| \leq \tilde{d}$*.*

To achieve some superior synchronization performances, the tracking error $\boldsymbol{e}_j = [e_{j,1}, e_{j,2}, \cdots, e_{j,n}]^{\mathrm{T}}$ is desired to satisfy the following performance constraints:

$$\begin{cases} -H_{ji}\varrho_{j,i}(t) < e_{j,i}(t) < \varrho_{j,i}(t), & e_{j,i}(0) \geq 0 \\ -\varrho_{j,i}(t) < e_{j,i}(t) < H_{ji}\varrho_{j,i}(t), & e_{j,i}(0) < 0 \end{cases} \tag{10.5}$$

where $i = 1, 2, \cdots, n$, $H_{ji} \in [0, 1]$; $\varrho_{j,i}(t)$ is a bounded smooth strictly positive decreasing function satisfying $\varrho_{j,i}(t) = (\varrho_{j,i0} - \varrho_{j,i\infty})\mathrm{e}^{-l_{ji}t} + \varrho_{j,i\infty}$, with $l_{ji} > 0$. The constant $\varrho_{j,i0} = \varrho_{j,i}(0)$ is selected such that (10.5) is satisfied at $t = 0$, i.e.,

$$\begin{cases} \varrho_{j,i0} > e_{j,i}(0), & e_{j,i}(0) \geq 0 \\ \varrho_{j,i0} > -e_{j,i}(0), & e_{j,i}(0) < 0 \end{cases}$$

$\varrho_{j,i\infty} = \lim_{t\to\infty} \varrho_{j,i}(t) > 0$ represents the maximum allowable size of $e_{j,i}(t)$ at the steady state and can be set arbitrarily small. The prespecified parameter value l_{ji} gives the smallest convergence speed of error $e_{j,i}$; the maximum overshoot of $e_{j,i}$ is prescribed less than $H_{ji}\varrho_{j,i}(0)$.

The control objective is that: for all $j \in \{m, s\}$, designing control torque $\boldsymbol{\tau}_j$ to stabilize the tracking error $\boldsymbol{e}_j(t)$, while ensuring the synchronization satisfy the desired performance (10.5).

10.3 ADAPTIVE CONTROLLER

This section provides the specific control design.

10.3.1 PPC strategy

To get the prescribed performance (10.5), the PPC scheme is employed. Under the PPC-based framework, according to the error $\boldsymbol{e}_j$ and its expected performance function $\boldsymbol{\varrho}_j = [\varrho_{j,1}, \varrho_{j,2}, \cdots, \varrho_{j,n}]^{\mathrm{T}}$, we first introduce the error transformation $\bar{\boldsymbol{e}}_j = [\bar{e}_{j,1}, \bar{e}_{j,2}, \cdots, \bar{e}_{j,n}]^{\mathrm{T}}$:

$$\bar{e}_{j,i}(t) = T_{j,i}\left(\frac{e_{j,i}(t)}{\varrho_{j,i}(t)}\right) = \begin{cases} \ln\left(\dfrac{H_{ji} + \dfrac{e_{j,i}(t)}{\varrho_{j,i}(t)}}{1 - \dfrac{e_{j,i}(t)}{\varrho_{j,i}(t)}}\right), & e_{j,i}(0) \geq 0 \\ \ln\left(\dfrac{1 + \dfrac{e_{j,i}(t)}{\varrho_{j,i}(t)}}{H_{ji} - \dfrac{e_{j,i}(t)}{\varrho_{j,i}(t)}}\right), & e_{j,i}(0) < 0 \end{cases}$$

where $i = 1, 2, \cdots, n$.

According to the PPC method, the desired performance of the error $\boldsymbol{e}_j(t)$ can be achieved by guaranteeing the boundedness of $\bar{\boldsymbol{e}}_j(t)$.

The derivative of the error transformation with respect to time can be written as

$$\dot{\bar{e}}_{j,i}(t) = \frac{\partial T_{j,i}\left(\dfrac{e_{j,i}(t)}{\varrho_{j,i}(t)}\right)}{\partial \dfrac{e_{j,i}(t)}{\varrho_{j,i}(t)}} \frac{1}{\varrho_{j,i}(t)} \left(\dot{e}_{j,i}(t) - \frac{\dot{\varrho}_{j,i}(t)}{\varrho_{j,i}(t)} e_{j,i}(t)\right)$$

Since the real-time change rate of the communication delay is difficult to obtain, in order to facilitate the design and analysis, the following auxiliary variables are introduced. Let $\bar{\boldsymbol{e}}^1_{vj} = [\bar{e}^1_{vj,1}, \bar{e}^1_{vj,2}, \cdots, \bar{e}^1_{vj,n}]^{\mathrm{T}}$ and $\bar{\boldsymbol{e}}^2_{vj} = [\bar{e}^2_{vj,1}, \bar{e}^2_{vj,2}, \cdots, \bar{e}^2_{vj,n}]^{\mathrm{T}}$, with

$$\bar{e}^1_{vj,i}(t) = -\frac{\partial T_{j,i}\left(\dfrac{e_{j,i}(t)}{\varrho_{j,i}(t)}\right)}{\partial \dfrac{e_{j,i}(t)}{\varrho_{j,i}(t)}} \frac{1}{\varrho_{j,i}(t)} \frac{\dot{\varrho}_{j,i}(t)}{\varrho_{j,i}(t)} e_{j,i}(t)$$

$$\bar{e}^2_{vj,i}(t) = \frac{\partial T_{j,i}\left(\dfrac{e_{j,i}(t)}{\varrho_{j,i}(t)}\right)}{\partial \dfrac{e_{j,i}(t)}{\varrho_{j,i}(t)}} \frac{1}{\varrho_{j,i}(t)} \dot{e}_{j,i}(t)$$

Then $\dot{\bar{\boldsymbol{e}}}_j(t) = \bar{\boldsymbol{e}}^1_{vj}(t) + \bar{\boldsymbol{e}}^2_{vj}(t)$. Denote

$$\boldsymbol{\mathcal{T}}_j(t) = \mathrm{diag}\left[\frac{\partial T_{j,1}\left(\dfrac{e_{j,1}(t)}{\varrho_{j,1}(t)}\right)}{\partial \dfrac{e_{j,1}(t)}{\varrho_{j,1}(t)}} \frac{1}{\varrho_{j,1}(t)}, \frac{\partial T_{j,2}\left(\dfrac{e_{j,2}(t)}{\varrho_{j,2}(t)}\right)}{\partial \dfrac{e_{j,2}(t)}{\varrho_{j,2}(t)}} \frac{1}{\varrho_{j,2}(t)}, \cdots, \frac{\partial T_{j,n}\left(\dfrac{e_{j,n}(t)}{\varrho_{j,n}(t)}\right)}{\partial \dfrac{e_{j,n}(t)}{\varrho_{j,n}(t)}} \frac{1}{\varrho_{j,n}(t)}\right]$$

then $\bar{\boldsymbol{e}}^2_{vj} = \boldsymbol{\mathcal{T}}_j \dot{\boldsymbol{e}}_j$.

10.3.2 Design of the control torque

For the teleoperation system (10.3), in the presence of time-varying communication delays and unknown kinematics/dynamics, in order to achieve prescribed performance control, this section proposes a switching control architecture, as shown in Fig. 10.1.

As shown in Fig. 10.1, it has the same control architecture at both master and slave sides. Only the position signal $\boldsymbol{X}_j$ and the velocity signal $\dot{\boldsymbol{X}}_j$ are transmitted between the master and slave robots. After the master and slave robots obtain the synchronization error, they will calculate the control input through a controller in

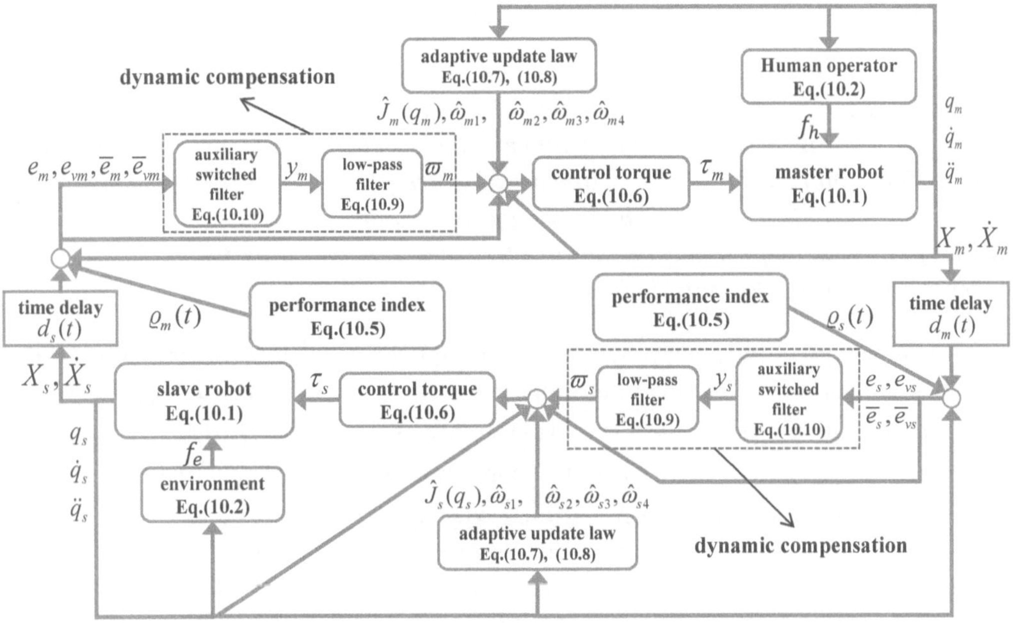

Figure 10.1 The block diagram of the proposed control algorithm.

the form of a "proportional + damping injection + compensator" structure. Different from the traditional prescribed performance control, in the case of time-varying communication delay, the algorithm in this chapter adds a switching filter based on the idea of dynamic compensation control. In order to reduce the chattering from the high-frequency switching of the signal, a low-pass filter is also introduced in the controller. To deal with unknown uncertainties, this chapter uses adaptive control techniques to estimate and compensate the unknown $\boldsymbol{J}_j(\boldsymbol{q}_j)$, ω_{j1}, ω_{j2}, ω_{j3} and ω_{j4}.

Specifically, the control torque $\boldsymbol{\tau}_j$ is designed in the form of "non-linear proportional + damping injection + compensator", i.e., namely

$$\begin{aligned}\boldsymbol{\tau}_j = {} & \boldsymbol{g}_{jo}(\boldsymbol{q}_j) - K_{j0}\dot{\boldsymbol{q}}_j - K_{j1}\hat{\boldsymbol{J}}_j^{\mathrm{T}}(\boldsymbol{q}_j)\boldsymbol{\mathcal{R}}_{j1}(\bar{\boldsymbol{e}}_j) - K_{j2}\hat{\boldsymbol{J}}_j^{\mathrm{T}}(\boldsymbol{q}_j)\boldsymbol{e}_j - \\ & K_{j3}\hat{\boldsymbol{J}}_j^{\mathrm{T}}(\boldsymbol{q}_j)\boldsymbol{\mathcal{R}}_{j2}(\boldsymbol{\varpi}_j) - K_{cj}\hat{\boldsymbol{J}}_j^{\mathrm{T}}(\boldsymbol{q}_j)\hat{\boldsymbol{J}}_j(\boldsymbol{q}_j)\dot{\boldsymbol{q}}_j - \hat{\omega}_{j1}\dot{\boldsymbol{q}}_j|\ddot{\boldsymbol{q}}_j|^2 - \\ & \hat{\omega}_{j2}\dot{\boldsymbol{q}}_j|\dot{\boldsymbol{q}}_j| - (\hat{\omega}_{j3} + \hat{\omega}_{j4})\dot{\boldsymbol{q}}_j \end{aligned} \tag{10.6}$$

where K_{j0}, K_{j1}, K_{j2}, K_{j3} and K_{cj} are some positive control gains. For any $k = 1, 2$ and $\boldsymbol{\eta} = [\eta_1, \eta_2, \cdots, \eta_n]^{\mathrm{T}} \in \mathbb{R}^n$, $\boldsymbol{\mathcal{R}}_{jk}(\boldsymbol{\eta}) = [\mathcal{R}_{j,k}(\eta_1), \mathcal{R}_{j,k}(\eta_2), \cdots, \mathcal{R}_{j,k}(\eta_n)]^{\mathrm{T}}$, where $\mathcal{R}_{j,k}(\cdot)$ is a strictly increasing, bounded, continuous, passing through the origin with continuous first derivative around the origin, such that $|\mathcal{R}_{j,k}(\eta_i)| \leq \min\{|\eta_i|, N_{ji}\}$ and $\mathcal{R}_{j,k}(-\eta_i) = -\mathcal{R}_{j,k}(\eta_i)$.

Remark 10.2. *In previous prescribed performance control works [33–35], the error transformation $\bar{\boldsymbol{e}}_j(t)$ is directly employed in the control torque. Given that the existences of model uncertainties and external disturbances in practical control, it is difficult and even impossible to push the tracking errors to asymptotically converge* *to zero. The existed unknown uncertainties may cause* $\frac{e_{j,i}(t)}{\varrho_{j,i}(t)} \geq 1$, *and further yield* $\bar{e}_{j,i}(t) = \infty$ *or invalid. In this case, on the one hand, the controllers designed in existing works* [33–35] *may face the effect of actuator saturation; on the other hand, it also reduces the action ability of other bounded components in the final control output. These may eventually lead to closed-loop control instability. To overcome this limitation, the controller (10.6) adds two smooth nonlinear bounded functions* $\mathcal{R}_{j1}(\cdot)$ *and* $\mathcal{R}_{j2}(\cdot)$. *This helps avoid actuator saturation and improves the practicability of prescribed performance control algorithms.*

The controller (10.6) consists of five parts: ① the gravity compensation $\boldsymbol{g}_{jo}(\boldsymbol{q}_j)$; ② the nonlinear proportional plus damping term $-K_{j0}\dot{\boldsymbol{q}}_j - K_{j1}\hat{\boldsymbol{J}}_j^{\mathrm{T}}(\boldsymbol{q}_j)\mathcal{R}_{j1}(\bar{\boldsymbol{e}}_j) - K_{j2}\hat{\boldsymbol{J}}_j^{\mathrm{T}}(\boldsymbol{q}_j)\boldsymbol{e}_j$; ③ the adaptive estimation compensation $-\hat{\omega}_{j1}\dot{\boldsymbol{q}}_j|\ddot{\boldsymbol{q}}_j|^2 - \hat{\omega}_{j2}\dot{\boldsymbol{q}}_j|\dot{\boldsymbol{q}}_j| - (\hat{\omega}_{j3} + \hat{\omega}_{j4})\dot{\boldsymbol{q}}_j$; ④ the adaptive Jacobian compensation $-K_{cj}\hat{\boldsymbol{J}}_j^{\mathrm{T}}(\boldsymbol{q}_j)\hat{\boldsymbol{J}}_j(\boldsymbol{q}_j)\dot{\boldsymbol{q}}_j$; ⑤ the auxiliary filter compensation term $-K_{j3}\hat{\boldsymbol{J}}_j^{\mathrm{T}}(\boldsymbol{q}_j)\mathcal{R}_{j2}(\boldsymbol{\varpi}_j)$.

The first component and the third component are, respectively, the grav- ity compensation term and the adaptive estimation compensation term, while the second component is the nonlinear proportional plus damping term. The fourth component $-K_{cj}\hat{\boldsymbol{J}}_j^{\mathrm{T}}(\boldsymbol{q}_j)\hat{\boldsymbol{J}}_j(\boldsymbol{q}_j)\dot{\boldsymbol{q}}_j$ is used to offset the residual item in the derivation of stability analysis. Such a design architecture is common and similar to the previous works. In controller (10.6), $\hat{\boldsymbol{J}}_j(\boldsymbol{q}_j)$ denotes the estimation for the unknown Jacobian matrix $\boldsymbol{J}_j(\boldsymbol{q}_j)$. From Property 10.1,

$$\dot{\hat{\boldsymbol{X}}}_j = \boldsymbol{Y}_j(\boldsymbol{q}_j, \dot{\boldsymbol{q}}_j)\hat{\boldsymbol{\theta}}_j = \hat{\boldsymbol{J}}_j(\boldsymbol{q}_j)\dot{\boldsymbol{q}}_j$$

where $\hat{\boldsymbol{\theta}}_j$ is the estimation of $\boldsymbol{\theta}_j$, while its update law is designed as

$$\dot{\hat{\boldsymbol{\theta}}}_j = \psi_j \boldsymbol{Y}_j^{\mathrm{T}}(\boldsymbol{q}_j, \dot{\boldsymbol{q}}_j)\tilde{\boldsymbol{X}}_j \tag{10.7}$$

where ψ_j is a positive constant, $\tilde{\boldsymbol{X}}_j = \boldsymbol{X}_j - \hat{\boldsymbol{X}}_j$. Note that, since $\boldsymbol{X}_j$ is measured by the physical sensors, $\tilde{\boldsymbol{X}}_j$ is available for the control design.

$\hat{\omega}_{j1}$, $\hat{\omega}_{j2}$, $\hat{\omega}_{j3}$ and $\hat{\omega}_{j4}$ are the estimations for ω_{j1}, ω_{j2}, ω_{j3} and ω_{j4}, respectively, whose adaptive update laws are given by

$$\begin{cases} \dot{\hat{\omega}}_{j1} = \psi_{j11}|\dot{\boldsymbol{q}}_j|^2|\ddot{\boldsymbol{q}}_j|^2 - \psi_{j12}\hat{\omega}_{j1} \\ \dot{\hat{\omega}}_{j2} = \psi_{j21}|\dot{\boldsymbol{q}}_j|^3 - \psi_{j22}\hat{\omega}_{j2} \\ \dot{\hat{\omega}}_{j3} = \psi_{j31}|\dot{\boldsymbol{q}}_j|^2 - \psi_{j32}\hat{\omega}_{j3} \\ \dot{\hat{\omega}}_{j4} = \psi_{j41}|\dot{\boldsymbol{q}}_j|^2 - \psi_{j42}\hat{\omega}_{j4} \end{cases} \tag{10.8}$$

where ψ_{ji1} and ψ_{ji2}, $i = 1, 2, 3, 4$, are some positive constants.

The fifth component is the auxiliary filter compensation term, which is the main difference in comparison with the previous works. $\boldsymbol{\varpi}_j(t)$is the state of the compensator, which is designed by a '2-level' switched filter framework including a low-pass filter and a conventional switched filter. Before the output of switched filter to act on the control torque, let it pass through a low-pass filter, which helps to reduce the control chattering caused by signal switching. Specifically, $\boldsymbol{\varpi}_j$ is designed by a low-pass filter, i.e.,

$$\dot{\boldsymbol{\varpi}}_j = -K_{j4}\boldsymbol{\varpi}_j + K_{j5}|\boldsymbol{y}_j|\boldsymbol{\mathcal{R}}(\boldsymbol{e}_j) \tag{10.9}$$

where $j \in \{m, s\}$, K_{j4} and K_{j5} are some constant positive scalars, $\boldsymbol{\mathcal{R}}(\boldsymbol{e}_j) = [\mathcal{R}(e_{j,1}), \mathcal{R}(e_{j,2}), \cdots, \mathcal{R}(e_{j,n})]^{\mathrm{T}}$ is used to replace the general sign function to avoid the control chattering, $\mathcal{R}(e_{j,i})$ $(i = 1, 2, \cdots, n)$ is a strictly increasing, bounded, continuous, passing through the origin with continuous first derivative around the origin, such that $|\mathcal{R}(e_{j,i})| \leq 1$ and $\mathcal{R}(-e_{j,i}) = -\mathcal{R}(e_{j,i})$. $\boldsymbol{y}_j$ is the state of switched filter. To give its definition, let us denote $\boldsymbol{e}_{cj} = [\boldsymbol{e}_j^{\mathrm{T}}, \bar{\boldsymbol{e}}_j^{\mathrm{T}}, \hat{\boldsymbol{\theta}}_j^{\mathrm{T}}, \tilde{\boldsymbol{X}}_j^{\mathrm{T}}]^{\mathrm{T}} \in \mathbb{R}^{n_{j1}}$.

Then based on the concept of dynamic compensation, $\boldsymbol{y}_j = [y_{j,1}, y_{j,2}, \cdots, y_{j,n_{j1}}]^{\mathrm{T}} \in \mathbb{R}^{n_{j1}}$ is designed as

$$\dot{\boldsymbol{y}}_j = -P_j(\boldsymbol{y}_j + \boldsymbol{\mathcal{K}}_j(t)\boldsymbol{e}_{cj}) + \boldsymbol{u}_j,\ j \in \{m, s\} \tag{10.10}$$

where

$$\begin{aligned}\boldsymbol{u}_j = -\boldsymbol{\mathcal{K}}_j(t)&\begin{bmatrix} \boldsymbol{e}_{vj} \\ \bar{\boldsymbol{e}}_{vj}^1 + \boldsymbol{\mathcal{T}}_j\boldsymbol{e}_{vj} \\ \dot{\hat{\boldsymbol{\theta}}}_j \\ \dot{\tilde{\boldsymbol{X}}}_j \end{bmatrix} - \\ &P_{cj}\mathrm{sgn}\left((\boldsymbol{y}_j + \boldsymbol{\mathcal{K}}_j(t)\boldsymbol{e}_{cj})^{\mathrm{T}}\boldsymbol{\mathcal{K}}_j(t)\bar{\boldsymbol{\mathcal{T}}}_j\begin{bmatrix} \boldsymbol{X}_{v d\ell}(t, d_\ell(t)) \\ \boldsymbol{X}_{v d\ell}(t, d_\ell(t)) \\ \boldsymbol{0}\end{bmatrix}\right) \times \\ &\boldsymbol{\mathcal{K}}_j(t)\bar{\boldsymbol{\mathcal{T}}}_j\begin{bmatrix} \boldsymbol{X}_{v d\ell}(t, d_\ell(t)) \\ \boldsymbol{X}_{v d\ell}(t, d_\ell(t)) \\ \boldsymbol{0}\end{bmatrix}\end{aligned} \tag{10.11}$$

$\boldsymbol{X}_{vd\ell}(t, d_\ell(t)) = \dot{\boldsymbol{X}}_\ell(\theta)|_{\theta=t-d_\ell(t)}$, $\boldsymbol{e}_{vj} = \dot{\boldsymbol{X}}_j - \boldsymbol{X}_{vd\ell}(t, d_\ell(t))$, $\ell \in \{m, s\}$ and $\ell \neq j$; P_j and P_{cj} are some positive constant gains; $\boldsymbol{0}$ is an appropriately dimensioned zero vector; $\bar{\boldsymbol{\mathcal{T}}}_j = \begin{bmatrix} \boldsymbol{I} & & \\ & \boldsymbol{\mathcal{T}}_j & \\ & & \mathbb{O} \end{bmatrix} \in \mathbb{R}^{n_{j1}\times n_{j1}}$, with $\boldsymbol{I} = \mathrm{diag}[1, 1, \cdots, 1] \in \mathbb{R}^n$, $\mathbb{O} \in \mathbb{R}^{(n_{j1}-2n)\times(n_{j1}-2n)}$ being a zero matrix; $\boldsymbol{\mathcal{K}}_j(t) = \mathrm{diag}[\kappa_{j,1}(t), \kappa_{j,2}(t), \cdots, \kappa_{j,n_{j1}}(t)]$ is the designed switching rule and is given by

$$\kappa_{j,i}(t) = \begin{cases} 1, & y_{j,i}(t)e_{cj,i}(t) > 0 \text{ or } y_{j,i}(t)e_{cj,i}(t) = 0,\ y_{j,i}(t^-)e_{cj,i}(t^-) \leq 0 \\ -1, & y_{j,i}(t)e_{cj,i}(t) < 0 \text{ or } y_{j,i}(t)e_{cj,i}(t) = 0,\ y_{j,i}(t^-)e_{cj,i}(t^-) > 0 \end{cases} \tag{10.12}$$

where $e_{cj,i}(t)$ is the ith element of $\boldsymbol{e}_{cj}(t) = [e_{cj,1}(t), e_{cj,2}(t), \cdots, e_{cj,n_{j1}}(t)]^{\mathrm{T}}$.

Remark 10.3. *The complete controller consists of (10.6)∼(10.10). In comparison with the previous works, the design difference of the novel design lies in the introduction of auxiliary filter subsystems (10.9) and (10.10).*

(1) The basic idea of the novel design is to use the concept of dynamic compensation. When the closed-loop teleoperation system is experiencing external disturbances, the adverse effects will be shown in the variations of the errors and estimations, including $\mathbf{e}_j$, $\bar{\mathbf{e}}_j$, $\hat{\boldsymbol{\theta}}_j$, $\tilde{\boldsymbol{X}}_j$ *and/or their derivatives. Based on the designs of (10.6), (10.9) and (10.10), their changes will be firstly applied to the switched filter subsystem (10.10), subsequently to the low-pass filter (10.9), and finally to the control torque (10.6).*

(2) Different from the traditional dynamic output feedback, the dynamic filter subsystems are given by a switching design. This novel design can make full use of the state information including the tracking error and the adaptive estimation. Moreover, due to the introduction of subsystem (10.10) the closed-loop system is a switched system. It also provides a great deal of freedom for the stability analysis, for example, the multiple Lyapunov functions method can be employed.

10.3.3 Complete closed-loop teleoperation system

The complete closed-loop teleoperation system consists of the robot dynamics (10.1), the operator/environment dynamics (10.2), the adaptive controller (10.6), the operator/environment dynamics (10.9), (10.10) and the update laws (10.7) and (10.8). Due to the existence of (10.10) the resulting closed-loop system belongs to a switched system.

Remark 10.4. *(10.10) includes two kinds of switching, i.e., the switching rule* $\boldsymbol{\mathcal{K}}_j(t)$ *and the sign function switching in (10.11). To use the analysis method for switched system, should be transformed into a standard switched system. Let us recall the modeling process shown in Chapter 3. From the definition,* $\kappa_{j,i}(t)$ $(i = 1, 2, \cdots, n_{j1})$ *switches between* 1 *and* -1*. Then,* $\boldsymbol{\mathcal{K}}_j(t) = \mathrm{diag}[\kappa_{j,1}(t), \kappa_{j,2}(t), \cdots, \kappa_{j,n_{j1}}(t)]$ *will have* $2^{n_{j1}}$ *states, i.e.,*

$$\begin{gathered}
\mathrm{diag}[-1,-1,-1,\cdots,-1] \in \mathbb{R}^{n_{j1}\times n_{j1}} \\
\mathrm{diag}[1,-1,-1,\cdots,-1] \in \mathbb{R}^{n_{j1}\times n_{j1}} \\
\mathrm{diag}[-1,1,-1,\cdots,-1] \in \mathbb{R}^{n_{j1}\times n_{j1}} \\
\mathrm{diag}[1,1,-1,\cdots,-1] \in \mathbb{R}^{n_{j1}\times n_{j1}} \\
\vdots \\
\mathrm{diag}[-1,1,1,\cdots,1] \in \mathbb{R}^{n_{j1}\times n_{j1}} \\
\mathrm{diag}[1,1,1,\cdots,1] \in \mathbb{R}^{n_{j1}\times n_{j1}}
\end{gathered}$$

which are numbered in turn as: 1, 2, $\cdots$, $2^{n_{j1}}$*. Denote the index set be* $\mathcal{S} = \{1, 2, \cdots, 2^{n_{j1}}\}$*. Then one can define the mapping* $\bar{r}_j : \boldsymbol{\mathcal{K}}_j(t) \to \mathcal{S}$ *as*

$$\bar{r}_j(\boldsymbol{\mathcal{K}}_j(t)) = i, \text{ if and only if } \boldsymbol{\mathcal{K}}_j(t) \text{ is in the } i\text{th state}$$

where $i \in \mathcal{S}$*. For the sake of simplifying the expression, let* $r_j^1(t) = \bar{r}_j(\boldsymbol{\mathcal{K}}_j(t))$*. Denote time sequence* $\{t_k^{j1}\}_{k\geq 0}$ *be the switching time of* $r_j^1(t)$*, with* $t_0^{j1} = t_0 = 0$*. Then* $r_j^1(t)$ *can be seen as a normal state-dependent switching signal. On the other hand, two modes* 1 *and* -1 *can be achieved by the sign function. Similar to the modeling of*

$r_j^1(t)$, *one can define another standard switching signal* $r_j^2 : \mathbb{R}_+ \to \mathcal{S}$ *such that* $r_j^2(t)$ *has a one-to-one correspondence with the value of the sign function, where we let* $\{t_k^{j2}\}_{k\geq 0}$ *be the switching times of* $r_j^2(t)$, *with* $t_0^{j2} = t_0 = 0$. *To deal with the existing two sets of switching, a virtual switching signal* $r_j : [t_0, \infty) \to \mathcal{S} \times \mathcal{S}$ *is introduced and defined as* $r_j(t) = (r_j^1(t), r_j^2(t))$, *where the switching time sequence is written as* $\{t_k^j\}_{k\geq 0}$. *Obviously,* $\{t_k^j\}_{k\geq 0} = \{t_k^{j1}\}_{k\geq 0} \cup \{t_k^{j2}\}_{k\geq 0}$. *Consequently, (10.10) can be rewritten as the following normal switched system in a standard form with respect to switching signal* $r_j(t)$.

$$\dot{\boldsymbol{y}}_j := h(\boldsymbol{y}_j, \boldsymbol{e}_{cj}, \boldsymbol{u}_j, r_j(t)) = -P_j(\boldsymbol{y}_j + \boldsymbol{\mathcal{K}}_j(t)\boldsymbol{e}_{cj}) + \boldsymbol{u}_j \tag{10.13}$$

Remark 10.5. *As shown in (10.10) and discussed in Remark 10.4, the switched filters are designed separately for the master robot and the slave robot. For the complete closed-loop system which includes both the master and slave robots, it is necessary to unify the definition of switching signals. Similarly, define the virtual switching signal* $r : [t_0, \infty) \to \mathcal{S} \times \mathcal{S} \times \mathcal{S} \times \mathcal{S}$ *with*

$$r(t) = (r_m(t), r_s(t))$$

Let $\{\bar{t}_k\}_{k\geq 0}$ *be the switching times, with* $\bar{t}_0 = t_0$. *Then* $\{\bar{t}_k\}_{k\geq 0} = \{t_k^m\}_{k\geq 0} \cup \{t_k^s\}_{k\geq 0}$. *In this chapter,* $r(t)$ *can be seen as the switching signal of the complete closed-loop system.*

Remark 10.6. *Denote* $\chi_{j1} = \boldsymbol{q}_j$, $\chi_{j2} = \dot{\boldsymbol{q}}_j$, $\chi_{j3} = \tilde{\boldsymbol{X}}_j$, $\chi_{j4} = \tilde{\omega}_{j1}$, $\chi_{j5} = \tilde{\omega}_{j2}$, $\chi_{j6} = \tilde{\omega}_{j3}$, $\chi_{j7} = \tilde{\omega}_{j4}$, $\chi_{j8} = \tilde{\boldsymbol{\theta}}_j$, $\chi_{j9} = \boldsymbol{\varpi}_j$, $\chi_{j10} = \boldsymbol{y}_j$, $\chi_{j11} = \boldsymbol{e}_j$, $\chi_{j12} = \bar{\boldsymbol{e}}_j$, $\chi_j = [\chi_{j1}^{\mathrm{T}}, \chi_{j2}^{\mathrm{T}}, \cdots, \chi_{j11}^{\mathrm{T}}, \chi_{j12}^{\mathrm{T}}]^{\mathrm{T}}$, *and let* $\chi = [\chi_m^{\mathrm{T}}, \chi_s^{\mathrm{T}}]^{\mathrm{T}}$. *Also denote* $\boldsymbol{\omega} = [\boldsymbol{f}_h^{*T}, \boldsymbol{f}_e^{*T}]^{\mathrm{T}}$. *Integrated the above definitions, there exists a nonlinear function* $\boldsymbol{H}_1(\cdot,\cdot,\cdot,\cdot,\cdot)$ *such that*

$$\dot{\chi} = \boldsymbol{H}_1(\chi, \chi(t - d_m(t)), \chi(t - d_s(t)), \boldsymbol{\omega}, r(t))$$

which means the complete closed-loop teleoperation system can be seen as a switched time delay system.

10.4 STABILITY ANALYSIS

This section presents the stability analysis of the complete closed-loop system. Given that the existences of the exogenous forces $\boldsymbol{f}_h^*$ and $\boldsymbol{f}_e^*$, the stability conclusion for the resulting closed-loop teleoperation system will be conducted in the framework of state-independent input-to-output stability.

In this respect, the state and output of a complete closed-loop system are first introduced, denoted as

$$\begin{aligned}\boldsymbol{x} = \Big[&\boldsymbol{q}_m^{\mathrm{T}}, \boldsymbol{q}_s^{\mathrm{T}}, \dot{\boldsymbol{q}}_m^{\mathrm{T}}, \dot{\boldsymbol{q}}_s^{\mathrm{T}}, \tilde{\omega}_{m1}, \tilde{\omega}_{s1}, \tilde{\omega}_{m2}, \tilde{\omega}_{s2}, \tilde{\omega}_{m3}, \tilde{\omega}_{s3}, \tilde{\omega}_{m4}, \tilde{\omega}_{s4}, \tilde{\boldsymbol{\theta}}_m^{\mathrm{T}}, \tilde{\boldsymbol{\theta}}_s^{\mathrm{T}},\\ &\boldsymbol{y}_m^{\mathrm{T}}, \boldsymbol{y}_s^{\mathrm{T}}, \boldsymbol{\varpi}_m^{\mathrm{T}}, \boldsymbol{\varpi}_s^{\mathrm{T}}, \boldsymbol{e}_m^{\mathrm{T}}, \boldsymbol{e}_s^{\mathrm{T}}, \bar{\boldsymbol{e}}_m^{\mathrm{T}}, \bar{\boldsymbol{e}}_s^{\mathrm{T}}, \tilde{\boldsymbol{X}}_m^{\mathrm{T}}, \tilde{\boldsymbol{X}}_s^{\mathrm{T}}\Big]^{\mathrm{T}} \in \mathbb{R}^{n_1}\\ \boldsymbol{z} = \Big[&\dot{\boldsymbol{q}}_m^{\mathrm{T}}, \dot{\boldsymbol{q}}_s^{\mathrm{T}}, \tilde{\omega}_{m1}, \tilde{\omega}_{s1}, \tilde{\omega}_{m2}, \tilde{\omega}_{s2}, \tilde{\omega}_{m3}, \tilde{\omega}_{s3}, \tilde{\omega}_{m4}, \tilde{\omega}_{s4}, \tilde{\boldsymbol{\theta}}_m^{\mathrm{T}}, \tilde{\boldsymbol{\theta}}_s^{\mathrm{T}},\\ &\boldsymbol{y}_m^{\mathrm{T}}, \boldsymbol{y}_s^{\mathrm{T}}, \boldsymbol{\varpi}_m^{\mathrm{T}}, \boldsymbol{\varpi}_s^{\mathrm{T}}, \boldsymbol{e}_m^{\mathrm{T}}, \boldsymbol{e}_s^{\mathrm{T}}, \bar{\boldsymbol{e}}_m^{\mathrm{T}}, \bar{\boldsymbol{e}}_s^{\mathrm{T}}, \tilde{\boldsymbol{X}}_m^{\mathrm{T}}, \tilde{\boldsymbol{X}}_s^{\mathrm{T}}\Big]^{\mathrm{T}} \in \mathbb{R}^{n_2}\end{aligned}$$

Further, define $\boldsymbol{x}_t : [-\bar{d}, 0] \rightarrow \mathbb{R}^{n_1}$ and $\boldsymbol{z}_t : [-\bar{d}, 0] \rightarrow \mathbb{R}^{n_2}$ by $\boldsymbol{x}_t = \{\boldsymbol{x}(t+\phi), \phi \in [-\bar{d}, 0]\} \in C([-\bar{d}, 0]; \mathbb{R}^{n_1})$ and $\boldsymbol{z}_t = \{\boldsymbol{z}(t+\phi), \phi \in [-\bar{d}, 0]\} \in C([-\bar{d}, 0]; \mathbb{R}^{n_2})$, respectively.

For the resulting closed-loop switched system, based on the multiple Lyapunov-Krosovskii functionals method, it holds

Theorem 10.1. *For the nonlinear teleoperation system (10.1), (10.6)∼(10.10), under the initial condition* $\varrho_{j,i0} > |e_{j,i}(0)|$ $(i = 1, 2, \cdots, n)$, *if for all* $j \in \{m, s\}$, *the constant positive scalars* K_{j0}, K_{j1}, K_{j2}, K_{j3}, K_{j4}, K_{j5}, K_{cj}, P_j, P_{cj}, ψ_{j11}, ψ_{j12}, ψ_{j21}, ψ_{j22}, ψ_{j31}, ψ_{j32}, ψ_{j41}, ψ_{j42} *and* ψ_j *satisfy*

$$\min_{\substack{j \in \{m,s\} \\ i = 1,2,3,4}} \left\{ \begin{array}{c} K_{j0} - \vartheta \\ P_j - \dfrac{nK_{j5}}{2} - \vartheta \\ K_{j4} - \dfrac{K_{j3} + K_{j5}}{2} - \vartheta \\ P_j - \dfrac{K_{j2}}{2} - \vartheta \\ P_j - \dfrac{K_{j1}}{2} - \vartheta \\ P_j\left(1 - \dfrac{1}{2\lambda_1}\right) - \vartheta \\ \dfrac{\psi_{ji2}}{2\psi_{ji1}} - \vartheta \\ \dfrac{\vartheta}{2\bar{d}} \end{array} \right\} > 0 \tag{10.14}$$

$$\begin{cases} P_{cj} \geq \tilde{d} \\ K_{cj} \geq \dfrac{K_{j1} + K_{j2} + K_{j3}}{2} \end{cases} \tag{10.15}$$

where $\lambda_1 > 0$, $\vartheta > 0$, *then for any bounded exogenous forces* $\boldsymbol{f}_h^*$ *and* $\boldsymbol{f}_e^*$, *the complete closed-loop teleoperation system is state-independent input-to-output stable (SIIOS). Under the hypotheses, the position tracking errors* $\mathbf{e}_m(t)$ *and* $\mathbf{e}_s(t)$ *maintain bounded, and evolve strictly within the predefined region shown in (10.5).*

Proof. For the complete closed-loop system, taking the Lyapunov-Krasovskii functionals as

$$V(\boldsymbol{x}_t, r(t)) = \sum_{k=1}^{6} V_k$$

where

$$V_1 = \sum_{j \in \{m,s\}} \frac{1}{2} \dot{\boldsymbol{q}}_j^{\mathrm{T}} \boldsymbol{M}_{jo}(\boldsymbol{q}_j) \dot{\boldsymbol{q}}_j$$

$$V_2 = \sum_{j \in \{m,s\}} \sum_{i=1}^{4} \frac{1}{2\psi_{ji1}} \tilde{\omega}_{ji} \tilde{\omega}_{ji}$$

$$V_3 = \sum_{j \in \{m,s\}} \frac{1}{2} \boldsymbol{\varpi}_j^{\mathrm{T}} \boldsymbol{\varpi}_j$$

$$V_4 = \sum_{j\in\{m,s\}} \frac{1}{2}\boldsymbol{s}_j^{\mathrm{T}}\boldsymbol{s}_j$$

$$V_5 = \sum_{j\in\{m,s\}} \frac{1}{2\psi_j}\tilde{\boldsymbol{\theta}}_j^{\mathrm{T}}\tilde{\boldsymbol{\theta}}_j + \frac{1}{2}\tilde{\boldsymbol{X}}_j^{\mathrm{T}}\tilde{\boldsymbol{X}}_j$$

$$V_6 = \frac{\vartheta}{2}\int_{-\bar{d}}^{0} \boldsymbol{z}_t^{\mathrm{T}}(\tau)\left(\frac{-\tau}{\bar{d}} + \frac{2(\tau+\bar{d})}{\bar{d}}\right)\boldsymbol{z}_t(\tau)\mathrm{d}\tau$$

where $\boldsymbol{s}_j = \boldsymbol{y}_j + \boldsymbol{\mathcal{K}}_j\boldsymbol{e}_{cj}$.

Firstly, along with system (10.3) and using Property 2.4, it has

$$\begin{aligned}
D^+V_1 &= \sum_{j\in\{m,s\}} \dot{\boldsymbol{q}}_j^{\mathrm{T}}\boldsymbol{M}_{jo}(\boldsymbol{q}_j)\ddot{\boldsymbol{q}}_j + \dot{\boldsymbol{q}}_j^{\mathrm{T}}\dot{\boldsymbol{M}}_{jo}(\boldsymbol{q}_j)\dot{\boldsymbol{q}}_j \\
&= \sum_{j\in\{m,s\}} \dot{\boldsymbol{q}}_j^{\mathrm{T}}\left(\boldsymbol{J}_j^{\mathrm{T}}(\boldsymbol{q}_j)\boldsymbol{f}_{he_j}^* + \boldsymbol{\tau}_j - \boldsymbol{\tau}_{jd} - \boldsymbol{g}_{jo}(\boldsymbol{q}_j)\right) \\
&\le \sum_{j\in\{m,s\}} \frac{1}{4}|\boldsymbol{f}_{he_j}^*|^2 - K_{j0}|\dot{\boldsymbol{q}}_j|^2 + \frac{K_{j1}}{2}|\bar{\boldsymbol{e}}_j|^2 + \frac{K_{j2}}{2}|\boldsymbol{e}_j|^2 - \\
&\quad \left(K_{cj} - \frac{K_{j1}+K_{j2}+K_{j3}}{2}\right)\dot{\boldsymbol{q}}_j^{\mathrm{T}}\hat{\boldsymbol{J}}_j^{\mathrm{T}}(\boldsymbol{q}_j)\hat{\boldsymbol{J}}_j(\boldsymbol{q}_j)\dot{\boldsymbol{q}}_j + \\
&\quad \frac{K_{j3}}{2}|\boldsymbol{\varpi}_j|^2 - \dot{\boldsymbol{q}}_j^{\mathrm{T}}\boldsymbol{\tau}_{jd} - \hat{\omega}_{j1}|\dot{\boldsymbol{q}}_j|^2|\ddot{\boldsymbol{q}}_j|^2 - \\
&\quad \hat{\omega}_{j2}|\dot{\boldsymbol{q}}_j|^3 - \hat{\omega}_{j3}|\dot{\boldsymbol{q}}_j|^2 - \hat{\omega}_{j4}|\dot{\boldsymbol{q}}_j|^2
\end{aligned}$$

where $\boldsymbol{f}_{he_m}^* = \boldsymbol{f}_h^*$, $\boldsymbol{f}_{he_s}^* = \boldsymbol{f}_e^*$. Since

$$\begin{aligned}
-\dot{\boldsymbol{q}}_j^{\mathrm{T}}\boldsymbol{\tau}_{jd} &\le \omega_{j1}|\dot{\boldsymbol{q}}_j||\ddot{\boldsymbol{q}}_j| + \omega_{j2}|\dot{\boldsymbol{q}}_j|^3 + \omega_{j3}|\dot{\boldsymbol{q}}_j|^2 + \omega_{j4}|\dot{\boldsymbol{q}}_j| \\
&\le \omega_{j1}|\dot{\boldsymbol{q}}_j|^2|\ddot{\boldsymbol{q}}_j|^2 + \omega_{j2}|\dot{\boldsymbol{q}}_j|^3 + \omega_{j3}|\dot{\boldsymbol{q}}_j|^2 + \omega_{j4}|\dot{\boldsymbol{q}}_j|^2 + \frac{1}{4}(\omega_{j1}+\omega_{j4})
\end{aligned}$$

then

$$\begin{aligned}
D^+V_1 \le \sum_{j\in\{m,s\}} &\frac{1}{4}|\boldsymbol{f}_{he_j}^*|^2 - K_{j0}|\dot{\boldsymbol{q}}_j|^2 + \frac{K_{j1}}{2}|\bar{\boldsymbol{e}}_j|^2 + \frac{K_{j2}}{2}|\boldsymbol{e}_j|^2 - \\
&\left(K_{cj} - \frac{K_{j1}+K_{j2}+K_{j3}}{2}\right)\dot{\boldsymbol{q}}_j^{\mathrm{T}}\hat{\boldsymbol{J}}_j^{\mathrm{T}}(\boldsymbol{q}_j)\hat{\boldsymbol{J}}_j(\boldsymbol{q}_j)\dot{\boldsymbol{q}}_j + \\
&\frac{K_{j3}}{2}|\boldsymbol{\varpi}_j|^2 + \tilde{\omega}_{j1}|\dot{\boldsymbol{q}}_j|^2|\ddot{\boldsymbol{q}}_j|^2 + \tilde{\omega}_{j2}|\dot{\boldsymbol{q}}_j|^3 + \tilde{\omega}_{j3}|\dot{\boldsymbol{q}}_j|^2 + \tilde{\omega}_{j4}|\dot{\boldsymbol{q}}_j|^2 + \frac{1}{4}(\omega_{j1}+\omega_{j4})
\end{aligned}$$

From (10.8), it exists

$$\begin{aligned}
D^+V_2 &= \sum_{j\in\{m,s\}}\sum_{i=1}^{4} \frac{1}{\psi_{ji1}}\tilde{\omega}_{ji}\dot{\hat{\omega}}_{ji} \\
&= \sum_{j\in\{m,s\}} -\tilde{\omega}_{j1}|\dot{\boldsymbol{q}}_j|^2|\ddot{\boldsymbol{q}}_j|^2 + \frac{\psi_{j12}}{\psi_{j11}}\tilde{\omega}_{j1}\hat{\omega}_{j1} - \tilde{\omega}_{j2}|\dot{\boldsymbol{q}}_j|^3 + \\
&\quad \frac{\psi_{j22}}{\psi_{j21}}\tilde{\omega}_{j2}\hat{\omega}_{j2} - \tilde{\omega}_{j3}|\dot{\boldsymbol{q}}_j|^2 + \frac{\psi_{j32}}{\psi_{j31}}\tilde{\omega}_{j3}\hat{\omega}_{j3} - \tilde{\omega}_{j4}|\dot{\boldsymbol{q}}_j|^2 + \frac{\psi_{j42}}{\psi_{j41}}\tilde{\omega}_{j4}\hat{\omega}_{j4}
\end{aligned}$$

$$\leq \sum_{j\in\{m,s\}} -\tilde{\omega}_{j1}|\dot{\boldsymbol{q}}_j|^2|\ddot{\boldsymbol{q}}_j|^2 - \tilde{\omega}_{j2}|\dot{\boldsymbol{q}}_j|^3 - (\tilde{\omega}_{j3}+\tilde{\omega}_{j4})|\dot{\boldsymbol{q}}_j|^2 -$$
$$\sum_{i=1}^{4}\frac{\psi_{ji2}}{2\psi_{ji1}}|\tilde{\omega}_{ji}|^2 + \sum_{i=1}^{4}\frac{\psi_{ji2}}{2\psi_{ji1}}|\omega_{ji}|^2$$

While based on (10.9),

$$\begin{aligned} D^+V_3 &= \sum_{j\in\{m,s\}} -K_{j4}|\boldsymbol{\varpi}_j|^2 + K_{j5}|\boldsymbol{y}_j|\boldsymbol{\varpi}_j^{\mathrm{T}}\boldsymbol{\mathcal{R}}(\boldsymbol{e}_j) \\ &\leq \sum_{j\in\{m,s\}} -K_{j4}|\boldsymbol{\varpi}_j|^2 + \frac{K_{j5}}{2}|\boldsymbol{\varpi}_j|^2 + \frac{K_{j5}}{2}|\boldsymbol{y}_j|^2|\boldsymbol{\mathcal{R}}(\boldsymbol{e}_j)|^2 \\ &\leq \sum_{j\in\{m,s\}} -\left(K_{j4}-\frac{K_{j5}}{2}\right)|\boldsymbol{\varpi}_j|^2 + \frac{nK_{j5}}{2}|\boldsymbol{y}_j|^2 \end{aligned}$$

On the other hand, according to the definition of $\boldsymbol{s}_j$ and (10.10), for any $t \in [\bar{t}_k, \bar{t}_{k+1})$, it holds

$$\dot{\boldsymbol{s}}_j = -P_j\boldsymbol{s}_j + \boldsymbol{u}_j + \boldsymbol{\mathcal{K}}_j\left[\dot{\boldsymbol{e}}_j^{\mathrm{T}}, \bar{\boldsymbol{e}}_{vj}^{2\mathrm{T}}, \boldsymbol{0}^{\mathrm{T}}\right]^{\mathrm{T}}$$

And moreover,

$$(|\boldsymbol{y}_j|^2 + |\boldsymbol{e}_{cj}|^2) \leq |\boldsymbol{s}_j| \leq 2(|\boldsymbol{y}_j|^2 + |\boldsymbol{e}_{cj}|^2) \tag{10.16}$$

Further for all $t \in [\bar{t}_k, \bar{t}_{k+1})$, $k \geq 0$,

$$\begin{aligned} D^+V_4 = \sum_{j\in\{m,s\}} \Bigg(& -P_j|\boldsymbol{s}_j|^2 - P_{cj}\left|\boldsymbol{s}_j^{\mathrm{T}}\boldsymbol{\mathcal{K}}_j\bar{\boldsymbol{\mathcal{T}}}_j\begin{bmatrix}\boldsymbol{X}_{vd\ell}(t,d_\ell(t))\\ \boldsymbol{X}_{vd\ell}(t,d_\ell(t))\\ \boldsymbol{0}\end{bmatrix}\right| + \\ & \dot{d}_\ell(t)\boldsymbol{s}_j^{\mathrm{T}}\boldsymbol{\mathcal{K}}_j\bar{\boldsymbol{\mathcal{T}}}_j\begin{bmatrix}\boldsymbol{X}_{vd\ell}(t,d_\ell(t))\\ \boldsymbol{X}_{vd\ell}(t,d_\ell(t))\\ \boldsymbol{0}\end{bmatrix}\Bigg) \end{aligned}$$

Given that $P_{cj} \geq \tilde{d}$, from Assumption 10.4

$$D^+V_4 \leq \sum_{j\in\{m,s\}} -P_j\left(|\boldsymbol{y}_j|^2 + |\boldsymbol{e}_j|^2 + |\bar{\boldsymbol{e}}_j|^2 + |\hat{\boldsymbol{\theta}}_j|^2 + |\tilde{\boldsymbol{X}}_j|^2\right)$$

In addition, derivation of V_5 and V_6 gives

$$\begin{aligned} D^+V_5 &= \sum_{j\in\{m,s\}} \tilde{\boldsymbol{X}}_j^{\mathrm{T}}\left(\mathbf{Y}_j(\boldsymbol{q}_j,\dot{\boldsymbol{q}}_j)\boldsymbol{\theta}_j - \mathbf{Y}_j(\boldsymbol{q}_j,\dot{\boldsymbol{q}}_j)\hat{\boldsymbol{\theta}}_j\right) - \tilde{\boldsymbol{\theta}}_j^{\mathrm{T}}\mathbf{Y}_j^{\mathrm{T}}(\boldsymbol{q}_j,\dot{\boldsymbol{q}}_j)\tilde{\boldsymbol{X}}_j \\ &= 0 \end{aligned}$$

and

$$\begin{aligned} D^+V_6 &= \vartheta\boldsymbol{z}^{\mathrm{T}}\boldsymbol{z} - \frac{\vartheta}{2}\boldsymbol{z}_t^{\mathrm{T}}(-\bar{d})\boldsymbol{z}_t(-\bar{d}) - \frac{\vartheta}{2\bar{d}}\int_{-\bar{d}}^{0}\boldsymbol{z}_t^{\mathrm{T}}(\tau)\boldsymbol{z}_t(\tau)\mathrm{d}\tau \\ &\leq \vartheta\boldsymbol{z}^{\mathrm{T}}\boldsymbol{z} - \frac{\vartheta}{2\bar{d}}\int_{-\bar{d}}^{0}\boldsymbol{z}_t^{\mathrm{T}}(\tau)\boldsymbol{z}_t(\tau)\mathrm{d}\tau \end{aligned}$$

Consequently, for all $t \in [\bar{t}_k, \bar{t}_{k+1})$, $k \geq 0$, there exists λ_1 such that

$$
\begin{aligned}
D^+V(\boldsymbol{x}_t, r(t)) \leq & \sum_{j\in\{m,s\}} \Bigg(- \left(K_{cj} - \frac{K_{j1}+K_{j2}+K_{j3}}{2}\right) \dot{\boldsymbol{q}}_j^{\mathrm{T}} \hat{\boldsymbol{J}}_j^{\mathrm{T}}(\boldsymbol{q}_j) \hat{\boldsymbol{J}}_j(\boldsymbol{q}_j)\dot{\boldsymbol{q}}_j - \\
& K_{j0}|\dot{\boldsymbol{q}}_j|^2 + \frac{1}{4}|\boldsymbol{f}_{he_j}^*|^2 + \frac{K_{j1}}{2}|\bar{\boldsymbol{e}}_j|^2 + \frac{K_{j2}}{2}|\boldsymbol{e}_j|^2 + \frac{K_{j3}}{2}|\boldsymbol{\varpi}_j|^2 - \\
& \sum_{i=1}^{4} \frac{\psi_{ji2}}{2\psi_{ji1}}|\tilde{\omega}_{ji}|^2 + \sum_{i=1}^{4} \frac{\psi_{ji2}}{2\psi_{ji1}}|\omega_{ji}|^2 + \frac{1}{4}(\omega_{j1}+\omega_{j4}) - \\
& \left(K_{j4} - \frac{K_{j5}}{2}\right)|\boldsymbol{\varpi}_j|^2 + \frac{nK_{j5}}{2}|\boldsymbol{y}_j|^2 - P_j\Big(|\boldsymbol{y}_j|^2 + |\boldsymbol{e}_j|^2 + \\
& |\bar{\boldsymbol{e}}_j|^2 + |\hat{\boldsymbol{\theta}}_j|^2 + |\tilde{\boldsymbol{X}}_j|^2\Big)\Bigg) + \vartheta \boldsymbol{z}^{\mathrm{T}}\boldsymbol{z} - \frac{\vartheta}{2\bar{d}}\int_{-\bar{d}}^{0} \boldsymbol{z}_t^{\mathrm{T}}(\tau)\boldsymbol{z}_t(\tau)\mathrm{d}\tau \\
\leq & \sum_{j\in\{m,s\}} \Bigg(-(K_{j0}-\vartheta)|\dot{\boldsymbol{q}}_j|^2 - (P_j - \vartheta)|\tilde{\boldsymbol{X}}_j|^2 - \left(P_j - \frac{nK_{j5}}{2} - \vartheta\right)|\boldsymbol{y}_j|^2 - \\
& \left(P_j - \frac{K_{j2}}{2} - \vartheta\right)|\boldsymbol{e}_j|^2 - \left(P_j - \frac{K_{j1}}{2} - \vartheta\right)|\bar{\boldsymbol{e}}_j|^2 - \\
& \left(K_{j4} - \frac{K_{j3}+K_{j5}}{2} - \vartheta\right)|\boldsymbol{\varpi}_j|^2 - \sum_{i=1}^{4}\left(\frac{\psi_{ji2}}{2\psi_{ji1}} - \vartheta\right)|\tilde{\omega}_{ji}|^2 - \\
& \left(P_j\frac{2\lambda_1 - 1}{2\lambda_1} - \vartheta\right)|\tilde{\boldsymbol{\theta}}_j|^2 + P_j(2\lambda_1 - 1)|\boldsymbol{\theta}_j|^2 + \frac{1}{4}(\omega_{j1}+\omega_{j4}) + \\
& \sum_{i=1}^{4}\frac{\psi_{ji2}}{2\psi_{ji1}}|\omega_{ji}|^2\Bigg) + \frac{1}{4}|\boldsymbol{f}_h^*|^2 + \frac{1}{4}|\boldsymbol{f}_e^*|^2 - \frac{\vartheta}{2\bar{d}}\int_{-\bar{d}}^{0}\boldsymbol{z}_t^{\mathrm{T}}(\tau)\boldsymbol{z}_t(\tau)\mathrm{d}\tau \\
\leq & -\mu\|\boldsymbol{z}_t\|_{M_2}^2 + \Xi
\end{aligned}
\tag{10.17}
$$

where Ξ and μ are, respectively, $\Xi = \sum_{j\in\{m,s\}} \Big(P_j(2\lambda_1 - 1)|\boldsymbol{\theta}_j|^2 + \frac{1}{4}(\omega_{j1}+\omega_{j4}) + \sum_{i=1}^{4}\frac{\psi_{ji2}}{2\psi_{ji1}}|\omega_{ji}|^2\Big) + \frac{1}{4}|\boldsymbol{f}_h^*|^2 + \frac{1}{4}|\boldsymbol{f}_e^*|^2$ and

$$
\mu = \min_{i,j}\left\{
\begin{array}{c}
K_{j0} - \vartheta,\ P_j - \dfrac{nK_{j5}}{2} - \vartheta \\
K_{j4} - \dfrac{K_{j3}+K_{j5}}{2} - \vartheta,\ P_j - \dfrac{K_{j2}}{2} - \vartheta \\
P_j - \dfrac{K_{j1}}{2} - \vartheta,\ P_j\left(1 - \dfrac{1}{2\lambda_1}\right) - \vartheta \\
\dfrac{\psi_{ji2}}{2\psi_{ji1}} - \vartheta,\ \dfrac{\vartheta}{2\bar{d}}
\end{array}
\right\} > 0
$$

If and only if $y_{j,i}(t)e_{cj,i}(t) = 0$ and $y_{j,i}(t^-)e_{cj,i}(t^-) > 0$, $\kappa_{j,i}(t)$ switches from 1 to -1. If and only if $y_{j,i}(t)e_{cj,i}(t) = 0$ and $y_{j,i}(t^-)e_{cj,i}(t^-) < 0$, $\kappa_{j,i}(t)$ switches from -1 to 1. Therefore for any $t = \bar{t}_{k+1}$, $k \geq 0$, it has

$$
V(\boldsymbol{x}_{\bar{t}_{k+1}}, r(\bar{t}_{k+1})) \leq V(\boldsymbol{x}_{\bar{t}_{k+1}}, r(\bar{t}_k)) \tag{10.18}
$$

In addition, from Property 2.1 and (10.16), and utilizing the fact that

$$\frac{1}{2}\int_{-\bar{d}}^{0} \boldsymbol{z}_t^{\mathrm{T}}(\tau)\left(\frac{-\tau}{\bar{d}}+\frac{2(\tau+\bar{d})}{\bar{d}}\right)\boldsymbol{z}_t(\tau)\mathrm{d}\tau \leq \int_{-\bar{d}}^{0} \boldsymbol{z}_t^{\mathrm{T}}(\tau)\boldsymbol{z}_t(\tau)\mathrm{d}\tau$$

there then exist positive constants α_1 and α_2 such that

$$\alpha_1|\boldsymbol{z}(t)|^2 \leq V(\boldsymbol{x}_t, r(t)) \leq \alpha_2\|\boldsymbol{z}_t\|_{M_2}^2 \tag{10.19}$$

Combining (10.17)∼(10.19), and using Lemma 2.4, the complete closed-loop system is SIIOS, with $\boldsymbol{f}_h^*$, $\boldsymbol{f}_e^*$, ω_{m1}, ω_{s1}, ω_{m2}, ω_{s2}, $\cdots$, ω_{m4}, ω_{s4}, $\boldsymbol{\theta}_m$ and $\boldsymbol{\theta}_s$ being the inputs, and $\boldsymbol{z}_t$ being the output. Thus for all $j \in \{m,s\}$, $\boldsymbol{e}_j$, $\bar{\boldsymbol{e}}_j$, $\tilde{\omega}_{j1}$, $\tilde{\omega}_{j2}$, $\tilde{\omega}_{j3}$, $\tilde{\omega}_{j4}$ and $\tilde{\boldsymbol{\theta}}_j$ remain bounded for any bounded input. In addition, the boundedness of $\bar{\boldsymbol{e}}_j$ implies that $\boldsymbol{e}_j$ meets the prescribed performance (10.5). □

Remark 10.7. *The conclusion in Theorem 10.1 can also be generalized to the joint space prescribed performance control of homogeneous teleoperation systems, and this generalization is straightforward. At this time, the tracking error is defined as*

$$\boldsymbol{e}_j = \boldsymbol{q}_j - \boldsymbol{q}_\ell(t-d_\ell(t))$$
$$\boldsymbol{e}_{vj} = \dot{\boldsymbol{q}}_j - \boldsymbol{q}_{vd\ell}(t, d_\ell(t))$$

where $\boldsymbol{q}_{vd\ell}(t, d_\ell(t)) = \dot{\boldsymbol{q}}_\ell(\theta)|_{\theta=t-d_\ell(t)}$.

Then the controller is designed by

$$\begin{aligned}\boldsymbol{\tau}_j = \boldsymbol{g}_{jo}(\boldsymbol{q}_j) - K_{j0}\dot{\boldsymbol{q}}_j - K_{j1}\boldsymbol{\mathcal{R}}_{j1}(\lambda_{j1}\bar{\boldsymbol{e}}_j) - K_{j2}\boldsymbol{e}_j - K_{j3}\boldsymbol{\mathcal{R}}_{j2}(\lambda_{j2}\boldsymbol{\varpi}_j)\\ - \hat{\omega}_{j1}\dot{\boldsymbol{q}}_j|\ddot{\boldsymbol{q}}_j|^2 - \hat{\omega}_{j2}\dot{\boldsymbol{q}}_j|\dot{\boldsymbol{q}}_j| - (\hat{\omega}_{j3}+\hat{\omega}_{j4})\dot{\boldsymbol{q}}_j\end{aligned} \tag{10.20}$$

At the same time, in the auxiliary switched filter (10.10), $\boldsymbol{u}_j$ *and* $\boldsymbol{e}_{cj}$ *are replaced by*

$$\begin{aligned}\boldsymbol{u}_j = -\boldsymbol{\mathcal{K}}_j(t)\begin{bmatrix}\boldsymbol{e}_{vj}\\ \bar{\boldsymbol{e}}_{vj}^1+\boldsymbol{\mathcal{T}}_j\boldsymbol{e}_{vj}\end{bmatrix} - \\ P_{cj}\mathrm{sgn}\left((\boldsymbol{y}_j+\boldsymbol{\mathcal{K}}_j(t)\boldsymbol{e}_{cj})^{\mathrm{T}}\boldsymbol{\mathcal{K}}_j(t)\bar{\boldsymbol{\mathcal{T}}}_j\begin{bmatrix}\boldsymbol{q}_{vd\ell}(t,d_\ell(t))\\ \boldsymbol{q}_{vd\ell}(t,d_\ell(t))\end{bmatrix}\right)\times \\ \boldsymbol{\mathcal{K}}_j(t)\bar{\boldsymbol{\mathcal{T}}}_j\begin{bmatrix}\boldsymbol{q}_{vd\ell}(t,d_\ell(t))\\ \boldsymbol{q}_{vd\ell}(t,d_\ell(t))\end{bmatrix}\end{aligned}$$

and

$$\boldsymbol{e}_{cj} = [\boldsymbol{e}_j^{\mathrm{T}}, \bar{\boldsymbol{e}}_j^{\mathrm{T}}]^{\mathrm{T}}$$

Repeating the analysis of Theorem 10.1, if

$$\min_{\substack{j\in\{m,s\}\\ i=1,2,3,4,5}}\left\{\begin{array}{c} K_{j0}-1-\vartheta \\ K_{j4}-\dfrac{K_{j3}+K_{j5}}{2}-\vartheta \\ P_j-\dfrac{nK_{j5}}{2}-\vartheta \\ P_j-\dfrac{K_{j2}}{2}-\vartheta \\ P_j-\dfrac{K_{j1}}{2}-\vartheta \\ P_j\left(1-\dfrac{1}{2\lambda_1}\right)-\vartheta \\ \dfrac{\psi_{ji2}}{2\psi_{ji1}}-\vartheta \\ P_{cj}-\tilde{d} \end{array}\right\}>0$$

the complete closed-loop system is SIIOS, with $\boldsymbol{z}=[\dot{\boldsymbol{q}}_m^{\mathrm{T}},\dot{\boldsymbol{q}}_s^{\mathrm{T}},\tilde{\omega}_{m1},\tilde{\omega}_{s1},\tilde{\omega}_{m2},\tilde{\omega}_{s2},\tilde{\omega}_{m3},\tilde{\omega}_{s3},\tilde{\omega}_{m4},\tilde{\omega}_{s4},\boldsymbol{y}_m^{\mathrm{T}},\boldsymbol{y}_s^{\mathrm{T}},\boldsymbol{\varpi}_m^{\mathrm{T}},\boldsymbol{\varpi}_s^{\mathrm{T}},\boldsymbol{e}_m^{\mathrm{T}},\boldsymbol{e}_s^{\mathrm{T}},\boldsymbol{e}_m^{\mathrm{T}},\boldsymbol{e}_s^{\mathrm{T}}]^{\mathrm{T}}$ *being the output.*

Remark 10.8. *By using the PPC strategy, the proposed switched controller achieves the high performance shown in (10.5), which ensures the tracking errors converge to a predefined arbitrarily small region, with convergence speed no less than a prespecified value, while exhibiting maximum overshoot less than some sufficiently small preassigned constant.*

10.5 SIMULATION STUDY

This section presents the simulation results. The used robot manipulators with revolute joints are given in Fig. 7.2, which includes one 2-DOF serial link manipulator and one 3-DOF planar arm, whose nominal dynamics are shown in (10.1) and can be seen in many works [45, 46].

For a simulation purpose, the external forces of human operator and environment are assumed to be modeled as

$$\left\{\begin{array}{l} \boldsymbol{f}_h=\boldsymbol{f}_h^*-\boldsymbol{B}_h(t)\dot{\boldsymbol{X}}_m-\boldsymbol{K}_h(t)(\boldsymbol{X}_m-\boldsymbol{X}_{m0}) \\ \boldsymbol{f}_e=\boldsymbol{f}_e^*+\boldsymbol{B}_e(t)\dot{\boldsymbol{X}}_s+\boldsymbol{K}_e(t)(\boldsymbol{X}_s-\boldsymbol{X}_{s0}) \end{array}\right. \tag{10.21}$$

where $\boldsymbol{B}_h=2+0.4\sin(2t)\sin(|\dot{\boldsymbol{q}}_m|)$; $\boldsymbol{K}_h=1+0.5\cos(5t)\sin(3|\boldsymbol{q}_m|)$; $\boldsymbol{B}_e=5+1.5\cos(3t)\cos(3|\dot{\boldsymbol{q}}_s|)$; $\boldsymbol{K}_e=3+0.6\sin(5t)\cos(|\boldsymbol{q}_s|)$; $\boldsymbol{X}_{m0}=\boldsymbol{X}_{s0}=[0,0]^{\mathrm{T}}$; the exogenous forces $\boldsymbol{f}_h^*$ and $\boldsymbol{f}_e^*$ are given as Fig. 10.2.

Remark 10.9. *The external forces (10.21) satisfy Assumption 10.2. For simplicity, let us take 2-DOF serial link manipulator as an example. As shown in*

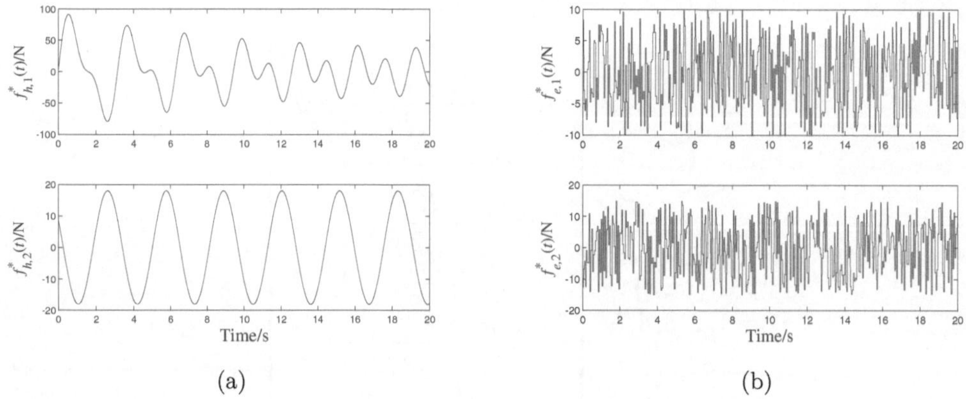

Figure 10.2 The exogenous forces. (a) The human operator exogenous force $\boldsymbol{f}_h^*(t)$. (b) The environment exogenous force $\boldsymbol{f}_e^*(t)$.

Fig. 7.2(a),

$$\boldsymbol{X} = \boldsymbol{h}(\boldsymbol{q}) = \begin{bmatrix} l_1\cos(q_1) + l_2\cos(q_1+q_2) \\ l_1\sin(q_1) + l_2\sin(q_1+q_2) \end{bmatrix}$$

$$\boldsymbol{J}(\boldsymbol{q}) = \begin{bmatrix} -l_1\sin(q_1) - l_2\sin(q_1+q_2) & -l_2\sin(q_1+q_2) \\ l_1\cos(q_1) + l_2\cos(q_1+q_2) & l_2\cos(q_1+q_2) \end{bmatrix}$$

Then

$$\dot{\boldsymbol{J}}(\boldsymbol{q}) = \begin{bmatrix} -l_1\cos(q_1)\dot{q}_1 - l_2\cos(q_1+q_2)(\dot{q}_1+\dot{q}_2) & -l_2\cos(q_1+q_2)(\dot{q}_1+\dot{q}_2) \\ -l_1\sin(q_1)\dot{q}_1 - l_2\sin(q_1+q_2)(\dot{q}_1+\dot{q}_2) & -l_2\sin(q_1+q_2)(\dot{q}_1+\dot{q}_2) \end{bmatrix}$$

Obviously, for any $\boldsymbol{X} \in \mathbb{R}^2$, $|\boldsymbol{h}(\boldsymbol{q})| \le \bar{k}_1$ and $|\boldsymbol{J}(\boldsymbol{q})\boldsymbol{X}| \le \bar{k}_2|\boldsymbol{X}|$, where $\bar{k}_1$ and $\bar{k}_2$ are bounded positive constants. In addition, it is easy to verify that

$$\begin{aligned} |\dot{\boldsymbol{J}}(\boldsymbol{q})\dot{\boldsymbol{q}}| &= \left(l_1^2\dot{q}_1^4 + 2l_1l_2\cos(q_2)\dot{q}_1^2(\dot{q}_1^2 + 2\dot{q}_1\dot{q}_2 + \dot{q}_2^2) + l_2^2(\dot{q}_1^2 + 2\dot{q}_1\dot{q}_2 + \dot{q}_2^2)^2\right)^{\frac{1}{2}} \\ &\le \left((l_1^2 + l_1l_2)\dot{q}_1^4 + (l_1l_2 + l_2^2)(\dot{q}_1^2 + 2\dot{q}_1\dot{q}_2 + \dot{q}_2^2)^2\right)^{\frac{1}{2}} \\ &\le \left((l_1^2 + l_1l_2)\dot{q}_1^4 + 4(l_1l_2 + l_2^2)(\dot{q}_1^2 + \dot{q}_2^2)^2\right)^{\frac{1}{2}} \\ &\le \left((l_1^2 + 5l_1l_2 + 4l_2^2)(\dot{q}_1^2 + \dot{q}_2^2)^2\right)^{\frac{1}{2}} \\ &= \left(l_1^2 + 5l_1l_2 + 4l_2^2\right)^{\frac{1}{2}} |\dot{\boldsymbol{q}}|^2 \\ &:= \bar{k}_3|\dot{\boldsymbol{q}}|^2 \end{aligned}$$

Then for any $j \in \{m, s\}$, due to the boundedness of $\boldsymbol{M}_j$, $\boldsymbol{B}_j$ and $\boldsymbol{K}_j$, there exist k_{j1}, k_{j2}, k_{j3} and k_{j4} such that

$$|\boldsymbol{\phi}_{he_j}(\boldsymbol{q}_j, \dot{\boldsymbol{q}}_j, \ddot{\boldsymbol{q}}_j)| \le k_{j1}|\ddot{\boldsymbol{q}}_j| + k_{j2}|\dot{\boldsymbol{q}}_j|^2 + k_{j3}|\dot{\boldsymbol{q}}_j| + k_{j4}$$

where $\boldsymbol{\phi}_{he_m}(\cdot,\cdot,\cdot) = \boldsymbol{\phi}_h(\cdot,\cdot,\cdot)$, $\boldsymbol{\phi}_{he_s}(\cdot,\cdot,\cdot) = \boldsymbol{\phi}_e(\cdot,\cdot,\cdot)$.

Similarly, this conclusion also holds for the three-degree-of-freedom planar manipulator. It can be seen that the conclusion is correct.

10.5.1 Stability verification with artificial delays

Given that the master and slave robots in many practical teleoperation systems are often heterogeneous, the first set simulation will consider the asymmetric teleoperation system. It is assumed that the master and slave robots are respectively the 2-DOF serial link manipulator and 3-DOF planar arm in Fig. 7.2.

In the simulation, the time delays are given in Fig. 10.3, where $\bar{d} = 1.8$, $\tilde{d} = 5$. For a simulation purpose, the disturbance and friction torques are assumed to be

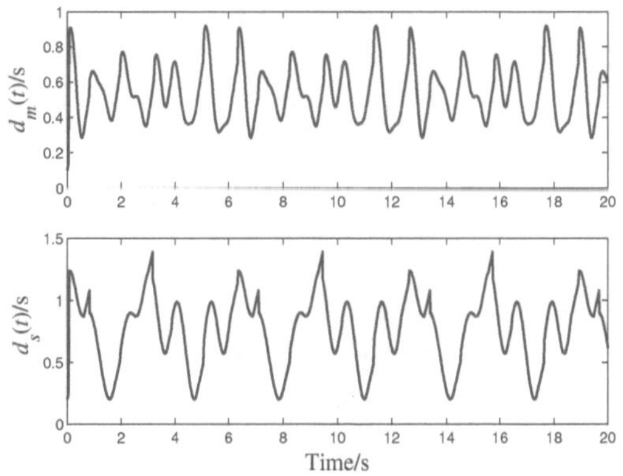

Figure 10.3 The communication delays $d_m(t)$ and $d_s(t)$.

$$\boldsymbol{f}_m(\boldsymbol{q}_m, \dot{\boldsymbol{q}}_m) = [0.2\dot{q}_{m,1} + 0.3\text{sgn}(\dot{q}_{m,1}), 0.6\dot{q}_{m,2} + 0.2\text{sgn}(\dot{q}_{m,2})]^{\mathrm{T}}$$

and

$$\boldsymbol{f}_s(\boldsymbol{q}_s, \dot{\boldsymbol{q}}_s) = [0.3\dot{q}_{s,1} + 0.2\text{sgn}(\dot{q}_{s,1}), 0.1\dot{q}_{s,2} + 0.2\text{sgn}(\dot{q}_{s,2}), 0.4\dot{q}_{s,3} + 0.1\text{sgn}(\dot{q}_{s,3})]^{\mathrm{T}}$$

and the dynamics uncertainties are assumed to be

$$\begin{aligned}
\Delta \boldsymbol{M}_m(\cdot) &= 0.1\sin(|\ddot{\boldsymbol{q}}_m|)\boldsymbol{M}_{mo}(\cdot)\\
\Delta \boldsymbol{C}_m(\cdot,\cdot) &= 0.15\cos(10t)\boldsymbol{C}_{mo}(\cdot,\cdot)\\
\Delta \boldsymbol{g}_m(\cdot) &= 0.2\sin(20t)\boldsymbol{g}_{mo}(\cdot)\\
\Delta \boldsymbol{M}_s(\cdot) &= 0.2\cos(10t)\sin(|\ddot{\boldsymbol{q}}_s|)\boldsymbol{M}_{so}(\cdot)\\
\Delta \boldsymbol{C}_s(\cdot,\cdot) &= 0.1\sin(10t)\boldsymbol{C}_{so}(\cdot,\cdot)\\
\Delta \boldsymbol{g}_s(\cdot) &= 0.15\sin(10t)\boldsymbol{g}_{so}(\cdot)
\end{aligned}$$

Then according to Theorem 10.1, the simulation results are shown in Figs. 10.4–10.8. Among them, Fig. 10.4 gives the response curve of $|\boldsymbol{z}(t)|$, where $|\boldsymbol{z}(t)|$ remains bounded, which is in accordance with the SIIOS analysis of Theorem 10.1. According to the definition of $|\boldsymbol{z}(t)|$, the joint velocities, the tracking errors and their transformations, and the adaptive estimation errors are also bounded. In fact, this can be seen from the results of Figs. 10.5 and 10.6. Figs. 10.5 and 10.6 give, respectively, the tracking errors and the adaptive estimations. As shown in Fig. 10.5, the tracking errors satisfy the prescribed performance (10.5), where the performance functions are set as $\varrho_{m,1}(t) = \varrho_{s,1}(t) = (3.4 - 0.06)e^{-t} + 0.06$, $\varrho_{m,2}(t) = \varrho_{s,2}(t) = (4.8 - 0.06)e^{-t} + 0.06$, $H_{m1} = H_{m2} = H_{s1} = H_{s2} = 0.9$. To better show the tracking performance, Fig. 10.7 gives the positions of the master and slave robots. Finally, the control torques are shown in Fig. 10.8.

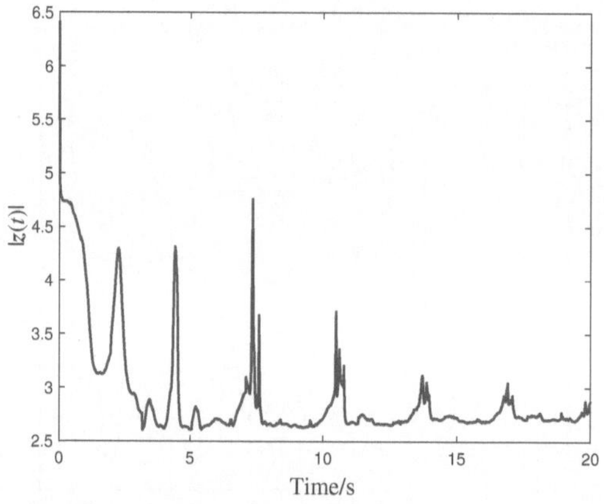

Figure 10.4 The response curve of the output $\boldsymbol{z}(t)$.

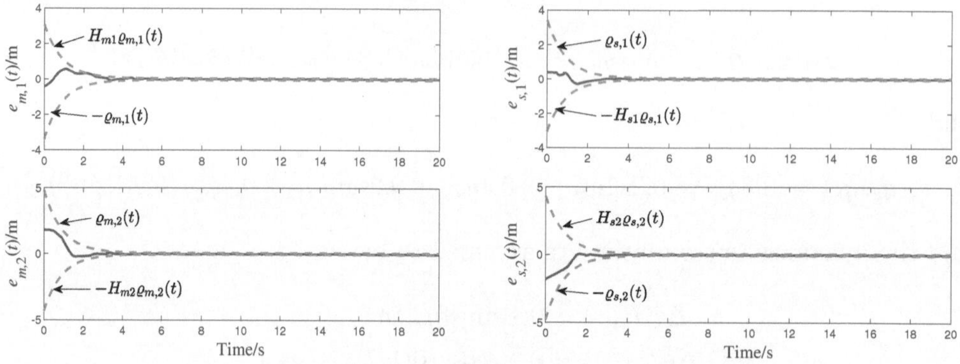

(a) (b)

Figure 10.5 The response curves of the tracking errors. (a) $\boldsymbol{e}_m(t)$. (b) $\boldsymbol{e}_s(t)$.

The control torques in Fig. 10.8 are relatively smooth. It is mainly because of the utilized low-pass filter (10.9). To verify the function of low-pass filter, let us consider the conventional switching-based control method developed in Chapter 5. In Chapter 5, the switched filter is directly used for the control torque. In this case, the simulation

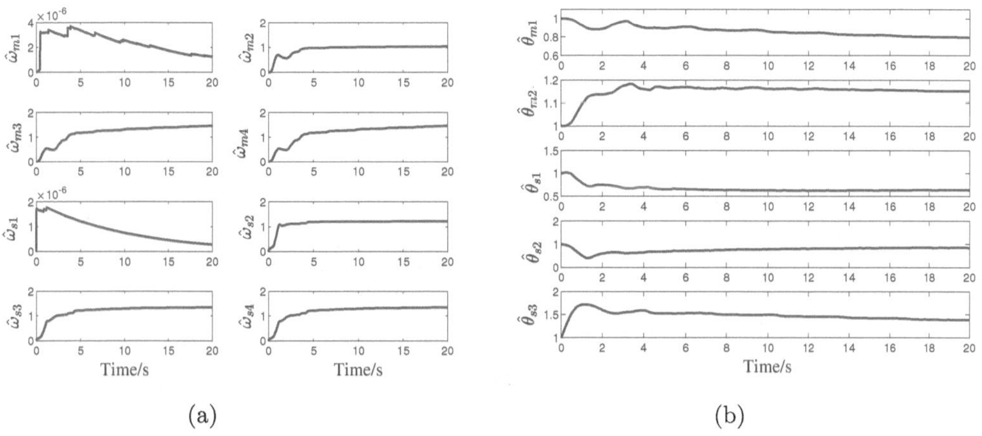

Figure 10.6 The adaptive estimations. (a) $\hat{\omega}_{j1}(t)$, $\hat{\omega}_{j2}(t)$, $\hat{\omega}_{j3}(t)$, $\hat{\omega}_{j4}(t)$, $j \in \{m, s\}$. (b) $\hat{\boldsymbol{\theta}}_j(t)$, $j \in \{m, s\}$.

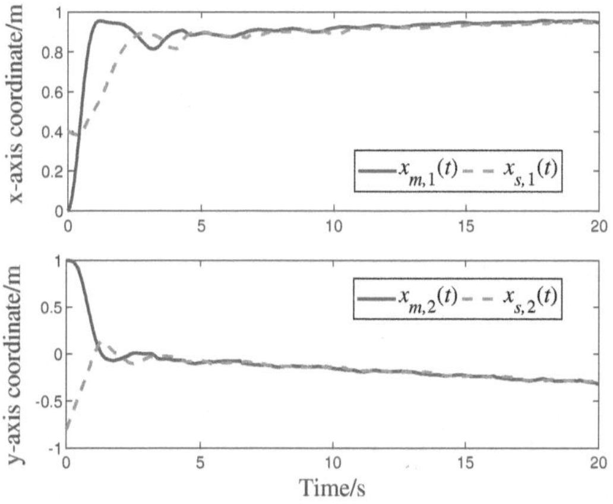

Figure 10.7 The end-effector positions of the master and slave robots.

results are shown in Figs. 10.9 and 10.10, which give, respectively, the control torques and the corresponding tracking errors. Compared Figs. 10.9 and 10.10 with Figs. 10.8 and 10.5, the low-pass filter can reduce the chattering caused by control switching when guaranteeing the desired performance.

10.5.2 Stability verification with practical delays

In Subsection 10.5.1, the effectiveness of the proposed method has been verified by two known artificial delays. To better verify the performance in practical implementation, in this subsection, the practical communication delays between two computers located respectively in Beijing and Shenzhen that are connected by transmitting data through public Internet will be used. The time delays can be seen in Fig.4 of our previous

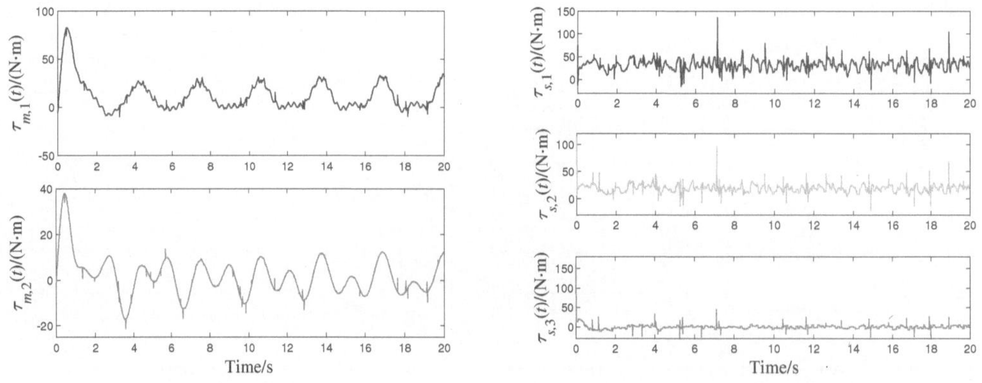

(a) (b)

Figure 10.8 The control inputs. (a) The control torque $\boldsymbol{\tau}_m(t)$ for the master robot. (b) The control torque $\boldsymbol{\tau}_s(t)$ for the slave robot.

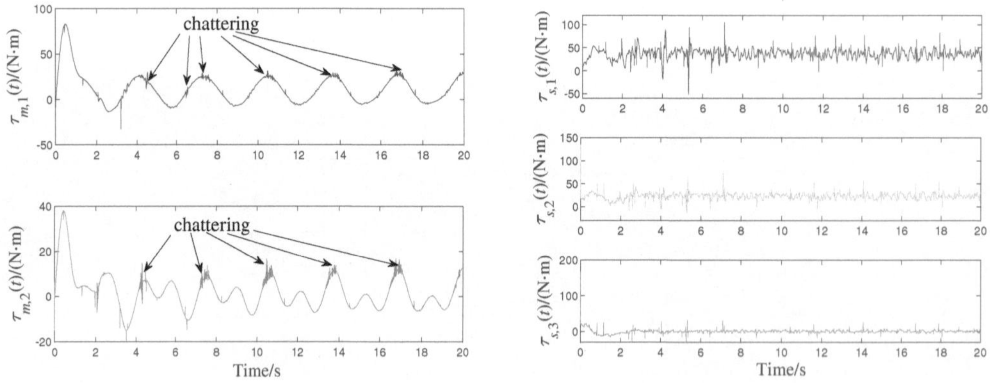

(a) (b)

Figure 10.9 The control inputs without using low-pass filters. (a) The control torque $\boldsymbol{\tau}_m(t)$ for the master robot. (b) The control torque $\boldsymbol{\tau}_s(t)$ for the slave robot.

work [47][1]. Since the delays cannot be known in advance, it is assumed that $\tilde{d} = 5$[2]. In addition, in order to better verify the effectiveness, the simulation also sets more stringent conditions. On the one hand, the operator's exogenous force input is simulated with a square wave signal. Specifically, let $\boldsymbol{f}_h^* = [0, f_{h,2}]^{\mathrm{T}}$, where $f_{h,2}$ is a periodic square wave from $-20\mathrm{N} \sim 20\mathrm{N}$ with period 3 s and initial state -20 N. On the other hand, more harsh model uncertainties will also be considered here, where

[1]Corresponds to the Fig. 7.4 in Chapter 7 of this book.

[2]The delays in Fig.4 [47] do not satisfy the Assumption 10.4. However, the robustness of the proposed method can still guarantee the stability.

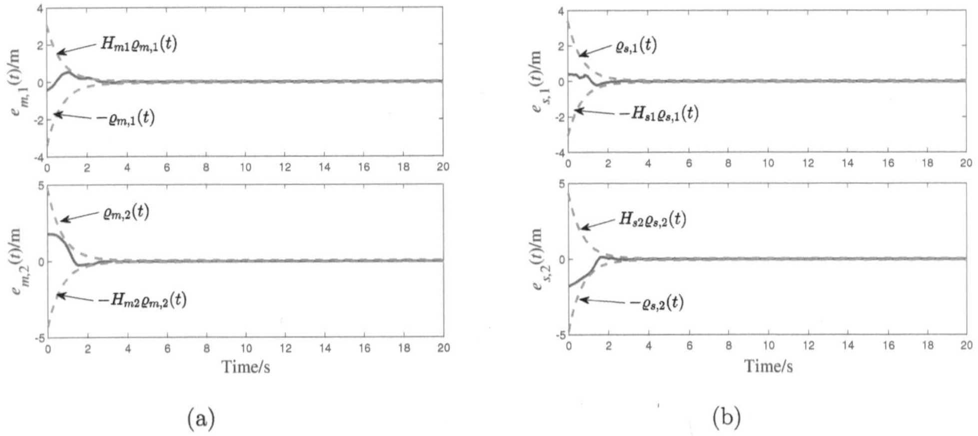

Figure 10.10 The response curves of the tracking errors without using low-pass filter. (a) Tracking error $\boldsymbol{e}_m(t)$. (b) Tracking error $\boldsymbol{e}_s(t)$.

it is assumed that

$$
\begin{aligned}
\Delta \boldsymbol{M}_m(\cdot) &= 0.2\sin(20t)\boldsymbol{M}_{mo}(\cdot) \\
\Delta \boldsymbol{C}_m(\cdot,\cdot) &= 0.3\cos(15t)\boldsymbol{C}_{mo}(\cdot,\cdot) \\
\Delta \boldsymbol{g}_m(\cdot) &= 0.3\sin(30t)\boldsymbol{g}_{mo}(\cdot) \\
\Delta \boldsymbol{M}_s(\cdot) &= 0.3\cos(20t)\sin(|\ddot{\boldsymbol{q}}_s|)\boldsymbol{M}_{so}(\cdot) \\
\Delta \boldsymbol{C}_s(\cdot,\cdot) &= 0.25\sin(20t)\boldsymbol{C}_{so}(\cdot,\cdot) \\
\Delta \boldsymbol{g}_s(\cdot) &= 0.3\sin(20t)\boldsymbol{g}_{so}(\cdot)
\end{aligned}
$$

Then under the same control gains as the ones in Subsection 10.5.1, the simulation results are shown in Figs. 10.11 and 10.12. It is shown that the proposed method

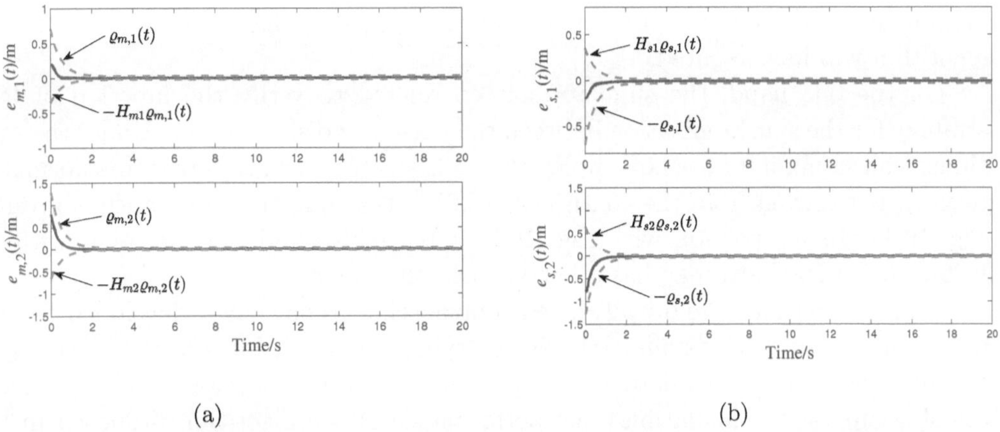

Figure 10.11 The response curves of the tracking errors with practical communication delays and a square human operator exogenous force. (a) Tracking error $\boldsymbol{e}_m(t)$. (b)Tracking error $\boldsymbol{e}_s(t)$.

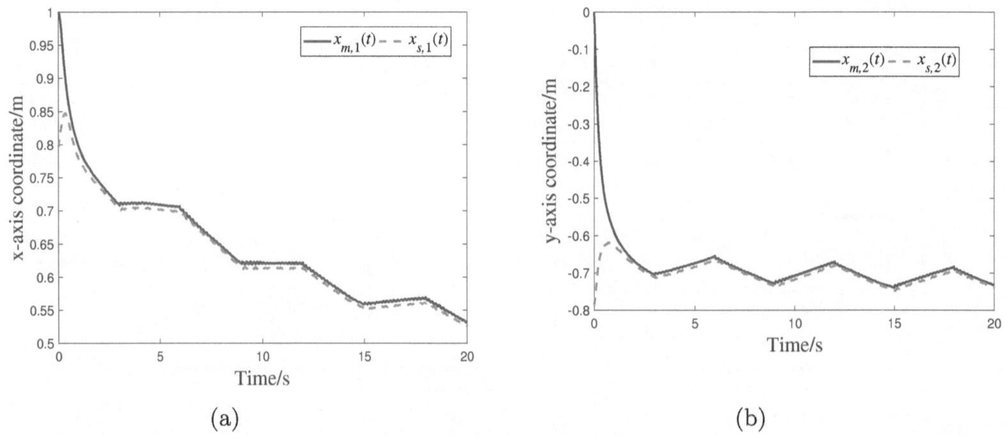

Figure 10.12 The end-effector positions of the master and slave robots with practical communication delays and a square human operator exogenous force. (a) x-axis coordinate. (b) y-axis coordinate.

can guarantee the desired performance of teleoperation system with the practical communication delays.

Remark 10.10. *Note that, since the practical delays are far less than the artificial delays in Fig. 10.3, in the following simulations, the communication delays still use the ones in Fig. 10.3.*

10.5.3 Comparisons on task performance

The PPC controller combined with the auxiliary switched filter method proposed in this chapter can realize the prescribed performance control of the teleoperation system under the time-varying delay, which is beyond the existing work. In this group of simulation, the control performance of the proposed algorithm and the existing algorithm will be compared.

On the one hand, the simulation is conducted to verify the function of PPC strategy for the synchronization between the master and slave robots. Under the same simulation setup, if it does not employ the PPC strategy, i.e., the error transformation term $\bar{\boldsymbol{e}}_j$ is removed from the controller (10.6) [3], the control performance is given as Fig. 10.13. In comparison with Fig. 10.5, it is easy to find the controller with the utilization of PPC strategy has a better task performance.

On the other hand, many advanced control schemes have been developed for teleoperation system with asymmetric time-varying delays. In Liu's work [36], the semi-autonomous task-space bilateral teleoperation control is developed. In the second set of comparative simulations, the performance of the algorithm proposed in this

[3]In this case, the control algorithm is the same as the design in Chapter 7, only subtask control is not considered.

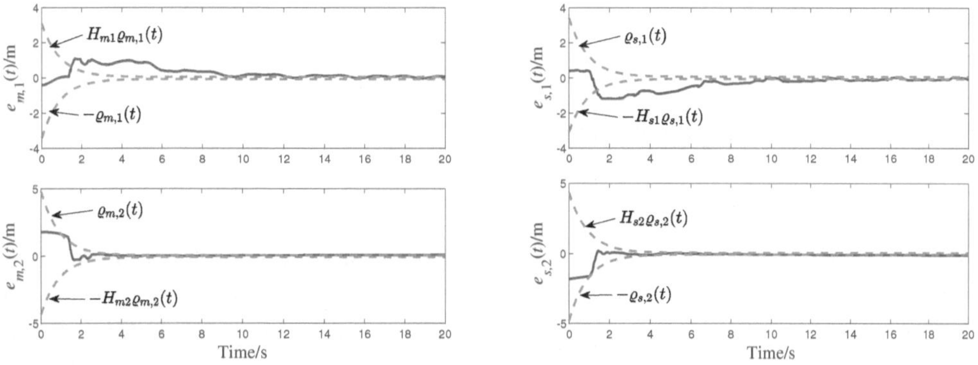

(a) (b)

Figure 10.13 The response curves of the tracking errors without using PPC strategy in the controller (10.6). (a) Tracking error $\boldsymbol{e}_m(t)$. (b) Tracking error $\boldsymbol{e}_s(t)$.

chapter and the Liu's algorithm [36] will be compared [4]. Under the same initial conditions, the control performance of Liu's method [36] is shown in Fig. 10.14[5]. Compared

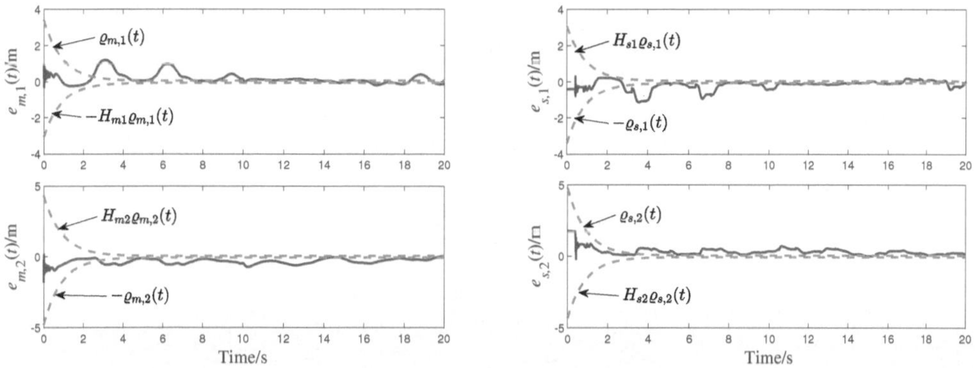

(a) (b)

Figure 10.14 The response curves of the tracking errors under the control of Liu's method [36]. (a) Tracking error $\boldsymbol{e}_m(t)$. (b) Tracking error $\boldsymbol{e}_s(t)$.

[4]Although the Liu's algorithm [36] gives the semi-autonomous control of the teleoperation system, since the joint velocity on the null space of the Jacobian matrix does not affect the motion of the task space. Although semi-autonomous control is not considered in this group of comparative simulations, it does not affect the results.

[5]Given that Liu et al. [36] only provide the non-adaptive control for teleoperation system with nominal dynamics when the external forces are non-passive, in this comparison study, it is also assumed that all of the kinematics and dynamics parameters are exactly known for their controller.

with Fig. 10.5, it is obvious that the developed algorithm of this chapter obtains a better task performance, although there is no any uncertainty in the control of Liu's method [36].

10.5.4 Comparison study with varying time delays

In this subsection, both the master and slave robots are chosen as the 2-DOF serial link manipulator.

In previous studies, some algorithms have been proposed for the joint space adaptive prescribed performance control of teleoperation systems [33–35]. However, they only consider the case of constant communication delay, the more complex and practical time-varying communication delay has not been studied, so there are limitations.

For a comparison purpose, firstly, let us take Yang's work [35] as an example. In Yang's work [35], a BLF-based adaptive controller is designed to guarantee the prescribed transient- and steady-state synchronization performance between the master and slave robots, which, however, only considers the constant communication delays. According to the Yang's control design [35], if the time delays are constant, its control performance is shown as Fig. 10.15(a), where the desired performance is achieved[6]. However, if the communication delays are time-varying (in the simulation, the varying time delays have the same upper bounds as the constant delays), under the same simulation setup, the closed-loop control performance is shown in Fig. 10.15(b), which is unstable. It shows that it is completely necessary to address the prescribed performance control for teleoperation system with varying time delays, which is actually one of the main contribution of this chapter.

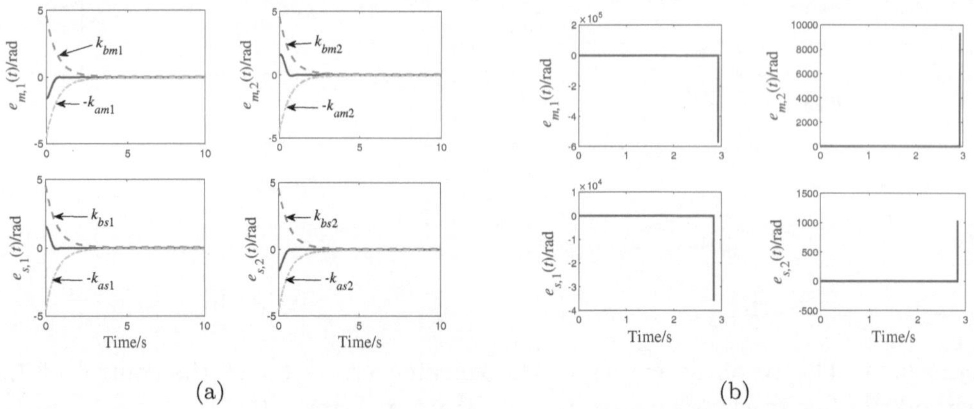

Figure 10.15 The control performance under the Yang's controller [35]. (a) Tracking errors under constant delays. (b) Tracking errors under varying time delays.

Since Yang's work [35] proposed a predefined performance control method based on barrier Lyapunov function, this chapter will also further consider the comparison between the proposed algorithm and the existing prescribed performance control

[6]In Fig. 10.15(a), k_{bji} and k_{aji} play the same roles as the $\varrho_{j,i}$ and $H_{ji}\varrho_{j,i}$, respectively.

methods. Let us take another work by Yang et al. [33] as a comparison. The developed method [33] considers the prescribed performance control of the teleoperation system with actuator saturation. To better verify the influence of time-varying delays, it is assumed there is no any model uncertainty and actuator saturation in the control of Yang's method [33]. For the simulation setup, the time delays are as the ones in Fig. 10.3, which have upper bounds 0.92 and 1.4 such that $d_m(t) \leq 0.92$ and $d_s(t) \leq 1.4$. According to Theorem 1 of Yang's work [33], the control gains can be set as $K_m = 6.093\ 9$, $K_s = 6.115$, $B_m = B_s = 1$, $k_{m1} = k_{s1} = k_{m2} = k_{s2} = k_{m3} = k_{s3} = 0$, $\lambda_m = 0.35$, $\lambda_s = 0.5$. The synchronization error constraints are set as $\varrho_{m,1}(t) = \varrho_{m,2}(t) = \varrho_{s,1}(t) = \varrho_{s,2}(t) = (4.570\ 8 - 0.05)e^{-t} + 0.05$. In the simulation, the slave robot robot is assumed to run in free motion, while external force applied on the master robot by human operator is $\boldsymbol{f}_h = \frac{\boldsymbol{f}_h^*}{5}$, where $\boldsymbol{f}_h^*$ is shown in Fig. 10.2. Since $\boldsymbol{f}_h$ is state-independent, it does not affect the stability of Yang's method [33]. First, under the same control gain, for constant communication delays $d_m(t) = 0.92$ and $d_s(t) = 1.4$, the Yang's controller [33] with the given gains can guarantee the stability of the closed-loop systems, and achieve the prescribed performance. However, when the time delays are time varying as shown in Fig. 10.3, the Yang's control scheme [33] will not work. In this case, the joint position tracking errors are shown in Fig. 10.16. It is inferred that the varying time delay has an adverse effect on the control performance of Yang's method [33].

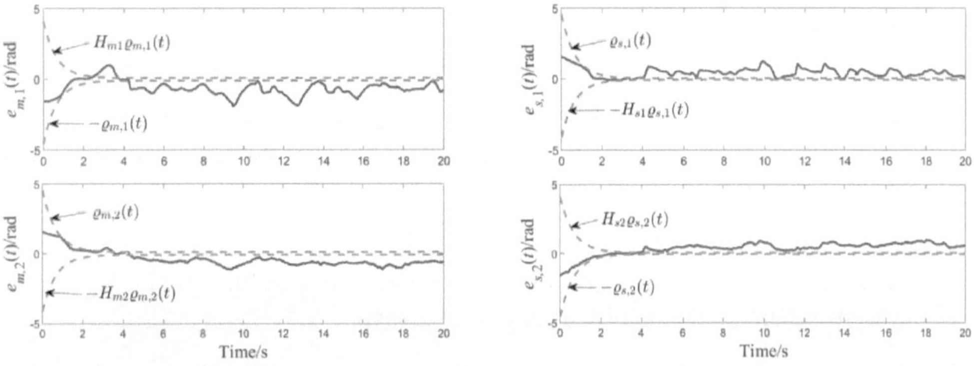

(a) (b)

Figure 10.16 The response curves of the tracking errors under the control of Yang's method [33]. (a) Tracking error $\boldsymbol{e}_m(t)$. (b) Tracking error $\boldsymbol{e}_s(t)$.

As a comparison, the developed algorithm of this chapter is also performed on the same teleoperation system. According to Remark 10.7, under the same simulation setup, the tracking performance is shown in Fig. 10.17. In comparison with Fig. 10.16, a better performance is achieved by our proposed control framework.

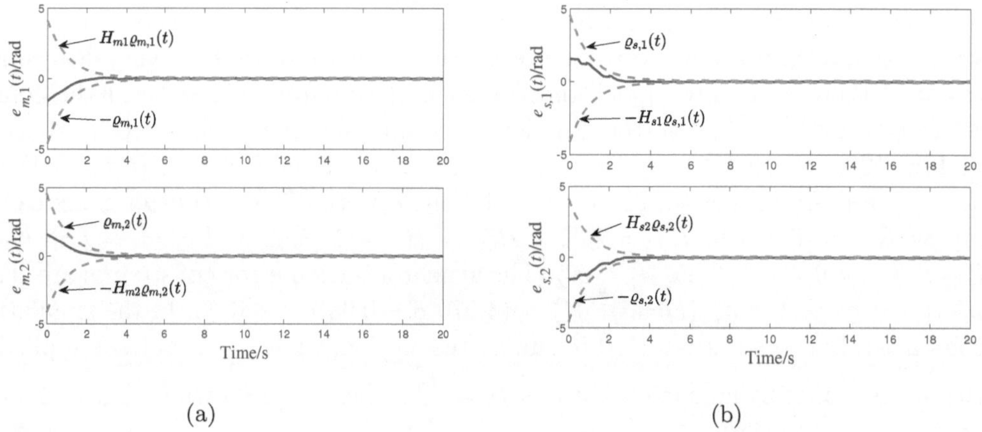

Figure 10.17 The response curves of the joint-space tracking errors under the developed control. (a) Tracking error $\boldsymbol{e}_m(t)$. (b) Tracking error$\boldsymbol{e}_s(t)$.

10.5.5 Discussions

This chapter investigates a switching-based control framework for improved task performance in teleoperation system with asymmetric time-varying delays. Its effectiveness is demonstrated by the simulation results shown in Subsection 10.5.1 and Subsection 10.5.2. Compared with the existing results, the main advantages and disadvantages lie in:

(1) Compared with the existing ordinary asymptotic control algorithms (such as the Liu's method [36] and the design in Chapter 7 of this book), this algorithm can ensure better synchronization performance, the comparison results can be seen Subsection 10.5.3.

(2) In comparison with the existing PPC- or BLF-based control schemes developed by Yang et al. [33,35], the proposed method is applicable to the teleoperation system in the presence of varying time delays.

(3) The issue of transparency is not addressed in this chapter. The complete absence of transparency analysis is a shortcoming of this chapter, which, however, will be considered in our ongoing work.

10.6 CONCLUSION

In this chapter, aiming at the high-precision control requirements of the teleoperation system, combined with the auxiliary switching filter control method, a new prescribed performance controller is proposed. Quantitative analysis and control of the task space (transient/steady-state) synchronization performance of the master and slave robots under time-varying communication delays is achieved while allowing the master and slave robots to interact with the operator/environment. On the one hand, compared with the existing predefined performance control algorithms, the new PPC control method proposed in this chapter can be applied to the time-varying communication delay situation, and allows the master and slave robots to interact with the operator and the environment, which is more in line with practical application scenarios. On the other hand, different from the previous design of most auxiliary switched filters, this chapter adopts a method similar to that proposed in Chapter 6 of this book, and introduces a double-layer filter compensator of "low-pass filter + switched filter", which reduces the control chattering caused by signal switching.

Bibliography

[1] P. F. Hokayem and M. W. Spong, "Bilateral teleoperation: an historical survey," *Automatica*, vol. 42, no. 12, pp. 2035–2057, 2006.

[2] P. M. Kebria, H. Abdi, M. M. Dalvand, A. Khosravi, and S. Nahavandi, "Control methods for Internet-based teleoperation systems: a review," *IEEE Transactions on Human-Machine Systems*, vol. 49, no. 1, pp. 32–46, 2018.

[3] C.-C. Hua and P. X. Liu, "Delay-dependent stability criteria of teleoperation systems with asymmetric time-varying delays," *IEEE Transactions on Robotics*, vol. 26, no. 5, pp. 925–932, 2010.

[4] Y.-C. Liu and N. Chopra, "Control of semi-autonomous teleoperation system with time delays," *Automatica*, vol. 49, no. 6, pp. 1553–1565, 2013.

[5] N. Chopra, M. W. Spong, and R. Lozano, "Synchronization of bilateral teleoperators with time delay," *Automatica*, vol. 44, no. 8, pp. 2142–2148, 2008.

[6] Z. Li, Y. Xia, and F. Sun, "Adaptive fuzzy control for multilateral cooperative teleoperation of multiple robotic manipulators under random network-induced delays," *IEEE Transactions on Fuzzy Systems*, vol. 22, no. 2, pp. 437–450, 2013.

[7] S. Islam, P. X. Liu, and A. El Saddik, "Teleoperation systems with symmetric and unsymmetric time varying communication delay," *IEEE Transactions on Instrumentation and Measurement*, vol. 62, no. 11, pp. 2943–2953, 2013.

[8] S. Islam, P. X. Liu, and A. El Saddik, "Nonlinear control for teleoperation systems with time varying delay," *Nonlinear Dynamics*, vol. 76, no. 2, pp. 931–954, 2014.

[9] Z. Chen, Y.-J. Pan, and J. Gu, "A novel adaptive robust control architecture for bilateral teleoperation systems under time-varying delays," *International Journal of Robust and Nonlinear Control*, vol. 25, no. 17, pp. 3349–3366, 2015.

[10] Z. Chen, Y.-J. Pan, and J. Gu, "Integrated adaptive robust control for multilateral teleoperation systems under arbitrary time delays," *International Journal of Robust and Nonlinear Control*, vol. 26, no. 12, pp. 2708–2728, 2016.

[11] Z. Li, L. Ding, H. Gao, G. Duan, and C.-Y. Su, "Trilateral teleoperation of adaptive fuzzy force/motion control for nonlinear teleoperators with communication random delays," *IEEE Transactions on Fuzzy Systems*, vol. 21, no. 4, pp. 610–624, 2012.

[12] Z. Li, X. Cao, Y. Tang, R. Li, and W. Ye, "Bilateral teleoperation of holonomic constrained robotic systems with time-varying delays," *IEEE Transactions on Instrumentation and Measurement*, vol. 62, no. 4, pp. 752–765, 2013.

[13] Y. Kang, Z. Li, X. Cao, and D. Zhai, "Robust control of motion/force for robotic manipulators with random time delays," *IEEE Transactions on Control Systems Technology*, vol. 21, no. 5, pp. 1708–1718, 2012.

[14] S. Qiu, Z. Li, W. He, L. Zhang, C. Yang, and C.-Y. Su, "Brain–machine interface and visual compressive sensing-based teleoperation control of an exoskeleton robot," *IEEE Transactions on Fuzzy Systems*, vol. 25, no. 1, pp. 58–69, 2016.

[15] Z. Li, C.-Y. Su, L. Wang, Z. Chen, and T. Chai, "Nonlinear disturbance observer-based control design for a robotic exoskeleton incorporating fuzzy approximation," *IEEE Transactions on Industrial Electronics*, vol. 62, no. 9, pp. 5763–5775, 2015.

[16] Z. Li and C.-Y. Su, "Neural-adaptive control of single-master–multiple-slaves teleoperation for coordinated multiple mobile manipulators with time-varying communication delays and input uncertainties," *IEEE Transactions on Neural Networks and Learning Systems*, vol. 24, no. 9, pp. 1400–1413, 2013.

[17] E. Nuno, R. Ortega, and L. Basanez, "An adaptive controller for nonlinear teleoperators," *Automatica*, vol. 46, no. 1, pp. 155–159, 2010.

[18] I. Sarras, E. Nuno, and L. Basanez, "An adaptive controller for nonlinear teleoperators with variable time-delays," *Journal of the Franklin Institute*, vol. 351, no. 10, pp. 4817–4837, 2014.

[19] P. Malysz and S. Sirouspour, "Nonlinear and filtered force/position mappings in bilateral teleoperation with application to enhanced stiffness discrimination," *IEEE Transactions on Robotics*, vol. 25, no. 5, pp. 1134–1149, 2009.

[20] I. G. Polushin, A. Tayebi, and H. J. Marquez, "Control schemes for stable teleoperation with communication delay based on ios small gain theorem," *Automatica*, vol. 42, no. 6, pp. 905–915, 2006.

[21] I. G. Polushin, P. X. Liu, and C.-H. Lung, "A control scheme for stable force-reflecting teleoperation over ip networks," *IEEE Transactions on Systems, Man, and Cybernetics, Part B (Cybernetics)*, vol. 36, no. 4, pp. 930–939, 2006.

[22] I. G. Polushin, P. X. Liu, and C.-H. Lung, "A force-reflection algorithm for improved transparency in bilateral teleoperation with communication delay," *IEEE/ASME Transactions on Mechatronics*, vol. 12, no. 3, pp. 361–374, 2007.

[23] I. G. Polushin, S. N. Dashkovskiy, A. Takhmar, and R. V. Patel, "A small gain framework for networked cooperative force-reflecting teleoperation," *Automatica*, vol. 49, no. 2, pp. 338–348, 2013.

[24] S.-J. Lee and H.-S. Ahn, "Controller designs for bilateral teleoperation with input saturation," *Control Engineering Practice*, vol. 33, pp. 35–47, 2014.

[25] F. Hashemzadeh, I. Hassanzadeh, and M. Tavakoli, "Teleoperation in the presence of varying time delays and sandwich linearity in actuators," *Automatica*, vol. 49, no. 9, pp. 2813–2821, 2013.

[26] J. T. Spooner, M. Maggiore, R. Ordonez, and K. M. Passino, *Stable adaptive control and estimation for nonlinear systems: neural and fuzzy approximator techniques*. John Wiley & Sons, 2004.

[27] Y.-C. Chang, "An adaptive H_∞ tracking control for a class of nonlinear multiple-input multiple-output (MIMO) systems," *IEEE Transactions on Automatic Control*, vol. 46, no. 9, pp. 1432–1437, 2001.

[28] G. Bianchini, R. Genesio, A. Parenti, and A. Tesi, "Global H_∞ controllers for a class of nonlinear systems," *IEEE Transactions on Automatic Control*, vol. 49, no. 2, pp. 244–249, 2004.

[29] C. P. Bechlioulis, Z. Doulgeri, and G. A. Rovithakis, "Neuro-adaptive force/position control with prescribed performance and guaranteed contact maintenance," *IEEE Transactions on Neural Networks*, vol. 21, no. 12, pp. 1857–1868, 2010.

[30] K. P. Tee, S. S. Ge, and E. H. Tay, "Barrier Lyapunov functions for the control of output-constrained nonlinear systems," *Automatica*, vol. 45, no. 4, pp. 918–927, 2009.

[31] A. K. Kostarigka and G. A. Rovithakis, "Adaptive dynamic output feedback neural network control of uncertain mimo nonlinear systems with prescribed performance," *IEEE Transactions on Neural Networks and Learning Systems*, vol. 23, no. 1, pp. 138–149, 2011.

[32] B. Ren, S. S. Ge, K. P. Tee, and T. H. Lee, "Adaptive neural control for output feedback nonlinear systems using a barrier Lyapunov function," *IEEE Transactions on Neural Networks*, vol. 21, no. 8, pp. 1339–1345, 2010.

[33] Y. Yang, C. Ge, H. Wang, X. Li, and C. Hua, "Adaptive neural network based prescribed performance control for teleoperation system under input saturation," *Journal of the Franklin Institute*, vol. 352, no. 5, pp. 1850–1866, 2015.

[34] Y. Yang, C. Hua, and X. Guan, "Synchronization control for bilateral teleoperation system with prescribed performance under asymmetric time delay," *Nonlinear Dynamics*, vol. 81, no. 1, pp. 481–493, 2015.

[35] Y. Yang, C. Hua, and X. Guan, "Finite time control design for bilateral teleoperation system with position synchronization error constrained," *IEEE Transactions on Cybernetics*, vol. 46, no. 3, pp. 609–619, 2015.

[36] Y.-C. Liu, "Task-space bilateral teleoperation systems for heterogeneous robots with time-varying delays," *Robotica*, vol. 33, no. 10, pp. 2065–2082, 2015.

[37] P. Malysz and S. Sirouspour, "A kinematic control framework for single-slave asymmetric teleoperation systems," *IEEE Transactions on Robotics*, vol. 27, no. 5, pp. 901–917, 2011.

[38] D.-H. Zhai and Y. Xia, "Adaptive fuzzy control of multilateral asymmetric teleoperation for coordinated multiple mobile manipulators," *IEEE Transactions on Fuzzy Systems*, vol. 24, no. 1, pp. 57–70, 2015.

[39] C.-C. Cheah, M. Hirano, S. Kawamura, and S. Arimoto, "Approximate jacobian control for robots with uncertain kinematics and dynamics," *IEEE Transactions on Robotics and Automation*, vol. 19, no. 4, pp. 692–702, 2003.

[40] C.-C. Cheah, C. Liu, and J.-J. E. Slotine, "Adaptive tracking control for robots with unknown kinematic and dynamic properties," *The International Journal of Robotics Research*, vol. 25, no. 3, pp. 283–296, 2006.

[41] H. Wang, "Passivity based synchronization for networked robotic systems with uncertain kinematics and dynamics," *Automatica*, vol. 49, no. 3, pp. 755–761, 2013.

[42] Y.-C. Liu and M.-H. Khong, "Adaptive control for nonlinear teleoperators with uncertain kinematics and dynamics," *IEEE/ASME Transactions on Mechatronics*, vol. 20, no. 5, pp. 2550–2562, 2015.

[43] A. Haddadi, K. Razi, and K. Hashtrudi-Zaad, "Operator dynamics consideration for less conservative coupled stability condition in bilateral teleoperation," *IEEE/ASME Transactions on Mechatronics*, vol. 20, no. 5, pp. 2463–2475, 2015.

[44] T. Tsuji, P. G. Morasso, K. Goto, and K. Ito, "Human hand impedance characteristics during maintained posture," *Biological Cybernetics*, vol. 72, no. 6, pp. 475–485, 1995.

[45] R. Kelly, V. S. Davila, and J. A. L. Perez, *Control of Robot Manipulators in Joint Space*. Springer Science & Business Media, 2005.

[46] L. Sciavicco and B. Siciliano, *Modelling and Control of Robot Manipulators*. Springer Science & Business Media, 2001.

[47] D.-H. Zhai and Y. Xia, "Adaptive control of semi-autonomous teleoperation system with asymmetric time-varying delays and input uncertainties," *IEEE Transactions on Cybernetics*, vol. 47, no. 11, pp. 3621–3633, 2016.

CHAPTER 11

Practical given performance control of robotic systems

TO improve the practical application ability of classical predefined performance control, this chapter investigates a switching-based prescribed-performance-like control approach for nonlinear teleoperation systems, which balances the performance and implementability of classical predefined performance control algorithm when its hard constraint is violated. With the help of auxiliary switched filter control technique, the proposed switched control is capable of guaranteeing the prescribed performance for the synchronization between the master and slave robots in simultaneous presence of unknown dynamic parameters and asymmetric time delays, regardless of whether the robots is in contact with the human operator/environment or not.

11.1 INTRODUCTION

In the previous chapters, for some occasions that require precise operation, such as surgical robot systems, the control needs to comprehensively consider indicators such as maximum overshoot, convergence rate, steady-state error, etc.; that is, it is necessary to control transient and steady-state performance at the same time. Many existing methods for robot control cannot take into account both aspects of performance when used in such teleoperation systems [1–14], so they face limitations. In this regard, in recent years, researchers have begun to try to apply predefined performance control methods such as PPC/BLF to teleoperation systems [15–18]. For the teleoperation system with constant communication time delay and the master and slave robots are both in free motion, Yang et al. propose the PPC/BLF-based predefined performance control [15–17], which realizes high-performance control with both transient performance and steady-state performance. In our previous work [18][1], using the PPC strategy combined with the auxiliary switched filter control method, the existing predefined performance control method is extended to the high-precision control of the teleoperation system with time-varying communication delay, and the master and slave robots are allowed to have contact with the operator/environment.

[1]It corresponds to the research in Chapter 10 of this book.

DOI: 10.1201/9781003382058-11

However, considering the complexity of real robotic systems, the above work still faces many difficulties when applied to the predefined performance control of practical teleoperation systems.

Generally, for the given tracking error $\boldsymbol{e}(t) = [e_1(t), e_2(t), \cdots, e_n(t)]^{\mathrm{T}} \in \mathbb{R}^n$, the mathematical expression of predefined performance is given as [17,19,20]

$$\begin{cases} -H_i\varrho_i(t) < e_i(t) < \varrho_i(t), & e_i(0) \geq 0 \\ -\varrho_i(t) < e_i(t) < H_i\varrho_i(t), & e_i(0) < 0 \end{cases} \quad \forall\, t \geq 0 \tag{11.1}$$

where $i = 1, 2, \cdots, n$, $H_i \in [0, 1]$, $\varrho_i(t)$ is the performance function and is a bounded smooth strictly positive decreasing function satisfying

$$\varrho_i(t) = (\varrho_{i0} - \varrho_{i\infty})e^{-l_i t} + \varrho_{i\infty}$$

where $l_i > 0$, $\varrho_{i0} = \varrho_i(0)$ is selected such that (11.1) is satisfied at $t = 0$, $\varrho_{i\infty} = \lim_{t\to\infty} \varrho_i(t) > 0$ represents the maximum allowable size of $e_i(t)$ at the steady state and can be set arbitrarily small, the maximum overshoot of $e_i(t)$ is prescribed less than $H_i\varrho_i(0)$. In the rest of this chapter, for simplicity of expression, let $H_i = 1$. To achieve the predefined performance (11.1), the PPC approach and the BLF strategy take different technical routes. Here, let us take the classical PPC scheme as an example. From the classical PPC schemes [19–21], the predefined performance shown by (11.1) can be satisfied by pursuing the boundedness of the transformed error $\bar{e}_i(t)$ as follows:

$$\bar{e}_i(t) = T_i\left(\frac{e_i(t)}{\varrho_i(t)}\right) \tag{11.2}$$

where $T_i(\cdot)$ is a smooth strictly increasing function, and defined as

$$T_i : (-1, 1) \to (-\infty, \infty)$$

From the definition of $T_i(\cdot)$, the expression (11.1) can be seen as a hard constraint that ensures $\bar{e}_i(t)$ nonsingular. In a classical PPC-based predefined performance control system [19–21], for all $t \geq t_0 = 0$, the hard constraint shown in formula (11.1) must be satisfied. This is very strict in practice.

On the one hand, as shown in Fig. 11.1, the performance function $\varrho_i(t)$ in (11.1) is strictly decreasing, which is dominated by the fixed parameters ϱ_{i0}, $\varrho_{i\infty}$ and l_i. On the other hand, due to the existences of disturbances and uncertainties, it is hard and even impossible to guarantee that the error is controlled to zero. Therefore, when $\varrho_i(t)$ has a small value, it is likely to occur that $|e_i(t)|$ is very close to $\varrho_i(t)$. At this time, if a disturbance is injected to the closed-loop system, there may arise the event $|e_i(t)| \geq \varrho_i(t)$. Here, the disturbances refer to the ones that are not taken into account by the control design. For example, for a PPC control system that does not take into account the actuator output capability, actuator saturation can be viewed as such a disturbance. Since $\dfrac{|e_i(t)|}{\varrho_i(t)}$ is close to 1, it can lead to a large $\bar{e}_i$, and cause a higher control torque output, but the actual actuator output capacity is always limited, and this higher torque output is likely to lead to actuator saturation. In this regard, the predefined performance (11.1) is difficult to maintain, and eventually the event $|e_i(t)| \geq \varrho_i(t)$ may occur. Another example is the problem of digital implementation of controllers. Many controllers are digitally implemented today, which face the problem of sampling control. In the process of digital implementation, assuming that $\{\bar{t}_k\}_{k\geq 0}$ is the sampling time sequence, then in $[\bar{t}_k, \bar{t}_{k+1})$, the input of the controller is $\boldsymbol{\tau}(t) = \boldsymbol{\tau}(\bar{t}_k)$. If a disturbance is injected during $(\bar{t}_k, \bar{t}_{k+1})$ and affects the control system, the control input $\boldsymbol{\tau}(\bar{t}_k)$ is impossible to adapt and adjust in time. In this case, the event $|e_i(t)| \geq \varrho_i(t)$ may also occur. Therefore, it can be said that the classical predefined performance control is sensitive to external disturbances. In the actual control, the event $|e_i(t)| \geq \varrho_i(t)$ is entirely possible.

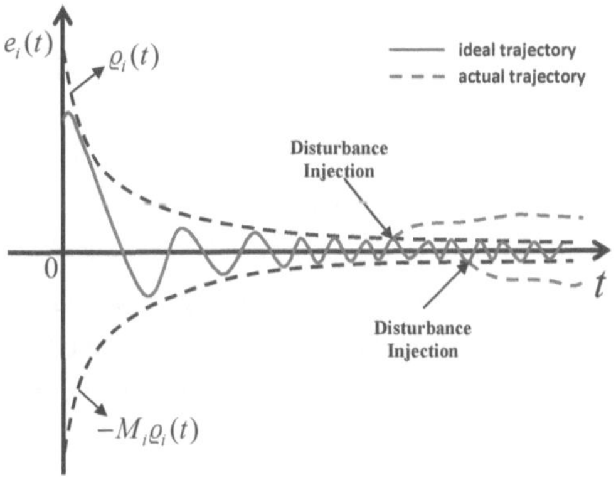

Figure 11.1 Graphical illustration of the synchronization performance of classical PPC method.

Mathematically, the event $|e_i(t)| \geq \varrho_i(t)$ will cause the equation (11.2) to appear singular. Once the singularity occurs, the introduced error transformation $\bar{e}_i(t)$ will be invalid, which may lead to the failure of the classical PPC algorithm. Similar problems also exist in classical BLF control systems. Existing achievements considering predefined performance control, for example, the references [15–17, 19–25] and Chapter 10 of this book, etc., do not consider the handling of the above-mentioned singular events, which inevitably leads to limited applications.

Specific to the teleoperation system, the network attack [26–28] existing in the communication process will lead to sudden changes in system performance, thus inducing $|e_i(t)| \geq \varrho_i(t)$. In addition, the robots need to maintain contact with the operator and/or the environment during most of working hours. The motions of both the master and slave robots are usually affected by the operator/environment forces. Those external forces can be regarded as exogenous disturbances that affect the synchronization between the master and slave robots, which can raise the stability issues in a teleoperation system with a classical PPC or BLF controller. For example, when the motion of a master robot which is manipulated by a human operator is suddenly changed and becomes very fast, the singular event $|e_i(t)| \geq \varrho_i(t)$ will arise if the slave robot cannot follow the motion in real time. In addition, if there exists an obstacle in the operation space, the singular event may also happen when the slave robot encounters obstruction. Although the predefined performance control schemes have been investigated for teleoperation system [15–17], all of them have assumed that the communication delays are constant and both of the master and slave robots run in free motion. The practical contact with operator/environment can be seen as one of the excitation sources for singular event $|e_i(t)| \geq \varrho_i(t)$ and has an adverse effect on the performance of the existing methods [15–17]. To show this, let us take the control design of Yang et al. [17] as example. When both the master and slave robots run in free motion and the communication delays are constant, the system is stable, as shown in Fig. 11.2(a). By contrast, if the robots contact with the human operator/environment, the system is unstable as in Fig. 11.2(b). Obviously, the algorithm proposed by Yang et al. [17] fails in this case. One of the key reasons for the instability of the system is that when a singular event $|e_i(t)| \geq \varrho_i(t)$ occurs, the error transformation $\bar{e}_i(t)$ is invalid, see Fig. 11.1.

Roughly speaking, the singular event $|e_i(t)| \geq \varrho_i(t)$ can be regarded as a fault of the classical PPC control system. But the problem considered in this chapter is different from the existing fault-tolerant tracking control with predefined performance. Some predefined performance fault-tolerant control methods are proposed to deal with actuator failures and improve control performance [29–33]. However, since the actuator faults they consider are only one of the disturbances that lead to the occurrence of singular events, they cannot avoid the occurrence of singular events or ensure the stability and control performance of the closed-loop system after the occurrence of singular events.

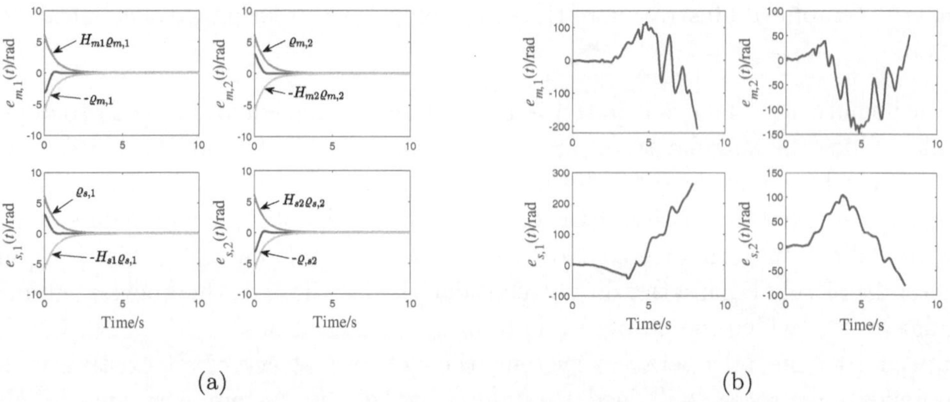

Figure 11.2 The tracking performance of Yang's method [17]. (a) Free motion case. (b) Contact with the operator and the environment.

Of course, disturbances that cause singular events can also be overcome or weakened by using disturbance rejection control techniques (such as disturbance observers, extended state observers, etc.) [13,34]. However, the existing anti-disturbance control technologies cannot completely offset the effects of these disturbances. In addition to the aforementioned digital implementation problems, there are also the following incentives. According to the conclusion of Guo et al. [34], the existing disturbance rejection control technology has some application limitations more or less, and can only deal with some specific disturbances, but cannot eliminate all disturbances. For example, Wang et al. propose a prescribed performance control algorithm based on an extended state observer for a class of nonlinear servomechanical systems, where the extended state observer is used to estimate and compensate for unknown disturbances [35]. However, according to Wang's work [35], it needs to limit the unknown perturbation to have a bounded rate of change, which faces great limitations in practical system applications.

According to the above discussion, it is necessary to address the stability problem when there arises the singular event $|e_i(t)| \geq \varrho_i(t)$ in a classical predefined performance control system. Regardless of these recent developments, however, no work has been available in the literature on addressing such control problem. It is, therefore, the purpose of this chapter is to fill such a gap.

To solve this problem, this chapter develops a switching-based prescribed performance-like control (PPLC) approach, which is applicable to teleoperation systems involving with varying time delays and unknown dynamics. The diagram of the proposed PPLC scheme is shown in Fig. 11.3. Its underlying idea is to introduce the

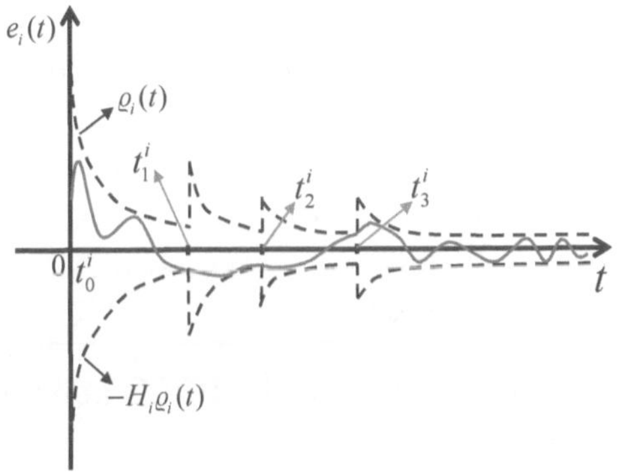

Figure 11.3 Graphical illustration of the synchronization performance of switching-based PPLC method.

switched performance function to guarantee $|e_i(t)| < \varrho_i(t)$ for all time. The algorithm keeps monitoring whether the singular event $|e_i(t)| \geq \varrho_i(t)$ occurs or not. Whenever there arises the singular event due to the disturbances that are not considered by control design, the performance function $\varrho_i(t)$ is adjusted accordingly. Then the rest can follow the classical PPC approach to achieve desired convergence. Although it is very intuitive, how to achieve such a control and guarantee the closed-loop stability is a big challenge. Specifically, the main difficulties and mathematical obstacles we are facing are highlighted as follows: ① How to choose an appropriate value for performance function $\varrho_i(t)$ at the moment when there arises the singular event? A large initial value of $\varrho_i(t)$ will lead to a large overshoot thereafter. A small initial value may cause the Zeno behavior and even instability. ② How to avoid the Zeno behavior in such system?[2] In other words, in addition to selecting the appropriate initial value of $\varrho_i(t)$, what other measures need to be synchronized to eliminate Zeno behavior? ③ How to obtain the stability conditions for the resulting closed-loop system? The piecewise performance function mechanism makes the closed-loop system model more complex. How to analyze the stability of the system is a challenge. In this chapter, these challenging issues are taken fully into account by developing a proposed PPLC algorithm with an effective stability analysis approach. A switching mechanism is introduced into the classical controller design and on this basis, a nonlinear proportional plus auxiliary compensation controller is designed, where a switched filter control technique is employed. By modeling the resulting closed-loop system as a special switched delayed system and using the piecewise Lyapunov-Krasovskii functional method, the state-independent input-to-output stability of closed-loop system is established, where a feasible range of the 'initial' values of piecewise continuous performance function is given to achieve a tradeoff between the closed-loop stability and overshoot. To sum up, the main contribution of this chapter can be summarized as follows: ① A switching-based adaptive controller is designed, which is capable of guaranteeing the control implementability and avoiding the occurrence of Zeno behavior. ② Sufficient conditions are obtained for the resulting closed-loop system by applying the piecewise Lyapunov-Krasovskii functional method. ③ The proposed switching-based PPLC approach has a wider range of applications than the classical PPC scheme. Specifically, when there arises the singular event, it can guarantee the closed-loop stability with a satisfactory performance. It is also capable of reducing to a classical PPC algorithm if there does not arise the singular event.

11.2 PROBLEM FORMULATION

For the sake of simplifying the expression, the PPLC approach is firstly de- veloped for networked robotic vehicle.

11.2.1 Dynamics of the robotic vehicle

Let us consider the following robotic vehicle:

$$\boldsymbol{M}(\boldsymbol{q})\ddot{\boldsymbol{q}} + \boldsymbol{C}(\boldsymbol{q},\dot{\boldsymbol{q}})\dot{\boldsymbol{q}} + \boldsymbol{g}(\boldsymbol{q}) + \boldsymbol{f}(\dot{\boldsymbol{q}}) + \boldsymbol{f}_d = -\boldsymbol{J}^{\mathrm{T}}(\boldsymbol{q})\boldsymbol{f}_\lambda + \boldsymbol{\tau} \tag{11.3}$$

where $\boldsymbol{q} \in \mathbb{R}^n$, $\boldsymbol{M}(\boldsymbol{q})$ and $\boldsymbol{C}(\boldsymbol{q},\dot{\boldsymbol{q}})$ are respectively the symmetric positive definite inertia matrix and coriolis/centrifugal matrix, $\boldsymbol{g}(\boldsymbol{q})$ is the gravitational torque, $\boldsymbol{f}(\dot{\boldsymbol{q}})$ is the friction, $\boldsymbol{f}_d$ is the bounded external disturbance, $\boldsymbol{\tau}$ is the control torque, $\boldsymbol{f}_\lambda$ is the contact force, and $\boldsymbol{J}(\boldsymbol{q})$ is the Jacobian matrix. In this chapter, the contact force $\boldsymbol{f}_\lambda$ is assumed to be bounded. In addition, for simplicity, the Jacobian matrix $\boldsymbol{J}(\boldsymbol{q})$ is assumed to be exactly known.

[2]The definition of Zeno behavior can be seen in Heymann's work [36] and here refers to the phenomenon that the number of occurrence of the event $|e_i(t)| \geq \varrho_i(t)$ is infinite in a finite amount of time.

The dynamics (11.3) satisfies the classical properties [37]:

Property 11.1. *There exist constant scalars* $\rho_1 > 0$ *and* $\rho_2 > 0$ *such that* $\rho_1 \boldsymbol{I} \leq \boldsymbol{M}(\boldsymbol{q}) \leq \rho_2 \boldsymbol{I}$, *where* $\boldsymbol{I}$ *is an identity matrix with suitable dimensions.*

Property 11.2. $\dot{\boldsymbol{M}}(\boldsymbol{q}) - 2\boldsymbol{C}(\boldsymbol{q}, \dot{\boldsymbol{q}})$ *is skew-symmetric.*

Property 11.3. *The dynamics model is linearly parameterizable. That is, for any* $\boldsymbol{z}_1, \boldsymbol{z}_2 \in \mathbb{R}^n$, *it has*

$$\boldsymbol{Y}(\boldsymbol{q}, \dot{\boldsymbol{q}}, \boldsymbol{z}_1, \boldsymbol{z}_2)\boldsymbol{\theta} = \boldsymbol{M}(\boldsymbol{q})\boldsymbol{z}_1 + \boldsymbol{C}(\boldsymbol{q}, \dot{\boldsymbol{q}})\boldsymbol{z}_2 - \boldsymbol{g}(\boldsymbol{q}) - \boldsymbol{f}(\dot{\boldsymbol{q}})$$

where $\boldsymbol{Y}(\boldsymbol{q}, \dot{\boldsymbol{q}}, \boldsymbol{z}_1, \boldsymbol{z}_2) \in \mathbb{R}^{m \times n}$ *is a regression matrix composed of known functions,* $\boldsymbol{\theta} \in \mathbb{R}^m$ *is a vector composed of unknown parameters.*

The robot system shown in (11.3) runs at the remote end. Its desired motion $(\boldsymbol{q}_d, \dot{\boldsymbol{q}}_d)$ is generated locally and is transmitted to the robot side through the communication network. In order to simplify the analysis, only the position tracking control is considered. To measure the tracking performance, the tracking error $\boldsymbol{e} = [e_1, e_2, \cdots, e_n]^{\mathrm{T}}$ is introduced:

$$\boldsymbol{e} = \boldsymbol{q} - \boldsymbol{q}_d(t - d(t))$$

where $d(t)$ is the communication delay, which satisfies

Assumption 11.1. *There exist bounded constant positive scalars* $\bar{d}$ *and* $\tilde{d}$ *such that* $0 \leq d(t) \leq \bar{d}$ *and* $|\dot{d}(t)| \leq \tilde{d}$.

11.2.2 Switching-based PPLC strategy

Since the control design cannot take into account all of the disturbances, it may occur $|e_i(t)| \geq \varrho_i(t)$, which will cause the invalid of a classical PPC based controller. To achieve a tradeoff between the control implementability and control performance, this chapter investigates a switching-based PPLC scheme. A switching-based PPLC system can be divided into two phases: initial phase (classical PPC phase) and PPLC phase. The closed-loop system starts from the initial phase. In this phase the control system is initialized as a PPC system. The controller keeps monitoring whether the event $|e_i(t)| \geq \varrho_i(t)$ occurs or not. If it never occurs, the proposed control performs the same performance as the classical PPC scheme. By choosing the proper performance function $\varrho_i(t)$ (or, equivalently, l_i, $\varrho_i(0)$ and $\varrho_{i\infty}$), the prescribed control performance can be achieved. Once the event $|e_i(t)| \geq \varrho_i(t)$ occurs, the control system will switch to the second phase. Let t_k^i denote the switching time. In this phase, the control is still regarded as a classical PPC with t_k^i being the new 'initial' time. The control flow is the same as the initial phase. Roughly speaking, this design is intuitive, while the technical difficulties lie in how to guarantee the closed-loop stability and the better synchronization performance without occurring the Zeno behavior.

Specifically, in the developed switching-based PPLC scheme, to guarantee the control implementability, the switched performance function is introduced as

$$\varrho_i(t) = (\varrho_i(t_k^i) - \varrho_{i\infty})\mathrm{e}^{-l_i(t-t_k^i)} + \varrho_{i\infty}, \ \forall t \in [t_k^i, t_{k+1}^i), \ k \geq 0 \tag{11.4}$$

where as shown in Fig. 11.3, $t_0^i = t_0 = 0$, the time sequence $\{t_k^i\}_{k\geq 0}$ is defined by

$$t_{k+1}^i = \min\{t|t > t_k^i, \text{ such that } |e_i(t)| \geq \varrho_i(t)\}$$

$\varrho_{i\infty} = \delta > 0$, $\delta > 0$ is a prespecified value which represents the steady-state convergence performance.

Remark 11.1. *Once $\varrho_i(t_k^i)$ is given, since l_i and $\varrho_{i\infty}$ are known, the value of $\varrho_i(t)$ can be calculated in real time. So using the measured error $e_i(t)$, the control algorithm can easily monitor whether the event $|e_i(t)| \geq \varrho_i(t)$ has occurred.*

It is known that in the classical PPC approach, the initial value $\varrho_i(t_k^i)$ should be selected such that the hard constraint (11.1) is satisfied at $t = t_k^i$. However, although the switching-based PPLC algorithm follows the classical PPC approach in every time interval $[t_k^i, t_{k+1}^i)$, the control should carefully take the 'initial' value $\varrho_i(t_k^i)$, since it is related to the control performance. In this case, (11.5) gives a feasible range of those "initial" values, and achieves a tradeoff between the control implementability and control performance.

$$\begin{cases} \varrho_i(0) > |e_i(0)| \\ \varrho_i(t_k^i) > \max\left\{ \dfrac{\varrho_i(t_k^{i-})}{|e_i(t_k^{i-})|}|e_i(t_k^i)|, |e_i(t_k^i)| \right\}, \ \forall\, k \geq 1 \end{cases} \tag{11.5}$$

When $\varrho_i(t_k^i)$ takes the value according to the above rules, the control can achieve a balance between closed-loop stability and overshoot.

Remark 11.2. *Because $\varrho_i(t_k^{i-}) > |e_i(t_k^{i-})|$, when $k \geq 1$, the 'initial' value in (11.5) can be further simplified as*

$$\varrho_i(t_k^i) > \frac{\varrho_i(t_k^{i-})}{|e_i(t_k^{i-})|}|e_i(t_k^i)|, \ \forall\, k \geq 1$$

Remark 11.3. *The motivation of the switching-based PPLC method proposed in this chapter with the idea of phased control is obvious, and the key point of the design is to choose the "initial" value $\varrho_i(t_k^i)$, $k \geq 1$. To achieve a balance between implementability and control performance, it cannot simply be selected as $\varrho_i(t_k^i) > |e_i(t_k^i)|$, but needs to satisfy (11.5).*

The synchronization error still needs to meet the piecewise constraint as shown in (11.1). Accordingly, the error transformation strategy is piecewisely designed as

$$\bar{e}_i(t) = T_i\left(\frac{e_i(t)}{\varrho_i(t)}\right), \ t \in [t_k^i, t_{k+1}^i) \tag{11.6}$$

where

$$T_i\left(\frac{e_i(t)}{\varrho_i(t)}\right) = \ln\left(\frac{1 + \dfrac{e_i(t)}{\varrho_i(t)}}{1 - \dfrac{e_i(t)}{\varrho_i(t)}} \right)$$

In what follows, let $\{t_k\}_{k\geq 0} = \bigcup_{i=1,2,\cdots,n}\{t_k^i\}_{k\geq 0}$, with $t_{k+1} > t_k$, $t_0 = 0$.

Remark 11.4. *According to the above design, the following conclusions are established at time t_k^i:*

(1) From (11.5) and the definition of $T_i(\cdot)$, $|\bar{e}_i(t_k^i)| \le |\bar{e}_i(t_k^{i-})|$.

(2) Since $\varrho_{i\infty} = \delta > 0$, considering the continuity of $e_i(t)$, it has $\bar{e}_i(t_k^i)\bar{e}_i(t_k^{i-}) > 0$.

Therefore, the task of control is: to design the control input $\boldsymbol{\tau}(t)$, so that the tracking error $\boldsymbol{e}(t)$ satisfies the desired (piecewised) prescribed performance (11.1), where the performance function is given by (11.4).

11.3 ADAPTIVE CONTROLLER

In view of the above control requirements, this section presents the specific PPLC controller design.

Let $\lambda > 0$ be a constant scalar. Denote

$$\begin{aligned}\boldsymbol{\eta} &= \dot{\boldsymbol{q}} + \lambda \boldsymbol{e} \\ \boldsymbol{e}_v &= \dot{\boldsymbol{q}} - \dot{\boldsymbol{q}}_d(\theta)|_{\theta=t-d(t)}\end{aligned}$$

For convenience, let $\boldsymbol{q}_{vd}(t, d(t)) = \dot{\boldsymbol{q}}_d(\theta)|_{\theta=t-d(t)}$. For the robot, $\boldsymbol{q}_{vd}(t, d(t))$ is the desired velocity signal it receives at time t. Then, $\dot{\boldsymbol{e}} = \boldsymbol{e}_v + \dot{d}(t)\boldsymbol{q}_{vd}(t, d(t))$. Substituting $\boldsymbol{\eta}$ into (11.3),

$$\begin{aligned}&\boldsymbol{M}(\boldsymbol{q})\dot{\boldsymbol{\eta}} + \boldsymbol{C}(\boldsymbol{q}, \dot{\boldsymbol{q}})\boldsymbol{\eta} \\ &\quad = -\boldsymbol{J}^{\mathrm{T}}(\boldsymbol{q})\boldsymbol{f}_\lambda + \boldsymbol{\tau} + \boldsymbol{M}(\boldsymbol{q})\lambda\left(\boldsymbol{e}_v + \dot{d}(t)\boldsymbol{q}_{vd}(t, d(t))\right) + \\ &\qquad \boldsymbol{C}(\boldsymbol{q}, \dot{\boldsymbol{q}})\lambda\boldsymbol{e} - \boldsymbol{g}(\boldsymbol{q}) - \boldsymbol{f}(\dot{\boldsymbol{q}}) - \boldsymbol{f}_d\end{aligned} \tag{11.7}$$

From Property 11.1 and Property 11.3,

$$\begin{cases}\boldsymbol{Y}(\boldsymbol{q}, \dot{\boldsymbol{q}}, \boldsymbol{e}, \boldsymbol{e}_v)\boldsymbol{\theta} = \boldsymbol{M}(\boldsymbol{q})\lambda\boldsymbol{e}_v + \boldsymbol{C}(\boldsymbol{q}, \dot{\boldsymbol{q}})\lambda\boldsymbol{e} - \boldsymbol{g}(\boldsymbol{q}) - \boldsymbol{f}(\dot{\boldsymbol{q}}) \\ |\boldsymbol{M}(\boldsymbol{q})\lambda\dot{d}(t)\boldsymbol{q}_{vd}(t, d(t))| \le \theta_{11}|\boldsymbol{q}_{vd}(t, d(t))|\end{cases}$$

where θ_{11} is an unknown bounded constant scalar. It is also assumed $|\boldsymbol{f}_d| \le \theta_{21}$, where $\theta_{21} > 0$ is an unknown bounded constant.

To achieve the control objective, the controller of this chapter is designed as

$$\begin{aligned}\boldsymbol{\tau} = &-\boldsymbol{Y}(\boldsymbol{q}, \dot{\boldsymbol{q}}, \boldsymbol{e}, \boldsymbol{e}_v)\hat{\boldsymbol{\theta}} - \hat{\theta}_1|\boldsymbol{q}_{vd}(t, d(t))|^2\boldsymbol{\eta} - \hat{\theta}_2\boldsymbol{\eta} - K_1\boldsymbol{\eta} \\ &- K_2\boldsymbol{\mathcal{R}}_1(\bar{\boldsymbol{e}}) - K_0\boldsymbol{J}^{\mathrm{T}}(\boldsymbol{q})\boldsymbol{J}(\boldsymbol{q})\boldsymbol{\eta} - K_{31}\boldsymbol{\varpi}_1 - K_{32}\boldsymbol{\mathcal{R}}_1(\boldsymbol{\varpi}_2)\end{aligned} \tag{11.8}$$

where K_0, K_1, K_2, K_{31} and K_{32} are some positive scalars. For any $\boldsymbol{\chi} = [\chi_1, \chi_2, \cdots, \chi_n]^{\mathrm{T}} \in \mathbb{R}^n$, $\boldsymbol{\mathcal{R}}_1(\boldsymbol{\chi}) = [\mathcal{R}_1(\chi_1), \mathcal{R}_1(\chi_2), \cdots, \mathcal{R}_1(\chi_n)]^{\mathrm{T}}$, where $\mathcal{R}_1(\cdot)$ is a strictly increasing, bounded, continuous, passing through the origin with continuous first derivative around the origin, such that $|\mathcal{R}_1(\chi_i)| \le \min\{|\chi_i|, N_i\}$ and $\mathcal{R}_1(-\chi_i) = -\mathcal{R}_1(\chi_i)$, N_i a bounded positive constant. Structurally, the controller (11.8) consists of four parts.

The first part is the adaptive estimation compensator, i.e., $-\boldsymbol{Y}(\boldsymbol{q},\dot{\boldsymbol{q}},\boldsymbol{e},\boldsymbol{e}_v)\hat{\boldsymbol{\theta}}-\hat{\theta}_1|\boldsymbol{q}_{vd}(t,d(t))|^2\boldsymbol{\eta}-\hat{\theta}_2\boldsymbol{\eta}$, where $\hat{\boldsymbol{\theta}}$, $\hat{\theta}_1$ and $\hat{\theta}_2$ are respectively the estimations of $\boldsymbol{\theta}$, $\theta_1:=\frac{\theta_{11}^2}{4\varepsilon_0}$ and $\theta_2:=\frac{\theta_{21}^2}{4\varepsilon_0}$, ε_0 is a positive constant scalar. In this chapter, the corresponding adaptive update laws are given as

$$\begin{cases}\dot{\hat{\boldsymbol{\theta}}}=\psi_1\boldsymbol{Y}^{\mathrm{T}}(\boldsymbol{q},\dot{\boldsymbol{q}},\boldsymbol{e},\boldsymbol{e}_v)\boldsymbol{\eta}-\psi_2\hat{\boldsymbol{\theta}}\\ \dot{\hat{\theta}}_1=\psi_3|\boldsymbol{q}_{vd}(t,d(t))|^2|\boldsymbol{\eta}|^2-\psi_4\hat{\theta}_1\\ \dot{\hat{\theta}}_2=\psi_5|\boldsymbol{\eta}|^2-\psi_6\hat{\theta}_2\end{cases} \tag{11.9}$$

where $\psi_1,\psi_2,\cdots,\psi_6$ are some positive constant scalars.

Secondly, $-K_1\boldsymbol{\eta}-K_2\boldsymbol{\mathcal{R}}_1(\bar{\boldsymbol{e}})$ is the proportional controller. Regarding the nonlinear saturation function used $\boldsymbol{\mathcal{R}}_1(\cdot)$, two points need to be emphasized. ① The function $\boldsymbol{\mathcal{R}}_1(\cdot)$ utilized here can avoid the Zeno behavior, where the theoretic analysis will be introduced in Remark 11.7. ② As shown in Fig. 11.1, due to the $e_i(t)$ in $\boldsymbol{e}(t)=[e_1(t),e_2(t),\cdots,e_n(t)]^T$ may be very close to the performance function $\varrho_i(t)$, the nonlinear saturating function $\mathcal{R}_1(\cdot)$ on the outside of $\bar{e}_i(t)$ will also be helpful for avoiding the actuator saturation caused by the high control torque $K_2\bar{e}_i(t)$.

Thirdly, $-K_0\boldsymbol{J}^{\mathrm{T}}(\boldsymbol{q})\boldsymbol{J}(\boldsymbol{q})\boldsymbol{\eta}$ is the additional compensator. It is used to offset the residual item in the derivation of stability analysis.

Lastly, $-K_{31}\boldsymbol{\varpi}_1-K_{32}\boldsymbol{\mathcal{R}}_1(\boldsymbol{\varpi}_2)$ is the auxiliary filter compensator, where $\boldsymbol{\varpi}_1$ and $\boldsymbol{\varpi}_2$ are the states of the developed auxiliary filters. For any $i=1,2,\cdots,n$, $\boldsymbol{\varpi}_1=[\varpi_{1,1},\varpi_{1,2},\cdots,\varpi_{1,n}]^{\mathrm{T}}$ and $\boldsymbol{\varpi}_2=[\varpi_{2,1},\varpi_{2,2},\cdots,\varpi_{2,n}]^{\mathrm{T}}$ are designed as

$$\begin{cases}\dot{\varpi}_{1,i}=-K_{41}\varpi_{1,i}+K_5|y_{1,i}|\mathcal{R}(e_i)\\ \dot{\varpi}_{2,i}=-K_{42}\varpi_{2,i}+K_6|y_{2,i}|\mathcal{R}(\bar{e}_i)\end{cases} \tag{11.10}$$

where $\mathcal{R}(\cdot)$ is consistent with the definition of $\mathcal{R}_1(\cdot)$, but the upper bound is fixed as $N_i=1$. For any $t\geq 0$, $y_{1,i}(t)$ is given as

$$\dot{y}_{1,i}=-K_7(y_{1,i}+\kappa_{1,i}(t)e_i)+u_{1,i} \tag{11.11}$$

with

$$\begin{aligned}u_{1,i}=&-\kappa_{1,i}(t)e_{v,i}-K_8\mathrm{sgn}\Big((y_{1,i}+\kappa_{1,i}(t)e_i)\kappa_{1,i}(t)q_{vd,i}(t,d(t))\Big)\times\\ &\kappa_{1,i}(t)q_{vd,i}(t,d(t))\end{aligned} \tag{11.12}$$

and for all $t\in[t_k^i,t_{k+1}^i)$, $k\geq 0$, $y_{2,i}(t)$ is given as

$$\dot{y}_{2,i}=-K_9(y_{2,i}+\kappa_{2,i}(t)\bar{e}_i)+u_{2,i} \tag{11.13}$$

with

$$u_{2,i} = -\kappa_{2,i}(t)\frac{\partial T_i\left(\frac{e_i}{\varrho_i}\right)}{\partial \frac{e_i}{\varrho_i}}\left(\frac{e_{v,i}}{\varrho_i} - \frac{e_i}{\varrho_i^2}\dot{\varrho}_i\right) -$$

$$K_{10}\text{sgn}\left((y_{2,i} + \kappa_{2,i}(t)\bar{e}_i)\kappa_{2,i}(t)\frac{\partial T_i\left(\frac{e_i}{\varrho_i}\right)}{\partial \frac{e_i}{\varrho_i}}\frac{1}{\varrho_i}q_{vd,i}(t, d(t))\right)\times$$

$$\kappa_{2,i}(t)\frac{\partial T_i\left(\frac{e_i}{\varrho_i}\right)}{\partial \frac{e_i}{\varrho_i}}\frac{1}{\varrho_i}q_{vd,i}(t, d(t)) \tag{11.14}$$

where $q_{vd,i}(t, d(t))$ is the i-th component of vector $\boldsymbol{q}_{vd}(t, d(t))$. For simplicity, $K_{41}, K_{42}, K_5, K_6, \cdots, K_{10}$ are some constant positive scalars. To make full use of the state information (synchronization error), similar to previous chapters, $\kappa_{1,i}(t)$ and $\kappa_{2,i}(t)$ are designed as two switching functions, i.e.,

$$\kappa_{1,i}(t) = \begin{cases} 1, & y_{1,i}(t)e_i(t) > 0 \text{ or } y_{1,i}(t)e_i(t) = 0,\ y_{1,i}(t^-)e_i(t^-) \le 0 \\ -1, & y_{1,i}(t)e_i(t) < 0 \text{ or } y_{1,i}(t)e_i(t) = 0,\ y_{1,i}(t^-)e_i(t^-) > 0 \end{cases} \tag{11.15}$$

$$\kappa_{2,i}(t) = \begin{cases} 1, & y_{2,i}(t)\bar{e}_i(t) > 0 \text{ or } y_{2,i}(t)\bar{e}_i(t) = 0,\ y_{2,i}(t^-)\bar{e}_i(t^-) \le 0 \\ -1, & y_{2,i}(t)\bar{e}_i(t) < 0 \text{ or } y_{2,i}(t)\bar{e}_i(t) = 0,\ y_{2,i}(t^-)\bar{e}_i(t^-) > 0 \end{cases} \tag{11.16}$$

Remark 11.5. *If t_k^i is a switching time of $\kappa_{2,i}(t)$, since $\kappa_{2,i}(t)$ switches from 1 to -1 if and only if $y_{2,i}(t)\bar{e}_i(t) = 0$ and $y_{2,i}(t^-)\bar{e}_i(t^-) > 0$, while it switches from -1 to 1 if and only if $y_{2,i}(t)\bar{e}_i(t) = 0$ and $y_{2,i}(t^-)\bar{e}_i(t^-) < 0$, it holds $|y_{2,i}(t_k^i) + \kappa_{2,i}(t_k^i)\bar{e}_i(t_k^i)| \le |y_{2,i}(t_k^{i-}) + \kappa_{2,i}(t_k^{i-})\bar{e}_i(t_k^{i-})|$. If t_k^i is not a switching time of $\kappa_{2,i}(t)$, because $|\bar{e}_i(t_k^i)| \le |\bar{e}_i(t_k^{i-})|$, and considering the continuity of $y_{2,i}(t)$, it still holds $|y_{2,i}(t_k^i) + \kappa_{2,i}(t_k^i)\bar{e}_i(t_k^i)| \le |y_{2,i}(t_k^{i-}) + \kappa_{2,i}(t_k^{i-})\bar{e}_i(t_k^{i-})|$.*

Remark 11.6. *In the control design scheme proposed in this chapter, the auxiliary filter system consists of (11.10), (11.11) and (11.13). Due to the introduction of $\kappa_{1,i}$ and $\kappa_{2,i}$ and sign functions, the auxiliary filter system is a switched dynamic system. In order to analyze this system using switching control theory, it is first necessary to model it in the form of a standard switched dynamic system. The modeling process is similar to the previous chapters of this book.*

(1) Let $\boldsymbol{\mathcal{K}}_1(t) = \text{diag}[\kappa_{1,1}(t), \kappa_{1,2}(t), \cdots, \kappa_{1,n}(t)] \in \mathbb{R}^{n\times n}$, $\boldsymbol{\mathcal{K}}_2(t) = \text{diag}[\kappa_{2,1}(t), \kappa_{2,2}(t), \cdots, \kappa_{2,n}(t)] \in \mathbb{R}^{n\times n}$. Denote $\{t_{1,k}^{\kappa_1}\}_{k\ge 0}$ and $\{t_{2,k}^{\kappa_2}\}_{k\ge 0}$ be the switching times of $\boldsymbol{\mathcal{K}}_1(t)$ and $\boldsymbol{\mathcal{K}}_2(t)$, respectively.

(2) *Since both* $\kappa_{1,i}$ *and* $\kappa_{2,i}$ *have two values, i.e.,* 1 *and* -1, $\boldsymbol{\mathcal{K}}_1(t)$ *and* $\boldsymbol{\mathcal{K}}_2(t)$ *will have* 2^n *states, i.e.,*

$$\begin{gathered}\mathrm{diag}[-1,-1,-1,\cdots,-1]\in\mathbb{R}^{n\times n}\\ \mathrm{diag}[1,-1,-1,\cdots,-1]\in\mathbb{R}^{n\times n}\\ \mathrm{diag}[-1,1,-1,\cdots,-1]\in\mathbb{R}^{n\times n}\\ \mathrm{diag}[1,1,-1,\cdots,-1]\in\mathbb{R}^{n\times n}\\ \vdots\\ \mathrm{diag}[-1,1,1,\cdots,1]\in\mathbb{R}^{n\times n}\\ \mathrm{diag}[1,1,1,\cdots,1]\in\mathbb{R}^{n\times n}\end{gathered}$$

which can be numbered in turn as: $1,2,\cdots,2^n$. *Also denote* $\mathcal{S}=\{1,2,\cdots,2^n\}$.

(3) *Then, define* $\bar{r}_1:\boldsymbol{\mathcal{K}}_1(t)\to\mathcal{S}$ *and* $\bar{r}_2:\boldsymbol{\mathcal{K}}_2(t)\to\mathcal{S}$ *by*

$$\begin{aligned}\bar{r}_1(\boldsymbol{\mathcal{K}}_1(t))=i,&\ \textit{iff}\ \boldsymbol{\mathcal{K}}_1(t)\ \textit{is in the}\ i\textit{-th state}\\ \bar{r}_2(\boldsymbol{\mathcal{K}}_2(t))=j,&\ \textit{iff}\ \boldsymbol{\mathcal{K}}_2(t)\ \textit{is in the}\ j\textit{-th state}\end{aligned}$$

For convenience, let $r_1^1(t):=\bar{r}_1(\boldsymbol{\mathcal{K}}_1(t))$, $r_2^1(t):=\bar{r}_2(\boldsymbol{\mathcal{K}}_2(t))$. *In this case,* $\{t_{1,k}^{\kappa_1}\}_{k\geq 0}$ *and* $\{t_{2,k}^{\kappa_2}\}_{k\geq 0}$ *are also the switching times of* $r_1^1(t)$ *and* $r_2^1(t)$, *respectively.*

(4) *The sign functions in (11.12) and (11.14) will also introduce two switching modes* 1 *and* -1. *Let us introduce* $\kappa_{r,1,i}(t)$ *and* $\kappa_{r,2,i}(t)$ *by:* $\kappa_{r,1,i}(t)=1$ *when the sign function in (11.12) takes value* 1, $\kappa_{r,1,i}(t)=2$ *when the sign function in (11.12) takes value* -1, $\kappa_{r,2,i}(t)=1$ *when the sign function in (11.14) takes value* 1, $\kappa_{r,2,i}(t)=2$ *when the sign function in (11.14) takes value* -1. *Denote* $\boldsymbol{\mathcal{K}}_{r,1}(t)=\mathrm{diag}[\kappa_{r,1,1}(t),\kappa_{r,1,2}(t),\cdots,\kappa_{r,1,n}(t)]$, $\boldsymbol{\mathcal{K}}_{r,2}(t)=\mathrm{diag}[\kappa_{r,2,1}(t),\kappa_{r,2,2}(t),\cdots,\kappa_{r,2,n}(t)]$. *Similarly, one can define two mappings* $\bar{r}_1^2:\boldsymbol{\mathcal{K}}_{r,1}(t)\to\mathcal{S}$ *and* $\bar{r}_2^2:\boldsymbol{\mathcal{K}}_{r,2}(t)\to\mathcal{S}$ *by*

$$\begin{aligned}\bar{r}_1^2(\boldsymbol{\mathcal{K}}_{r,1}(t))=i,&\ \textit{iff}\ \boldsymbol{\mathcal{K}}_{r,1}(t)\ \textit{is in the}\ i\textit{-th state}\\ \bar{r}_2^2(\boldsymbol{\mathcal{K}}_{r,2}(t))=j,&\ \textit{iff}\ \boldsymbol{\mathcal{K}}_{r,2}(t)\ \textit{is in the}\ j\textit{-th state}\end{aligned}$$

Denote $r_1^2(t):=\bar{r}_1^2(\boldsymbol{\mathcal{K}}_{r,1}(t))$, $r_2^2(t):=\bar{r}_2^2(\boldsymbol{\mathcal{K}}_{r,2}(t))$, *and let* $\{t_{1,k}\}_{k\geq 0}$ *and* $\{t_{2,k}\}_{k\geq 0}$ *denote, respectively, the switching times of* $r_1^2(t)$ *and* $r_2^2(t)$, *with* $t_{1,k}=t_{2,k}=0$.

(5) *Defining virtual switching signals* $r_1:[t_0,\infty)\to\mathcal{S}\times\mathcal{S}$ *and* $r_2:[t_0,\infty)\to\mathcal{S}\times\mathcal{S}$ *by* $r_1(t)=(r_1^1(t),r_1^2(t))$, $r_2(t)=(r_2^1(t),r_2^2(t))$. *Let* $\{t_k^{\kappa_1}\}_{k\geq 0}=\{t_{1,k}^{\kappa_1}\}_{k\geq 0}\bigcup\{t_{1,k}\}_{k\geq 0}$, $\{t_k^{\kappa_2}\}_{k\geq 0}=\{t_{2,k}^{\kappa_2}\}_{k\geq 0}\bigcup\{t_{2,k}\}_{k\geq 0}$. *In this case, the subsystems* $\boldsymbol{y}_1=[y_{1,1},y_{1,2},\cdots,y_{1,n}]^{\mathrm{T}}$ *and* $\boldsymbol{y}_2=[y_{2,1},y_{2,2},\cdots,y_{2,n}]^{\mathrm{T}}$ *in (11.11) and (11.13) can be modeled as two standard switched systems with* $r_1(t)$ *and* $r_2(t)$ *being the switching signals, respectively.*

(6) *Let* $r(t)=(r_1(t),r_2(t))\in\mathcal{S}\times\mathcal{S}\times\mathcal{S}\times\mathcal{S}$ *represent the virtual switching signal of the complete closed-loop system. Its switching times are given by* $\{t_k^{r}\}_{k\geq 0}=\{t_k^{\kappa_1}\}_{k\geq 0}\bigcup\{t_k^{\kappa_2}\}_{k\geq 0}$. *Based on the above analysis, the complete closed-loop system composed of subsystems (11.7)~(11.11) and (11.13) is a switched system, with* $r(t)$ *being the switching signal.*

11.4 STABILITY ANALYSIS

Since the contact force $\boldsymbol{f}_\lambda$ may be non-passive, the stability will be performed in the sense of state-independent input-to-output stable (SIIOS). Denote

$$\boldsymbol{x} = [\boldsymbol{q}^{\mathrm{T}}, \boldsymbol{\eta}^{\mathrm{T}}, \tilde{\boldsymbol{\theta}}^{\mathrm{T}}, \tilde{\theta}_1, \tilde{\theta}_2, \boldsymbol{\varpi}_1^{\mathrm{T}}, \boldsymbol{\varpi}_2^{\mathrm{T}}, \boldsymbol{y}_1^{\mathrm{T}}, \boldsymbol{y}_2^{\mathrm{T}}, \boldsymbol{e}^{\mathrm{T}}, \bar{\boldsymbol{e}}^{\mathrm{T}}]^{\mathrm{T}} \in \mathbb{R}^{n_1}$$
$$\boldsymbol{z} = [\boldsymbol{\eta}^{\mathrm{T}}, \tilde{\boldsymbol{\theta}}^{\mathrm{T}}, \tilde{\theta}_1, \tilde{\theta}_2, \boldsymbol{\varpi}_1^{\mathrm{T}}, \boldsymbol{\varpi}_2^{\mathrm{T}}, \boldsymbol{y}_1^{\mathrm{T}}, \boldsymbol{y}_2^{\mathrm{T}}, \boldsymbol{e}^{\mathrm{T}}, \bar{\boldsymbol{e}}^{\mathrm{T}}]^{\mathrm{T}} \in \mathbb{R}^{n_2}$$

For the complete closed-loop system, the state and output are defined by $\boldsymbol{x}_t = \{\boldsymbol{x}(t+\phi), \phi \in [-\bar{d}, 0]\} \in C([-\bar{d}, 0]; \mathbb{R}^{n_1})$ and $\boldsymbol{z}_t = \{\boldsymbol{z}(t+\phi), \phi \in [-\bar{d}, 0]\} \in C([-\bar{d}, 0]; \mathbb{R}^{n_2})$. Then, the following conclusions are obtained.

Proposition 11.1. *For the complete closed-loop system, taking the Lyapunov-Krasovskii functionals as*

$$V(\boldsymbol{x}_t, r(t)) = \sum_{i=1}^{6} V_i(\boldsymbol{x}_t, r(t)) \tag{11.17}$$

with

$$\begin{aligned}
V_1(\boldsymbol{x}_t, r(t)) &= \frac{1}{2}\boldsymbol{\eta}^{\mathrm{T}}\boldsymbol{M}(\boldsymbol{q})\boldsymbol{\eta} \\
V_2(\boldsymbol{x}_t, r(t)) &= \frac{1}{2\psi_1}\tilde{\boldsymbol{\theta}}^{\mathrm{T}}\tilde{\boldsymbol{\theta}} + \frac{1}{2\psi_3}\tilde{\theta}_1^{\mathrm{T}}\tilde{\theta}_1 + \frac{1}{2\psi_5}\tilde{\theta}_2^{\mathrm{T}}\tilde{\theta}_2 \\
V_3(\boldsymbol{x}_t, r(t)) &= \frac{1}{2}\boldsymbol{\varpi}_1^{\mathrm{T}}\boldsymbol{\varpi}_1 + \frac{1}{2}\boldsymbol{\varpi}_2^{\mathrm{T}}\boldsymbol{\varpi}_2 \\
V_4(\boldsymbol{x}_t, r(t)) &= \frac{1}{2}\boldsymbol{s}_1^{\mathrm{T}}\boldsymbol{s}_1 \\
V_5(\boldsymbol{x}_t, r(t)) &= \frac{1}{2}\boldsymbol{s}_2^{\mathrm{T}}\boldsymbol{s}_2 \\
V_6(\boldsymbol{x}_t, r(t)) &= \frac{\vartheta}{2}\int_{-\bar{d}}^{0} \boldsymbol{z}_t^{\mathrm{T}}(\tau)\left(\frac{-\tau}{\bar{d}} + \frac{2(\tau+\bar{d})}{\bar{d}}\right)\boldsymbol{z}_t(\tau)\mathrm{d}\tau
\end{aligned}$$

where $\boldsymbol{s}_1 = \boldsymbol{y}_1 + \boldsymbol{\mathcal{K}}_1\boldsymbol{e}$, $\boldsymbol{s}_2 = \boldsymbol{y}_2 + \boldsymbol{\mathcal{K}}_2\bar{\boldsymbol{e}}$, ϑ is a positive constant. If the positive control gains K_1, K_2, K_{31}, K_{32}, K_{41}, K_{42}, K_5, K_6, $\cdots$, K_{10}, ψ_1, ψ_2, $\cdots$, ψ_5 and ψ_6 satisfy

$$\left\{\begin{array}{l}
\min\left\{\begin{array}{c}
K_1 - \dfrac{K_2 + K_{31} + K_{32}}{2} - \vartheta \\
\psi_2 - 2\psi_1\vartheta,\ \psi_4 - 2\psi_3\vartheta,\ \psi_6 - 2\psi_5\vartheta \\
K_{41} - \dfrac{K_{31} + K_5}{2} - \vartheta,\ K_{42} - \dfrac{K_{32} + K_5}{2} - \vartheta \\
K_7 - \dfrac{K_5}{2} - \vartheta,\ K_9 - \dfrac{K_2}{2} - \vartheta \\
K_9 - \dfrac{K_6}{2} - \vartheta
\end{array}\right\} > 0 \\
K_0 \geq \varepsilon > 0,\ K_8 \geq \tilde{d},\ K_{10} \geq \bar{d}
\end{array}\right. \tag{11.18}$$

then for any $t \in [t_k^r, t_{k+1}^r)$, it holds

$$V(\boldsymbol{x}_t, r(t)) \leq \mathrm{e}^{-\nu_1(t-t_k^r)}V(\boldsymbol{x}_{t_k^r}, r(t_k^r)) + \nu_2\|\boldsymbol{f}_{\lambda[0,\infty)}\|_\infty^2 + \Delta_2 \tag{11.19}$$

where ν_1, ν_2 and Δ_2 are bounded positive constants.

Proof. The proof of Proposition 11.1 is similar to our previous works [9,10]. However, since the switching-based PPLC controller applies two sets of switching rules, its stability analysis is more complicated. The detailed proof process is as follows.

Along with the subsystem (11.7), there exists $\varepsilon > 0$ such that

$$\begin{aligned}
&D^{+}V_1(\boldsymbol{x}_t, r(t)) \\
&= \frac{1}{2}\boldsymbol{\eta}^{\mathrm{T}}\dot{\boldsymbol{M}}(\boldsymbol{q})\boldsymbol{\eta} + \boldsymbol{\eta}^{\mathrm{T}}\boldsymbol{M}(\boldsymbol{q})\dot{\boldsymbol{\eta}} \\
&= \boldsymbol{\eta}^{\mathrm{T}}\Big(- K_1\boldsymbol{\eta} - K_2\boldsymbol{\mathcal{R}}_1(\bar{\boldsymbol{e}}) - K_{31}\boldsymbol{\varpi}_1 - K_{32}\boldsymbol{\mathcal{R}}_1(\boldsymbol{\varpi}_2) - \boldsymbol{Y}(\boldsymbol{q}, \dot{\boldsymbol{q}}, \boldsymbol{e}, \boldsymbol{e}_v)\hat{\boldsymbol{\theta}} - \\
&\quad \hat{\theta}_1|\boldsymbol{q}_{vd}(t, d(t))|^2\boldsymbol{\eta} - \hat{\theta}_2\boldsymbol{\eta} - \boldsymbol{f}_d + \boldsymbol{Y}(\boldsymbol{q}, \dot{\boldsymbol{q}}, \boldsymbol{e}, \boldsymbol{e}_v)\boldsymbol{\theta} + \\
&\quad \boldsymbol{M}(\boldsymbol{q})\lambda\dot{d}(t)\boldsymbol{q}_{vd}(t, d(t)) - \boldsymbol{J}^{\mathrm{T}}(\boldsymbol{q})\boldsymbol{f}_{\lambda} - K_0\boldsymbol{J}^{\mathrm{T}}(\boldsymbol{q})\boldsymbol{J}(\boldsymbol{q})\boldsymbol{\eta}\Big) \\
&\le -\left(K_1 - \frac{K_2 + K_{31} + K_{32}}{2}\right)|\boldsymbol{\eta}|^2 + \frac{K_2}{2}|\boldsymbol{\mathcal{R}}_1(\bar{\boldsymbol{e}})|^2 + \frac{K_{31}}{2}|\boldsymbol{\varpi}_1|^2 + \\
&\quad \frac{K_{32}}{2}|\boldsymbol{\mathcal{R}}_1(\boldsymbol{\varpi}_2)|^2 + \theta_{21}|\boldsymbol{\eta}| - \hat{\theta}_2|\boldsymbol{\eta}|^2 + \boldsymbol{\eta}^{\mathrm{T}}\boldsymbol{Y}(\boldsymbol{q}, \dot{\boldsymbol{q}}, \boldsymbol{e}, \boldsymbol{e}_v)\tilde{\boldsymbol{\theta}} + \\
&\quad \theta_{11}|\boldsymbol{q}_{vd}(t, d(t))||\boldsymbol{\eta}| - \hat{\theta}_1|\boldsymbol{q}_{vd}(t, d(t))|^2|\boldsymbol{\eta}|^2 - \\
&\quad (K_0 - \varepsilon)\boldsymbol{\eta}^{\mathrm{T}}\boldsymbol{J}^{\mathrm{T}}(\boldsymbol{q})\boldsymbol{J}(\boldsymbol{q})\boldsymbol{\eta} + \frac{1}{4\varepsilon}|\boldsymbol{f}_{\lambda}|^2 \\
&\le -\left(K_1 - \frac{K_2 + K_{31} + K_{32}}{2}\right)|\boldsymbol{\eta}|^2 + \frac{K_2}{2}|\boldsymbol{\mathcal{R}}_1(\bar{\boldsymbol{e}})|^2 + \frac{K_{31}}{2}|\boldsymbol{\varpi}_1|^2 + \\
&\quad \frac{K_{32}}{2}|\boldsymbol{\varpi}_2|^2 + \boldsymbol{\eta}^{\mathrm{T}}\boldsymbol{Y}(\boldsymbol{q}, \dot{\boldsymbol{q}}, \boldsymbol{e}, \boldsymbol{e}_v)\tilde{\boldsymbol{\theta}} + 2\varepsilon_0 + \tilde{\theta}_1|\boldsymbol{q}_{vd}(t, d(t))|^2|\boldsymbol{\eta}|^2 + \\
&\quad \tilde{\theta}_2|\boldsymbol{\eta}|^2 + \frac{1}{4\varepsilon}|\boldsymbol{f}_{\lambda}|^2
\end{aligned}$$

Along with the subsystem (11.9), it has

$$\begin{aligned}
D^{+}V_2(\boldsymbol{x}_t, r(t)) &= -\frac{1}{\psi_1}\tilde{\boldsymbol{\theta}}^{\mathrm{T}}\dot{\hat{\boldsymbol{\theta}}} - \frac{1}{\psi_3}\tilde{\theta}_1^{\mathrm{T}}\dot{\hat{\theta}}_1 - \frac{1}{\psi_5}\tilde{\theta}_2^{\mathrm{T}}\dot{\hat{\theta}}_2 \\
&= -\tilde{\boldsymbol{\theta}}^{\mathrm{T}}\boldsymbol{Y}^{\mathrm{T}}(\boldsymbol{q}, \dot{\boldsymbol{q}}, \boldsymbol{e}, \boldsymbol{e}_v)\boldsymbol{\eta} + \frac{\psi_2}{\psi_1}\tilde{\boldsymbol{\theta}}^{\mathrm{T}}\hat{\boldsymbol{\theta}} + \frac{\psi_4}{\psi_3}\tilde{\theta}_1^{\mathrm{T}}\hat{\theta}_1 - \\
&\quad \tilde{\theta}_1^{\mathrm{T}}|\boldsymbol{q}_{vd}(t, d(t))|^2|\boldsymbol{\eta}|^2 - \tilde{\theta}_2^{\mathrm{T}}|\boldsymbol{\eta}|^2 + \frac{\psi_6}{\psi_5}\tilde{\theta}_2^{\mathrm{T}}\hat{\theta}_2 \\
&\le -\tilde{\boldsymbol{\theta}}^{\mathrm{T}}\boldsymbol{Y}^{\mathrm{T}}(\boldsymbol{q}, \dot{\boldsymbol{q}}, \boldsymbol{e}, \boldsymbol{e}_v)\boldsymbol{\eta} - \tilde{\theta}_1^{\mathrm{T}}|\boldsymbol{q}_{vd}(t, d(t))|^2|\boldsymbol{\eta}|^2 - \\
&\quad \tilde{\theta}_2^{\mathrm{T}}|\boldsymbol{\eta}|^2 - \frac{\psi_2}{2\psi_1}|\tilde{\boldsymbol{\theta}}|^2 + \frac{\psi_2}{2\psi_1}|\boldsymbol{\theta}|^2 - \frac{\psi_4}{2\psi_3}|\tilde{\theta}_1|^2 + \\
&\quad \frac{\psi_4}{2\psi_3}|\theta_1|^2 - \frac{\psi_6}{2\psi_5}|\tilde{\theta}_2|^2 + \frac{\psi_6}{2\psi_5}|\theta_2|^2
\end{aligned}$$

Along with the subsystem (11.10),

$$
\begin{aligned}
D^+V_3(\boldsymbol{x}_t, r(t)) &= \sum_{i=1}^{n} \varpi_{1,i} \left(-K_{41}\varpi_{1,i} + K_5|y_{1,i}|\mathcal{R}(e_i)\right) + \\
&\quad \sum_{i=1}^{n} \varpi_{2,i} \left(-K_{42}\varpi_{2,i} + K_6|y_{2,i}|\mathcal{R}(\bar{e}_i)\right) \\
&\leq \sum_{i=1}^{n} \Bigg(-\left(K_{41} - \frac{K_5}{2}\right)|\varpi_{1,i}|^2 + \frac{K_5}{2}|y_{1,i}|^2 - \\
&\quad \left(K_{42} - \frac{K_6}{2}\right)|\varpi_{2,i}|^2 + \frac{K_6}{2}|y_{2,i}|^2 \Bigg) \\
&= -\left(K_{41} - \frac{K_5}{2}\right)|\boldsymbol{\varpi}_1|^2 - \left(K_{42} - \frac{K_6}{2}\right)|\boldsymbol{\varpi}_2|^2 + \\
&\quad \frac{K_5}{2}|\boldsymbol{y}_1|^2 + \frac{K_6}{2}|\boldsymbol{y}_2|^2
\end{aligned}
$$

For any $t \in [t_k^{\kappa_1}, t_{k+1}^{\kappa_1})$, from subsystem (11.11), for all $i = 1, 2, \cdots, n$, one has

$$
\begin{aligned}
\dot{s}_{1,i} = {} & -K_7 s_{1,i} + \kappa_{1,i}(t)\dot{d}(t)q_{vd,i}(t, d(t)) - \\
& K_8 \mathrm{sgn}\left(s_{1,i}\kappa_{1,i}(t)q_{vd,i}(t, d(t))\right) \dot{\kappa_{1,i}}(t) q_{vd,i}(t, d(t))
\end{aligned}
$$

and further when $K_8 \geq \tilde{d}$,

$$
\begin{aligned}
D^+V_4(\boldsymbol{x}_t, r(t)) &\leq -K_7|\boldsymbol{s}_1|^2 - K_8 \sum_{i=1}^{n} |s_{1,i}\kappa_{1,i}(t)q_{vd,i}(t, d(t))| + \\
&\quad \dot{d}(t) \sum_{i=1}^{n} s_{1,i}\kappa_{1,i}(t)q_{vd,i}(t, d(t)) \\
&\leq -K_7\left(|\boldsymbol{y}_1|^2 + |\boldsymbol{e}|^2\right)
\end{aligned}
$$

Denote $\{t_{k_j}\}_{k_j} = [t_k^{\kappa_2}, t_{k+1}^{\kappa_2}) \bigcap \{t_k\}_{k \geq 0}$. Then for any $t \in [t_k^{\kappa_2}, t_{k+1}^{\kappa_2}) \setminus \{t_{k_j}\}_{k_j}$, from (11.13),

$$
\begin{aligned}
\dot{s}_{2,i} = {} & -K_9 s_{2,i} + \kappa_{2,i}(t) \frac{\partial T_i\left(\dfrac{e_i}{\varrho_i}\right)}{\partial \dfrac{e_i}{\varrho_i}} \frac{1}{\varrho_i} \dot{d}(t) q_{vd,i}(t, d(t)) - \\
& K_{10}\mathrm{sgn}\left(s_{2,i}\kappa_{2,i}(t) \frac{\partial T_i\left(\dfrac{e_i}{\varrho_i}\right)}{\partial \dfrac{e_i}{\varrho_i}} \frac{1}{\varrho_i} q_{vd,i}(t, d(t)) \right) \times \\
& \kappa_{2,i}(t) \frac{\partial T_i\left(\dfrac{e_i}{\varrho_i}\right)}{\partial \dfrac{e_i}{\varrho_i}} \frac{1}{\varrho_i} q_{vd,i}(t, d(t))
\end{aligned}
$$

and further when $K_{10} \geq \tilde{d}$, one can obtain

$$\begin{aligned}
D^+V_5(\boldsymbol{x}_t, r(t)) &= \sum_{i=1}^{n} s_{2,i}\dot{s}_{2,i} \\
&= \sum_{i=1}^{n} -K_9|s_{2,i}|^2 + s_{2,i}\kappa_{2,i}(t)\frac{\partial T_i\left(\dfrac{e_i}{\varrho_i}\right)}{\partial \dfrac{e_i}{\varrho_i}}\frac{1}{\varrho_i}\dot{d}(t)q_{vd,i}(t,d(t)) - \\
&\quad K_{10}\left|s_{2,i}\kappa_{2,i}(t)\frac{\partial T_i\left(\dfrac{e_i}{\varrho_i}\right)}{\partial \dfrac{e_i}{\varrho_i}}\frac{1}{\varrho_i}q_{vd,i}(t,d(t))\right| \\
&\leq \sum_{i=1}^{n} -K_9|s_{2,i}|^2 + \tilde{d}\left|s_{2,i}\kappa_{2,i}(t)\frac{\partial T_i\left(\dfrac{e_i}{\varrho_i}\right)}{\partial \dfrac{e_i}{\varrho_i}}\frac{1}{\varrho_i}q_{vd,i}(t,d(t))\right| - \\
&\quad K_{10}\left|s_{2,i}\kappa_{2,i}(t)\frac{\partial T_i\left(\dfrac{e_i}{\varrho_i}\right)}{\partial \dfrac{e_i}{\varrho_i}}\frac{1}{\varrho_i}q_{vd,i}(t,d(t))\right| \\
&\leq -K_9(|\boldsymbol{y}_2|^2 + |\bar{\boldsymbol{e}}|^2)
\end{aligned}$$

When $t = t_{k_j}$, from Remark 11.5,

$$V_5(\boldsymbol{x}_{t_{k_j}}, r(t_{k_j})) \leq V_5(\boldsymbol{x}_{t_{k_j}^-}, r(t_{k_j}^-)) \tag{11.20}$$

Finally, it also holds

$$\begin{aligned}
D^+V_6(\boldsymbol{x}_t, r(t)) &= \vartheta \boldsymbol{z}^{\mathrm{T}}\boldsymbol{z} - \frac{\vartheta}{2}\boldsymbol{z}_t^{\mathrm{T}}(-\bar{d})\boldsymbol{z}_t(-\bar{d}) - \frac{\vartheta}{2\bar{d}}\int_{-\bar{d}}^{0}\boldsymbol{z}_t^{\mathrm{T}}(\tau)\boldsymbol{z}_t(\tau)\mathrm{d}\tau \\
&\leq \vartheta \boldsymbol{z}^{\mathrm{T}}\boldsymbol{z} - \frac{\vartheta}{2\bar{d}}\int_{-\bar{d}}^{0}\boldsymbol{z}_t^{\mathrm{T}}(\tau)\boldsymbol{z}_t(\tau)\mathrm{d}\tau
\end{aligned}$$

Denote $\{t_{k_i}\}_{k_i} = [t_k^r, t_{k+1}^r) \bigcap \{t_k\}_{k\geq 0}$. Then for all $t \in [t_k^r, t_{k+1}^r) \setminus \{t_{k_i}\}_{k_i}$, combining the above formulas, we can get

$$\begin{aligned}
&D^+V(\boldsymbol{x}_t, r(t)) \\
&\leq -\left(K_1 - \frac{K_2 + K_{31} + K_{32}}{2}\right)|\boldsymbol{\eta}|^2 - \frac{\psi_2}{2\psi_1}|\tilde{\boldsymbol{\theta}}|^2 - \frac{\psi_4}{2\psi_3}|\tilde{\theta}_1|^2 - \\
&\quad \frac{\psi_6}{2\psi_5}|\tilde{\theta}_2|^2 - K_7|\boldsymbol{e}|^2 - K_9|\bar{\boldsymbol{e}}|^2 - \left(K_7 - \frac{K_5}{2}\right)|\boldsymbol{y}_1|^2 - \\
&\quad \left(K_{41} - \frac{K_{31} + K_5}{2}\right)|\boldsymbol{\varpi}_1|^2 - \left(K_{42} - \frac{K_{32} + K_6}{2}\right)|\boldsymbol{\varpi}_2|^2 - \\
&\quad K_9|\boldsymbol{y}_2|^2 + \frac{K_2}{2}|\boldsymbol{\mathcal{R}}_1(\bar{\boldsymbol{e}})|^2 + 2\varepsilon_0 + \frac{\psi_2}{2\psi_1}|\boldsymbol{\theta}|^2 + \frac{\psi_4}{2\psi_3}|\theta_1|^2 + \frac{\psi_6}{2\psi_5}|\theta_2|^2 + \\
&\quad \frac{1}{4\varepsilon}|\boldsymbol{f}_\lambda|^2 + \frac{K_6}{2}|\boldsymbol{y}_2|^2 + \vartheta\boldsymbol{z}^{\mathrm{T}}\boldsymbol{z} - \frac{\vartheta}{2\bar{d}}\int_{-\bar{d}}^{0}\boldsymbol{z}_t^{\mathrm{T}}(\tau)\boldsymbol{z}_t(\tau)\mathrm{d}\tau
\end{aligned}$$

$$
\begin{aligned}
&\leq -\left(K_1 - \frac{K_2+K_{31}+K_{32}}{2} - \vartheta\right)|\boldsymbol{\eta}|^2 - \left(\frac{\psi_2}{2\psi_1} - \vartheta\right)|\tilde{\boldsymbol{\theta}}|^2 - \\
&\left(\frac{\psi_4}{2\psi_3} - \vartheta\right)|\tilde{\theta}_1|^2 - \left(\frac{\psi_6}{2\psi_5} - \vartheta\right)|\tilde{\theta}_2|^2 - (K_7 - \vartheta)|\boldsymbol{e}|^2 - \\
&\left(K_9 - \frac{K_2}{2} - \vartheta\right)|\bar{\boldsymbol{e}}|^2 - \left(K_7 - \frac{K_5}{2} - \vartheta\right)|\boldsymbol{y}_1|^2 - \\
&\left(K_9 - \frac{K_6}{2} - \vartheta\right)|\boldsymbol{y}_2|^2 - \left(K_{41} - \frac{K_{31}+K_5}{2} - \vartheta\right)|\boldsymbol{\varpi}_1|^2 - \\
&\left(K_{42} - \frac{K_{32}+K_6}{2} - \vartheta\right)|\boldsymbol{\varpi}_2|^2 - \frac{\vartheta}{2\bar{d}}\int_{-\bar{d}}^{0} \boldsymbol{z}_t^{\mathrm{T}}(\tau)\boldsymbol{z}_t(\tau)\mathrm{d}\tau + \\
&2\varepsilon_0 + \frac{\psi_2}{2\psi_1}|\boldsymbol{\theta}|^2 + \frac{\psi_4}{2\psi_3}|\theta_1|^2 + \frac{\psi_6}{2\psi_5}|\theta_2|^2 + \frac{1}{4\varepsilon}|\boldsymbol{f}_\lambda|^2 \\
&\leq -\mu_1|\boldsymbol{z}|^2 - \frac{\vartheta}{2\bar{d}}\int_{-\bar{d}}^{0} \boldsymbol{z}_t^{\mathrm{T}}(\tau)\boldsymbol{z}_t(\tau)\mathrm{d}\tau + \Delta \\
&\leq -\mu_2\|\boldsymbol{z}_t\|_{M_2}^2 + \Delta
\end{aligned}
$$

where $\Delta = \frac{1}{4\varepsilon}\|\boldsymbol{f}_{\lambda[0,\infty)}\|_\infty^2 + 2\varepsilon_0 + \frac{\psi_2}{2\psi_1}|\boldsymbol{\theta}|^2 + \frac{\psi_4}{2\psi_3}|\theta_1|^2 + \frac{\psi_6}{2\psi_5}|\theta_2|^2$, $\mu_2 = \min\left\{\mu_1, \frac{\vartheta}{2\bar{d}}\right\}$,

$$
\mu_1 = \min\left\{
\begin{array}{c}
K_1 - \dfrac{K_2+K_{31}+K_{32}}{2} - \vartheta \\
\dfrac{\psi_2}{2\psi_1} - \vartheta,\ \dfrac{\psi_4}{2\psi_3} - \vartheta,\ \dfrac{\psi_6}{2\psi_5} - \vartheta \\
K_7 - \dfrac{K_5}{2} - \vartheta,\ K_9 - \dfrac{K_2}{2} - \vartheta \\
K_{41} - \dfrac{K_{31}+K_5}{2} - \vartheta \\
K_{42} - \dfrac{K_{32}+K_6}{2} - \vartheta \\
K_9 - \dfrac{K_6}{2} - \vartheta
\end{array}
\right\}
$$

Considering the fact that $\frac{1}{2}\int_{-\bar{d}}^{0} \boldsymbol{z}_t^{\mathrm{T}}(\tau)\left(\frac{-\tau}{\bar{d}} + \frac{2(\tau+\bar{d})}{\bar{d}}\right)\boldsymbol{z}_t(\tau)\mathrm{d}\tau \leq \int_{-\bar{d}}^{0} \boldsymbol{z}_t^{\mathrm{T}}(\tau)$ $\boldsymbol{z}_t(\tau)\mathrm{d}\tau$, there exist positive constants α_1 and α_2 such that

$$
\alpha_1|\boldsymbol{z}(t)|^2 \leq V(\boldsymbol{x}_t, r(t)) \leq \alpha_2\|\boldsymbol{z}_t\|_{M_2}^2 \tag{11.21}
$$

Thus for all $t \in [t_k^r, t_{k+1}^r) \setminus \{t_{k_i}\}_{k_i}$,

$$
D^+V(\boldsymbol{x}_t, r(t)) \leq -\frac{\mu_2}{\alpha_2}V(\boldsymbol{x}_t, r(t)) + \Delta \tag{11.22}
$$

and further there exists $\varepsilon \in (0,1)$ such that

$$
V(\boldsymbol{x}_t, r(t)) \geq \frac{\alpha_2}{\mu_2}\varepsilon\Delta \Rightarrow D^+V(\boldsymbol{x}_t, r(t)) \leq -(1-\varepsilon)\frac{\mu_2}{\alpha_2}V(\boldsymbol{x}_t, r(t))
$$

For simplicity, denote $\mu = (1-\varepsilon)\frac{\mu_2}{\alpha_2}$, $\Delta_1 = \frac{\alpha_2}{\mu_2}\varepsilon\Delta$.

Without loss of generality, it is assumed that the number of the times in $\{t_{k_i}\}_{k_i}$ is N (from the analysis of the subsequent Remark 11.7, it can be seen that N exists and is bounded). Then the interval $[t_k^r, t_{k+1}^r)$ will be divided into $N+1$ small intervals by $\{t_{k_i}\}_{k_i}$, i.e., $[t_k^r, t_{k_1})$, $[t_{k_1}, t_{k_2})$, $[t_{k_2}, t_{k_3})$, $\cdots$, $[t_{k_N}, t_{k+1}^r)$. From (11.20) and (11.22), when $t \in [t_k^r, t_{k_1})$,

$$V(\boldsymbol{x}_t, r(t)) \leq \mathrm{e}^{-\mu(t-t_k^r)} V(\boldsymbol{x}_{t_k^r}, r(t_k^r)) + \Delta_1$$

when $t \in [t_{k_1}, t_{k_2})$,

$$V(\boldsymbol{x}_t, r(t)) \leq \mathrm{e}^{-\mu(t-t_{k_1})} V(\boldsymbol{x}_{t_{k_1}}, r(t_{k_1}) + \Delta_1$$

when $t \in [t_{k_N}, t_{k+1}^r)$,

$$V(\boldsymbol{x}_t, r(t)) \leq \mathrm{e}^{-\mu(t-t_{k_N})} V(\boldsymbol{x}_{t_{k_N}}, r(t_{k_N}) + \Delta_1$$

at time $t = t_{k_i}$, $i = 1, 2, \cdots, N$,

$$V(\boldsymbol{x}_{t_{k_i}}, r(t_{k_i})) \leq V(\boldsymbol{x}_{t_{k_i}^-}, r(t_{k_i}^-))$$

From Remark 11.5, when $t \in [t_k^r, t_{k_1})$,

$$V(\boldsymbol{x}_t, r(t)) \leq \mathrm{e}^{-\mu(t-t_k^r)} V(\boldsymbol{x}_{t_k^r}, \mathfrak{p}_k) + \Delta_1$$

when $t \in [t_{k_1}, t_{k_2})$,

$$V(\boldsymbol{x}_t, r(t)) \leq \mathrm{e}^{-\mu(t-t_k^r)} V(\boldsymbol{x}_{t_k^r}, \mathfrak{p}_k) + (\mathrm{e}^{-\mu(t-t_{k_1})} + 1)\Delta_1$$

when $t \in [t_{k_2}, t_{k_3})$,

$$V(\boldsymbol{x}_t, r(t)) \leq \mathrm{e}^{-\mu(t-t_k^r)} V(\boldsymbol{x}_{t_k^r}, \mathfrak{p}_k) + (\mathrm{e}^{-\mu(t-t_{k_1})} + \mathrm{e}^{-\mu(t-t_{k_2})} + 1)\Delta_1$$

when $t \in [t_{k_N}, t_{k+1}^r)$,

$$V(\boldsymbol{x}_t, r(t)) \leq \mathrm{e}^{-\mu(t-t_k^r)} V(\boldsymbol{x}_{t_k^r}, \mathfrak{p}_k) + (\sum_{i=1}^{N} \mathrm{e}^{-\mu(t-t_{k_i})} + 1)\Delta_1$$

Hence, for any $t \in [t_k^r, t_{k+1}^r)$,

$$V(\boldsymbol{x}_t, r(t)) \leq \mathrm{e}^{-\mu(t-t_k^r)} V(\boldsymbol{x}_{t_k^r}, \mathfrak{p}_k) + (N+1)\Delta_1$$

□

According to the conclusion of Proposition 11.1, the following closed-loop stability criterion can be obtained.

Theorem 11.1. *Under the proposed switching-based PPLC algorithm, the resulting closed-loop system is SIIOS. Specifically, the output signal of the closed-loop system satisfies*

$$|\boldsymbol{z}(t)| \leq \beta(\|\boldsymbol{z}_0\|, t) + \gamma(\|\boldsymbol{f}_{\lambda[0,t)}\|_\infty) + \Delta_4, \ \forall t \geq t_k^r$$

where $\beta(\cdot,\cdot) \in \mathcal{KL}$, $\gamma(\cdot) \in \mathcal{K}$, Δ_4 *is a bounded constant positive scalar. The specific values are*

$$\beta(s,t) = \alpha_2^{\frac{1}{2}} \mathrm{e}^{-\frac{\nu_1}{2}t} s$$

$$\gamma(s) = \left(\left(\sum_{i=1}^{k} \mathrm{e}^{-\nu_1 (t - t_{k-i}^r)} + 1 \right) \nu_2 \right)^{\frac{1}{2}} s$$

$$\Delta_4 = \left(\left(\sum_{i=1}^{k} \mathrm{e}^{-\nu_1 (t - t_{k-i}^r)} + 1 \right) \Delta_2 \right)^{\frac{1}{2}}$$

Then, $\mathbf{e}(t)$, $\bar{\mathbf{e}}(t)$, $\tilde{\boldsymbol{\theta}}$, $\tilde{\theta}_1$ *and* $\tilde{\theta}_2$ *will remain bounded. The boundedness of* $\bar{\mathbf{e}}(t)$ *also means that the error* $\mathbf{e}(t)$ *satisfies the piecewise prescribed performance defined by (11.1), (11.4) and (11.5).*

Proof. The proof of Theorem 11.1 will be based on the Proposition 11.1, and use the multiple Lyapunov-Krasovskii functionals method [38–40]. The specific proof is as follows.

Denote $\Delta_3 = \nu_2 \|\boldsymbol{f}_{\lambda[0,\infty)}\|_\infty^2 + \Delta_2$, from Proposition 11.1, for any $t \in [t_k^r, t_{k+1}^r)$,

$$V(\boldsymbol{x}_t, r(t)) \leq \mathrm{e}^{-\nu_1 (t - t_k^r)} V(\boldsymbol{x}_{t_k^r}, r(t_k^r)) + \Delta_3$$

Since $\kappa_{1,i}(t)$ switches from 1 to -1 if and only if $y_{1,i}(t)e_i(t) = 0$ and $y_{1,i}(t^-)e_i(t^-) > 0$, while it switches from -1 to 1 if and only if $y_{1,i}(t)e_i(t) = 0$ and $y_{1,i}(t^-)e_i(t^-) < 0$, and $\kappa_{2,i}(t)$ switches from 1 to -1 if and only if $y_{2,i}(t)e_i(t) = 0$ and $y_{2,i}(t^-)e_i(t^-) > 0$, while it switches from -1 to 1 if and only if $y_{2,i}(t)e_i(t) = 0$ and $y_{2,i}(t^-)e_i(t^-) < 0$, for all $t = t_{k+1}^r$, $k \geq 0$, one can get

$$V(\boldsymbol{x}_{t_{k+1}^r}, r(t_{k+1}^r)) \leq V(\boldsymbol{x}_{t_{k+1}^r}, r(t_k^r))$$

Then for any $t \geq t_k^r$,

$$\begin{aligned}
V(\boldsymbol{x}_t, r(t)) &\leq \mathrm{e}^{-\nu_1 (t - t_k^r)} V(\boldsymbol{x}_{t_k^r}, r(t_k^r)) + \Delta_3 \\
&\leq \mathrm{e}^{-\nu_1 (t - t_k^r)} V(\boldsymbol{x}_{t_k^r}, r(t_{k-1}^r)) + \Delta_3 \\
&\leq \mathrm{e}^{-\nu_1 (t - t_k^r)} \left(\mathrm{e}^{-\nu_1 (t_k^r - t_{k-1}^r)} V(\boldsymbol{x}_{t_{k-1}^r}, r(t_{k-1}^r)) + \Delta_3 \right) + \Delta_3 \\
&= \mathrm{e}^{-\nu_1 (t - t_{k-1}^r)} V(\boldsymbol{x}_{t_{k-1}^r}, r(t_{k-1}^r)) + (\mathrm{e}^{-\nu_1 (t - t_{k-1}^r)} + 1)\Delta_3 \\
&\leq \mathrm{e}^{-\nu_1 (t - t_{k-2}^r)} V(\boldsymbol{x}_{t_{k-2}^r}, r(t_{k-2}^r)) + \left(\sum_{i=1}^{2} \mathrm{e}^{-\nu_1 (t - t_{k-i}^r)} + 1 \right) \Delta_3 \\
&\ \ \vdots \\
&\leq \mathrm{e}^{-\nu_1 (t - t_0^r)} V(\boldsymbol{x}_{t_0^r}, r(t_0^r)) + \left(\sum_{i=1}^{k} \mathrm{e}^{-\nu_1 (t - t_{k-i}^r)} + 1 \right) \Delta_3
\end{aligned}$$

From (11.21), since $t_0^r = t_0 = 0$, it has

$$|z(t)|^2 \leq \alpha_2 \mathrm{e}^{-\nu_1 t}\|z_0\|_\infty^2 + \left(\sum_{i=1}^{k} \mathrm{e}^{-\mu(t-t_{k-i}^r)} + 1\right)\Delta_3$$

and further

$$|z(t)| \leq \alpha_2^{\frac{1}{2}} \mathrm{e}^{-\frac{\nu_1}{2}t}\|z_0\| + \left(\left(\sum_{i=1}^{k} \mathrm{e}^{-\nu_1(t-t_{k-i}^r)} + 1\right)\Delta_2\right)^{\frac{1}{2}} + \left(\left(\sum_{i=1}^{k} \mathrm{e}^{-\nu_1(t-t_{k-i}^r)} + 1\right)\nu_2\right)^{\frac{1}{2}} \|f_{\lambda[0,\infty)}\|_\infty$$

By causality,

$$|z(t)| \leq \alpha_2^{\frac{1}{2}} \mathrm{e}^{-\frac{\nu_1}{2}t}\|z_0\| + \left(\left(\sum_{i=1}^{k} \mathrm{e}^{-\nu_1(t-t_{k-i}^r)} + 1\right)\Delta_2\right)^{\frac{1}{2}} + \left(\left(\sum_{i=1}^{k} \mathrm{e}^{-\nu_1(t-t_{k-i}^r)} + 1\right)\nu_2\right)^{\frac{1}{2}} \|f_{\lambda[0,t)}\|_\infty$$

which means that $z(t)$ is bounded.

Given that $z = [\eta^{\mathrm{T}}, \tilde{\theta}^{\mathrm{T}}, \tilde{\theta}_1, \tilde{\theta}_2, \varpi_1^{\mathrm{T}}, \varpi_2^{\mathrm{T}}, y_1^{\mathrm{T}}, y_2^{\mathrm{T}}, e^{\mathrm{T}}, \bar{e}^{\mathrm{T}}]^{\mathrm{T}}$, the boundedness of z implies that the synchronization error e, the estimation errors $\tilde{\theta}$, $\tilde{\theta}_1$, $\tilde{\theta}_2$ and $\bar{e}$ are bounded. Moreover, the boundedness of $\bar{e}$ also implies the synchronization error e meets the piecewise constraint (11.1), (11.4) and (11.5). □

Remark 11.7. ***(Zeno Behavior Analysis)*** *When the control gains satisfy (11.18), the switching of performance function $\varrho(t) = [\varrho_1(t), \varrho_2(t), \cdots, \varrho_n(t)]^{\mathrm{T}}$ would not occur the Zeno behavior.*

On the one hand, similar to the Proposition 11.1, let us take the Lyapunov-Krasovskii functionals as

$$V(x_{1t}, r(t)) = V_1(x_{1t}, r(t)) + V_2(x_{1t}, r(t)) + V_4(x_{1t}, r(t)) + \frac{1}{2}\varpi_1^{\mathrm{T}}\varpi_1 + \frac{\vartheta}{2}\int_{-\bar{d}}^{0} z_{1t}^{\mathrm{T}}(\tau)\left(\frac{-\tau}{\bar{d}} + \frac{2(\tau+\bar{d})}{\bar{d}}\right) z_{1t}(\tau)\mathrm{d}\tau$$

where $x_1 = [q^{\mathrm{T}}, \eta^{\mathrm{T}}, \tilde{\theta}^{\mathrm{T}}, \tilde{\theta}_1, \tilde{\theta}_2, \varpi_1^{\mathrm{T}}, y_1^{\mathrm{T}}, e^{\mathrm{T}}]^{\mathrm{T}}$, $z_1 = [\eta^{\mathrm{T}}, \tilde{\theta}^{\mathrm{T}}, \tilde{\theta}_1, \tilde{\theta}_2, \varpi_1^{\mathrm{T}}, y_1^{\mathrm{T}}, e^{\mathrm{T}}]^{\mathrm{T}}$.

Considering the bounded nonlinear saturation function $\mathcal{R}_1(\cdot)$ placed outside $\bar{e}(t)$ and ϖ_2, when doing the similar proof as Proposition 11.1 and Theorem 11.1, there exists $\beta_1 \in \mathcal{KL}$, $\gamma_1 \in \mathcal{K}$ and bounded constant positive scalar Δ_5 such that

$$|z_1(t)| \leq \beta_1(\|z_{10}\|, t) + \gamma_1(\|f_{\lambda[0,t)}\|_\infty) + \Delta_5$$

which means that $|e(t)|$ does not suddenly change to infinity in a bounded time. Here, the control gain is consistent with Proposition 11.1.

On the other hand, as shown in (11.5), for any $i = 1, 2, \cdots, n$ *and* $k \geq 0$*, the performance function* $\varrho_i(t)$ *satisfies* $\varrho_i(t_k^i) > |e_i(t_k^i)|$*, and its decay rate is bounded at every time* t_k^i*.*

From this, it can be inferred that $\lim_{k\to\infty} t_k^i = \infty$*. In fact, if there is a bounded* $T_0 < \infty$ *such that* $\lim_{k\to\infty} t_k^i = T_0 < \infty$*, then there will be* $\lim_{k\to\infty}(t_{k+1}^i - t_k^i) = 0$*. This means that the event* $|e_i(t)| \geq \varrho_i(t)$ *occurs a very short time after* t_k^i*. However, since* $\varrho_i(t_k^i) > |e_i(t_k^i)|$*, it further means that the increasing rate of* $|e_i(t)|$ *and/or the decay rate of* $\varrho_i(t)$ *will reach infinity at time* t_k^i*. This contradicts the above analysis. That is, in any bounded time, the number of event* $|e_i(t)| \geq \varrho_i(t)$ *will be limited. Therefore, the Zeno behavior will not occur in the closed-loop system.*

Remark 11.8. *The proof of Proposition 11.1 uses the conclusion of Remark 11.7, so theoretically, the Remark 11.7 should be stated before the Proposition 11.1. However, since the proof of Remark 11.7 is similar to that of the Proposition 11.1 and Theorem 11.1, in this chapter, for convenience, Remark 11.7 is placed after them. It should be noted that although the analysis of Remark 11.7 is similar to the proof of Proposition 11.1, it does not need to consider the time series* $\{t_k\}_{k\ geq0}$*, it will not contradict the proof of Proposition 11.1.*

Remark 11.9. *In the classical PPC control system, the event* $|e_i(t)| \geq \varrho_i(t)$ *will cause the singularity of the logarithmic function used in the error transformation in the PPC algorithm, which will further lead to control failure. In order to ensure the control implementability and control performance, this chapter proposes a switching-based PPLC method. Compared with the classical PPC method, the algorithm proposed in this chapter has the following characteristics.*

(1) *Due to the introduction of the switched performance function, the switched-based PPLC method is not a PPC control in the classical sense.*

(2) *Under the PPLC framework, only when the singular event* $|e_i(t)| \geq \varrho_i(t)$ *occurs, the switching mechanism of the performance function will be triggered, see (11.4). Therefore, in an ideal case, the PPLC method will degenerate into the classical PPC control.*

(3) *The PPLC method proposed in this chapter avoids the Zeno behavior, and can ensure the implementability of the control and provide better control performance than a controller without any PPC strategy when the singular event* $|e_i(t)| \geq \varrho_i(t)$ *occurs.*

Remark 11.10. *The complete controller consists of (11.8)∼ (11.16). The control gains in (11.18) can be solved by the following steps. ① Choose the appropriate* λ*,* ε*,* ϑ*,* K_2*,* K_{31}*,* K_{32}*,* K_5*,* K_6*,* ψ_1*,* ψ_3 *and* ψ_5*. ② According to the characteristics of the actual communication channel, initialize* $\bar{d}$ *and* $\tilde{d}$*. ③ From (11.18), calculate* K_0*,* K_1*,* K_{41}*,* K_{42}*,* K_7*,* K_8*,* K_9*,* K_{10}*,* ψ_2*,* ψ_4 *and* ψ_6*.*

11.5 APPLICATION TO TELEOPERATION SYSTEM

Consider the following teleoperation system consisting of master and slave robots with n degrees of freedom,

$$\begin{cases} \boldsymbol{M}_m(\boldsymbol{q}_m)\ddot{\boldsymbol{q}}_m + \boldsymbol{C}_m(\boldsymbol{q}_m, \dot{\boldsymbol{q}}_m)\dot{\boldsymbol{q}}_m + \boldsymbol{g}_m(\boldsymbol{q}_m) + \\ \qquad \boldsymbol{f}_m(\dot{\boldsymbol{q}}_m) + \boldsymbol{f}_{dm} = \boldsymbol{J}_m^{\mathrm{T}}(\boldsymbol{q}_m)\boldsymbol{f}_h + \boldsymbol{\tau}_m \\ \boldsymbol{M}_s(\boldsymbol{q}_s)\ddot{\boldsymbol{q}}_s + \boldsymbol{C}_s(\boldsymbol{q}_s, \dot{\boldsymbol{q}}_s)\dot{\boldsymbol{q}}_s + \boldsymbol{g}_s(\boldsymbol{q}_s) + \\ \qquad \boldsymbol{f}_s(\dot{\boldsymbol{q}}_s) + \boldsymbol{f}_{ds} = -\boldsymbol{J}_s^{\mathrm{T}}(\boldsymbol{q}_s)\boldsymbol{f}_e + \boldsymbol{\tau}_s \end{cases} \tag{11.23}$$

where subscripts $\{m, s\}$ denote the master robot and the slave robot, respectively, and subscripts $\{h, e\}$ denote the human operator and the environment, respectively. For all $j = m, s$, $\boldsymbol{f}_j(\dot{\boldsymbol{q}}_j)$ is the frication torque, $\boldsymbol{f}_{dj}$ is the bounded external disturbance, the rest of the components are the same as (2.11).

In the teleoperation scenario, the positions of the master and slave robots are expected to converge to each other. In this case, the synchronization errors in master side and in slave side are, respectively, defined as

$$\begin{aligned} \boldsymbol{e}_m &= \boldsymbol{q}_m - \boldsymbol{q}_s(t - d_s(t)) \\ \boldsymbol{e}_s &= \boldsymbol{q}_s - \boldsymbol{q}_m(t - d_m(t)) \end{aligned}$$

where $d_m(t)$ and $d_s(t)$ are the time delays in the forward and backward communication channels, and are assumed to satisfy the Assumption 11.1.

Corollary 11.1. *For the master and slave robot subsystems represented by (11.23), under the same assumptions and PPLC design as in Section 11.3, the tracking errors $\boldsymbol{e}_m$ and $\boldsymbol{e}_s$ and their corresponding error transformations will remain bounded. Further, $\boldsymbol{e}_m$ and $\boldsymbol{e}_s$ can meet the piecewise constraint (11.1), (11.4) and (11.5).*

11.6 SIMULATION STUDY

The proposed design will be verified by a teleoperation system, which consists of two 3-DOF planar arms. The dynamics models of the master and slave robots are shown in (11.23). The physical parameters of robots can be found in Sciaviicco's work [41]. For teleoperation system, the frication models are assumed to be

$$\begin{aligned} \boldsymbol{f}_m(\dot{\boldsymbol{q}}_m) &= [0.3\dot{q}_{m,1} + 0.2\mathrm{sgn}(\dot{q}_{m,1}), 0.1\dot{q}_{m,2} + 0.2\mathrm{sgn}(\dot{q}_{m,2}), \\ &\qquad 0.4\dot{q}_{m,3} + 0.1\mathrm{sgn}(\dot{q}_{m,3})]^{\mathrm{T}} \\ \boldsymbol{f}_s(\dot{\boldsymbol{q}}_s) &= [0.3\dot{q}_{s,1} + 0.2\mathrm{sgn}(\dot{q}_{s,1}), 0.1\dot{q}_{s,2} + 0.2\mathrm{sgn}(\dot{q}_{s,2}), 0.4\dot{q}_{s,3} + 0.1\mathrm{sgn}(\dot{q}_{s,3})]^{\mathrm{T}} \end{aligned}$$

The bounded disturbances are

$$\begin{aligned} \boldsymbol{f}_{dm} &= [0.6\sin(|\dot{\boldsymbol{q}}_m|), 0.3\cos(2|\dot{\boldsymbol{q}}_m|), 0.5\sin(2t)\cos(2|\dot{\boldsymbol{q}}_m|)]^{\mathrm{T}} \\ \boldsymbol{f}_{ds} &= [0.3\cos(5|\dot{\boldsymbol{q}}_s|), 0.6\sin(0.5|\dot{\boldsymbol{q}}_s|), 0.2\cos(6t)\cos(7|\dot{\boldsymbol{q}}_s|)]^{\mathrm{T}} \end{aligned}$$

The communication delays are shown in Fig. 11.4, with $\tilde{d} = 5$. In addition, the human and environment forces are given as Fig. 11.5.

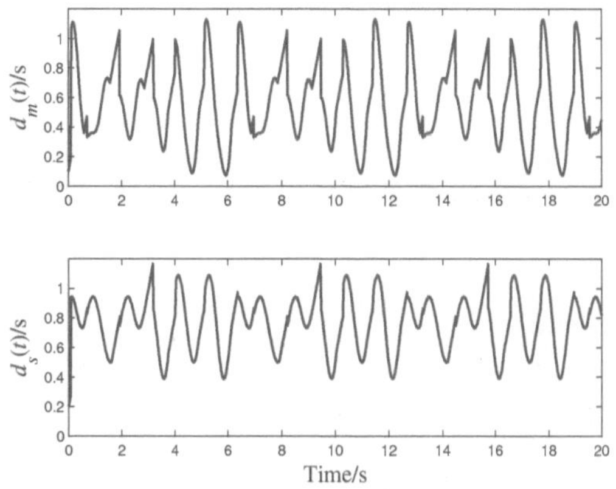

Figure 11.4 The communication delays $d_m(t)$ and $d_s(t)$.

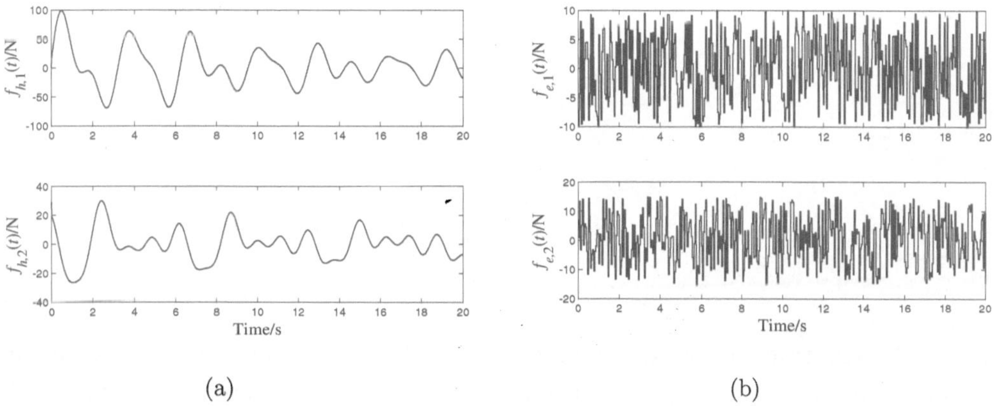

Figure 11.5 The human operator and environment forces. (a) The human force $\boldsymbol{f}_h(t)$. (b) The environment force $\boldsymbol{f}_e(t)$.

11.6.1 Stability verification

First, the closed-loop stability is demonstrated. It takes $\delta = 0.05$, where δ is defined in (11.4). From Corollary 11.1, the simulation results are given as Figs. 11.6–11.8. Among them, Fig. 11.6 gives the synchronization performance between the master and slave robots. It can be seen that the switching-based PPLC method proposed in this chapter has obtained satisfactory tracking performance, and the tracking error also satisfies the performance constraints defined by (11.1), (11.4) and (11.5). In Fig. 11.6, the occurrence of singular events can be observed, which is mainly caused by the injection of non-zero external force and the digital realization of the controller. Figs. 11.7 and 11.8 give the trajectories of adaptive estimation of unknown parameters and the control torques, respectively.

In Fig. 11.6, the proposed switching mechanism for performance function has been triggered. From the aforementioned discussion, the key issue for this switching

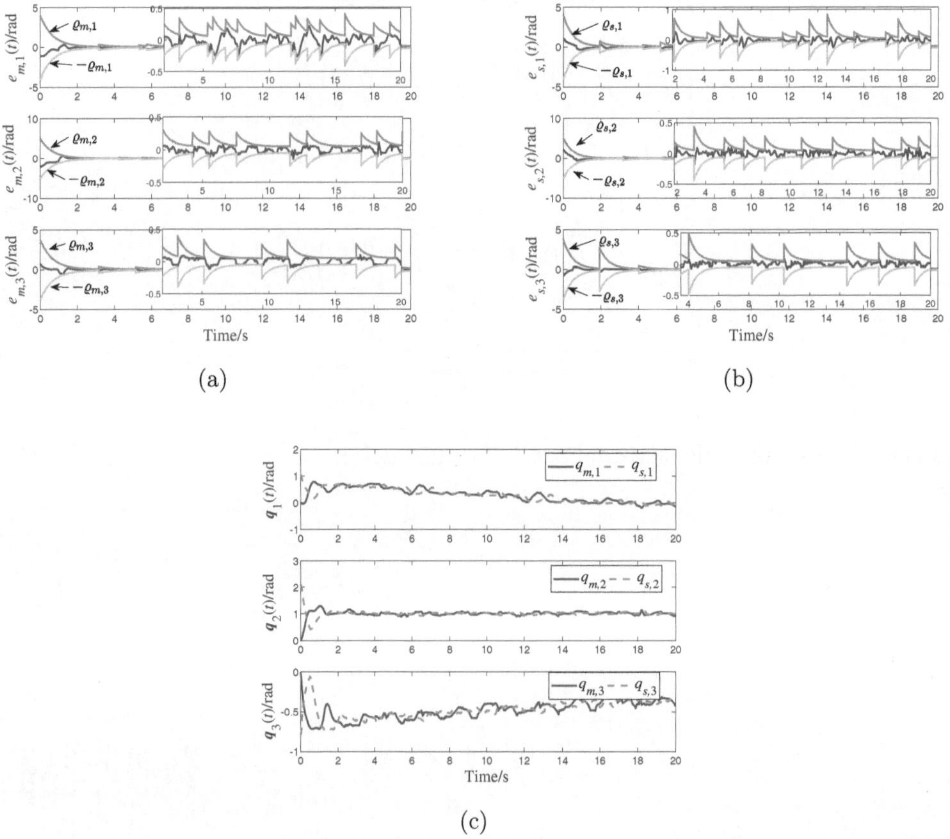

Figure 11.6 The tracking performance between the master and slave robots. (a) Synchronization error $\boldsymbol{e}_m(t)$. (b) Synchronization error $\boldsymbol{e}_s(t)$. (c) Joint positions.

design is to avoid the Zeno behavior, which has been overcome by this chapter. To verify the effectiveness, let us remove the nonlinear saturating function $\boldsymbol{\mathcal{R}}_1(\cdot)$ from the control torque (11.8) and just guarantee $\varrho_i(t_k^i) > |e_i(t_k^i)|$ in (11.5). Then under the same simulation setup, the synchronization performance is shown in Fig. 11.9. In Fig. 11.9(a), after $t > 5.56s$ and $t > 5.686s$, it is observed that $\varrho_{m,i}(t)$ and $\varrho_{s,i}(t)$, $i = 1, 2, 3$, are almost strictly increasing, respectively. In fact, the increasing properties of $\varrho_{m,i}(t)$ and $\varrho_{s,i}(t)$ mean that the events $|e_{mi}(t)| \geq \varrho_{mi}(t)$ and $|e_{si}(t)| \geq \varrho_{si}(t)$ keep occurring. Further, from Fig. 11.9, the trajectories of $\varrho_{m,i}(t)$ and $\varrho_{s,i}(t)$ go infinity at time $t = 5.758s$ and $t = 5.69s$ respectively, which mean that the numbers of the occurrence of the events $|e_{m,i}(t)| \geq \varrho_{m,i}(t)$ and $|e_{s,i}(t)| \geq \varrho_{s,i}(t)$ are infinite over a very short period at $t = 5.758s$ and $t = 5.69s$ respectively, i.e., there arises the Zeno behavior. It can also be seen from Fig. 11.9 that when the Zeno behavior occurs, the closed-loop system is unstable. Therefore, the Zeno behavior needs to be avoided, and the control design proposed in this book avoids the Zeno behavior, see Remark 11.7 for details.

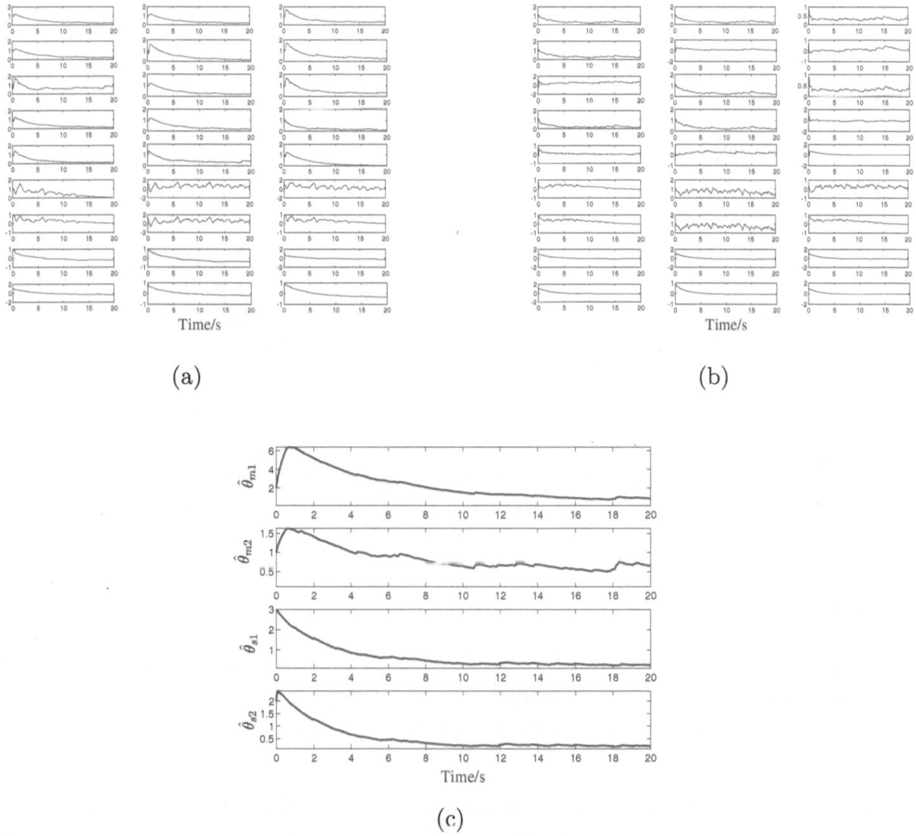

Figure 11.7 The adaptive estimations. (a) $\hat{\boldsymbol{\theta}}_m(t)$, (b) $\hat{\boldsymbol{\theta}}_s(t)$, (c) $\hat{\theta}_{m1}(t)$, $\hat{\theta}_{m2}(t)$, $\hat{\theta}_{s1}(t)$ and $\hat{\theta}_{s2}(t)$.

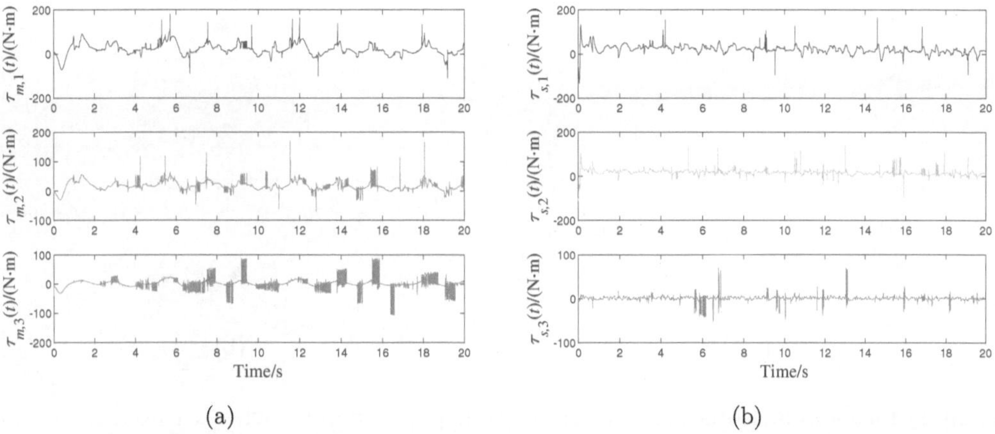

Figure 11.8 The control inputs. (a) $\boldsymbol{\tau}_m(t)$, (b) $\boldsymbol{\tau}_s(t)$.

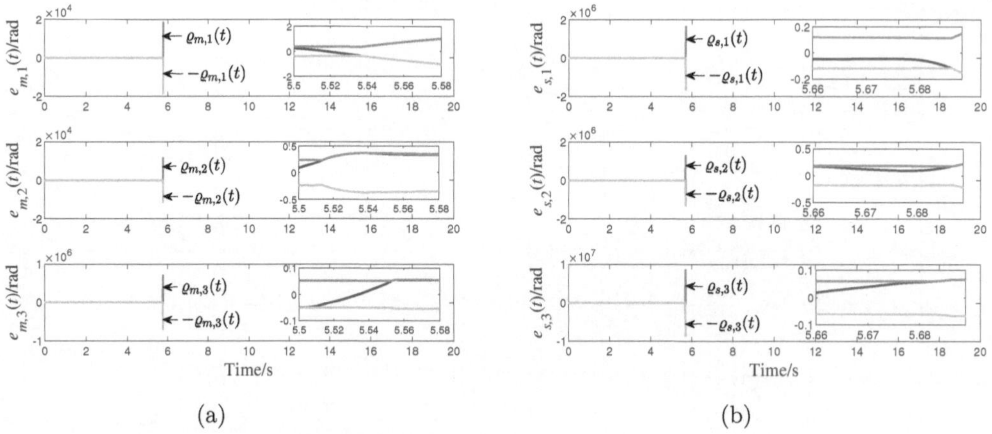

Figure 11.9 The tracking performance between the master and slave robots with Zeno behavior. Synchronization error (a) $\boldsymbol{e}_m(t)$, (b) $\boldsymbol{e}_s(t)$.

11.6.2 Function verification of PPLC

From Remark 11.9, the switching-based PPLC is not a PPC algorithm in the classical sense, which looks like that it deviates from the original control objective of classical PPC algorithm. Therefore, there is doubt whether the switching-based PPLC strategy can improve the control performance as the classical PPC approach.

To answer this question, first, let us remove the developed PPLC strategy. Under the same simulation setup, the synchronization performance is given in Fig. 11.10.

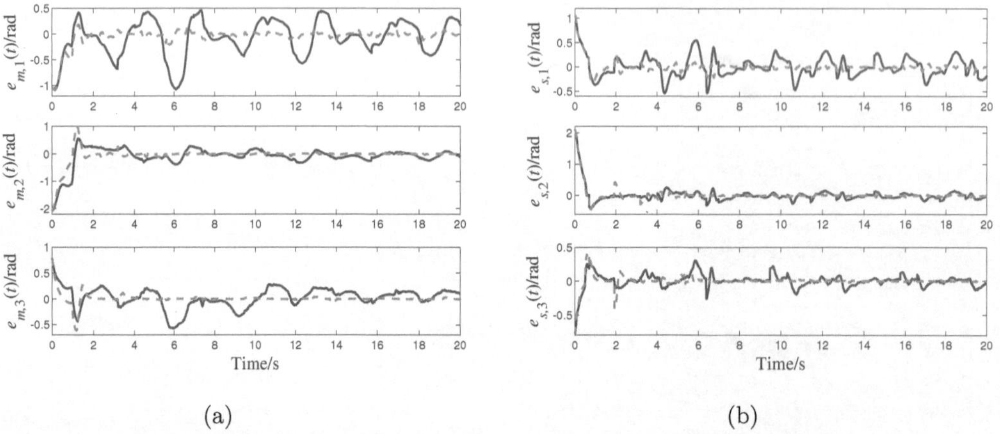

Figure 11.10 Comparisons on synchronization performances with/without switching-based PPLC scheme (The blue solid lines and red dotted lines are, respectively, the error trajectories without and with using switching-based PPLC scheme). Synchronization error (a) $\boldsymbol{e}_m(t)$ and (b) $\boldsymbol{e}_s(t)$.

It is shown that the proposed switching-based PPLC algorithm has improved the synchronization performance.

As pointed out in Remark 11.9, if the singular event $|e_i(t)| \geq \varrho_i(t)$ does not occur, i.e., in ideal situation, the proposed switching-based PPLC algorithm is capable of reducing to a classical PPC approach. In this simulation, the nonzero external forces need to bear the main responsibility for those singularities. To simulate the ideal situation, similar to the existing works [15–17], let the master and slave robots run in free motion, then the synchronization performance is given in Fig. 11.11, where the switching mechanism is not triggered, which is in accordance with the analysis of Remark 11.9.

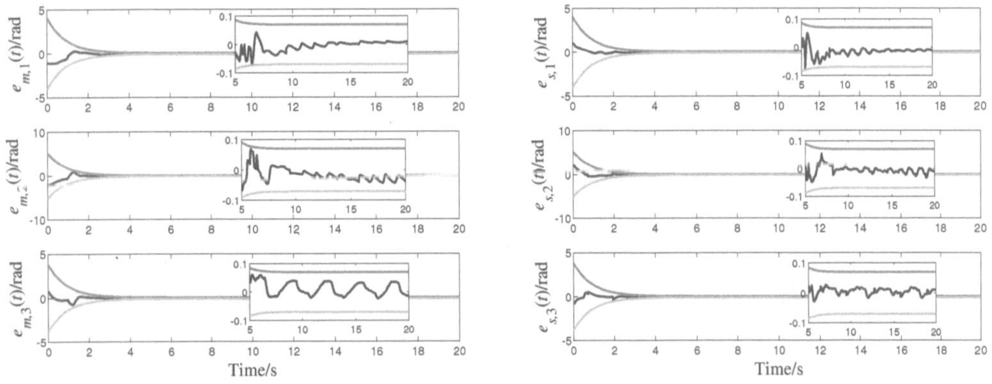

(a) (b)

Figure 11.11 Synchronization performance when the switching-based PPLC-based controller runs in ideal situation. Synchronization error (a) $\boldsymbol{e}_m(t)$, (b)$\boldsymbol{e}_s(t)$.

Since it has no Zeno behavior, the proposed PPLC approach can retain the original features of classical PPC algorithm. On the one hand, from the comparison shown in Fig. 11.10, one can find that the proposed PPLC algorithm can improve convergence speed and overshoot. On the other hand, similar to the classical PPC algorithm, under the switching-based PPLC approach, the value of steady-state error can be changed by taking different δ in (11.4). For example, taking $\delta = 0.05$ in the first simulation, when taking $\delta = 0.001$, under the same simulation conditions, the simulation results are shown in Fig. 11.12. It can be seen that the PPLC algorithm can obtain a smaller convergence error when choosing a smaller δ. To show this effect more intuitively, let us introduce the following performance index:

$$p(e(t), \delta) = \frac{T(e(t), \delta)}{total\ simulation\ time} \times 100\%$$

where $T(e(t), \delta)$ denotes all times such that the $|e(t)|$ with δ is less than or equal to the $|e(t)|$ with $\delta = 0.05$. For all $i = 1, 2, 3$, $j = m, s$, and those $\delta_2 < \delta_1 < 0.05$, if $p(e_{j,i}(t), \delta_1) > 50\%$ and $p(e_{j,i}(t), \delta_2) > p(e_{j,i}(t), \delta_1)$, we can claim that the switching-based PPLC approach is capable of improving the steady-state error. Under the same

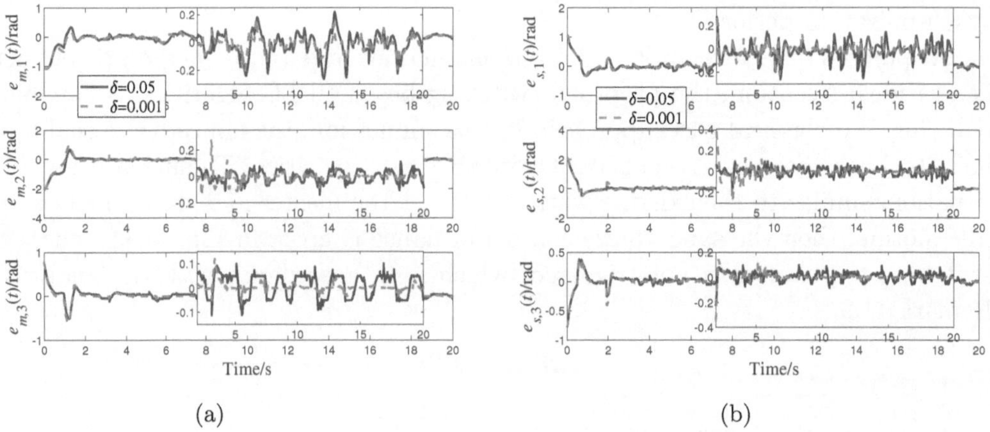

Figure 11.12 Comparison of the tracking performance of the proposed algorithm under different values of δ (the solid line is the error response curve with the value $\delta = 0.05$, and the dotted line is the error response curve with the value $\delta = 0.001$). Synchronization error (a) $\boldsymbol{e}_m(t)$, (b) $\boldsymbol{e}_s(t)$.

control hypotheses, for different δ, the values of $p(e_{m,i}(t), \delta)$ and $p(e_{s,i}(t), \delta)$ are given as Table 11.1.

Table 11.1 $p(e(t), \delta)$ with different δ

$p(e(t), \delta)$	$\delta = 0.01$	$\delta = 0.001$	$\delta = 0.0002$
$e(t) = e_{m1}(t)$	51.7280%	68.3880%	83.2855%
$e(t) = e_{m2}(t)$	68.7570%	78.9875%	82.2740%
$e(t) = e_{m3}(t)$	81.4070%	87.7720%	91.6490%
$e(t) = e_{s1}(t)$	55.5980%	60.6095%	79.7370%
$e(t) = e_{s2}(t)$	65.5885%	68.8950%	70.7490%
$e(t) = e_{s3}(t)$	79.7160%	82.0555%	89.5425%

11.6.3 Comparison studies

Compared with the existing classical predefined performance control, the PPLC method proposed in this chapter has made a great improvement, and can guarantee the implementability and control performance of the controller when the singular event $|e_{j,i}(t)| \geq \varrho_{j,i}(t)$ occurs. This group of simulations will give the comparisons of the PPLC method and the existing algorithm with predefined performance.

Firstly, the classical PPC strategy is used to replace the switching-based PPLC. In this case, the classical PPC algorithm [18] (corresponds to Chapter 10 of this book) is utilized. Given that it may occur the singular event in this case, for all $j = m, s$ and $i = 1, 2, 3$, the simulation takes $\varrho_{j,i\infty} = 0.05$ and $l_{j,1} = 0.1$, $l_{j,2} = 0.3$, $l_{m,3} = 0.3$ and $l_{s,3} = 0.2$. [3] The synchronization performance between the master and slave robots is shown in Fig. 11.13. To compare the performance, the following indicators

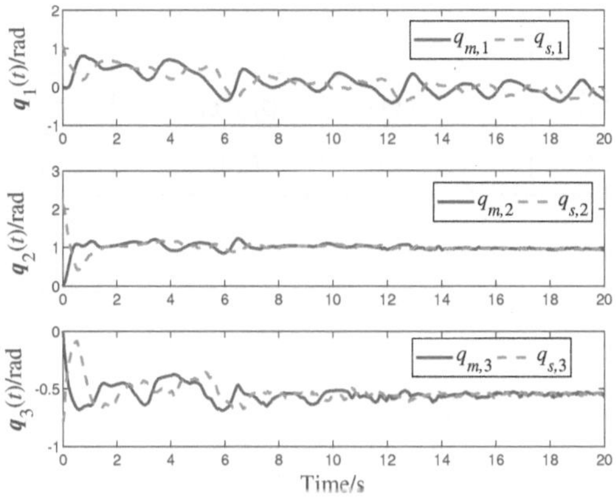

Figure 11.13 Synchronization performance of master and slave robots by using the classic PPC algorithm [18].

are introduced:

$$p_{cp}(e(t)) = \frac{T_{cp}(e(t))}{\text{total simulation time}} \times 100\%$$
$$p_{sp}(e(t)) = \frac{T_{sp}(e(t))}{\text{total simulation time}} \times 100\%$$

where $T_{cp}(e(t))$ denotes all times such that $|e(t)|$ with using classical PPC is less than or equal to $|e(t)|$ without using any PPC, and $T_{sp}(c(t))$ denotes all times such that $|e(t)|$ with using switching-based PPLC is less than or equal to $|e(t)|$ without using any PPC. Under the same simulation setup, the comparison results are shown in Table 11.2, where the performance of classical PPC algorithm is worse than the switching-based PPLC strategy. Generally speaking, when we implement the classical PPC, from (11.1), a better synchronization performance can be obtained by decreasing $\varrho_{j,i\infty}$ and/or increasing $l_{j,i}$, $j = m, s$, $i = 1, 2, 3$. Unfortunately, due to $\bar{e}_{j,i}(t)$ becomes invalid when $|e_{j,i}(t)| \geq \varrho_{j,i}(t)$, such effort is failed in the given simulation setup, as shown in Fig. 11.14. [4]

Secondly, the developed switching-based PPLC algorithm has a better performance when the robots keep contact with the human operator and/or environment.The classical PPC-based adaptive neural network control has been developed by Yang et al. for teleoperation system with constant delays [15]. The comparison study is performed between this chapter and the Yang's work [15]. To be more fair, it is

[3] Those parameters can help the PPC controller achieve the best performance in our simulation.

[4] To guarantee that the controller is implementable, the classical PPC strategy is removed from the corresponding joint's controller when $|e_{j,i}(t)| \geq \varrho_{j,i}(t)$.

Table 11.2 $p_{cp}(e(t))$ and $p_{sp}(e(t))$

$e(t) = e_{m,1}(t)$		$e(t) = e_{m,2}(t)$		$e(t) = e_{m,3}(t)$	
$p_{cp}(e(t))$	$p_{sp}(e(t))$	$p_{cp}(e(t))$	$p_{sp}(e(t))$	$p_{cp}(e(t))$	$p_{sp}(e(t))$
64.441 0	87.723 0	88.801 5	93.612 0	82.130 5	88.967 5
$e(t) = e_{s,1}(t)$		$e(t) = e_{s,2}(t)$		$e(t) = e_{s,3}(t)$	
$p_{cp}(e(t))$	$p_{sp}(e(t))$	$p_{cp}(e(t))$	$p_{sp}(e(t))$	$p_{cp}(e(t))$	$p_{sp}(e(t))$
62.130 5	84.028 0	84.520 0	90.492 5	77.280 0	80.095 5

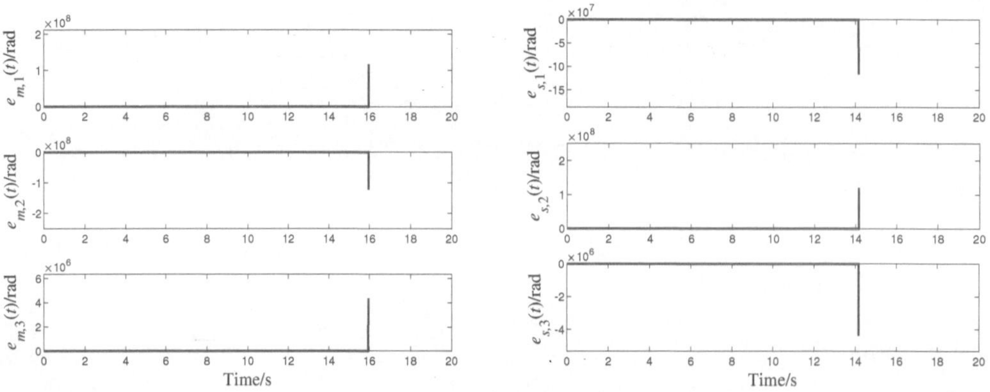

Figure 11.14 The synchronization errors when using the classical PPC algorithm in our controller with small $\varrho_{j,i\infty} = 0.01$ and large $l_{j,i} = 0.3$. Synchronization error (a) $\boldsymbol{e}_m(t)$ and (b) $\boldsymbol{e}_s(t)$.

assumed that the robots' dynamics are exactly known and the communication delays are constant, where the time delays are set to be $T_m = 0.92$ and $T_s = 1.4$. Similarly, consider that $\bar{e}_{j,i}$ is invalid when the event $|e_{j,i}(t)| \geq \varrho_{j,i}(t)$ occurs, when the control is implemented, the PPC strategy will be removed when a singular event occurs. Under the external forces shown in Fig. 11.5, the simulation results are given in Fig. 11.15, while the comparison results are given in Fig. 11.16. It is obvious that the proposed PPLC scheme has a better performance than the classical PPC algorithm [15] when the robots keep contact with the human operator and/or environment.

Another comparison study is performed between the PPLC approach of this chapter and the BLF-based scheme [17]. In the BLF-based framework [17], a sliding mode incorporated with BLF strategy control design is developed for the finite-time control of teleoperation system with constant delays. Note that, the proposed BLF-based control can only handle the case that both the master and slave robots run in free motion, as shown in Fig. 11.17. In practices, however, the robots inevitably need to contact with the human operator and/or environment. Unfortunately, if we put some external forces on the master and slave robots (for example, as the ones in Fig. 11.5),

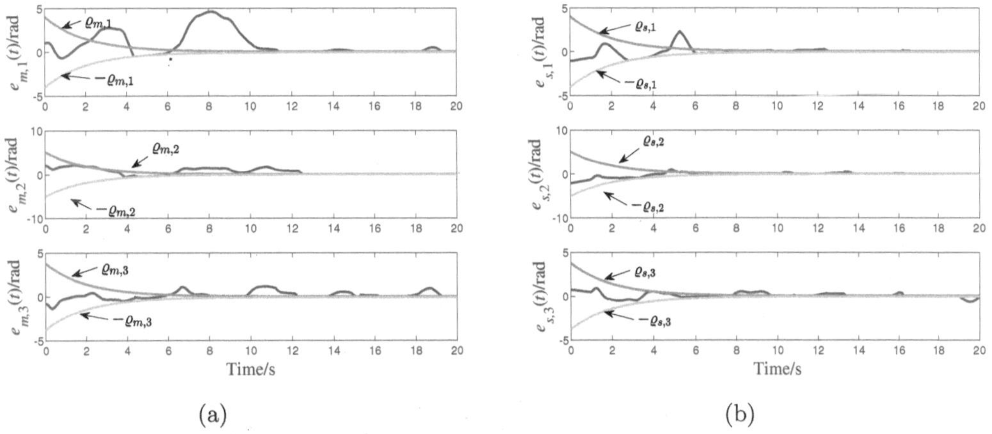

Figure 11.15 Comparisons with the classical PPC algorithm [15]. Synchronization error (a) $\boldsymbol{e}_m(t)$, (b) $\boldsymbol{e}_s(t)$.

the closed-loop system with the BLF-based control becomes unstable, which is shown in Fig. 11.18.

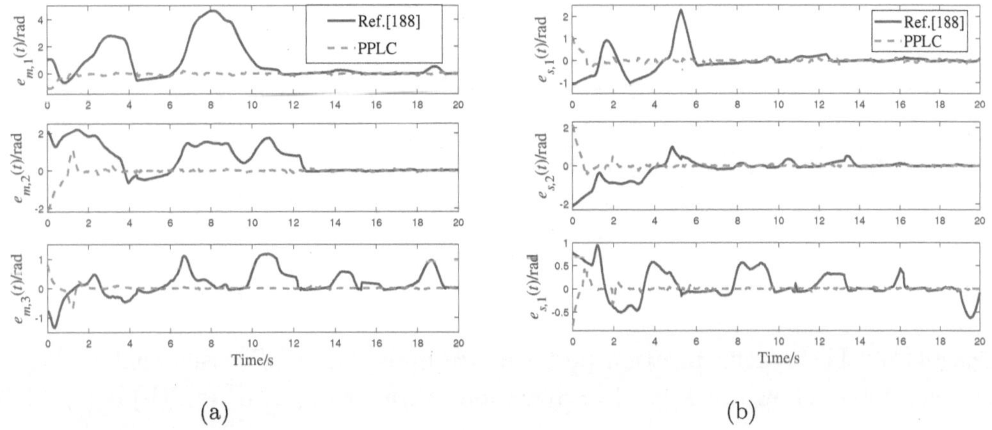

Figure 11.16 The synchronization performance under the classical PPC method [15] with external forces shown in Fig. 11.5. Synchronization error (a) $\boldsymbol{e}_m(t)$, (b) $\boldsymbol{e}_s(t)$.

Thirdly, the proposed switching-based PPLC approach is applicable to the teleoperation system with varying time delays, which has been demonstrated by the simulation in previous simulations. However, the existing control methods [15–17] can only handle the constant delays. The time-varying components in time delay also have a negative effect on the effectiveness of those methods. To show this, let us take the BLF-based method [17] as an example. In the varying time delays case, the synchronization performance is shown in Fig. 11.19, which is unstable.

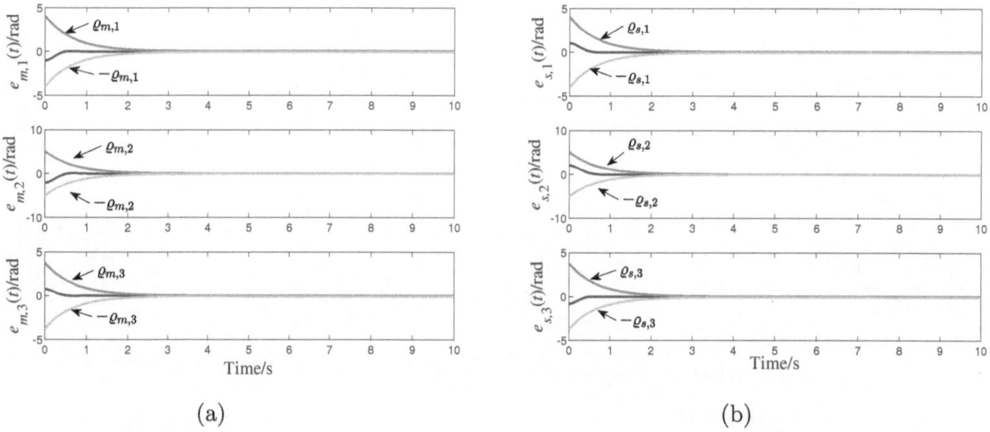

Figure 11.17 The synchronization performance under the BLF-based control [17] when the robots run in free motion case. Synchronization error (a) $\boldsymbol{e}_m(t)$, (b) $\boldsymbol{e}_s(t)$.

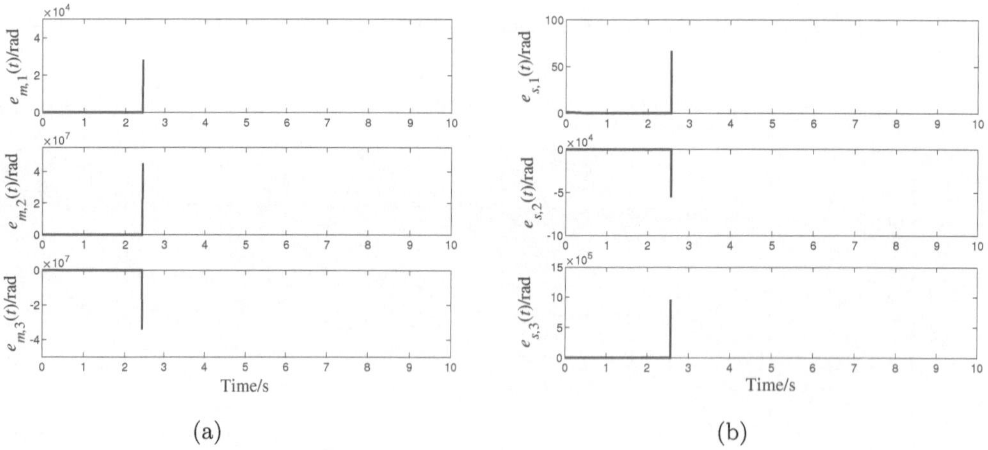

Figure 11.18 The synchronization performance under the BLF-based control [17] with external forces shown in Fig. 11.5. Synchronization error (a) $\boldsymbol{e}_m(t)$, (b) $\boldsymbol{e}_s(t)$.

11.6.4 Discussions

From the above simulation, one can conclude that:

- The developed switching-based PPLC cannot avoid the occurrence of singular event $|e_{j,i}(t)| \geq \varrho_{j,i}(t)$. Alternately, it just provides a solution for engineer to achieve the control when there arises such singular event. A better performance is obtained in comparison with the controllers without using any PPC strategy.
- The switching-based PPLC approach has also inherited the merits of the classical PPC. Moreover, since the classical predefined performance control (including the classical PPC scheme and the BLF-based framework) is invalid when it

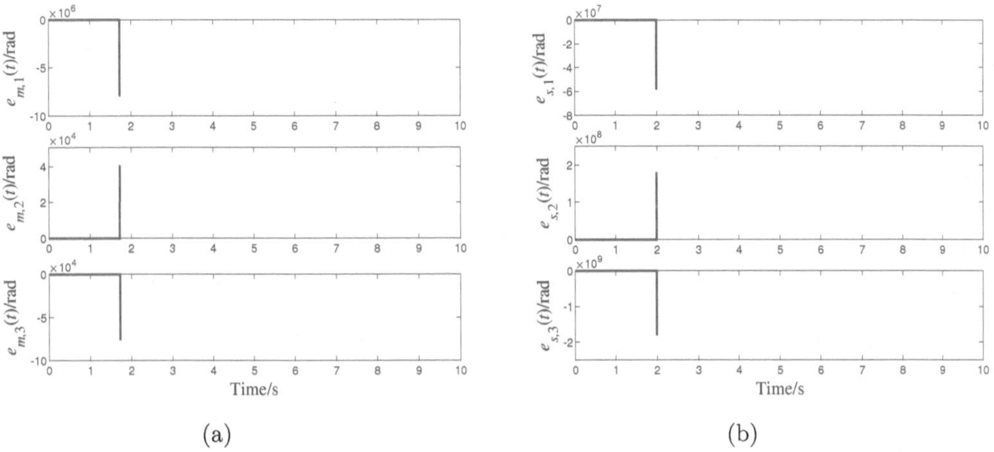

Figure 11.19 The synchronization performance under the BLF-based control [17] with varying time delays. Synchronization error (a) $\boldsymbol{e}_m(t)$ and (b) $\boldsymbol{e}_s(t)$.

occurs $|e_{j,i}(t)| \geq \varrho_{j,i}(t)$, the proposed approach expands the application range of classical predefined performance control.

11.7 CONCLUSION

In this chapter, a new switching-based PPLC method is proposed for teleoperation systems in complex networks and external environments, which is used to ensure the implementability and control performance of the controller when there is a violation of hard constraints in the classical predefined performance control strategy. The basic idea of the proposed switching-based PPLC method is to introduce a switching mechanism to the classical prescribed performance strategy, specifically, to passively introduce a switched performance function so that the hard constraints are satisfied at any time. Combined with the auxiliary switched filter control method proposed in the previous chapters, the control design proposed in this chapter can guarantee the control performance of the closed-loop teleoperation system in the presence of model uncertainty and time-varying communication delay, regardless of whether the robot contacts with the operator and/or contacts with the environment. When the hard constraints are never violated, the new switching-based prescribed performance-like controller will have exactly the same control performance as the classical prescribed performance controller. Once the hard constraint violation event occurs, the classical prescribed performance control will have a singular phenomenon and cannot work. In this case, the proposed switching performance function mechanism will work to ensure the smooth implementation of the control algorithm and provide good control performance. Compared with the classical predefined performance control, the proposed PPLC method further improves the environmental adaptability of the controller and helps to improve the precise operation ability of the robot in complex and dynamic environments.

Bibliography

[1] P. F. Hokayem and M. W. Spong, "Bilateral teleoperation: an historical survey," *Automatica*, vol. 42, no. 12, pp. 2035–2057, 2006.

[2] M. Ferre, R. Aracil, C. Balaguer, M. Buss, and C. Melchiorri, *Advances in Telerobotics*. Springer, 2007.

[3] E. Nuno, L. Basanez, and R. Ortega, "Passivity-based control for bilateral teleoperation: a tutorial," *Automatica*, vol. 47, no. 3, pp. 485–495, 2011.

[4] E. Nuno, R. Ortega, and L. Basanez, "An adaptive controller for nonlinear teleoperators," *Automatica*, vol. 46, no. 1, pp. 155–159, 2010.

[5] Z. Li and C.-Y. Su, "Neural-adaptive control of single-master–multiple-slaves teleoperation for coordinated multiple mobile manipulators with time-varying communication delays and input uncertainties," *IEEE Transactions on Neural Networks and Learning Systems*, vol. 24, no. 9, pp. 1400–1413, 2013.

[6] N. Chopra, M. W. Spong, and R. Lozano, "Synchronization of bilateral teleoperators with time delay," *Automatica*, vol. 44, no. 8, pp. 2142–2148, 2008.

[7] Y.-C. Liu and N. Chopra, "Control of semi-autonomous teleoperation system with time delays," *Automatica*, vol. 49, no. 6, pp. 1553–1565, 2013.

[8] I. G. Polushin, S. N. Dashkovskiy, A. Takhmar, and R. V. Patel, "A small gain framework for networked cooperative force-reflecting teleoperation," *Automatica*, vol. 49, no. 2, pp. 338–348, 2013.

[9] D.-H. Zhai and Y. Xia, "Adaptive fuzzy control of multilateral asymmetric teleoperation for coordinated multiple mobile manipulators," *IEEE Transactions on Fuzzy Systems*, vol. 24, no. 1, pp. 57–70, 2015.

[10] D.-H. Zhai and Y. Xia, "Adaptive control for teleoperation system with varying time delays and input saturation constraints," *IEEE Transactions on Industrial Electronics*, vol. 63, no. 11, pp. 6921–6929, 2016.

[11] F. Hashemzadeh, I. Hassanzadeh, and M. Tavakoli, "Teleoperation in the presence of varying time delays and sandwich linearity in actuators," *Automatica*, vol. 49, no. 9, pp. 2813–2821, 2013.

[12] D. Sun, S. Hu, X. Shao, and C. Liu, "Global stability of a saturated nonlinear PID controller for robot manipulators," *IEEE Transactions on Control Systems Technology*, vol. 17, no. 4, pp. 892–899, 2009.

[13] H. Yang, X. You, Z. Liu, and F. Sun, "Extended-state-observer-based adaptive control for synchronisation of multi-agent systems with unknown nonlinearities," *International Journal of Systems Science*, vol. 46, no. 14, pp. 2520–2530, 2015.

[14] L. Wang, T. Chai, and L. Zhai, "Neural-network-based terminal sliding-mode control of robotic manipulators including actuator dynamics," *IEEE Transactions on Industrial Electronics*, vol. 56, no. 9, pp. 3296–3304, 2009.

[15] Y. Yang, C. Hua, and X. Guan, "Synchronization control for bilateral teleoperation system with prescribed performance under asymmetric time delay," *Nonlinear Dynamics*, vol. 81, no. 1, pp. 481–493, 2015.

[16] Y. Yang, C. Ge, H. Wang, X. Li, and C. Hua, "Adaptive neural network based prescribed performance control for teleoperation system under input saturation," *Journal of the Franklin Institute*, vol. 352, no. 5, pp. 1850–1866, 2015.

[17] Y. Yang, C. Hua, and X. Guan, "Finite time control design for bilateral teleoperation system with position synchronization error constrained," *IEEE Transactions on Cybernetics*, vol. 46, no. 3, pp. 609–619, 2015.

[18] D.-H. Zhai and Y. Xia, "A novel switching-based control framework for improved task performance in teleoperation system with asymmetric time-varying delays," *IEEE Transactions on Cybernetics*, vol. 48, no. 2, pp. 625–638, 2017.

[19] C. P. Bechlioulis, Z. Doulgeri, and G. A. Rovithakis, "Neuro-adaptive force/position control with prescribed performance and guaranteed contact maintenance," *IEEE Transactions on Neural Networks*, vol. 21, no. 12, pp. 1857–1868, 2010.

[20] A. K. Kostarigka and G. A. Rovithakis, "Adaptive dynamic output feedback neural network control of uncertain mimo nonlinear systems with prescribed performance," *IEEE Transactions on Neural Networks and Learning Systems*, vol. 23, no. 1, pp. 138–149, 2011.

[21] C. P. Bechlioulis and G. A. Rovithakis, "Robust partial-state feedback prescribed performance control of cascade systems with unknown nonlinearities," *IEEE Transactions on Automatic Control*, vol. 56, no. 9, pp. 2224–2230, 2011.

[22] B. Ren, S. S. Ge, K. P. Tee, and T. H. Lee, "Adaptive neural control for output feedback nonlinear systems using a barrier Lyapunov function," *IEEE Transactions on Neural Networks*, vol. 21, no. 8, pp. 1339–1345, 2010.

[23] K. P. Tee, S. S. Ge, and E. H. Tay, "Barrier Lyapunov functions for the control of output-constrained nonlinear systems," *Automatica*, vol. 45, no. 4, pp. 918–927, 2009.

[24] K. P. Tee, B. Ren, and S. S. Ge, "Control of nonlinear systems with time-varying output constraints," *Automatica*, vol. 47, no. 11, pp. 2511–2516, 2011.

[25] W. He and S. S. Ge, "Vibration control of a flexible beam with output constraint," *IEEE Transactions on Industrial Electronics*, vol. 62, no. 8, pp. 5023–5030, 2015.

[26] Y. Dong, N. Gupta, and N. Chopra, "On content modification attacks in bilateral teleoperation systems," in *Proceedings of the American Control Conference*. IEEE, 2016, pp. 316–321.

[27] T. Bonaci, J. Herron, T. Yusuf, J. Yan, T. Kohno, and H. J. Chizeck, "To make a robot secure: An experimental analysis of cyber security threats against teleoperated surgical robots," *arXiv preprint arXiv:1504.04339*, 2015.

[28] Y. Dong, N. Gupta, and N. Chopra, "False data injection attacks in bilateral teleoperation systems," *IEEE Transactions on Control Systems Technology*, vol. 28, no. 3, pp. 1168–1176, 2019.

[29] D. Zhai, L. An, J. Li, and Q. Zhang, "Adaptive fuzzy fault-tolerant control with guaranteed tracking performance for nonlinear strict-feedback systems," *Fuzzy Sets and Systems*, vol. 302, pp. 82–100, 2016.

[30] S. J. Yoo, "Fault-tolerant control of strict-feedback non-linear time-delay systems with prescribed performance," *IET Control Theory & Applications*, vol. 7, no. 11, pp. 1553–1561, 2013.

[31] Y. Xu, S. Tong, and Y. Li, "Prescribed performance fuzzy adaptive fault-tolerant control of non-linear systems with actuator faults," *IET Control Theory & Applications*, vol. 8, no. 6, pp. 420–431, 2014.

[32] M. Chen, X. Liu, and H. Wang, "Adaptive robust fault-tolerant control for nonlinear systems with prescribed performance," *Nonlinear Dynamics*, vol. 81, no. 4, pp. 1727–1739, 2015.

[33] G. Gao, J. Wang, and X. Wang, "Prescribed-performance fault-tolerant control for feedback linearisable systems with an aircraft application," *International Journal of Control*, vol. 90, no. 5, pp. 932–949, 2017.

[34] L. Guo and S. Cao, "Anti-disturbance control theory for systems with multiple disturbances: A survey," *ISA Transactions*, vol. 53, no. 4, pp. 846–849, 2014.

[35] S. Wang, X. Ren, J. Na, and T. Zeng, "Extended-state-observer-based funnel control for nonlinear servomechanisms with prescribed tracking performance," *IEEE Transactions on Automation Science and Engineering*, vol. 14, no. 1, pp. 98–108, 2016.

[36] M. Heymann, F. Lin, G. Meyer, and S. Resmerita, "Analysis of zeno behaviors in a class of hybrid systems," *IEEE Transactions on Automatic Control*, vol. 50, no. 3, pp. 376–383, 2005.

[37] R. Kelly, V. S. Davila, and J. A. L. Perez, *Control of Robot Manipulators in Joint Space*. Springer Science & Business Media, 2005.

[38] D. Liberzon, *Switching in Systems and Control*. Springer, 2003.

[39] Z. Sun and S. S. Ge, *Stability Theory of Switched Dynamical Systems.* Springer, 2011.

[40] M. S. Branicky, "Multiple Lyapunov functions and other analysis tools for switched and hybrid systems," *IEEE Transactions on Automatic Control*, vol. 43, no. 4, pp. 475–482, 1998.

[41] L. Sciavicco and B. Siciliano, *Modelling and Control of Robot Manipulators.* Springer Science & Business Media, 2001.